Nonlocal Quantum Field Theory and Stochastic Quantum Mechanics

Fundamental Theories of Physics

A New International Book Series on the Fundamental Theories of Physics: Their Clarification, Development and Application

Nonlocal Quantum Field Theory and Stochastic Quantum Mechanics

by

Khavtgain Namsrai

Institute of Physics and Technology, Academy of Sciences,
Mongolian People's Republic, and
Joint Institute for Nuclear Research, Dubna, U.S.S.R.

D. Reidel Publishing Company

A MEMBER OF THE KLUWER ACADEMIC PUBLISHERS GROUP

Dordrecht / Boston / Lancaster / Tokyo

Library of Congress Cataloging-in-Publication Data
Namsrai, Khavtgain, 1943–
Nonlocal quantum field theory and stochastic quantum mechanics.

(Fundamental theories of physics)
Bibliography: p.
Includes index.
1. Quantum field theory. 2. Stochastic processes. 3. Space and time. 4. Electromagnetic interactions. 5. Weak interactions (Nuclear physics) I. Title. II. Series.
QC174.45.N33 1986 530.1'43 85-25617
ISBN-13: 978-94-010-8513-7 e-ISBN-13: 978-94-009-4518-0
DOI: 10.1007/978-94-009-4518-0

Published by D. Reidel Publishing Company,
P.O. Box 17,3300 AA Dordrecht, Holland.

Sold and distributed in the U.S.A. and Canada
by Kluwer Academic Publishers,
190 Old Derby Street, Hingham, MA 02043, U.S.A.

In all other countries, sold and distributed
by Kluwer Academic Publishers Group,
P.O. Box 322, 3300 AH Dordrecht, Holland.

Dedicated
to my father
Damdiny Khavtgai

Table of Contents

Preface

In this book we attempt to give a physical foundation for the concept of nonlocality and to introduce the fundamental length in physics by the hypothesis of stochasticity of space-time in the microworld. The structure of space-time has always played an important role in physical theory. It is used to simplify concepts connected with symmetry and to give understanding about a great many complicated phenomena in a unified way, in terms of a few simple principles. For example, Einstein's courageous idea of unification of space and time was in fact a starting point for the general theory of gravitation and became the cornerstone of modern relativistic quantum theory. At present, further development of high energy physical experiments and theories dictates a deeper level in the understanding of the structure of space-time and its properties at small distances. However, our concepts of space-time are confirmed experimentally to be valid to distances of the order 10^{-15}–10^{-16} cm.

Among the proposed properties of space-time at small distances [for example, the concept of superspace and pregeometry, lattice (discrete) and cellular structures of space-time, higher-dimensional geometry, etc.] an important role is played by the idea of the stochastic character of space-time. This idea is based on the fact that the quantum fluctuations in geometry are inescapable if one believes in the quantum principle and Einstein's theory. A possible deeper connection between the concept of nonlocality and the hypothesis of space-time stochasticity, originally due to D. I. Blokhintsev, stimulates the *physical foundation* for the *nonlocal quantum field theories* constructed by G. Wataghin, A. Pais and G. E. Uhlenbeck, H. Yukawa, G. V. Efimov, and others.

This book is devoted to the construction of a self-consistent model of nonlocal quantum field theory and stochastic quantum mechanics based on the hypothesis of space-time stochasticity and nonlocality in the microworld. In that case the occurrence of form-factors in the theory (i.e., *violation of the concept of locality* at small distances) and the *random behavior of a physical system* are caused by the stochastic nature of space-time on a small scale. The averaging of any fields independent of their nature (i.e., mass, spin, charge, etc.)

over this stochastic space-time leads to the nonlocal fields considered by G. V. Efimov. In other words, stochasticity of space-time (after being averaged on a large scale) as a self-memory makes the theory nonlocal. This allows one to consider in a unified way the effect of stochasticity (or nonlocality) in all physical processes. Moreover, the universal character of this hypothesis of space-time at small distances enables us to re-interpret the dynamics of stochastic particles and to study some important problems of the theory of stochastic processes [such as the relativistic description of diffusion, Feynman-type processes, and the problem of the origin of self-turbulence in the motion of free particles within nonlinear (stochastic) mechanics]. In this direction our approach (Part II) may be useful in recent developments of the stochastic interpretation of quantum mechanics and fields due to E. Nelson, D. Kershaw, I. Fényes, F. Guerra, de la Peña-Auerbach, J.-P. Vigier, M. Davidson, and others. In particular, as shown by N. Cufaro Petroni and J.-P. Vigier, within the discussed approach, a *causal action-at-distance interpretation* of a series of experiments by A. Aspect and his co-workers indicating a possible *nonlocality property of quantum mechanics*, may also be obtained. Aspect's results have recently inspired a great interest in different nonlocal theories and models devoted to an understanding of the implications of this nonlocality.

This book consists of two parts. The first part is an attempt to give the basic principles of nonlocal theory of quantized fields and to construct the theory of electromagnetic and four-fermion weak interactions of leptons from the stochastic space-time point of view. In the second part, we present to the reader the main ideas of stochastic quantum mechanics and stochastic quantization methods for field theory. The core of the book is based on the original results obtained during the years (1877–83) in collaboration with Professor G. V. Efimov and his students; it covers nonlocal theory of quantized fields, the relativistic and nonrelativistic dynamics of stochastic particles, stochastic quantum mechanics and quantum fields. Since the book is addressed primarily to theoretical and mathematical physicists, it emphasizes a conceptual structure – the definition and formulation of the problem and techniques of solution. Since the emphasis differs from that of conventional physics texts, students may find this book useful as a supplement to their texts.

In both parts we include the mathematical fundamentals of quantum field theory and stochastic processes: generalized function and functional methods, and their applications in quantum field theory, the basic concepts of probability theory and random processes, and some new developments of field theory and stochastic quantization methods. We have organized the book so that it will be useful to new students as well as to experienced researchers. It is neither self-contained nor complete, but it intends to develop the central ideas, to explain the main results of a physical and mathematical nature, and to provide an introduction to the related literature.

Physicists concerned with condensed matter may be interested in the discussion of the stochastic solution to the classical Yang–Mills equation and of vacuum structure; the behavior of vacuum energy density in the strong coupling limit; the mechanism of vacuum tunneling phenomena in the non-Abelian gauge theories; and some problems of cosmic rays (their acceleration mechanism and the energy spectrum, and the ratio of the intensities of the electron component to the proton component at the same energy level). The central matters are functional integral techniques, stochastic quantization and space-time metric fluctuational methods.

I would like to thank Professor G. V. Efimov (JINR) for helpful and stimulating discussions, valuable comments, and for teaching me the principles of scientific work. The author is truly indebted to Professors N. N. Bogolubov and V. A. Meshcheryakov for their hospitality at the Joint Institute for Nuclear Research, Dubna. I wish to thank Doctors M. Dineykhan (Mongolian Academy of Sciences, Ulan-Bator, Mongolia) and P. Exner (JINR and Nuclear Centre, Charles University, Prague, Czechoslovakia) for useful comments and help in preparing the manuscript. I would also like to thank Professors Ch. Tseren and Sh. Tsegmid (Academy of Sciences, Ulan-Bator, Mongolia); N. Sodnom, E. Damdinsuren, and O. Lkhavga (Mongolian State Univ. Ulan-Bator, Mongolia); B. M. Barbashov, V. G. Kadyshevsky and A. N. Sisakhyan (JINR); V. Ya. Fainberg and D. A. Kirzhnits (Lebedev Physical Institute, Moscow, USSR), Ya. A. Smorodinskii (Kurchatov Institute, Moscow, USSR); N. E. Turin (High Energy Institute, Serpukhov, USSR); G. M. Zinovjev (Institute of Theoretical Physics, Kiev, USSR); P. N. Bogolubov (Institute for Nuclear Research, Academy of Sciences, Moscow, USSR); Hans-J. Treder (German Academy of Sciences, Babelsberg, DDR); Alwyn van der Merwe (Univ. of Denver, Colorado, USA); and J.-P. Vigier (Henri Poincaré Institute, Paris, France) for constant interest in my work and valuable comments and encouragement in my investigations.

The author wishes to thank D. Reidel's referees for their valuable comments, and Miss D. Bond and G. G. Sandukovskaya who have corrected the grammar and style of the English version of the manuscript.

Finally, I wish to thank my mother S. Myadag, sister Kh. Tseveg, wife Dz. Tserendulam, and children Nyamtseren and Tsedevsuren for the warmth of their love and support.

KHAVTGAIN NAMSRAI

Dubna-Ulan-Bator
September, 1984.

Notations used in the book

To introduce the general notation, all four-component vectors are chosen real. The metric is given by the Minkowski one, $g_{\mu\nu}$ ($\mu, \nu = 0, 1, 2, 3$), for which

$$g_{\mu\nu} = 0 \qquad \text{for} \quad \mu \neq \nu,$$

$$g_{00} = -g_{11} = -g_{22} = -g_{33} = 1.$$

The product of two four-vectors p and q with components

$$p = (p_0, \mathbf{p}) = (p_0, p_1, p_2, p_3), \qquad q = (q_0, \mathbf{q}) = (q_0, q_1, q_2, q_3)$$

is defined as

$$(pq) = p_\mu q_\nu g_{\mu\nu} = p_\mu q_\mu = p_0 q_0 - \mathbf{pq} = p_0 q_0 - p_1 q_1 - p_2 q_2 - p_3 q_3.$$

Summation is carried out by repeating indices, omitting the symbol of summation.

Derivatives will be noted as follows:

$$i\frac{\partial}{\partial x_\mu} = i\,\partial_\mu = p_\mu \qquad (\mu = 0, 1, 2, 3),$$

$$\Box = p^2 = -\partial_0^2 + \partial^2 = -\partial^2/\partial x_0^2 + \partial^2/\partial x_1^2 + \partial^2/\partial x_2^2 + \partial^2/\partial x_3^2,$$

$$\partial^s = \partial_{\mu_1} \cdots \partial_{\mu_s} = \partial^s/\partial x_{\mu_1} \cdots \partial x_{\mu_s}.$$

The equivalent notation

$$f(x) = f(x_0, \mathbf{x}) = f(x_0, x_1, x_2, x_3)$$

will be used for a function $f(x)$ defined in four-dimensional space-time. Sometimes $x_0 = t$ will be used.

The following equivalent notations

$$\int \mathrm{d}^4 x f(x) = \int \mathrm{d}x_0 \int \mathrm{d}\mathbf{x} f(x_0, \mathbf{x}) = \int \mathrm{d}x_0 \int \mathrm{d}^3 x f(x_0, \mathbf{x})$$

$$= \int \mathrm{d}x_0 \int \mathrm{d}x_1 \int \mathrm{d}x_2 \int \mathrm{d}x_3 f(x_0, x_1, x_2, x_3)$$

are used for integrals of some function $f(x)$ over all four-dimensional space-time.

Four- and three-dimensional δ-function will be read

$$\delta^{(4)}(x) = \delta(x) = \delta(x_0)\,\delta^{(3)}(x) = \delta(x_0)\,\delta(\mathbf{x}) = \delta(x_0)\,\delta(x_1)\,\delta(x_2)\,\delta(x_3),$$

omitting upper indices (3) and (4).

The Dirac γ_μ-matrices are chosen in the form:

$$\gamma_\mu = (\gamma_0, \gamma),\ (\mu = 0, 1, 2, 3), \qquad \gamma_\mu\gamma_\nu + \gamma_\nu\gamma_\mu = 2g_{\mu\nu},$$

$$\gamma_5 = i\gamma_0\gamma_1\gamma_2\gamma_3, \qquad \gamma_5^2 = 1,$$

$$0_\mu = \gamma_\mu(1+\gamma_5), \qquad \hat{p} = p_\mu\gamma_\mu = p_0\gamma_0 - \mathbf{p}\gamma.$$

In this book the unit system (with rare exceptions) in which the light velocity c and the Planck constant $\hbar$ divided by 2π ($\hbar = h/2\pi$) are both equal to unity, i.e., $\hbar = c = 1$, is used. In this unit system, energy, momentum and mass have the dimension of inverse length and time $x_0 = t$ has the dimension of length.

Open sets in the real n-dimensional space $\mathbb{R}^n$ will be denoted by G, Ω, Γ, etc., and $\bar{G}, \bar{\Omega}, \bar{\Gamma}$, etc., are then their closure in $\mathbb{R}^n$. In this notation $S = \bar{G}\backslash G$ means boundary of the set G.

Sometimes the symbols $\forall$ and $\exists$ will be used, which mean: $\forall$ – 'for any' and $\exists$ – "there exist". For example, $\forall\ a \in \Gamma$, $f(a) = 1$ means 'for any a belonging to Γ, $f(a) = 1$ always', and $\exists\ a \in \Gamma$, $f(a) = 1$ means 'there exists an a belonging to Γ, for which $f(a) = 1$'.

The symbol $[a]$ means the enteger part of the number a, for example, $[5] = 5$, $[5.3] = 5$, $[\pi] = 3$.

All other mathematical symbols and notations used in the book should be self-explanatory.

Part I

Nonlocal Quantum Field Theory

Part I is devoted to the construction of a nonlocal theory of quantized fields and presents systematically the results of studies in this direction. In Chapter 1, a method is formulated for introducing stochasticity into space-time with indefinite metric (i.e., Minkowski space) and of averaging in it. This method allows us to solve the problems of conserving relativistic invariance and of eliminating ultraviolet divergences in constructing quantum field theory from a stochastic space-time point of view. A class of test functions and form factors (or generalized functions) arising from the averaging prescription in a measured stochastic space-time is also considered.

Chapter 2 deals with the formulation and proof of basic principles (the quantization problem, the causality condition and unitarity) of the relativistic quantum field theory in the case of nonlocal interactions. The physical meaning of the form factors is discussed too. It is shown that the finite and unitary S-matrix describing the nonlocal interactions of quantized fields is a solution of the Cauchy problem for the evolution equation (Schrödinger equation in the interaction picture at imaginary time) with retardation.

The finite S-matrix obeying the basic theoretical principles is constructed for the case of the electromagnetic and four-fermion weak interactions (Chapters 3 and 4). The different characteristics of electromagnetic and weak processes are calculated and restrictions on the magnitude of the fundamental length are obtained.

Functional methods and their applications in quantum field theory are considered in Chapter 5. Here attention is focussed on functional analysis of Grassmann algebra and on a definition of the functional integrals in quantum field theory. Vacuum energy estimates have been obtained in the φ_d^4-nonlocal model and the anharmonic oscillator case.

Chapter 1

Foundation of the Nonlocal Model of Quantized Fields

1.1. Introduction

The fundamental problems of quantum field theory (QFT) are a unified description of particle physics phenomena and a strict mathematical formulation of the basic principles of the theory. One of the directions in which these problems are investigated, is called the axiomatic approach in QFT. Bogolubov, Medvedev and Polivanov's (BMP) formulation, for example, is based on the variational derivative method of the S-matrix and on the causality condition. Another field approach is connected with names of Garding, Wightman, Haag, Lehmann, Symanzik and Zimmermann, etc. The mathematical fundamentals of QFT were given in a series of monographs [see, for example, Bogolubov *et al.* (1958), Streater and Wightman (1964), Jost (1965), and Bogolubov *et al.* (1975)].

Progress achieved in QFT in the last decade is connected with the developments of *non-Abelian gauge field theories* (in particular, the one introduced by Yang and Mills, 1954) and with the *constructive quantum field theory* advanced by Glimm and Jaffe and their followers. The former have opened a way to the unification of all interactions existing in nature, in particular, the link between electromagnetism and the weak interaction (made known by S. Weinberg, 1967, 1980; A. Salam, 1968, 1980; and Sh. Glashow, 1961, 1980) in what is now being called 'electroweak' interaction. In the electroweak theory, the weak and electromagnetic interactions are described in a unified way in terms of an exact but spontaneously broken gauge symmetry. Evidence for the discovery of the $W^{\pm}$, Z^0-bosons with the masses $m_W = 81 \pm 2\,\mathrm{GeV}$ and $m_{Z^0} = 93 \pm 2\,\mathrm{GeV}$ (which coincide exactly with the theoretical predictions), which are indispensable to the electroweak theory, was presented in 1983 by two large experimental groups called UA1 and UA2 working at CERN.

The constructive quantum field theory has been enriched by new powerful methods which rely on the ideas from Euclidean field theory and use probabilistic techniques and concepts. A central role is played by the Euclidean–Markov field $\varphi(f)$, introduced by Schwinger (1959), Nakano (1959) and Symanzik (1966, 1969). Further, Nelson (1971, 1973) described the Euclidean boson quantum field theory as a Markov process: the Euclidean fields are random variables and the Green functions are expectations of products of them (see Part II). In Nelson's formulation, the *Markov property is crucial* (Chapter 6) for the reconstruction of the physical theory from the Euclidean theory and for the study of the Euclidean theory itself. Osterwalder and Schrader (1973, 1975) and Glaser (1974) established the necessary and sufficient conditions for Euclidean Green's functions to define a unique Wightman field theory. Thus the Euclidean methods in QFT have generated a constructive Euclidean field theory which is the synthesis of the Nakano–Schwinger Euclidean field theory with the Feynman–Wiener functional integral theory at an imaginary time and, on a very deep level, represents the joining of QFT (at least, the theory of scalar boson fields) with the theory of the Markov process. This interrelation allows one to insert the theory of boson fields into classical statistical mechanics.

Recently, a profound connection between quantum theory and the theory of stochastic processes has been established, more explicitly with the developments towards stochastic mechanics and its relations with quantum mechanics, originally proposed by Fényes (1952) and Kershaw (1964), neatly formulated by Nelson (1966, 1967), and successively extended by many authors: de la Peña-Auerbach and Cetto (1975), Dankel (1970), Caubet (1976), Dohrn and Guerra (1978), Guerra and Ruggiero (1973), Guerra and Loffredo (1980), Vigier (1982), Davidson (1979–82), Lee (1980), Yasue (1979), and others. This problem concerns Part II of the book.

Historically old approaches devoted to the construction of a finite theory of quantized fields free from *ultraviolet divergences* exist. These approaches began from the early stages of the development of field theory and were based on the analysis of the fundamental principles (causality condition, locality and properties of geometry on a small scale, etc.) of modern local QFT at small distances (or, equivalently, at very high energies). The *locality* condition (the commutation rules), e.g., is one of the fundamental principles of the local QFT. More clearly, this means that the commutator of operators of physical fields disappears outside the light cone. On the other hand, this property of *locality* ensures the independence of events separated by spacelike intervals, i.e. the *causality condition* in space-time (usually called microcausality). A strict formulation of the microcausality condition in QFT has been given by Bogolubov and co-workers (see Bogolubov and Shirkov, 1980).

We present here (Part I) a mathematical method of constructing a nonlocal theory of quantized fields by means of the hypothesis of *space-time stochasticity.* Our model belongs to a class of approaches in which one assumes a modification of the usual (local) theory of quantized fields, based on the analysis of the principle of locality, and postulates that a possible violation of locality at small distances may be conditioned by intrinsic problems of QFT, like the ultraviolet divergences, the problem of electron self-energy, etc. This problem appears especially inevitable as soon as we want to describe spread-out (*extended*) *objects* (particles) within QFT.

In attempts to reach a consistent construction of QFT with a reasonable locality (macrocausality, i.e., maybe a slight violation of locality at small distances) condition, a distinguished role is played by nonlocal QFTs (Wataghin, 1934; Pais and Uhlenbeck, 1950; Yukawa, 1950; Kirzhnits, 1966; Efimov, 1977; etc.). Moreover, constructing a nonlocal theory of quantized fields is stimulated by the latest results from a series of experiments (Aspect *et al.*, 1982a, b) designed to test a fundamental aspect of quantum theory, which show that:

(a) quantum theory holds good as far as we can tell,
(b) Bell's inequality is violated,

or, in a more concrete form, one may say that these experiments establish: (a) the non-existence of a local hidden variable, (b) the existence of quantum superluminal *connections* between the action of the independent parts of a measuring device separated by *space-like intervals.* The last fact tells us that quantum phenomena possess a *nonlocality property.* In other words, there appears to be some form of 'nonlocal' effect, an effect that Einstein called a 'spooky action at a distance'. These results of Aspect have indicated that quantum mechanics is adequate even for those situations where nonlocal effects might seem to be necessary and have recently inspired great interest in different nonlocal theories and models devoted to an understanding of the implications of this nonlocality.

There exist two different approaches to the construction of nonlocal theory. Supporters of the first approach assume that in nature there exists some new *fundamental constant* of the dimension of length, together with such constants as the *velocity of light c* and the *Planck constant ħ*. Further, they assume that, in a domain characterized by this new universal length l, one must expect a principle change in our concepts of physical world and, in particular, the concept of space-time and locality (causality).

Some possibilities of introducing the concept of fundamental length in physics were considered by Heisenberg (1936); Born (1949); Pais and Uhlenbeck (1950); Yukawa (1950); Takano (1961, 1967); Markov (1958, 1959); Tamm (1965); Kirzhnits (1966); Blokhintzev (1973b); Fubini (1974); Ginzburg (1975); Efimov (1977a); Ehrlich (1976, 1978); Cheon (1978, 1979); Hsu and Mac (1979); Lacroix (1979); Kadyshevsky (1980); and Brout *et al.* (1980), where earlier references concerning this problem are cited.

The second approach is based on the assumption that the parameter l is not a universal fundamental constant, but characterizes only the domain of *nonlocal interaction* (or the size of *extended particles*, say, electrons) of the considered quantum fields. In particular, such an opinion is the basis of Efimov's (1977a) construction of a nonlocal theory. Thus, the parameter l, with the dimension of length, arises inevitably in any attempt to introduce non-locality into the theory. Recently, high-precision measurements (Bailey *et al.*, 1979; Van Dyck *et al.*, 1977, 1979; Robiscoe, 1968; Robiscoe and Shyn, 1970) in atomic physics, for example, measurements of the anomalous magnetic moment (AMM) of muons (electrons) and of the Lamb shift, have given the following restrictions on the parameter $l: l \lesssim 10^{-15}$ cm ($l \lesssim 10^{-14}$ cm) and $l \lesssim 10^{-13}$ cm, respectively (see Chapter 3, and Efimov, 1977a; Hsu and Mac, 1979). For discussion of various theoretical contributions, where earlier references can be found, as well as for a review comparing the theory and experiments, see Brodsky and Drell (1970); Lautrup *et al.* (1972); Calmet *et al.* (1977); Kinoshita (1979), and Levine *et al.* (1982). From the high-energy experimental data on $e^+ e^-$ physics, it follows that $l \lesssim 10^{-16}$ cm (Flügge, 1980; Wolf, 1980; Beron *et al.*, 1978; Barber *et al.*, 1979a, b, 1980; Bartel *et al.*, 1980; Berger *et al.*, 1980, and Ting, 1982). In other words, this means that the *leptons* are *point-like particles* with radii smaller than 10^{-16} cm. Recently, experiments have been designed to probe distances up to 10^{-17} cm, which allow one to test some theoretical estimates of the value of $l \lesssim 3 \times 10^{-5}$ fermi (Bracci *et al.*, 1983) and predictions about the structure of the leptons with radii $l \sim 4 \times 10^{-17}$ cm (Li, 1982).

The problem of constructing the nonlocal theory satisfying the basic principles of QFT, like the Lorentz covariance, finiteness, unitarity, causality and gauge invariance, is now solved (Efimov, 1972, 1977a). One assumes in this theory that neutral particles (for example, photons and neutrinos) are 'carriers' of *nonlocality*, while the charged fields are considered to be local. However, this way of introducing nonlocality into the theory does not remove all *ultraviolet divergences* from the perturbation series for the *S-matrix*. There are divergences in the so-called *vacuum polarization* diagrams constructed by using propagators of the charged particles. Usually, in order to remove the divergences in these diagrams, one uses the modified *Pauli–Villars regularization* (see Bogolubov and Shirkov, 1980; Slavnov, 1974; Efimov, 1972). This method of regularization of singular functions is to be understood as a formal procedure only and has no definite physical meaning.

It is generally accepted at present that the essence of mathematical methods for removing divergences is more or less explicitly connected with the properties of space-time at small distances, or with the very nature of high-energy interactions which is inherent to all types of interactions. Thus, we believe that the method of eliminating ultraviolet divergences must be the *same for all types of diagrams* and must have a *clear physical meaning*.

In this book we give a *proper foundation* for the concept of *nonlocality* and *stochasticity* which appears in physics and presents a basis for a nonlocal theory of quantized fields for scalar, electromagnetic and four-fermion weak interactions (Part I), and for stochastic mechanics and stochastic field theory (Part II) within the hypothesis of the space-time stochasticity in the microworld. We would like to introduce into our consideration the *stochastic space-time* $\mathbb{R}^4(\hat{x})$ with

$$\hat{x} = (x_0 + ib_4, \mathbf{x} + \mathbf{b}), \tag{1.1}$$

where $x_\mu = (x_0 = ct, \mathbf{x})$ is the regular part of components $\hat{x}$, and

$b_E = (b_4 = c\tau, \mathbf{b})$ is some random vector with a *measure* $w(b_E^2/l^2)$ obeying the conditions

$$\int dw(b_E^2/l^2) = 1, \qquad dw(b_E^2/l^2) \geqslant 0, \tag{1.2}$$

and where l is some universal length (a scale of errors). In our case, the universal length l characterizes physically a certain domain within which the existing space-time concepts and causality conditions may be violated and the stochastic properties or nonlocality in the metrics can be manifested if they exist.

It appears that the *field obtained by averaging out the space-time* $\mathbb{R}^4(\hat{x})$ is the *nonlocal field* considered by Efimov (1966, 1977a) (see Section 1.3). Equivalence of these approaches leads to the following hypothesis: the origin of the form factors, which change the electromagnetic (Efimov, 1972) and weak (Efimov *et al.*, 1973) potentials at small distances, and the properties of *vacuum polarization* may be connected with the stochasticity of space-time on the microscale. Thus, in our model, apart from the universal length l, there exists a functional arbitrariness connected with the choice of a measure $w(b_E^2/l^2)$ or of a form of weak and electromagnetic potentials at small distances (see Section 2.3). This situation allows us to interpret our approach as a phenomenological scheme having unknown parameters in the theory. Therefore, our model belongs to the second-class approaches to constructing nonlocal theories mentioned above.

Thus, two ideas, the introduction of stochasticity into spacetime in the microworld and the method of averaging (which allows a transition to a large scale of space-time) in it, lead to the construction of a QFT with nonlocal interaction, satisfying the basic principles of the theory. The first part of this book is devoted to the study of these problems.

1.2. Stochastic Space-Time

The structure of space-time and the physical phenomena within it, enter inseparably into human cognition, and their inter-relations are those of dialectical unity. The simple mechanical motions and the majority of biological and chemical processes, including human activity, are confined to the concept of Euclidean flat three-dimensional space. To order events in this space to each point-like event $\mathscr{P}$, three numbers $\mathbf{r} = (x, y, z)$ are attributed, i.e., the co-ordinates of this event; the passage from one co-ordinate system to another is fulfilled by the *Galilean transformation:*

$$\mathbf{r} = \mathbf{r}' + \mathbf{V}t, \quad t = t',$$

where $\mathbf{V}$ is the relative velocity of the two co-ordinate inertial systems.

In the course of the advance of human cognition to deep levels of the structure of matter, the conception of space-time has been defined more and more precisely and is still gaining further generalizations and developments. For example, in the first years of our century, Einstein and Minkowski *united* space and time as a single entity and introduced the new concept of a space-time manifold – the pseudo-Euclidean four-dimensional space, which was a starting point for the general theory of relativity and modern relativistic QFT. In the first case, the Newtonian gravitation is due to the *curvature* of this space-time manifold. In the Einstein theory of gravitation the ordering (or arithmetization) of the events is fulfilled by means of certain *agreements* (Poincaré, 1898; Einstein, 1950; Fridman, 1965; also see Blokhintzev, 1973b) based on the principle of the *universal constant velocity of light* and on an assumption about the existence of a '*standard*' *clock* (Marzke and Wheeler, 1964; Misner, Thorne and Wheeler, 1973). This arithmetization leads to *Minkowski space* with the *indefinite metric*

$$ds^2 = c^2\,dt^2 - d\mathbf{x}^2 = dx_0^2 - d\mathbf{x}^2. \tag{1.3}$$

Transformation conserving the metric (1.3) and connecting co-ordinates of events in different co-ordinate systems is the *Poincaré–Lorentz transformation*:

$$x = \frac{x' + Vt'}{(1 - V^2/c^2)^{1/2}}, \qquad y = y'. \qquad z = z', \qquad t = \frac{t' + (V/c^2)x'}{(1 - V^2/c^2)^{1/2}}.$$

Further, the discovery of the quantum phenomena was initiated by introducing an infinite-dimensional functional space, the Hilbert space, on which the states of quantum mechanical systems are defined. For the description of the subnucleon processes, it was necessary to introduce, simultaneously with the usual space-time, supplementary so-called internal abstract spaces connected with quantum numbers (spin, isospin, strangeness, charge, etc.) of the elementary particles. Roughly speaking, the 'ordering' of events in these spaces is reduced to group transformations and to their representations. This abstract concept of spaces turned out to be very useful and tenacious of life in classifications of the elementary particles and in the unified description of all their types of interactions.

The theoretical and experimental successes of high-energy physics now dictate a deeper level of understanding of the structure of space-time and its properties at small distances. In the near future, distances up to 10^{-17} cm will be probed by high-energy experiments glided at CERN, Fermilab., Serpukhov, etc. Moreover, as has been mentioned above, the intrinsic problems of QFT and the hypothesis of *fundamental length*, and recent results in Einstein–Podolsky–Rosen–Bohm–type experiments on single photon pairs, establishing *nonlocality* in quantum mechanics, give a strong stimulus to the investigation of the space-time structure of matter.

The concepts of superspace and pregeometry (Wheeler, 1964; Akama, 1981; Terazawa, 1981), the submicroscopic scale and *lattice* (*discrete*) structure of space-time (Wilson, 1974; Kogut and Sussking, 1975; Vialtsev, 1965) and stochastic (or fluctuational) and *higher-dimensional geometry* (Snyder, 1947) may become the arena of future physical theories. Friedberg and Lee (1983) and Cole (1972), for example, discussed discrete quantum mechanics and a *cellular space-time structure*, respectively. Finkelstein (1969, 1972, 1974) published a series of papers developing a theory including space-time structure – the so-called *space-time code*. As it has been supposed by Wheeler, the electrical and nuclear charges give evidence for the presence of a submicroscale structure of space-time resembling a *foam-like* structure which is on the whole homogeneous.

It is assumed that *quantum fluctuations in geometry* are inescapable, if one believes in the quantum principle and Einstein's theory. According to Misner, Thorne and Wheeler (1973, pp. 1193, 1201) 'The fluctuations widen the narrow swath cut through super-space by the classical history of the geometry. In other words, the geometry is not deterministic, even though it looks so at the everyday scale of observation.... The quantum fluctuations in the geometry of space are so great at small distances that even the topology fluctuates, makes 'wormholes', and traps lines of force.' Thus, it seems, the fluctuations in the geometry make a multiple connected topology which provides a natural description for the electric charge as electric lines of force trapped in the topology of a multiple connected space.

Thus, we believe that among the above-mentioned hypothesis of space-time properties, the idea of the stochastic character of space-time at small distance will play the most important role. In particular, as we shall see below, this idea leads to the concept of *nonlocality* and thus it gives the very nature of nonlocal QFT.

Some attempts were undertaken to construct quantum field theory in a stochastic space-time (Markov, 1959; Komar and Markov, 1959; Takano, 1961, 1967; Ingraham, 1967; Blokhintsev, 1973a, 1975 and references therein). A stochastic space-time, which can be used in constructing theories of elementary particles, was first considered by March (1934, 1937), Markov (1958) and Yukawa (1966). Recently, this problem has been discussed by Blokhintsev (1975), Roy Choudhury and Roy (1980), Cerofolini (1980), Prugovečki (1984). Frederick (1976) and Roy (1979) considered mathematical spaces with stochastic metric and quantized domain, respectively. A mathematical definition of linear space with random metric, mainly in the Euclidean case, was given by Menger (1942, 1949); Sherstnev (1963) and Schweizer (1967). Prugovečki's (1978) papers are devoted to the construction of the relativistic kinematics of massive and massless particles in a stochastic phase space.

In contradistinction to Markov, Takano, Frederick, Menger, Sherstnev, Schweizer and others, we do not start here from metric space with stochastic metric, but we consider the behavior of dynamic systems in space-time with a measure (or distribution), say in the space-time $\mathbb{R}^4$ $(\hat{x})$ with stochastic components (1.1) and measure (1.2). This view is also the basis of Prugovečki's statement: 'If we, however, accept the uncertainty principle and yet retain

Einstein's operational definition for the structure of space-time, then we are forced into describing space-time in stochastic terms ...' (Brooke, Guz and Prugovečki, 1981, p. 1731). This proposed way has also another justification: *stochasticity in physics* appears more naturally on the stage of arithmetization (or all reality is subject to an intrinsic stochasticity inherent in the measurement process caused by stochasticity of space-time) rather than on the metrical stage. The *stochastic metric* is introduced on a more deeper level, where one should take into account the gravitational effects connected with introducing stochasticity of space-time into physics. The study of latter problems is, however, beyond the scope of this book, although some initial idea in this direction will be discussed in Section 11.7 of the Part II.

Of course, stochasticity of space-time and random metric are connected with each other as a single entity concept; for instance, any space, the points of which have the measure, reduces to the space with random (or stochastic) metric. Indeed, in our case of the space-time $\mathbb{R}^4(\hat{x})$, the *metric averaged* is given by

$$s_R^2 = \langle s^2 \rangle_{\mathbb{R}^4(\hat{x})} = \int d^4 b_E\, w(b_E^2/l^2)[(x_0 + ib_4)^2 - (\mathbf{x} + \mathbf{b})^2] = x_0^2 - \mathbf{x}^2 - l^2. \quad (1.4)$$

On the other hand, this result can be rewritten in the form

$$s_R^2 = s_0^2 - a^2, \qquad s_0^2 = x_0^2 - \mathbf{x}^2, \quad (1.5)$$

where a is a random number with a measure $w(a/l)$ such that

$$\int da\, w(a/l) = 1, \qquad \int da\, aw(a/l) = 0, \qquad \int da\, a^2 w(a/l) = l^2.$$

We see that modification of the metric form (1.5) may be understood geometrically as the introduction of the *fifth dimension* which is conditioned by the probability measure. A metric form of the type of (1.4) (we call it nonlocal metric form) will be considered also (Section 11.6).

An attempt to construct a quantum field theory, based on the metric form (1.4) [or (1.5)] was done by Markov and Takano.

Now we concretize the form of the space-time $\mathbb{R}^4(\hat{x})$ and explain the following question: why the points of space $\mathbb{R}^4(\hat{x})$ have to be of the form (1.1) and the reason for it. The answer to this question will inevitably depend on the dimension of space and the operation of the arithmetization of events. In the nonrelativistic case (space dimension is three), it is sufficient to assume that space is stochastic with respect to the spatial component $\mathbf{x} \to \hat{\mathbf{x}} = \mathbf{x} + \mathbf{b}$, where $\mathbf{b}$ is a universal stochastic variable, subject to a *probability measure* with $w(\mathbf{b}) = w(|\mathbf{b}|^2)$, for example,

$$w(\mathbf{b}) = (2\pi l^2)^{-3/2} \exp(-|\mathbf{b}|^2/2l^2). \quad (1.6)$$

This form of the measure follows from the homogeneity and isotropy of space.

In the relativistic case, such an operation requires some explanation. In the relativistic theory the space in which physical processes are investigated is the Minkowski space, i.e., the space $\mathbb{R}^4(\hat{x})$ must be the Minkowski space. The *indefinite metric* of this space leads to a number of specific problems, which are not encountered in Euclidean space.

These difficulties peculiar to physical space are associated with the requirements of *invariance* and the condition of *normalization* of the probability of some value (in particular, the measure $w(b_\mu)$ itself) of the interval $b^2 = b_0^2 - \mathbf{b}^2$ in a space with indefinite metric (for more detail, see Blokhintsev, 1973b). Thus, the invariance requirement means that the measure $w(b_\mu)$ of the vector b_μ must be a function of the interval $b^2 = b_\mu b^\mu$, and the normalization condition gives the equation

$$\int dw(b_\mu b^\mu) = 1.$$

The simultaneous fulfillment of these two requirements for $w(b_\mu)$ is actually *impossible* in Minkowski space. The above difficulties can be avoided by making the following assumptions:

1. The stochastic nature of the space $\mathbb{R}^4(\hat{x})$ is manifested in the Euclidean domain of the variables $\hat{x}_E = x_E + b_E$.

2. Physical quantities are regarded as functions of the *complex time* $t + i\tau$ in the limit of microscale, where the stochasticity of space-time may be manifested [τ is a stochastic variable which ensures the hypothesis of a stochastic nature of the *Euclidean space* $\mathbb{E}^4(\hat{\mathbf{x}}, \tau)$ instead of the Minkowski space $\mathbb{R}^4(\hat{\mathbf{x}}, t)$]. A justification of the importance of the method with $t \to t + i\tau$ in quantum field theory and quantum mechanics can be found in the papers by Alebastrov and Efimov (1974) and Davidson (1978), respectively (see also Jannussis and Papatheou, 1982). Thus, in our model the real points of the space-time $\mathbb{R}^4(\hat{x})$ consist of two parts (1.1), where $b_4 = c\tau$, and any physical quantity $F(\cdot)$ in $\mathbb{R}^4(\hat{x})$ depends on arguments of the type $x_0 + ib_4, \mathbf{x} + \mathbf{b}$ with the measure $w(b_E^2/l^2)$, i.e., $F(\cdot) = F(x_0 + ib_4, \mathbf{x} + \mathbf{b})$.

1.3. The Method of Averaging in the Stochastic Space-Time and Nonlocality

Since, in accordance with our assumption, the real points of the space $\mathbb{R}^4(\hat{\mathbf{x}})$ are stochastic, these points cannot be used as the basis for a coordinate system, and for the same reason derivatives with respect to them cannot be formulated. However, the space of common experience (i.e., the laboratory frame) is nonstochastic on a large scale. The stochastic nature of space is manifested only in the microscopic world. Therefore, one can take a macroscale of nonstochastic space mathematically continued from a stochastic microregion.

Such a mathematical construction ensures a nonstochastic space to which the stochastic physical space can be referred (see also Frederick, 1976). In our case, this mathematical construction reduces to *averaging with respect to the measure* $w(b_E^2/l^2)$ at each point $\hat{x} = (x_0 + ic\tau, \mathbf{x} + \mathbf{b})$ of the space $\mathbb{R}^4(\hat{x})$. Thus, by the physical quantity $F(\mathbf{x}, t)$, we understand $\langle F(\hat{x})\rangle_{\mathbb{R}^4(\hat{x})}$

$$F(\mathbf{x}, t) = \langle F(\hat{x})\rangle_{\mathbb{R}^4(\hat{x})} = \int \mathrm{d}^4 b_E\, w(b_E^2/l^2) F(x_0 + ib_4, \mathbf{x} + \mathbf{b}), \tag{1.7}$$

averaged with respect to the measure $w(b_E^2/l^2)$ at a given time t.

It should be noted that with a suitable choice of the function $w(b_E^2/l^2)$ the method of averaging (1.7), which ensures the transition from the small to the large scale, leads to an *entirely new object*. In particular, a scalar field $\varphi(\hat{x})$ after averaging in $\mathbb{R}^4(\hat{x})$ acquires the following form:

$$\begin{aligned}\varphi_R(x) = \langle \varphi(\hat{x})\rangle_R &= \int \mathrm{d}^4 b_E\, w(b_E^2/l^2)\varphi(x_0 + ib_4, \mathbf{x} + \mathbf{b}) \\ &= (2\pi)^{-2} \int \mathrm{d}^4 p\, \mathrm{e}^{ipx}\, \tilde{K}(p^2 l^2)\tilde{\varphi}(p),\end{aligned} \tag{1.8}$$

where

$$\begin{aligned}\tilde{K}(p^2 l^2) &= \int \mathrm{d}^4 b_E \exp(-i\mathbf{pb} - p_0 b_4) w(b_E^2/l^2) \\ &= (2\pi)^2 \int_0^\infty \mathrm{d}r\, r^2 w(r^2/l^2) J_1(r\sqrt{-p^2})(-p^2)^{-1/2}.\end{aligned} \tag{1.9}$$

Here $p^2 = p_0^2 - \mathbf{p}^2$, and $J_1(x)$ is the *Bessel function.* Further we are interested only in the class of measures $w(b_E^2/l^2)$ for which, $\tilde{K}(z)$ in (1.9) *are entire analytical functions* of the variable z with a finite order of growth $\infty > \rho \geqslant \frac{1}{2}$ and which decrease rapidly enough when $z = p^2 \to -\infty$ (i.e., in the Euclidean direction). Thus, expression (1.8) may be rewritten in the other representation by means of a *generalized function* (or distribution) (see Section 1.4)

$$\varphi_R(x) = \int \mathrm{d}^4 y K(x - y)\varphi(y) = \int \mathrm{d}^4 p\, \mathrm{e}^{ipx}\, \tilde{K}(p^2 l^2)\tilde{\varphi}(p), \tag{1.10}$$

where $K(x)$ is some generalized function, the *Fourier transform* of which is given by formula (1.9). The following representation is valid for $K(x)$:

$$K(x) = K(l^2\square)\,\delta^{(4)}(x) = \sum_{n=0}^{\infty} \frac{c_n}{(2n)!}(l^2\square)^n\,\delta^{(4)}(x), \quad \square = -\frac{\partial^2}{\partial x_0^2} + \frac{\partial^2}{\partial \mathbf{x}^2} \tag{1.11}$$

and the coefficients $\{c_n\}$ depend on the particular choice of the measure $w(b_E^2/l^2)$. In quantum field theory, such an object, say $\varphi_R(x)$ has been investigated carefully by Efimov (1968, 1977a) from the point of view of generalized functions $K(x)$ whose space-time properties depend essentially on different sequences of the coefficients $\{c_n\}$. Efimov showed that an object constructed by means of such generalized functions $K(x)$ is smeared (*nonlocal*) in space-time. Thus, the relativistic invariant generalized functions $K(x)$ are a basis of a correct description of *extended* objects (see Chapter 2). In this case, roughly speaking, the parameter l may be identified with the size of an extended object (a particle).

Thus, within the framework of our scheme all fields: neutral and charged are extended (nonlocalized) over the space $\mathbb{R}^4(\hat{x})$. This allows one to take into account in a unified way the effect of stochasticity (or nonlocality) in all physical processes. At this point, even the *space-time metric* $g_{\mu\nu}(x)$ is changed and becomes *nonlocal*:

$$g_{\mu\nu}(x) \to G_{\mu\nu}(x) = \int \mathrm{d}^4 y K(x-y) g_{\mu\nu}(y). \tag{1.12}$$

Some consequences of this interesting possibility will be discussed below in Chapter 11.

Now let us calculate the *causal Green functions* of the *extended field* $\varphi_R(x)$ (bearing in mind that the *T-ordering* symbol concerns the field $\varphi(\hat{x})$, where $\varphi(\hat{x})$ is a scalar field with a mass m) by the following formula:

$$\begin{aligned} D_c(x_1 - x_2) &= \langle 0 | T[\varphi_R(\hat{x}_1)\varphi_R(\hat{x}_2)] | 0 \rangle \\ &= \iint \mathrm{d}^4 b_{1\mathrm{E}}\, \mathrm{d}^4 b_{2\mathrm{E}}\, w(b_{1\mathrm{E}}^2/l^2) w(b_{2\mathrm{E}}^2/l^2) \times \\ &\quad \times \langle 0 | T\{\varphi(x_{10} + ib_{14}, \mathbf{x}_1 + \mathbf{b}_1)\varphi(x_{20} + ib_{24}, \mathbf{x}_2 + \mathbf{b}_2)\} | 0 \rangle \\ &= \iint \mathrm{d}^4 b_{1\mathrm{E}}\, d^4 b_{2\mathrm{E}}\, w(b_{1\mathrm{E}}^2/l^2) w(b_{2\mathrm{E}}^2/l^2) i^{-1} (2\pi)^{-4} \times \\ &\quad \times \int \mathrm{d}^4 p (m^2 - p^2 - i\varepsilon)^{-1} \times \\ &\quad \times \exp[ip_0(x_{10} - x_{20} + ib_{14} - ib_{24}) - i\mathbf{p}(\mathbf{x}_1 - \mathbf{x}_2 + \mathbf{b}_1 - \mathbf{b}_2)] \\ &= i^{-1}(2\pi)^{-4} \int \mathrm{d}^4 p\, \mathrm{e}^{ip(x_1 - x_2)} [\tilde{K}(p^2 l^2)]^2 (m^2 - p^2 - i\varepsilon)^{-1}, \end{aligned} \tag{1.13}$$

where $\tilde{K}(p^2 l^2)$ is given by formula (1.9).

Summarizing this section we see that according to the hypothesis of the stochasticity of space-time we come to the Efimov nonlocal theory differing only in that the causal Green function $S(\hat{p})$ of any charged particle is replaced by (in the momentum representation):

$$S(\hat{p}) \to S_R(\hat{p}) = V(-p^2 l^2) S(\hat{p}),$$

where

$$V(-p^2 l^2) = |\tilde{K}(p^2 l^2)|^2, \qquad (\hat{p} = \gamma_\mu p_\mu),$$

for which the *Mellin representation* (see Section 2.4.4)

$$V(-p^2 l^2) = \frac{1}{2i} \int_{-\beta+i\infty}^{-\beta-i\infty} d\zeta \frac{v(\zeta)}{\sin \pi\zeta} l^{2\zeta}(m^2 - p^2 - i\varepsilon)^{\zeta} \quad (0 < \beta < 1)$$

is valid. The form of the functions $V(-p^2 l^2)$ and $v(\zeta)$ depends on the form of the *measure* $w(b_E^2/l^2)$. For example (Namsrai, 1981c), let

$$w_m(y) = \begin{cases} \pi^{-2} l^{-4} (\frac{1}{2} ml\pi)^{-1/2} \dfrac{m^2 l^2 (1-y^2)^{-1/4}}{4(\sin ml/ml - \cos ml)} J_{-1/2}(ml(1-y^2)^{1/2}), & 0 \leqslant y < \\ 0 & y \geqslant \end{cases}$$

Then we obtain

$$V_b = [\tilde{K}_m(p^2 l^2)]^2 = \left[\frac{m^2 l^2}{\sin ml/ml - \cos ml}\right]^2 [(m^2 - p^2) l^2]^{-2} \times$$

$$\times \left\{\frac{\sin[(m^2 - p^2) l^2]^{1/2}}{[(m^2 - p^2) l^2]^{1/2}} - \cos[(m^2 - p^2) l^2]^{1/2}\right\}^2, \tag{1.14}$$

$$v_b(\zeta) = 9 \times 2^{4+2\zeta}(2\zeta^2 + 7\zeta + 5)/\Gamma(7 + 2\zeta). \tag{1.15}$$

Here m is some parameter $(m^2 l^2 \ll 1)$ which can be identified with the particle mass. Notice that the form factor (1.14) in the case $m = 0$ describes the extended electron as a uniformly charged ball of radius l (see Section 2.3 and Efimov, 1977a). Now let us write the *Gaussian measure*:

$$w_G(y^2) = 4\pi^{-2} 1^{-4} \exp(-2y^2), \qquad 0 \leqslant y < \infty.$$

The function $V_G(-p^2 l^2)$ corresponding to this measure acquires the following form

$$V_G(-p^2 l^2) = \exp[(p_0^2 - \mathbf{p}^2) l^2/4]. \tag{1.16}$$

The main restrictions in the choice of *form factors* $V(-p^2 l^2)$ as *entire analytic functions* arise from the fundamental theoretical principle of the theory, i.e., from unitarity and causality (see Chapter 2 and Alebastrov and Efimov, 1973, 1974).

The physical meaning of form factors $V(-p^2 l^2)$ consist in changing a form of the potential between interacting fields (for example, the Coulomb and Yukawa laws) at small distances, and in making the theory finite in each order of the perturbation series of the theory in coupling constant.

1.4. The Class of Test Functions and Generalized Functions

In this section we give some elementary definitions of test functions and generalized functions (or distributions), and explain their main differences in the cases of local and nonlocal quantum field theories. Our exposition carried out below is based on the works due to Bogolubov, Logunov and Todorov (1975), Vladimirov (1979) and Efimov (1977a).

1.4.1. *Introduction*

The concept of generalized functions is the generalized classical definition of the function. This generalization arose from physical requirements and enabled some idealized concepts to be defined as a density of the material point, and of the point-like charge or dipole etc. On the other hand, the concept of generalized functions is based on the fact that it is practically impossible to measure the value of a physical quantity at a given point, and that one can only measure its averaged value in the relatively small surroundings of this point. As a result, it is usually assumed that the limit of sequences of these averaged values may be understood as a value of the physical quantity under consideration at a given point.

For example, in order to explain the above-mentioned situation we attempt to define the density generated by the material point with the mass $m = 1$. Assuming that this point is the origin of the co-ordinate system, we distribute this mass uniformly inside the sphere with the radius of ε and the centre at the point 0. As a result we obtain the averaged density $\rho_\varepsilon(x)$

$$\rho_\varepsilon(x) = \begin{cases} (4\pi\varepsilon^3/3)^{-1}, & \text{if } |x| < \varepsilon, \\ 0, & \text{if } |x| > \varepsilon. \end{cases}$$

However we are interested in the density at $\varepsilon = +0$. Further, we denote the following function

$$\delta(x) = \begin{cases} +\infty, & \text{if } x = 0, \\ 0, & \text{if } x \neq 0, \end{cases} \tag{1.17}$$

as a pointwise limit of sequence of the averaged densities $\rho_\varepsilon(x)$ at $\varepsilon \to +0$. It should be required that

$$\int \delta(x)\, dx = 1.$$

However, for the function $\delta(x)$ defined by the formula (1.17)

$$\int \delta(x)\, dx = 0.$$

This means that this function is not to generate mass and therefore it cannot be understood as density. Thus, a pointwise limit of sequence of the averaged densities does not fulfil our purpose. Now we define another limit, the so-called *weak limit*, of the sequence of the averaged densities $\rho_\varepsilon(x)$. It is easily verified that for any continuous function $f(x)$

$$\lim_{\varepsilon \to +0} \int \rho_\varepsilon(x) f(x)\, dx = f(0). \tag{1.18}$$

This formula denotes that the weak limit of the sequence of the functions $\rho_\varepsilon(x)$ for $\varepsilon \to +0$ is the functional $f(0)$ (but not function!) giving to each continuous function $f(x)$ the number $f(0)$ – its value at the point $x = 0$. We define this functional as the density $\delta(x)$ – the *famous Dirac δ-function*. Thus, we can write

$$\rho_\varepsilon(x) \underset{\text{weak}}{\Longrightarrow} \delta(x), \quad \text{for } \varepsilon \to +0.$$

This limit is given by formula (1.18) for any $f(x)$. The value of the functional δ onto the function $f(x)$ is the number $f(0)$ which will be denoted as

$$(\delta, f) = f(0). \tag{1.19}$$

We see that functional δ generates total mass. Indeed,

$$\int dx\, \delta(x) = (\delta, 1) = 1$$

at the point $x = 0$ [the second equality follows from (1.19)].

Thus, the density corresponding to the point-like distribution of masses cannot be described within the classical concept of the function, and for its description objects of a more general mathematical nature are needed – linear (continuous) functionals. Before defining the functionals we consider the spaces of test functions.

1.4.2. *Space of Test Functions*

From the example of δ-function we see that this function is defined by means of continuous functions as a linear (continuous) functional on these functions. It is said that continuous functions are the *test functions* for δ-function. This point of view takes as its basis the definition of any generalized function as a linear continuous functional onto sets of sufficiently well-defined so-called test functions. It is clear that with the smaller set of test functions there exists larger linear continuous functionals. On the other hand, the reserve of the test functions must be sufficiently larger.

First, some definitions are needed. Let $\mathscr{U}$ be the linear space in which the commutative and associative addition is defined as:

1. $u + v = v + u.$
2. $u + (v + w) = (u + v) + w$

and also

$$u + 0 = u, \qquad u + (-u) = 0$$

for each $u, v, w \in \mathcal{U}$ and $0 \in \mathcal{U}$. Moreover,

$$1 \cdot u = u, \qquad \lambda(\mu u) = (\lambda\mu)u$$

and

$$(\lambda + \mu)u = \lambda u + \mu u, \qquad \lambda(u + v) = \lambda u + \lambda v$$

for any $u, v \in \mathcal{U}$ and some (real and complex) numbers λ and μ.

The real function $p(u)$ defined on $\mathcal{U}$ is called the *norm* if the following conditions are fulfilled:

(i) for any number λ, $p(\lambda u) = |\lambda| p(u)$,
(ii) $p(u + v) \leqslant p(u) + p(v)$ (triangle inequality),
(iii) if $p(u) = 0$, then $u = 0$.

From the conditions (i) and (ii) follows non-negativity of the norm

$$0 = p(u - u) \leqslant p(u) + p(-u) = 2p(u).$$

The functions satisfying only conditions (i) and (ii) are called the *semi-norms*.

If the norm p is given in $\mathcal{U}$, then one can define the distance between any two elements u and $v \in \mathcal{U}$ as $p(u - v)$. We say that the *sequence* $(u_1, \ldots, u_n, \ldots)$ converges to the limit u if the distance between u_n and u tends to zero when $n \to \infty$, i.e., if

$$\lim_{n \to \infty} p(u_n - u) \to 0.$$

Convergence defined by such manner is sometimes called the *convergence over the norm* or the *strong convergence.*

Linear space $\mathcal{U}$ in which the convergence is given by the norm $p(u)$ is called the *normalized space.* As usual, the norm $p(u)$ of the element u denotes as $p(u) = \|u\|$. For example, the Hilbert space $L_2[a, b]$ of all complex functions $f(x)$ with the norm

$$(f, f) = \|f\|^2 = \int_a^b \mathrm{d}x f^*(x) f(x)$$

is *linear normalized space.* Other linear normalized space is the space $\mathbb{C}([a, b])$ of continuous functions on the interval $[a, b]$ with the norm

$$p(u) = \operatorname*{Sup}_{x \in [a, b]} |u(x)|.$$

As a further example, we consider the space $\mathbb{C}(\sigma, \rho, n)$ of complex-valued functions of n-real variables $x = x_1, \ldots, x_n$, having continuous partial derivatives up to the order σ inclusively, and decreasing no more slowly than $|x|^{-\rho}$ together with all derivatives at infinity. In other words, for the functions

$u(x)$ from $\mathbb{C}(\sigma, \rho, n)$ all the products of the type

$$x^\alpha D^\beta u(x), \quad |\alpha| \leqslant \rho, \quad |\beta| \leqslant \sigma \tag{1.20}$$

are bounded, where

$$x^\alpha = x_1^{\alpha_1} x_2^{\alpha_2} \cdots x_n^{\alpha_n}, \qquad D^\beta = \frac{\partial^{\beta_1 + \cdots + \beta_n}}{\partial x_1^{\beta_1} \cdots \partial x_n^{\beta_n}}, \qquad |\alpha| = \alpha_1 + \cdots + \alpha_n, \tag{1.21}$$

The norm in the space $\mathbb{C}(\sigma, \rho, n)$ is given by

$$p_{\rho\sigma}(u) = \max_{\substack{|\alpha| \leqslant \rho \\ |\beta| \leqslant \sigma}} \operatorname{Sup}_{x \in \mathbb{R}^n} |x^\alpha D^\beta u(x)|. \tag{1.22}$$

The spaces of the type of $\mathbb{C}(\sigma, \rho, n)$ play an important role in the theory of the generalized functions.

DEFINITION 1.1. Let

$$p_1(u) \leqslant p_2(u) \leqslant \cdots \leqslant p_\sigma(u) \leqslant \cdots$$

be increasing the sequence of norms co-ordinated in pairs in the linear space $\mathcal{U}$. Space $\mathcal{U}$ is called *countably normalized* (with the norms $\{p_\sigma(u)\}$), if the convergence in it is defined as follows. The sequence $\{u_\nu\} \in \mathcal{U}$ converges to the element $u \in \mathcal{U}$, if $\lim_{\nu \to \infty} p_\sigma(u_\nu - u) = 0$, for any σ.

We give some examples of countably normalized spaces:

(1) Space $D(G)$. Let G be the finite region, i.e., the bounded open set in the n-dimensional real space $\mathbb{R}^n$. Denote $D(G)$, the set of infinitely smooth functions (i.e., functions having continuous partial derivatives of all orders) in $\mathbb{R}^n$, tending to zero outside the region G. Define in $D(G)$ the countable system norm q_σ by the formula

$$q_0(u) = p_{00}(u) = \operatorname{Sup}_x |u(x)|, \ldots, q_\sigma(u) = p_{0\sigma}(u) = \max_{|\alpha| \leqslant \sigma} \operatorname{Sup}_x |D^\alpha u(x)|,$$

where symbol D^α is given by formula (1.21).

The particular case of space $D(G)$ is space $D(a)$ of infinitely smooth functions of the single real variable x, tending to zero outside the interval $-a < x < a$. For example, the function $u(x)$ defined by the equality

$$u_a(x) = \begin{cases} \exp[a^2/(x^2 - a^2)] & \text{for } -a < x < a \ (a > 0) \\ 0 & \text{for } \quad |x| \geqslant a \end{cases}$$

belongs to space $D(a)$.

(2) Space S[or $S(\mathbb{R}^n)$] consists of all infinitely smooth functions in the n-dimensional real space $\mathbb{R}^n$, which decrease rapidly any polynomials of $(x_1^2 + x_2^2 + \cdots + x_n^2)^{-1/2}$ together with all partial derivatives at $\|x\| \to \infty$, i.e.,

$$S = S(\mathbb{R}^n) = \mathbb{C}(\infty, \infty, \mathbb{R}^n).$$

In other words, for these functions all the norms (1.22) take finite values. We will define the convergence in S by the countable system norms

$$p_\sigma(u)=p_{\sigma\sigma}(u)=\max_{\substack{|\alpha|\leqslant\sigma\\|\beta|\leqslant\sigma}}\operatorname{Sup}_{x\in\mathbb{R}^n}|x^\alpha D^\beta u(x)|,\quad \sigma=0,1,2,\ldots \tag{1.23}$$

In particular, Hermit–Chebyshev functions, and in general, all functions of the type

$$P(x_1,\ldots,x_n)\exp(-x_1^2/a_1^2-\cdots-x_n^2/a_n^2)$$

may be used as functions of S-space, where $P(.)$ is an arbitrary polynomial.

(3) Space $S(G)$ is defined as a set of functions from S, which tend to zero outside (the generally considered unbounded) region G. The convergence in $S(G)$ is given by the same system norm (1.23). We assume a possibility $G=\mathbb{R}^n$, so that the space S is a particular case of space $S(G)$. Space $D(G)$ is also a specific case of spaces of the type $S(G)$ for the bounded regions.

1.4.3. *Linear Functionals and Generalized Functions*

1.4.3a. *General Definition.* Numerical function defined in linear space $\mathcal{U}$ is called *functional.* The functional $F(u)$ is called *linear*, if for any $u, v\in\mathcal{U}$ and any numbers α and β

$$F(\alpha u+\beta v)=\alpha F(u)+\beta F(v). \tag{1.24}$$

For example, the integral over the space of integrable functions on the some interval is the linear functional. However, the norm is nonlinear functional. The functional $F(u)$ is *continuous* if from the convergence of sequence $\{u_n\}\in\mathcal{U}$ to $u\in\mathcal{U}$ follows the convergence of sequence $F(u_n)$ to $F(u)$. In particular, if the convergence in the space $\mathcal{U}$ is given by the norm $p(u)$ (i.e., if $\mathcal{U}$ is the normalized space), then it is said that the linear functional $F(u)$ is continuous (or bounded) in $\mathcal{U}$. If there exists a positive number C_F such as that for any $u\in\mathcal{U}$, the following inequality

$$|F(u)|\leqslant C_F\,p(u) \tag{1.25}$$

holds. The set of all *linear functionals* in normalized space $\mathcal{U}$ is in turn linear space, and we call it the space conjugate to $\mathcal{U}$ and denote it through $\mathcal{U}'$. In $\mathcal{U}'$ one can also define the norm by the formula

$$p'(F)=\operatorname*{Sup}_{p(u)\leqslant 1}|F(u)|. \tag{1.26}$$

In other words, $p'(F)$ is the smallest of constants C_F for which the inequality (1.25) is valid.

The space $\mathcal{U}'$ conjugate to the normalized space $\mathcal{U}$, with the convergence defined by the norm (1.26) is called *full normalized space* irrespective of whether the initial space $\mathcal{U}$ is full or not.

Now we turn to the study of linear functionals in the countably normalized space topology which is given by system neighborhoods: the neighborhood of any element $u \in \mathcal{U}$ being

$$U_{\sigma\varepsilon}(u) = \{v, p_\sigma(v - u) < \varepsilon\},$$

where ε and σ are positive and natural numbers, respectively. This formula means that the element v belongs to the neighborhood $U_{\sigma\varepsilon}(u)$, if $p_\sigma(v-u)<\varepsilon$.

The linear functional $F(u)$ is to be continuous in $\mathcal{U}$, if for any $\varepsilon>0$ one can show the neighborhood $U_{\sigma\delta}(0)$ of zero in which $|F(u)| < \varepsilon$. In this neighborhood $p_\sigma(u) < \delta$ for any element $u \in \mathcal{U}$.

We have already seen that the space conjugate to the normalized space is itself full normalized space. However, the space $\mathcal{U}'$ conjugate to the countably normalized space $\mathcal{U}$ has a structure different from the structure of the space $\mathcal{U}$. In this case construction of the space $\mathcal{U}'$ can be described as follows:

If the functional $F(u)$ is bounded in the neighborhood of zero $p_\sigma(u) < \delta$, then this functional is continuous with respect to the norm $p_\sigma(u)$, i.e., the inequality of the type of (1.25) holds

$$|F(u)| \leqslant Cp_\sigma(u). \tag{1.27}$$

The smallest σ for which this inequality takes place, is called the *order of the functional* $F(u)$. We see that any functional from $\mathcal{U}'$ has a finite order.

Now we take some examples for functionals:

(1) Space $D'(a)$ conjugate to countably normalized space $D(a)$ consists of the functionals of the type

$$F(u) = \int_{-a}^{a} u^{(\sigma)}(x)\, \mathrm{d}\mu(x),$$

where $\sigma = 0, 1, 2, \ldots$ and $\mu(x)$ is a function with bounded variation (for example, see Gel'fand and Shilov, 1968).

(2) Space S' consists of all functionals of the type

$$F(u) = \int_{\mathbb{R}^n} \mathrm{d}^n x f(x) D^\alpha u(x),$$

where D^α is given by formula (1.21), and $f(x)$ is a continuous function of the polynomial growth.

DEFINITION 1.2. A *generalized function* is called any linear continuous functional over the countably normalized space S defined above, i.e., any element of space S'. The functions of space S are called *test functions*.

It is clear that the concept of the generalized functions depends on the choice of the *initial* (linear topological) *space* of the test functions. For example, instead of S we would take $D(G)$ as a base of test space. L. Schwartz (1957, 1959) defined the generalized functions as the continuous functionals on space D of all finite and infinite differentiable functions (D is the union of the spaces $D(G)$, when the region G is changed). Functions disappearing outside some finite region of space are called *finite functions. Closure* of points set on which a continuous function $u(x) \neq 0$ is called the *support* of this function.

We retain the term 'generalized function' for the functionals from S', while the functionals from D' we call the distributions due to L. Schwartz (1957, 1959), who calls the generalized functions from S' 'distributions tempérées' and Gel'fand and Shilov (1964, 1968) call these generalized functions of 'tempered growth'. It is verifiable that any generalized function is distribution while the opposite assertion is not, i.e., the inclution $S' \subset D'$ takes place.

Another definition of the generalized functions as classes of fundamental sequences is given by Bogolubov *et al.* (1975).

Now we give some basic properties of the generalized functions.

1.4.3b. *Transformation of the Arguments and Differentiation of the Generalized Functions.* Mutually synonymous transformation $y = \varphi(x)$ $[y_1 = \varphi_1(x), \ldots, y_n = \varphi_n(x), x = (x_1, \ldots, x_n)]$ or $x = \varphi^{-1}(y)$ of the space $\mathbb{R}^n$ onto itself leads to

$$(f(\varphi^{-1}(y)), u(y)) = \int_{\mathbb{R}^n} \mathrm{d}^n x f^*(x) u(\varphi(x)) |J(\varphi)| = (f(x), |J(\varphi)| u(\varphi(x))) \quad (1.28)$$

for any $u(x) \in D$ and f is a usual (local integrable) function. Here

$$J(\varphi) = \begin{vmatrix} \dfrac{\partial \varphi_1}{\partial x_1} & \cdots & \dfrac{\partial \varphi_1}{\partial x_n} \\ \vdots & \cdots & \vdots \\ \dfrac{\partial \varphi_n}{\partial x_1} & \cdots & \dfrac{\partial \varphi_n}{\partial x_n} \end{vmatrix} \neq 0.$$

When $f(x)$ is the distribution, the right-hand side of (1.28) is understood as the definition of the distribution $f(\varphi^{-1}(y))$. By means of the particular case $y_1 = x_1 - \Delta x_1, y_2 = x_2, \ldots, y_n = x_n$, it is easy to verify that

$$(\partial f/\partial x_k, u) \equiv -(f, \partial u/\partial x_k) \quad (1.29)$$

in the limit $\Delta x_1 \to 0$. Here we use the fact that in accordance with the continuity of the functional f the right-hand side of (1.28) in this particular case tends to the limit $-(f, \partial u/\partial x_1)$ at $\Delta x_1 \to 0$. By definition we call this limit the partial derivative of the generalized function $f(x)$ with respect to the variable x_1. We notice that the operation of the differentiation is continuous in S', i.e., the following equality takes place

$$\lim_{n \to \infty} (\partial f_n/\partial x, u) = (\partial f/\partial x, u)$$

for any $u \in S$ and $\lim_{n\to\infty}(f_n, u) = (f, u)$. These simpler properties of the differentiation operation in S and S' are not fulfilled in the case of normalized spaces. Therefore, the countably-normalized spaces play an important role in the theory of the generalized functions.

Further, we notice that

$$\partial^2 f/\partial x_i\,\partial x_k = \partial^2 f/\partial x_k\,\partial x_i, \tag{1.30}$$

i.e., for the generalized functions the order of the differentiation has no significance.

We consider some examples for the generalized functions of the single variable, which may be defined as derivative of usual local integrable functions:

(1) The derivative of the well-known discontinuous function

$$\theta(x) = \begin{cases} 1 & \text{for } x > 0, \\ 0 & \text{for } x < 0 \end{cases} \tag{1.31}$$

equals $\delta(x)$. Indeed, in accordance with definition (1.29) and formula (1.19) we have

$$\left(\frac{d\theta}{dx}, u\right) = -\int_{-\infty}^{\infty} dx\theta(x)\frac{du}{dx} = -\int_0^{\infty} dx\,\frac{du}{dx} = u(0) = (\delta, u). \tag{1.32}$$

(2) The functional $(d/dx)\ln|x|$ coincides with the principal value of $1/x$ in Cauchy's sense; in other words,

$$\left(\frac{d\ln|x|}{dx}, u(x)\right) = \mathscr{P}\int_{-\infty}^{\infty} dx\,\frac{u(x)}{x}, \tag{1.33}$$

where $\mathscr{P}$ is the symbol of the principal value of the integral. Indeed, by definition (1.29), we obtain

$$\left(\frac{d}{dx}\ln|x|, u(x)\right) = -\int_{-\infty}^{\infty} dx\ln|x|\cdot u'(x) = \lim_{\varepsilon\to+0}\left\{\int_{-\infty}^{-\varepsilon} dx\frac{u(x)}{x} + \int_{\varepsilon}^{\infty} dx\frac{u(x)}{x}\right\}$$

$$= \int_0^{\infty} dx\frac{u(x) - u(-x)}{x} = \mathscr{P}\int_{-\infty}^{\infty} dx\frac{u(x)}{x}, \quad (u'(x) \equiv du/dx).$$

Defined thus, the generalized function $d\ln|x|/dx$ is denoted as $1/x$ [or sometimes as $\mathscr{P}(1/x)$]. We notice that the usual function $1/x$ is not locally integrable (in the neighborhood of the point $x = 0$) and it, therefore, is not identical with the generalized function $1/x$. However, these two functions coincide at $x \neq 0$.

Analogously, we define the function $1/x^2$ as a derivative of the generalized function $-1/x$:

$$\left(\frac{1}{x^2}, u(x)\right) = \left(\frac{1}{x}, u'(x)\right) = \int_0^{\infty} dx\frac{u'(x) - u'(-x)}{x}$$

$$= \int_0^\infty dx \frac{u(x) + u(-x) - 2u(0)}{x^2}. \tag{1.34}$$

In general, we assume by the definition

$$x^{-n} = \frac{(-1)^{n-1}}{(n-1)!} \frac{d^n}{dx^n} \ln|x|, \quad n = 1, 2, \dots. \tag{1.35}$$

where the derivative must be understood in the sense of the definition of the differentiation in S'.

(3) The usual locally integrable function $\ln(x + i0)$ defines as

$$\ln(x + i0) = \lim_{y \to +0} \ln(x + iy) = \ln|x| + i \lim_{y \to +0} \arg(x + iy)$$
$$= \ln|x| + i\pi\theta(-x), \tag{1.36}$$

where $\theta(x)$ is given by formula (1.31). The derivative of the generalized function (1.36) is denoted as $1/(x + i0)$. In accordance with (1.32), (1.35) and (1.36) we have

$$\frac{1}{x + i0} \equiv \frac{d}{dx} \ln(x + i0) = \frac{1}{x} - i\pi\, \delta(x), \tag{1.37}$$

or in the general case

$$(x + i0)^{-n} = \frac{(-1)^{n-1}}{(n-1)!} \frac{d^n}{dx^n} \ln(x + i0) = x^{-n} + i\pi \frac{(-1)^n}{(n-1)!} \delta^{n-1}(x). \tag{1.38}$$

and

$$(x - i0)^{-n} = \frac{(-1)^{n-1}}{(n-1)!} \frac{d^n}{dx^n} \ln(x - i0) = x^{-n} + i\pi \frac{(-1)^{n-1}}{(n-1)!} \delta^{n-1}(x). \tag{1.39}$$

According to formula (1.29) derivatives of the delta-function on the right-hand sides of equations (1.38) and (1.39) are given by the expression

$$(\delta^n, u) = (-1)^n u^{(n)}(0), \qquad \delta^n = \frac{d^n}{dx^n} [\delta(x)].$$

1.4.3c. *The Fourier Transform of Generalized Functions.* We consider space $\mathbb{R}^n$ with the indefinite metric, i.e., the interval between two 'events' x and y as given by

$$s^2 = g_{ij} x_i y_j = x_1 y_1 + \cdots + x_l y_l - x_{l+1} y_{l+1} - \cdots - x_n y_n \tag{1.40}$$

and construct the Fourier transform with respect to this bilinear form.

LEMMA 1.1. *The Fourier transform of the test function* $u(x) \in S$ *is*

$$\tilde{u}(p) = Fu(x) = (2\pi)^{-n/2} \int_{\mathbb{R}^n} d^n x \, e^{ipx} \, u(x) \tag{1.41}$$

and is also the test function ($\tilde{u}(p) \in S$). *Here the bilinear form xp is given as (1.40). We see that formula (1.41) gives the mutually synonymous and continuous mapping of space* S *onto itself.*

The inverse transform of (1.41) is given by

$$u(x) = \tilde{F}\tilde{u}(p) = (2\pi)^{-n/2} \int_{\mathbb{R}^n} d^n p \, e^{-ipx} \, \tilde{u}(p); \tag{1.42}$$

which possesses the same properties as F. The Fourier transform therefore realizes an *isomorphism* S on S.

We define the Fourier transform of the generalized function $f(x)$ as the linear functional $\tilde{f}(p)$ on the space of the Fourier images of the test functions $u(x)$ by the formula

$$(\tilde{f}(p), \tilde{u}(p)) = (f(x), u(x)), \tag{1.43}$$

where $\tilde{u}(p)$ is connected with $u(x)$ by formula (1.41). Formula (1.43) may also be considered as a definition of inverse Fourier transform, if the generalized function $\tilde{f}(p)$ is given then its inverse image $f(x)$ determines its equality (1.43).

According to Lemma 1.1 the Fourier transform $\tilde{f}(p)$ of the generalized function $f(x)$ ($f(x) \in S'$) is also the generalized function of the same space S'. Conversely, any generalized function $\tilde{f}(p) \in S'$ has the inverse Fourier image $f(x) \in S'$. Thus, the following theorem corresponding to the Lemma 1.1 takes place.

THEOREM 1.1. *The Fourier transform of the generalized functions defined by formula* (1.43) *is the isomorphism of space* S' *onto itself. We recall that the* isomorphism *of the linear topological space* $\mathcal{U}_1$ *onto the same space* $\mathcal{U}_2$ *is called the mutually synonymous and mutually continuous mapping* $\mathcal{U}_1$ *onto* $\mathcal{U}_2$, *conserving linear operations.*

We notice that Lemma 1.1 and Theorem 1.1 do not hold for the spaces D and D' of finite test functions and distributions.

For example, we give Fourier transforms of some generalized functions.

$$F\theta(t) = (2\pi)^{-1/2} \int dt \theta(t) \, e^{i\omega t} = \frac{(2\pi)^{-1/2} i}{\omega + i0},$$

$$F\delta^{(n)}(x) = (2\pi)^{-n/2} \int_{\mathbb{R}^n} d^n x \delta^{(n)}(x) \, e^{ipx} = (2\pi)^{-n/2},$$

$$\tilde{F}\left(\frac{e^{ipa}}{(2\pi)^{n/2}}\right) = (2\pi)^{-n} \int_{\mathbb{R}^n} d^4 p \, e^{ip(a-x)} = \delta^{(n)}(x - a),$$

where $\delta^{(n)}(x) = \delta(x_1)\ldots\delta(x_n)$.

1.4.3d. *Multiplication of the Generalized Functions by a Smooth Function and Their Convolution.* The product of the generalized functions represents one of the most important problems in the theory of generalized functions. Since this is a nonlinear operation,. the product of the two pairs of the generalized functions cannot be defined in the usual manner. It is easy to verify that it is impossible to give multiplication which would be associative. Indeed,

$$\frac{1}{x}(x\,\delta(x)) = \frac{1}{x}0 = 0 \neq \left(\frac{1}{x}x\right)\delta(x) = \delta(x). \tag{1.44}$$

Nevertheless, there exists a wide class of functions for which one can define their product with the generalized functions from S' by the natural manner. This class is defined as follows.

It is said that the function $\varphi(x)$ is *multiplier* in space S of test functions if from $u(x)\in S$ it follows that $\varphi(x)u(x)\in S$. The space of all multipliers we denote as Q_M. It is clear that if $\varphi(x)$ is infinitely differentiable and a polynomially bounded function of x (together with all its derivatives), then $\varphi(x)$ is the multiplier in S.

If $\varphi(x)$ is the multiplier the product of $\varphi(x)$ by the generalized function $f\in S'$ is given by the formula

$$(\varphi(x)f, u(x)) = (f, \varphi^*(x)u(x)) \tag{1.45}$$

for any $u(x)\in S$. By definition we assume that the product is commutative, i.e., $\varphi f = f\varphi$.

In some cases one can define the product of two generalized functions. However, construction of the general theory in this direction is not complete and encounters definite difficulties. The production problem of generalized functions is crucial in quantum field theory. Since the latter always deals with the generalized functions, definition of those products requires new mathematical methods and specific regularization procedures. This problem will be partially discussed in Section 1.4.4 (see also Sections 2.2.2 and 2.4.4).

We now consider the operation of convolution. This operation is widely used in the theory of the generalized functions. The *convolution* for two usual functions $f(p)$ and $g(p)$ is given by the expression

$$f(p) * g(p) = \int_{\mathbb{R}^n} d^n q f(p-q)g(q) = \int_{\mathbb{R}^n} d^n q f(q)g(p-q). \tag{1.46}$$

As in the case of the multiplication operation the convolution is not defined for any pair of the generalized functions. Using the second equality in (1.46) the convolution of the generalized function f with the test function $u(p)\in S$ is defined as:

$$f * u(p) = (f(q), u(p-q)) = \int_{\mathbb{R}^n} d^n q f(q)u(p-q). \tag{1.47}$$

The first equality in (1.47) is the definition of the convolution as the operation of the functional $f(q)$ onto the test function $u(p-q)$ considering the latter as a function of q at fixed p. Clearly the convolution (1.47) is infinitely differentiable and a polynomially bounded function of p (together with all its derivatives). In other words, if $f \in S'$ and $u \in S$ then $f * u(p) \in Q_M$, i.e., the convolution (1.47) is the multiplier. Generally speaking, the function $f * u$ does not belong to the space S of test functions.

Now we go over to the problem of division and to the concept of the support of the generalized function. First, we consider the *division problem* which is inverse in operation to multiplication and leads to the study of the equation

$$\varphi(x) f = g, \tag{1.48}$$

where $g \in S'$ and $\varphi(x) \in Q_M$ are the given functions, and f is an unknown generalized function. In the case when $\varphi(x) \neq 0$ for any x, and it goes to zero not too rapidly (i.e., the function $\varphi^{-1}(x)$ is also a multiplier), equation (1.48) is solved elementarily. The problem of division becomes complicated if the function $\varphi(x)$ has zeros. Consider here only the case of a single independent variable x and assume that $\varphi(x)$ in (1.48) has a discrete set of zeros of a finite order. By this assumption the division problem in the space $S'(\mathbb{R}^1)$ is in essence reduced to the solution of simpler equations of the type of (1.48):

$$xf = g. \tag{1.49}$$

Solution of which has the form

$$(f, u) = Cu(0) + (g, u_1(x)),$$

where $u(x)$ and $u_1(x)$ belongs to $S(\mathbb{R}^1)$. While the general solution of the homogeneous equation

$$xf_0 = 0$$

is

$$(f_0, u) = Cu(0), \quad \text{i.e.,} \quad f_0 = C\delta(x) \tag{1.50}$$

Further, by induction it is easily shown that the more general equation

$$x^m f = g \tag{1.51}$$

always has a solution with respect to $f \in S'$, where $g \in S'$ and m is any natural number. An arbitrary factor in this solution is produced by a general solution of the homogeneous equation

$$x^m f_0 = 0,$$

that is

$$f_0 = \sum_{\nu=0}^{m-1} \frac{C_\nu}{\nu!} \cdot \delta^\nu(x), \quad \delta^\nu = \frac{d^\nu}{dx^\nu} \delta(x), \tag{1.52}$$

where C_ν are arbitrary constants (for details, see Bogolubov *et al.*, 1975).

Further we discuss other questions of interest including the *local properties* of the generalized functions. As opposed to the usual functions which are given in each point of some set, the generalized functions are determined whole as values of the functional on the space of the test functions. Generally speaking, they do not have definite values at separate points. It is said that the generalized functions f and g coincide in the region (or in the open set) G if for any test function $u(x)$ with the support on G the following equality $(f, u) = (g, u)$ takes place. Thus, by definition the set of points on which the generalized function f turns to zero, is *open*. Complement of this set to the whole space $\mathbb{R}^n$ is called the *support* of the generalized function f.

THEOREM 1.2. *Let f be the generalized function located in the origin of coordinate system (i.e., $f(x) = 0$, for $x \neq 0$). Then $f(x)$ is expressed by a finite linear combination of δ-function and its derivatives $\delta^v(x)$. This assertion is valid for both the cases of the generalized functions of many variables and of the distributions from $D'(\mathbb{R}^n)$, i.e., if $f \in D'(\mathbb{R}^n)$;*

$$f(x) = \sum_{|\alpha| \leqslant N} C_\alpha D^\alpha \delta(x), \tag{1.53}$$

where N is the order of f and C_α are some constants.

Proof. The proof of this theorem is based on formula (1.52) and on the fact that any generalized function has a finite order (see Gel'fand and Shilov, 1968).

1.4.4. *Generalized Functions of Quantum Field Theory*

Initial objects of the local quantum field theory are singular functions, for example:

- the causal Green function $\Delta_c(x)$ $(x = x_0, \mathbf{x})$ or the propagator of the particle with mass m,
- the positive-frequency part $\Delta_{(-)}(x)$ of the Pauli–Jordan function $\Delta(x - y) = [\varphi(x), \varphi(y)]$, where $\varphi(x)$ is the field operator.

Explicit forms of these functions in the case of the scalar field $\varphi(x)$, are given by the following formulas (for detail, see Bogolubov and Shirkov, 1980):

$$\Delta_{(-)}(x - y) = \langle 0|\varphi(x)\varphi(y)|0\rangle, \qquad \Delta_c(x - y) = \langle 0|T[\varphi(x)\varphi(y)]|0\rangle,$$

$$\Delta_{(-)}(x) = (2\pi)^{-3}\int d^4p\, e^{-ipx}\,\theta(p_0)\,\delta(p^2 - m^2) = \frac{i}{4\pi}\varepsilon(x_0)\delta(\lambda) +$$

$$+ \frac{m}{8\pi\sqrt{\lambda}}\theta(\lambda)\,[N_1(m\sqrt{\lambda}) - i\varepsilon(x_0)J_1(m\sqrt{\lambda})] +$$

$$+ \frac{m}{4\pi^2\sqrt{-\lambda}}\theta(-\lambda)K_1(m\sqrt{-\lambda}) \tag{1.54}$$

and

$$\Delta_c(x) = \tilde{F}((2\pi)^{-2} i^{-1} (m^2 - p^2 - i\varepsilon)^{-1})$$

$$\equiv (2\pi)^{-4} i^{-1} \int d^4 p\, e^{-ipx} (m^2 - p^2 - i\varepsilon)^{-1}$$

$$= \frac{1}{4\pi i} \delta(\lambda) - \frac{m}{8\pi i} \frac{\theta(\lambda)}{\sqrt{\lambda}} [J_1(m\sqrt{\lambda}) - iN_1(m\sqrt{\lambda})] +$$

$$+ \frac{m}{4\pi^2} \frac{\theta(-\lambda)}{\sqrt{-\lambda}} K_1(m\sqrt{-\lambda}), \qquad (1.55)$$

where $\lambda = x_0^2 - x_1^2 - x_2^2 - x_3^2$,

$$K_1(z) = -\frac{\pi}{2} H_1^{(1)}(iz) = \frac{\pi}{2}(J_1(iz) + iN_1(iz)),$$

and J_1, N_1, $H_1^{(1)}$, and K_1 are the Bessel, Neiman, Hankel and Kel'vin functions, respectively. Notice that in the case of QFT space-time is the Minkowski, i.e., $R = \mathbb{R}^4$.

We see that these functions have a higher singularity of the type of $\delta(\lambda)$, $\theta(\lambda)$, $\lambda^{-1/2}$ on the light cone $\lambda = 0$. These singularities are connected with the ultraviolet divergences in the theory. Removal of the latter leads, in fact, to regularization of the singular functions $\Delta_{(-)}(x)$ and $\Delta_c(x)$. On the other hand, the regularization procedure of the generalized functions $\Delta_{(-)}(x)$ and $\Delta_c(x)$ is crucial for their multiplication, convolution, etc. Due to the presence of the δ-function in expressions (1.54) and (1.55) for $\Delta_{(-)}(x)$ and $\Delta_c(x)$ the definition of their multiplication: $\Delta_c^2(x)$ and $[\Delta_{(-)}(x)]^2$ is not clear. This operation is connected with the multiplication properties of the generalized functions (for the δ-function: $\delta^2(x) = C\delta(x)$, where C is an arbitrary constant) and requires some regularization procedure (the Pauli–Villars regularization is often used) giving finite values for the regularized functions $\Delta_c^r(x)$ and $\Delta_{(-)}^r(x)$, i.e., $\Delta_c^r(0) < \infty$ and $\Delta_{(-)}^r(0) < \infty$. For example, in the above-mentioned case (Section 1.3), the regularization procedure played by the method of averaging in stochastic space-time.

However, it should be noted that the situation for $\Delta_{(-)}(x)$ is radically different from the case of $\Delta_c(x)$. It transpires that the product of two positive-frequency parts $\Delta_{(-)}(x)$ may be defined as the *Fourier transform* of the convolution

$$(\tilde{\Delta}_{(-)}^M * \tilde{\Delta}_{(-)}^m)(p) = \int d^4 q\, \tilde{\Delta}_{(-)}^M(q)\, \tilde{\Delta}_{(-)}^m(p - q)$$

$$= (2\pi)^{-2} \int d^4 q \theta(q_0) \theta(p_0 - q_0)\, \delta(q^2 - M^2)\, \delta((p - q)^2 - m^2)$$

$$= (8\pi p^2)^{-1}\theta(p_0)\theta(p^2 - (M+m)^2) \times$$
$$\times \{[p^2 - (M+m)^2][p^2 - (M-m)^2]\}^{1/2}. \qquad (1.56)$$

Here, integration was carried out in an elementary manner: choosing the axis q_0 along the time-like vector p and introducing new variables q^2, ω and $\mathbf{n}$; $q_0 = -\omega$, $\mathbf{q} = (\omega^2 - q^2)^{1/2}\,\mathbf{n}$, where $\mathbf{n}^2 = 1$, and after integration over q^2 and angles giving the three-dimensional unity vector $\mathbf{n}$, we obtain expression (1.56). We notice that the function (1.56) is usually continuous and bounded at infinity.

On the contrary, in the case of $\Delta_c(x)$ its product contains a *finite number* of arbitrary constants which are caused by the products of the type of $\delta^2(x)$ and $\delta(x)\theta(x)$. The reader may locate the general theory of such productions in Bogolubov and Parasiuk's (1957) work.

Thus, two main functions $\Delta_c(x)$ and $\Delta_{(-)}(x)$ are the generalized functions and therefore all the *coefficient functions* in the expansion of the S-matrix in a series of normal products of the field operator $\varphi(x)$

$$S = \sum_{n=0}^{\infty} \frac{1}{n!} \int d^4x_1 \cdots \int d^4x_n S_n(x_1, \ldots, x_n){:}\ \varphi(x_1)\cdots\varphi(x_n){:}$$

or *Wightman's* (1972) *functions*:

$$W_n(x_1, \ldots, x_n) = \langle 0 \mid \varphi(x_1)\cdots\varphi(x_n) \mid 0\rangle, \quad (x = x_0, \mathbf{x})$$

become also the generalized functions. Since the functions $S_n(x_1, \ldots, x_n)$ and $W_n(x_1, \ldots, x_n)$ are expressed by the product of $\Delta_c(x)$ and $\Delta_{(-)}(x)$, respectively. For example, we consider the structure of the functions $W_n(x_1, \ldots, x_n)$ which are constructed by $W_2(x, y) = \langle 0 \mid \varphi(x)\varphi(y) \mid 0\rangle$. According to the translation invariance, $W_2(x, y)$ represents in fact only a function of difference $y - x$, which we have denoted by $\Delta_{(-)}$. This function must possess the following conditions:

(1) $\Delta_{(-)}(\Lambda x) = \Delta_{(-)}(x)$ for all the Lorentz transformations $\Lambda \in L_+$.
(2) Support of $\Delta_{(-)}$ belongs to the future light cone $\bar{V}_+$.
(3) $\iint d^4x\, d^4y f(x)\, \Delta_{(-)}(y - x) f(y) \geqslant 0$ for all $f \in L(\mathbb{R}^2)$.
(4) $\Delta_{(-)}(x) = \Delta_{(-)}(-x)$, if vector x is space-like.
(5) If, besides $\langle 0|\varphi(x)|0\rangle = 0$, then
(6) $\lim_{\lambda\to\infty} \Delta_{(-)}(\lambda x) = 0$, if x is the space-like vector.

Conditions (1)–(4) and (6) are the consequences of the *Lorentz covariance, spectrality, positive definiteness, locality* and *group property* in *Garding and Wightman's axioms*, respectively. We recall that the generalized function $\tilde{f}(p)$ $(p = p_0, \mathbf{p})$,

$$\tilde{f}(p) = (2\pi)^{-2} \int d^4p\, e^{ipx} f(x)$$

is invariant under the proper Lorentz transformation in $\mathbb{R}^4(p)$ if and only if the Fourier image $f(x)$ is invariant under these transformations in $\mathbb{R}^4(x)$. If the Lorentz invariant generalized function $\tilde{f}(p)$ is located at the point $p = 0$ then a similar formula to (1.53) takes place

$$\tilde{f}(p) = \sum_{0 \leqslant |\alpha| \leqslant n_f} C_\alpha D^\alpha \delta^{(4)}(p), \qquad \delta^{(4)}(p) = \delta(p_0)\,\delta(\mathbf{p}).$$

Any generalized function satisfying conditions (1)–(3) and (6) can be represented in the form (*Källen–Lehmann representation*)

$$\Delta_{(-)}(x) = \int_{m^2 > 0} \mathrm{d}\mu(m^2)\, \Delta_{(-)}(x; m^2),$$

where

$$\Delta_{(-)}(x; m^2) = (4\pi)^{-1} \int \mathrm{d}^3 p E^{-1}(\mathbf{p}, m^2) \exp[iE(\mathbf{p}, m^2)t - i\mathbf{p}\mathbf{x}],$$

$$E(\mathbf{p}, m^2) = (|\mathbf{p}|^2 + m^2)^{1/2}$$

and μ is a polynomial (positive) bounded measure. In the case when $\mathrm{d}\mu(m^2) = \delta(m^2 - m_0^2)$, a field associated with the functions $W_n(x_1, \ldots, x_n) = [x_1 \cdots x_n]$ is called a *free field* with mass m_0. Here

$$[x_1 \cdots x_n] = 0 \quad \text{if } n \text{ is odd},$$

$$[x_1 \cdots x_{2n}] = \sum [x_{i_1} x_{j_1}] \cdots [x_{i_n} x_{j_n}] \tag{1.57}$$

where $[xy] = \Delta_{(-)}(y - x)$. A summation is carried out over all $(2n!)/2^n n!$ permutations of the numbers $i_1, \ldots, i_n$; $j_1, \ldots, j_n$ of $\{1, \ldots, 2n\}$ such that $i_1 < \cdots < i_n$ and $i_1 < j_1, \ldots, i_n < j_n$.

Thus, we see above that in QFT basic objects of study are different classes of the generalized functions. The type of classes depends essentially on the physical principles of the theory. In the local QFT concept of locality and microcausality condition plays an important role, according to which the character of the generalized functions and type of test functions are defined. For example, the principle of locality (or microcausality condition) is connected with a definition of the local properties of the test and generalized functions, and requires among them the existence of functions with *bounded support*. In other words, from the mathematical point of view, the concept of locality is connected with the possibility of giving a definition of the *support of a linear functional* on some space of test functions. In turn the concept of functional support is directly associated with the existence of some subspaces of functions in a test function space. The functions of these subspaces can be used as an instrument allowing the study of the space-time property of functionals in QFT. Usually, as such functions one can consider the functions of D-space of infinite-differentiable functions with bounded supports. By means of these functions the space-time properties of the functional are investigated, for example, the commutator of *Heisenberg's field* $\varphi(x)$:

$$[\varphi(x), \varphi(y)]_- = ? \tag{1.58}$$

or the *causality condition* for the S-matrix in the Bogolubov form

$$\frac{\delta}{\delta\varphi(x)}\left(\frac{\delta S}{\delta\varphi(y)} S^+\right) = ? \tag{1.59}$$

The requirement of strict *locality* due to Jaffe (1966) is directly formulated in such a way that the space of test functions always contains *functions with bounded supports*. According to the concept of locality QFT, as usually formulated, is based on the assumption that all its generalized functions belong to the spaces $\mathscr{D}'$ or $S' \subset \mathscr{D}'$. The choice of the space S of test functions has been dictated on the basis that the spaces S and S' go over themselves under the Fourier transform (see Section 1.4.3c), which in turn reflects the symmetry between x- and p-spaces in QFT. Moreover, we concluded from that that the space S' contains a sufficiently larger reserve of the functions (in particular, the δ-function and all its derivatives, and all the usual polynomial bounded functions etc.).

It is important to notice that the *tempered distributions* become traditional in the study of the local QFT and that the concept of locality is introduced here in the simplest and most natural manner.

Lately it has become clearer that this class of tempered distributions is not always an adequate instrument. It transpires that the class of *local distributions* can be *extended* essentially. The most significant results were obtained by Meiman (1964) and Jaffe (1966). It is noteworthy that the definition of the *concept* of microcausality by Meiman and *strict localizability* by Jaffe takes first place in the papers of these authors. They obtained different classes of test functions depending on the definitions introduced. Meinman's and Jaffe's idea was to choose the 'minimum' class of test functions, meaning that, when introducing a *certain definition* of *locality*, only those distributions which satisfy the definition introduced, must be defined on the 'minimum' class of test functions. The other distributions must not be defined with the whole class of test functions. The requirement of this 'minimality' permits one to obtain the important physical consequences from this purely mathematical hypothesis following from the definition of locality alone. We may now consider this situation in more detail. It should be noted that the assumption of the class of the generalized functions, is an independent hypothesis which has nontrivial physical consequences. An important corollary of this hypothesis is the *polynomial boundedness* of the S-matrix elements in the momentum space. We show that the growth of the generalized function in the p-space cannot be arbitrary if we want to conserve the local properties in the x-space, in particular, the microcausality principle. Without loss of generality, we consider the function $f(t)$ of the single variable t. The function $f(t)$ is a generalized function, i.e., which is a linear functional on some (for the present undetermined) space $\mathscr{U}$ of the test functions $u(t)$. In order to study the local properties of the function (in particular, the assumption that $f(t) = 0$ for $t < 0$, such type of assertion enters into the formulation of the *microcausality condition*) we assume that the space $\mathscr{U}$ must contain sufficiently larger reserves of test functions. It transpires that the *local properties* of the function $f(t)$ are connected with an acceptable growth of its Fourier image $\tilde{f}(\omega)$. It can be verified according to the following example.

Let the space $\mathscr{U}'_a$ of the functions $\tilde{f}(\omega)$ contains the generalized functions from S' and at the same time all usual functions having the behavior as $e^{b\omega}$ at infinity, where $|b| < a$ ($a > 0$ is fixed). The space $\mathscr{U}'_a$ can be regarded as the space of the *linear functionals* over the countably normalized space $\mathscr{U}_a$ of infinitely differentiable functions $u(\omega)$ with the norms:

$$p_n(u) = \max_{k \leqslant n} \ \operatorname*{Sup}_{\omega} \left| e^{a|\omega|} \frac{d^k u(\omega)}{d\omega^k} \right|, \qquad n = 0, 1, 2, \ldots \tag{1.60}$$

It is easy verify that $D \subset \mathscr{U}_a \subset S$. From the boundedness of the norms (1.60) it follows that the *Fourier transform* $\tilde{u}(t)$ of the functions from $\mathscr{U}_a$ is analytic in the band

$$-a < \operatorname{Im} t < a \tag{1.61}$$

and decreases within this band rapidly any negative power of $|t|$ at infinity. We denote this space of Fourier images through $\tilde{\mathscr{U}}_a$. It is impossible to define the local properties of the generalized functions in $\tilde{\mathscr{U}}_a$ since they do not contain finite functions. Moreover, in the case of the generalized functions $f \in \tilde{\mathscr{U}}'_a$ it is irrelevant to say that in the given interval they coincide with some usual function $\varphi(t)$, in particular, the assertion of the type of $f = 0$, for $t < 0$ loses its significance.

Indeed, for the linear functionals over the space $\tilde{\mathscr{U}}_a$ there exists no natural definition of the support in the absence of nonzero-finite test functions in $\tilde{\mathscr{U}}_a$.

As an example, we consider the series

$$f(t) = \sum_{n=0}^{\infty} \frac{b^n}{n!} \delta^n(t), \qquad \delta^n(t) = \frac{d^n}{dt^n} \delta(t), \tag{1.62}$$

for $0 < b < a$. This series converges in the space $\mathscr{U}'_a$ because the series

$$(f, u) = \sum_{n=0}^{\infty} \frac{1}{n!} u^{(n)}(0)(-b)^n = u(-b), \qquad u^{(n)}(t) = \frac{d^n}{dt^n} u(t) \tag{1.63}$$

is the Taylor series for the analytic function $u(t)$ inside its region of the analyticity ($b < a$).

Furthermore, in spite of the fact that every term of the series (1.62) is located at zero (and, in particular, that it disappears along the negative semi-axis t) the *support* of its *sum* is all points of any closed contour C_b surrounding the point $-b$ and belonging inside the band (1.61), since, according to the Cauchy formula:

$$u(-b) = \frac{1}{2\pi i}\int_{C_b} dz\,\frac{u(z)}{z+b}.$$

On the other hand, as mentioned above (Section 1.4.3) the series (1.62) [or (1.63)] diverges with respect to convergence in S' in accordance with the fact that the generalized functions from S' possess definite local properties.

Thus, the question arises: what is the widest class of the generalized functions which one can use to speak about their local properties? By means of the Fourier transform, the answer to this question may be roughly formulated as follows: the Fourier-image $\tilde{f}(\omega)$ of 'localizable' functions $f(t)$ must increase *slowly any exponent.* We have already noted that the existence of the dense set of the finite functions among the test functions in the x-space ensures the possibility of the usual definition of the local properties of the generalized functions. From the previous example for the space $\mathscr{U}'_a$, we have seen that if $\tilde{f}(\omega)$ can increase exponentially, then it is impossible to define the local properties of $f(t)$ synonymously. Jaffe (1967) found necessary and sufficient conditions for a function of the type of

$$g(z) = \sum_{\nu=0}^{\infty} c_\nu z^\nu, \quad c_\nu \geqslant 0,$$

and that the space $\tilde{S}(g)$ of Fourier transforms of the test functions from $S(g)$ contained finite functions of x-variables ($x = x_0, \mathbf{x}$). It appears that for this case it is both sufficient and necessary that the integral

$$\int_1^\infty dw\, w^{-2} \ln g(w^2) < \infty \tag{1.64}$$

converges, for example, the function $g(w^2) \sim \exp[(w/\ln w)(\ln\ln w)^{-1-\varepsilon}]$, $\varepsilon > 0$ satisfies this condition. Meiman (1964) considered a more general problem, the essence of which concerns the attempt to define the local properties of the generalized functions (in particular, the concept of their support) without having finite test functions. In this more general case one can go to the class of the generalized functions satisfying the weaker restrictions than (1.64), namely, which lead to the function increasing not rapidly any linear exponent at infinity:

$$|f(p)| \leqslant C_\varepsilon e^{\varepsilon|p|} \quad \text{for any } \varepsilon > 0. \tag{1.65}$$

In contradistinction to the local theory, nonlocal QFT allows stronger growth than (1.65) for the generalized functions in p-space. We will now discuss this problem.

The development of the nonlocal QFT and of the theory with the non-polynomial interaction Lagrangian has required attracting spaces of *analytical functions among which functions with finite supports are absent.* However, the study of the causality conditions and the locality of the theory is needed to construct mathematical methods devoted to the investigation of space-time characteristics of operator-valued generalized functions of the types of (1.58) and (1.59) in the case when they represent analytical functionals, i.e., they are defined only in the spaces of analytic functions. This problem has been considered in detail by Efimov (1968) (see also Fainberg and Soloviev, 1977, 1978; and Soloviev, 1980).

An answer to all the supplied problems connected with the definition of local and nonlocal generalized functions and the corresponding spaces of test functions can be obtained by studying the space-time properties of generalized functions of the type (1.11) which have arisen from the results of averaging in the stochastic space-time $\mathbb{R}^4(\hat{x})$, as was done previously in Section 1.3. We see that the function $K(x)$ (1.11) is the generalized form of the well-known local Dirac δ-function.

We say that the *generalized function* (1.11) is given in some test function space $\mathscr{U}$, if for any $f \in \mathscr{U}$ the functional

$$(K, f) = \int \mathrm{d}^4 x K(x) f(x) = \sum_{n=0}^{\infty} \frac{c_n}{(2n)!} \Box^n f(x)|_{x=0} < \infty \tag{1.66}$$

is well-defined, i.e., the obtained series converges absolutely. Here c_n are some, generally speaking, complex number coefficients, and we insert the parameter l into c_n. Passing to the momentum space in (1.66), we obtain

$$(K, f) = \int \mathrm{d}^4 p \tilde{K}(p^2) \tilde{f}(p) < \infty, \tag{1.67}$$

where

$$\tilde{K}(p^2) = \sum_{n=0}^{\infty} \frac{c_n}{(2n)!} (p^2)^n \tag{1.68}$$

and $\tilde{f}(p)$ is the Fourier transform of $f(x)$. In other words, the generalized function (1.11) is given on $\mathscr{U}$ if series (1.68) defines the function $\tilde{K}(p^2)$ for all p^2 and the integral (1.67) converges for any $f(x) \in \mathscr{U}$. Both conditions (1.66) and (1.67) are equivalent. The space-time properties of the generalized function (1.11) are as usual investigated for different sequences of the coefficients $\{c_n\}$, which correspond to different generalized functions.

1.4.5. *The Class of Test Functions in the Nonlocal Case*

We choose the space $\mathscr{E}$ of all entire analytic functions $f(z) = f(z_0, \mathbf{z})$ depending on four variables $z_\mu = x_\mu + iy_\mu$ in the capacity of the class of our test functions. We define in space $\mathscr{E}$ the countable sequence of the norms

$$\|f\|_m = \max_{|z_\mu| < m} |f(z_0, \mathbf{z})|.$$

We will say that the sequence of functions $\{f_n(z)\}$, where $f_n(z) \in \mathscr{E}$, converges correctly if it converges uniformly in any bounded region of real values of arguments $(x_0, \mathbf{x})$.

LEMMA 1.2. *If each function of the sequence* $\{\mathrm{f}_n(\mathrm{z})\}$, *where* $\mathrm{f}_n(\mathrm{z}) \in \mathscr{E}$ *is bounded according to norm* $\|\mathrm{f}_n\|_m$ *by the constant* M *and this sequence converges correctly, then its limit* f(z) *belongs to the space* $\mathscr{E}$ *and the norm of* f(z) *does not surpass* M.

This lemma is the other formulation of Vitaly's theorem concerning the convergence of analytic function sequences (Titchmarsh, 1939). Therefore, the norms introduced are compatible and the space $\mathscr{E}$ is complete and perfect.

DEFINITION 1.3. The sequence $\{f_n(z)\}$, where $f_n(z) \in \mathscr{E}$, converges in the region $\tilde{G} \subset \mathbb{C}^4$ if it converges uniformly in the closed region $\tilde{G}$.

The space $\mathscr{E}$ is not closed, relative to this convergence definition, but we will make use of it for another definition of the support of a functional on the space $\mathscr{E}$.

The space of all linear continuous functionals defined on the space $\mathscr{E}$ is denoted by $\mathscr{E}'$. The explicit form of such functionals can be found according to the usual procedure (Gel'fand and Shilov, 1968; Kantorovich and Akilov, 1964):

$$(F, f) = \int_{|z_\mu| < m} d\mu_F(z) f(z), \quad F \in \mathscr{E}', \tag{1.69}$$

where $\mu_F(z)$ is a complex and completely additive measure in the region $|z_\mu| < m$. This formula with all possible m gives the general form of *linear continous functions* on the space $\mathscr{E}$.

Let us introduce the concept of *functional support*, and give the following:

DEFINITION 1.4. The region $\tilde{G} \subset \mathbb{C}^4$ is said to be the *support* of the functional $F \in \mathscr{E}'$ if this functional can be extended continuously on the space $\mathscr{E}(\tilde{G})$ of functions, which are analytical in the closed region $\tilde{G}$. If the region $\tilde{G}$ is a point $z = (z_0, \mathbf{z})$ then *functional* is said to be *local.*

It follows from given definitions that the sequence of numbers (F, f_n) converges to zero if the support of the functional F is a region $\tilde{G}$ and $\{f_n\}$ where $f_n \in \mathscr{E}$ is any sequence which converges to zero in $\tilde{G}$.

DEFINITION 1.5. It is said that the compact $B \subset \tilde{G}$ is the defining set of the functional $F \in \mathscr{E}'(\tilde{G})$, if for any open $\tilde{\omega} \supset B$ the functional F is continued on $\mathscr{E}(\tilde{\omega})$, i.e., if $\forall \tilde{\omega} \supset B \exists B_{\tilde{\omega}}$ is compact in $\tilde{\omega}$ and $\exists C_{\tilde{\omega}}$ such that $\forall f \in \mathscr{E}(\tilde{\omega})$,

$$|(F, f)| \leqslant C_{\tilde{\omega}} \sup_{B_{\tilde{\omega}}} |f|.$$

If $F \in \mathscr{E}'(\tilde{G})$, then there exists at least one compact defining F and by the

Hahn–Banach theorem it follows (since $\mathscr{E}(\tilde{G})$ is the closed subspace of the space $\mathbb{C}(\tilde{G})$ of continuous functions on $\tilde{G}$) that there exist compact $B \subset \tilde{G}$ and the measure μ with the support in B, such that

$$(F, f) = \int_{\tilde{G}} f \,\mathrm{d}\mu, \quad \forall f \in \mathscr{E}(\tilde{G}).$$

Some further definitions are needed here. We denote by $\mathbb{R}^{4n}$ and $\mathbb{C}^{4n}$ the spaces of real and complex n vectors x_μ and $z_\mu = x_\mu + iy_\mu$, $\mu = 0,1,2,3$. Let $G, \Gamma, \Omega, \ldots$, and $\tilde{G}, \tilde{\Gamma}, \tilde{\Omega}, \ldots$ be open sets in $\mathbb{R}^4$ and $\mathbb{C}^4$, respectively. For denoting different spaces of functions we use the following symbols:

$\mathscr{U}$ is some arbitrary space of functions;

$\mathscr{U}'$ is the space of linear functionals defined on $\mathscr{U}$;
$C(q) = C(q, \mathbb{R}^{4n})$ is the space of functions $f(x_1, \ldots, x_n)$ defined in $\mathbb{R}^{4n}$, where $\{f(x_1, \ldots, x_n)\}$ are q times differentiable and continuous functions together with all partial derivatives up to qth order inclusively;

$C(q, r) = C(q, r, \mathbb{R}^{4n})$ is the space of functions of $C(q)$ for which all the following products

$$\left| x_{i_1 \nu_1} \cdots x_{i_s \nu_s} \frac{\partial^p f(x_1, \ldots, x_n)}{\partial x_{j_1 \mu_1} \cdots \partial x_{j_p \mu_p}} \right| < \infty$$

are finite ($s = 0, 1, \ldots, r$; $p = 0, 1, \ldots, q$; $\nu, \mu = 0, 1, 2, 3$);
$L_p = L_p(\mathbb{R}^{4n})$ is the space of the functions for which

$$\int \mathrm{d}^4 x_1 \cdots \int \mathrm{d}^4 x_n |f(x_1, \ldots, x_n)|^p < \infty;$$

Z is the space of all entire functions for which

$$\int \mathrm{d}^4 x_1 \cdots \int \mathrm{d}^4 x_n |f(x_1 + iy_1, \ldots, x_n + iy_n)| < \infty$$

for any $y_1, \ldots, y_n$;

$\mathscr{E} = \mathscr{E}(\mathbb{C}^{4n})$ is the space of all entire functions;
$\mathscr{E}(\tilde{\Omega})$ and $\mathscr{E}(\dot{\Omega})$ are the spaces of analytic functions in regions $\tilde{\Omega}$ and Ω, respectively;
$\mathscr{E}_\alpha$ is the space of entire functions of the finite order of growth α, i.e., for any $g(z) \in \mathscr{E}_\alpha$ there exist such positive numbers b and B, as

$$|g(z_1, \ldots, z_n)| \leqslant B \exp\left[b \sum_i |z_i|^\alpha \right].$$

The space Z_α ($\alpha \geqslant 1$) consists of all those entire functions $f(z_1, \ldots, z_n)$ of n complex variables, which satisfy the following requirements:

(1) For any $f(z_1, \ldots, z_n) \in Z_\alpha$, there exist such positive numbers $C > 0$ and

$A_j > 0$ $(j = 1, \ldots, n)$, that

$$|f(z_1, \ldots, z_n)| \leqslant C \exp\left[\sum_{j=1}^{n} A_j |z_j|^\alpha\right].$$

(2) For any $y_1, \ldots, y_n$

$$\int d^4x_1 \cdots \int d^4x_n |f(x_1 + iy_1, \ldots, x_n + iy_n)| < \infty.$$

The number α is chosen depending on the interaction Lagrangian [exactly, on the measure $w(b_E^2/l^2)$] under consideration and the way of introducing nonlocality into the theory.

On the space Z_α $(\alpha \geqslant 1)$ one defines all nonlocal generalized functions $K(x)$, for which the Fourier transform $\tilde{K}(p^2)$ is an entire analytic function in the p^2-plane, the order of which is

$$\rho < \tfrac{1}{2}\gamma = \alpha/2(\alpha - 1).$$

Thus, $Z_\alpha = Z \cap \mathscr{E}_\alpha$ is nontrivial space for $\alpha \geqslant 1$. Z_α^β is the space of entire functions such that for any $g(z) \in Z_\alpha^\beta$ there exist some positive numbers b, c, and B, that

$$|g(z)| = |g(x_1 + iy_1, \ldots, x_n + iy_n)| \leqslant B \exp\left\{\sum_i [-c|x_i|^\beta + b|y_i|^\alpha]\right\}.$$

The space Z_α^β is nontrivial in the cases: $\alpha = 1$, $\beta < 1$ and $\alpha > 1$, $\alpha \geqslant \beta > 0$. The spaces Z_α^β coincide actually with the spaces S_α^β investigated by Gel'fand and Shilov (1968).

Further, we denote by $\tilde{\mathscr{U}}$, $\tilde{S}$, $\tilde{D}$, $\tilde{Z}$, and $\tilde{Z}_\alpha$ the spaces of the Fourier transforms of the functions belonging to $\mathscr{U}$, $S, \ldots,$ and Z_α, respectively. For example,

$$\tilde{f}(p) = (2\pi)^{-2} \int d^4x\, e^{ipx} f(x) \in \tilde{\mathscr{U}}$$

if $f(x) \in \mathscr{U}$, and we have $\tilde{S} = S$, $\tilde{D} = Z_1 \cap S$.

The space $\tilde{Z}_\alpha$, which is the space of Fourier transforms of functions $f \in Z_\alpha$ consists of the differentiable functions $\tilde{f}(p_1, \ldots, p_n)$ which satisfy the condition:

(3) There always exist positive numbers C and B_j $(j = 1, 2, \ldots, n)$ that

$$|\tilde{f}(p_1, \ldots, p_n)| \leqslant C \exp\left\{-\sum_{j=1}^{n} B_j |p_j|^\gamma\right\},$$

where $\gamma = \alpha/(\alpha - 1) > 1$.

Let us now introduce some definitions concerning the *entire analytic functions* of the single complex variable $z = x + iy \in \mathbb{C}^1 = \mathbb{R}^1 + i\mathbb{R}^1$.

DEFINITION 1.6. Let $g(z)$ be an entire function and

$$M_g(r) = \max_{|z|=r} |g(z)|$$

then the number

$$\rho = \lim_{r \to \infty} \frac{\ln \ln M_g(r)}{\ln r}$$

is called the *order* of the *entire function* $g(z)$. If the order of the entire function $g(z)$ equals ρ, it means that, roughly speaking, $g(z)$ increases as $\exp[|z|^\rho]$. All entire functions increasing slowly with respect to $\exp[|z|^\varepsilon]$ (for any $\varepsilon > 0$ chosen arbitrary small) are called entire functions of zero order $\rho = 0$, but all functions increasing rapidly with respect to $\exp[|z|^N]$ for any $N > 0$ are the functions of infinite order $\rho = \infty$.

DEFINITION 1.7. Let $h(r)$ be the monotonously increasing function for the sufficiently large r. If

$$\lim_{r \to \infty} \ln M_g(r)/h(r) = 1,$$

then we say that the exact order of $g(z)$ is $h(r)$. If $h(r) = \sigma r^\rho$, then the number σ is called the *type of the function* $g(z)$.

The function $\lambda(r)$ is called slowly changing or slow, if

$$\lim_{r \to \infty} r\lambda'(r)/\lambda(r) = 0, \quad \lambda' = \mathrm{d}\lambda(r)/\mathrm{d}r$$

for example, $\lambda(r) = (\ln r)^\alpha$, $\forall \alpha$ or $\lambda(r) = \ln \ln r$. The concept of the exact order of growth of the function $g(z)$ yields a finer gradation order of growth. The entire functions with the exact orders $h(r) = \sigma r^\rho$ and $h(r) = \sigma r^\rho \cdot \lambda(r)$, where the function $\lambda(r)$ is slowly growing have the same order ρ.

Let

$$g(z) = \sum_{n=0}^{\infty} g_n z^n$$

be the entire function represented in the form of series. Let us introduce

$$m_g(r) = \max_{n \geqslant 0} |g_n| r^n.$$

Then, the following relation is valid in the limit $r \to \infty$

$$\ln M_g(r) \sim \ln m_g(r),$$

whence $h(r) \sim \ln m_g(r)$.

This relation allows one to estimate with sufficient accuracy an order of growth of an entire function.

DEFINITION 1.8. The following Fourier representations hold for the entire functions. If $h(r) = \sigma r$ for $g(r)$, then

$$g(z) = \int d\mu_g(\zeta)\, e^{iz\zeta}, \tag{1.70}$$

where $\mu_g(r)$ is the complex-valued absolutely additive measure in the complex ζ-plane, located in the region $|\zeta| < \sigma + \varepsilon$ $(\varepsilon > 0)$. This representation was proved by Gel'fand and Shilov (1968).

DEFINITION 1.9. If $g(z) \in \mathscr{E}_h$ [$\mathscr{E}_h$ is the space of the entire functions such that for any $g(z) = g(z_1, \ldots, z_n)$ there exist such numbers $\delta > 0$ and $B > 0$, that

$$|g(z_1, \ldots, z_n)| \leqslant B \exp\left\{\sum_j h[(1-\delta)|z_j|],\ Z_h = Z \cap \mathscr{E}_h\right\},$$

where $h(r) > 0$, $h'(r) > 0$, $h''(r) > 0$, then the representation (1.70) holds when $\mu_g(\zeta)$ is the complex-valued absolutely additive measure in the complex ζ-plane for which there exists such $\varepsilon > 0$ that

$$\int |d\mu_g(\zeta)| \exp\{H[(1+\varepsilon)|\zeta|]\} < \infty. \tag{1.71}$$

Here

$$H(s) = \max_{t \geqslant 0}\, [st - h(t)]. \tag{1.72}$$

The representation (1.71) belongs to Efimov (1970). If $h(r) = \sigma r^\alpha$, then

$$H(s) = s^{\alpha/(\alpha-1)} (\alpha - 1)(\alpha^\alpha \sigma)^{-1/(\alpha-1)}.$$

1.4.6. *The Class of Generalized Functions in the Nonlocal Case*

We study now the space-time properties of the relativistic invariant generalized functions of type (1.11) or (1.66), for different sequences of coefficients $\{c_n\}$. We will distinguish only the following possibilities (for detail see Efimov, 1977a)

(1) $c_n = 0$, for any $n > n_0 \geqslant 0$ (n_0 is arbitrary),

$$\text{(2)}\quad \overline{\lim_{n\to\infty}}\, (c_n)^{1/n} = 0, \tag{1.73}$$

$$\text{(3)}\quad \overline{\lim_{n\to\infty}}\, (c_n)^{1/n} = l^2. \tag{1.74}$$

Thus let us consider the functional

$$(K, f)(x) = \int d^4x K(x) f(x) = \sum_{n=0}^{\infty} \frac{c_n}{(2n)!} \Box^n f(x). \tag{1.75}$$

If there exist functions with bounded support in the space $\mathcal{U}$ of the test functions to which the function $f(x)$ belongs, then one can give the following definition.

DEFINITION 1.10. The generalized function

$$K(x) = \sum_{n=0}^{\infty} \frac{c_n}{(2n)!} \Box^n \delta^{(4)}(x) \tag{1.76}$$

is called *local* and is located at the point $x = 0$, if $(K, f) = 0$ for any $f(x) \in \mathcal{U}$, a support of which does not contain the point $x = 0$.

We assume that a sufficiency of functions with a bounded support exist in the space $\mathcal{U}$. It means that if some function $g(x)$ is locally integrable and

$$\int d^4x \; f(x)g(x) = 0$$

for any $f(x) \in D(G) \cap \mathcal{U}$, then $g(x) \equiv 0$ almost everywhere. We now consider the generalized function $K(x)$ for three cases:

Case 1. Local distributions when $c_n = 0$, $n > n_0 \geqslant 0$. For the existence of the functional (1.75) it is sufficient to require the functions of the test space $2n_0$ times differentiable with respect to all arguments. Thus, we arrive at the spaces of the type $C(q)$, where $q = 2n_0$. If one considers such $K(x)$ for which n_0 is arbitrary but is finite for any $K(x)$, then the functions of the test space must be infinitely differentiable and among those there must be the functions with a bounded support. So, we come to the spaces D and S. The spaces D, S, and $C(q, r) \subset C(q)$ are investigated in detail on the basis of which usual local quantum field theory is constructed. In this case the generalized functions $K(x)$ satisfy Definition 1.10. Therefore, they are local and are usually called distributions of *tempered order of growth*.

Case 2. Local distributions when $\overline{\lim}_{n\to\infty} (c_n)^{1/n} = 0$. The distributions of this type in the case of single variable function were considered first by Meiman (1964) and next by Soloviev (1971). We now define the conditions for which the functional (1.75) exists. At first we carry out some formal transformations. We make use of the following integral representation for the operator $\Box$:

$$\Box^n = \left(-\frac{\partial^2}{\partial x_0^2} + \frac{\partial^2}{\partial \mathbf{x}^2} \right)^n = a_n \int_{b^2 \leqslant 1} d^4 b \left(\frac{ib_4}{4} \frac{\partial}{\partial x_0} + \mathbf{b} \frac{\partial}{\partial \mathbf{x}} \right)^{2n}, \tag{1.77}$$

where

$$a_n = \pi^{-2} \cdot 2^{2n} n!(n+1)!/(2n)!, \; \lim_{n\to\infty} |a_n|^{1/n} = 1. \tag{1.78}$$

The integration in (1.77) is performed over the Euclidean globe $b^2 = b_1^2 + b_2^2 + b_3^2 + b_4^2 = \mathbf{b}^2 + b_4^2 \leqslant 1$. Let us introduce the function of a complex variable ξ:

$$W(\xi) = \sum_{n=0}^{\infty} c_n a_n \xi^{-2n-1}. \tag{1.79}$$

It follows from conditions (1.73) and (1.78) that $W(\xi)$ has no singularities in the complex ξ-plane except for the essential singularity at $\xi = 0$ only. Therefore,

$$c_n a_n = \frac{1}{2\pi i} \oint d\xi\, \xi^{2n} W(\xi),$$

where the integration is performed over any closed contour around the point $\xi = 0$. Making use of (1.77) and (1.79), the functional (1.75) can be written in the form

$$(K, f)(x) = \sum_{n=0}^{\infty} \frac{1}{(2n)!} \frac{1}{2\pi i} \oint d\xi\, \xi^{2n} W(\xi) \times$$

$$\times \int_{b^2 \leqslant 1} d^4 b \left(ib_4 \frac{\partial}{\partial x_0} + \mathbf{b} \frac{\partial}{\partial \mathbf{x}} \right)^{2n} f(x)$$

$$= \frac{1}{2\pi i} \oint d\xi\, W(\xi) \int_{b^2 \leqslant 1} d^4 b \sum_{n=0}^{\infty} \frac{\xi^n}{n!} \left(ib_4 \frac{\partial}{\partial x_0} + \mathbf{b} \frac{\partial}{\partial \mathbf{x}} \right)^n f(x_0, \mathbf{x}). \tag{1.80}$$

In the last term the summation is performed over all n because the integral over b of the odd powers of $(ib_4\, \partial/\partial x_0 + \mathbf{b}\, \partial/\partial \mathbf{x})$ is identically equal to zero. Observing that there is a displacement operator with respect to the arguments $(x_0, \mathbf{x})$ in (1.80), one can get

$$(K, f)(x) = \frac{1}{2\pi i} \oint d\xi\, W(\xi) \int_{b^2 \leqslant 1} d^4 b\, f(x_0 + ib_4 \xi, \mathbf{x} + \mathbf{b}\xi). \tag{1.81}$$

This expression is well defined because of $f(x) \in \mathscr{E}_0$ or Z_0.

The integration over ξ can be performed over a contour with an arbitrary small radius because $W(\xi)$ has the only singularity at the point $\xi = 0$. Therefore, the functional (1.81) can be extended continuously to the space of functions which are analytic at the point $x = (x_0, \mathbf{x})$. Hence, the generalized functions (1.76) with the condition (1.73) are local according to Definitions 1.4 and 1.5.

The Fourier transform of the generalized functions (1.76) can be written in the form

$$\tilde{K}(p^2) = (K, e^{ipz}) = \frac{1}{2\pi i} \oint d\xi\, W(\xi) \int_{b^2 \leqslant 1} d^4 b \exp(-p_0 b_4 \xi - i\mathbf{p}\mathbf{b}\xi)$$

$$= \frac{1}{2\pi i} \oint d\xi\, W(\xi) 2\pi^2 J_1(\xi(-p^2)^{1/2}) \frac{(-p^2)^{-1/2}}{\xi}. \tag{1.82}$$

The function $\tilde{K}(p^2)$ is an entire analytic function of the order $\rho < \frac{1}{2}$ in the complex p^2-plane, i.e. [see formula (1.65)],

$$|\tilde{K}(p^2)| \leqslant C_\varepsilon \exp(\varepsilon\sqrt{|p|^2})$$

for any $\varepsilon > 0$. According to the theory of entire analytic functions it is known that for these functions there is not any direction in the complex p^2-plane, along which they could decrease. It means that they cannot play the role of the *cutoff functions* in the construction of the perturbation theory for the S-matrix, if we want to make use of the quasi-local arbitrariness in constructing the coefficient functions $S_n(x_1, \ldots, x_n)$ in the perturbation expansion of the S-matrix.

The 'minimum' space of test functions in this case forms the class $\mathscr{E}_0$ of functions which are analytic with respect to each argument $z_\mu = x_\mu + iy_\mu$ in some band $|y_\mu| < \mathrm{d}_\mu$ where d_μ depends on the function $f(z)$ and can be an arbitrary quantity. The fixed band $|y_\mu| < \mathrm{d}_0$, where all functions $f \in \mathscr{E}$ are analytic, does not exist. In this way we got the relativistic generalization of Meiman's class C_0.

We notice that local distributions, when $\Sigma(1/n)(c_n)^{1/2n} < \infty$, correspond to the strict locality due to Jaffe (1966). In this case the theory will be strictly local, if the space of test functions contains functions with a bounded support.

Case 3. Nonlocal generalized functions when $\overline{\lim}_{n\to\infty} (c_n)^{1/n} = l^2$, where l is some parameter. We consider functional $(K,f)(x)$ for which the formula (1.75) is valid. Let us make use of the integral representation (1.77) and introduce the function $W_l(\xi)$ of the complex variable ξ according to (1.79). In this case it follows from conditions (1.74) and (1.78) that the function $W_l(\xi)$ is analytic in the complex plane ξ outside the circle $|\xi| = l$. Inside this circle it has some singularities, the positions of which depend on an explicit form of the coefficients c_n. Therefore,

$$c_n a_n = \frac{1}{2\pi i} \oint_{|\xi| > l} \mathrm{d}\xi\, \xi^{2n} W_l(\xi).$$

Repeating the calculations of Case 2, one can obtain for the functional (1.75) with condition (1.74) the following representation:

$$(K, f)(x) = \frac{1}{2\pi i} \int_{l < |\xi| < l + \varepsilon} \mathrm{d}\xi\, W_l(\xi) \int_{b^2 \leqslant 1} \mathrm{d}^4 b\, f(x_0 + ib_4\xi, \mathbf{x} + \mathbf{b}\xi). \quad (1.83)$$

This expression is well defined for any $f(x) \in \mathscr{E}$.

Let us demonstrate that the representation (1.83) is a *relativistic covariant* on $\mathscr{E}$. We have to prove that the integral in (1.83) over the Euclidean space is covariant under co-ordinate transformations from the homogeneous Lorentz group, i.e.,

$$\int_{b^2 \leqslant 1} d^4b f(\Lambda_{0\mu}(x_\mu + b_\mu\xi), \Lambda_{i\mu}(x_\mu + b_\mu\xi))$$

$$= \int_{b^2 \leqslant 1} d^4 bf(\Lambda_{0\mu} x_\mu + b_0 \xi, \Lambda_{i\mu} x_\mu + b_i \xi), \quad (1.84)$$

where Λ is a representation of the homogeneous Lorentz group L. The summation is performed over $\mu = 0, 1, 2, 3$. Here, we have introduced the notation $b_0 = ib_4$ and $i = 1,2,3$.

Any $\Lambda \in L$ can be represented in the form $\Lambda = \Lambda_1 \Lambda_2 \Lambda_3$ where Λ_1 and Λ_3 are usual rotations and Λ_2 is the purely Lorentz transformation along third axes. The integral (1.84) is obviously covariant with respect to rotations in the three-dimensional Euclidean space. Therefore, it will be sufficient for us to demonstrate the covariance of the representation (1.84) under the purely Lorentz transformation Λ_2. Then, the relation (1.84) can be written

$$\int_{b^2 \leqslant 1} d^4b\, f((x_0 + ib_4\xi) \operatorname{ch} \theta + (x_3 + b_3\xi) \operatorname{sh} \theta, x_1 + b_1\xi, x_2 + b_2\xi,$$

$$(x_0 + ib_4\xi) \operatorname{sh} \theta + (x_3 + b_3\xi) \operatorname{ch} \theta) =$$

$$\int_{b^2 \leqslant 1} d^4b\, f(x_0 \operatorname{ch} \theta + x_3 \operatorname{sh} \theta + ib_4\xi, x_1 + b_1 \xi,$$

$$x_2 + b_2 \xi, x_0 \operatorname{sh} \theta + x_3 \operatorname{ch} \theta + b_3 \xi), \quad (1.85)$$

where $\operatorname{th} \theta = v/c$, $l < |\xi| < l + \varepsilon$. Let us go over to the polar co-ordinates in the plane (b_4, b_3) and put $b_4 = r \cos \varphi$, $b_3 = r \sin \varphi$. Then, the left-hand side of (1.85) can be written as

$$\int_0^{2\pi} d\varphi \int_0^1 dr\, r \iint_{b_1^2 + b_2^2 = 1 - r^2} db_1\, db_2 f(x'_0 + ir\xi \cos(\varphi + i\theta),$$

$$x_1 + b_1\xi, x_2 + b_2\xi, x'_3 + r\xi \sin(\varphi + i\theta)), \quad (1.86)$$

where $x'_0 = x_0 \operatorname{ch} \theta + x_3 \operatorname{sh} \theta$, $x'_3 = x_0 \operatorname{sh} \theta + x_3 \operatorname{ch} \theta$. In so far as the function $f(z_0, \mathbf{z})$ is entirely analytic with respect to the arguments z_μ it will be the entirely analytic function with respect to the complex argument φ. In doing so the integrand in (1.86) is periodic along the real axis in the complex plane φ with the period 2π. Therefore, in (1.86) one can displace the integration contour over φ from 0 to 2π by any purely imaginary number, i.e.,

$$\int_0^{2\pi} d\varphi\, A(\varphi) = \int_{i\theta}^{i\theta + 2\pi} d\varphi\, A(\varphi) = \int_0^{2\pi} d\varphi\, A(\varphi - i\theta), \quad (1.87)$$

where $A(\varphi)$ is the integrand in (1.86). Thus, the right-hand side of (1.85) is equal to the left and our assertion is proved.

Now let us study the space-time properties of the generalized functions $K(x)$ in (1.83), i.e., consider the possible defining sets of the functional (K, f) according to Definition 1.5. In the representation (1.83) one can give a somewhat definite element of the Lorentz group $\Lambda = \Lambda(\theta, \varphi, \psi)$ with the fixed parameters θ, φ, and ψ, and then the functional (K, f) $(x = 0)$ has the following equivalent form:

$$(K, f) = \frac{1}{2\pi i} \oint_{l \leqslant |\xi| \leqslant l + \varepsilon} \mathrm{d}\xi \, W_l(\xi) \int_{b^2 \leqslant 1} \mathrm{d}^4 b f(\Lambda_{0\mu} b_\mu \xi, \Lambda_{j\mu} b_\mu \xi). \tag{1.88}$$

This functional can be extended to the space $\mathscr{E}(\tilde{\Omega}_l(\theta, \varphi, \psi))$ of functions which are analytic in the region

$$\tilde{\Omega}_l(\theta, \varphi, \psi) = \{z: |\Lambda_{0\mu} z_\mu|^2 + |\Lambda_{1\mu} z_\mu|^2 + |\Lambda_{2\mu} z_\mu|^2 + |\Lambda_{3\mu} z_\mu|^2 < l^2 + \varepsilon\}.$$

In particular,

$$\tilde{\Omega}_l(0, 0, 0) = \{z: |z_0|^2 + |z_1|^2 + |z_2|^2 + |z_3|^2 < l^2 + \varepsilon\},$$
$$\tilde{\Omega}_l(\theta.0,0) = \{z: |z_0 \operatorname{ch} \theta + z_3 \operatorname{sh} \theta|^2 + |z_1|^2 + |z_2|^2 + |z_3 \operatorname{ch} \theta + z_0 \operatorname{sh} \theta|^2 < l^2 + \varepsilon\}.$$

Projections of these regions onto the plane (x_0, x_3) are shown in Figure 1.

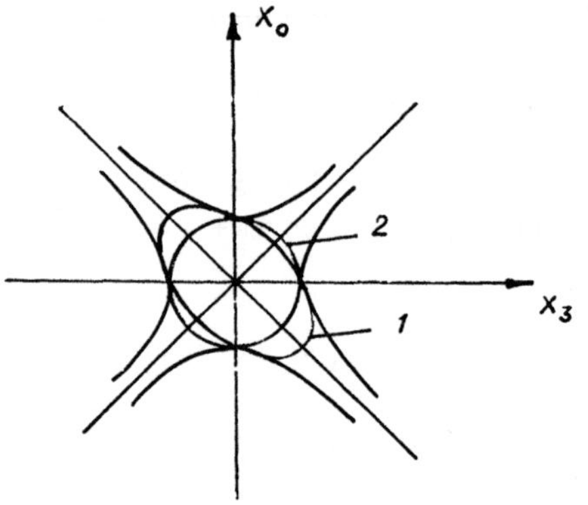

Fig. 1. The projections of some possible defining sets of the functional (1.88) on the $(x_0\, x_3)$-plane. 1 is the projection $\tilde{\Omega}_l(\theta, 0, 0)$ and 2 is the projection $\tilde{\Omega}_l(0, 0, 0)$.

Thus, the sets $\tilde{\Omega}_l(\theta, \varphi, \psi)$ are defining for the function (1.88); there is none among them, that can easily be seen, which would be contained in all the others. Therefore, a support of the generalized function $K(x)$ cannot be given synonymously. In this case the Fourier transform of $K(x)$ acquires the form

$$\tilde{K}(p^2) = (K, \mathrm{e}^{ipz}) = \frac{1}{2\pi i} \oint_{l \leqslant |\xi| < l + \varepsilon} \mathrm{d}\xi \, W_l(\xi) \int_{b^2 \leqslant 1} \mathrm{d}^4 b \exp(-p_0 b_4 \xi - i\mathbf{pb}\xi)$$

$$=\frac{1}{2\pi i}\oint_{l<|\xi|<l+\varepsilon} \mathrm{d}\xi\, W_l(\xi)2\pi^2 J_1(\xi(-p^2)^{1/2})(-p^2)^{-1/2}/\xi. \qquad (1.89)$$

From this, it follows that the function $\tilde{K}(p^2)$ is an entire analytic function in the complex p^2-plane and has the order of growth $\rho = \frac{1}{2}$ and type l, and for which the following estimation:

$$|\tilde{K}(p^2)| \leqslant C_\varepsilon \exp[(1+\varepsilon)(|p|^2)^{1/2}] \qquad (1.90)$$

is valid. Such functions may have only one direction in the complex plane along which they may decrease. This is just the case which will be used below.

The 'minimum' space of test functions in this case is the class $\mathscr{E}_l$ of functions which are analytic in the region $\Gamma_l \subset \mathbb{C}^4$. The point $z \in \Gamma_l$, if $-l^2 \leqslant y_0^2 - \mathbf{y}^2 \leqslant l^2$.

Chapter 2

The Basic Problems of Nonlocal Quantum Field Theory

2.1. Nonlocality and the Interaction Lagrangian

In this chapter we present the main ideas of constructing the nonlocal theory of quantized fields. Our basic postulation states that in quantum field theory nonlocality arising from the stochastic nature of space-time is manifested only in interaction processes between fields. In other words, it means that we must introduce nonlocality into the interaction Lagrangians of different fields. This proposal has some basis. Indeed, if we maintain the Einstein relation between momentum and energy $E = (\mathbf{p}^2 + m^2)^{1/2}$ to be valid for extended fields constructed by means of the relativistic invariant distributions $K(l^2\Box)$ (1.11), then they will satisfy free field equations, for example, in the case of the one-component scalar field $\varphi_R(x) = \phi(x)$:

$$(\Box - m^2)\phi(x) = 0. \tag{2.1}$$

According to our construction, the solution to equation (2.1) has the form

$$\phi(x) = (2\pi)^{-3/2} \int \mathrm{d}^4 p\, \mathrm{e}^{-ipx}\, \delta(p^2 - m^2)\tilde{K}(p^2 l^2)\, \tilde{\varphi}(p). \tag{2.2}$$

From this it is easy to see that due to the form factor $\tilde{K}(p^2l^2)$ and the function $\delta(p^2 - m^2)$, the extended field (2.2) is, indeed, free and local (the normalization condition $\tilde{K}(m^2 l^2) = 1$ is used here). Therefore, the free field case does not give new results in our proposed scheme.

Thus *nonlocality is introduced into interaction Lagrangians* $\mathscr{L}_{\text{in}}(x)$ while free Lagrangians $\mathscr{L}_0(x)$ of systems are considered to be local. Within this proposal, the self-consistent nonlocal model of QFT was formulated by Efimov (1977a) in the framework of the Bogolubov, Medvedev and Polivanov (1958) axiomatics (see Bogolubov and Shirkov, 1980).

The outline of the construction of the S-matrix in the nonlocal theory consists of the following. The general axioms are formulated, then a certain nonlocal Lagrangian of the classical field is considered. For example, in the case of the scalar field $\varphi(x)$ the total Lagrangian of this kind can be

$$\mathscr{L}(x) = \tfrac{1}{2}\varphi(x)(\square - m^2)\varphi(x) - g\left\{\int d^4y K(x-y)\varphi(y)\right\}^4, \tag{2.3}$$

where $K(x-y)$ is our nonlocal form factor (generalized function). In order to construct the S-matrix in perturbation theory, the *correspondence principle* is used, which states that for *infinitesimal* coupling constant g the S-matrix has the form

$$S = 1 - ig\int d^4x \colon\left\{\int d^4y K(x-y)\varphi(y)\right\}^4\colon \tag{2.4}$$

where $\varphi(x)$ is the quantized scalar field which satisfies equation (2.1). In constructing the highest perturbation orders, we then introduce the nonlocality method averaged in (1.8), according to which the propagator of the scalar particle changes to the following [see formula (1.13)]:

$$\frac{1}{m^2 - k^2 - i\varepsilon} \to \frac{[\tilde{K}(k^2l^2)]^2}{m^2 - k^2 - i\varepsilon} = V(-k^2l^2)\frac{1}{m^2 - k^2 - i\varepsilon} \tag{2.5}$$

in the momentum space $\mathbb{R}^4(k)$. Further, there is indicated the class of functions to which the form factor $V(-k^2l^2)$ belongs (see Chapter 1), and an intermediate regularization (see below) is introduced in the framework of which the finite S-matrix is constructed. At the end of the calculations one can remove an intermediate regularization, and thus prove that the S-matrix satisfies all general axioms: unitarity, causality, covariance, gauge invariance and so on. In the nonlocal theory one should construct a special method of quantization of nonlocal classical fields of type (2.3).

We present in the following two sections a canonical quantization method of nonlocal classical fields, and the physical meaning of the form factors $K(l^2\square)$ of the theory. In Section 2.4 we formulate the causality condition and unitarity for the nonlocal S-matrix in Bogolubov–Efimov axioms and prove the unitarity of the S-matrix in a functional form. Further, in the case of QFT with nonlocal interactions, we obtain the Schrödinger-type dynamical equation and show that the finite and unitary S-matrix describing nonlocal interactions of quantized fields is a solution of the *Cauchy problem* for this evolution equation with *retardation.*

2.2. Quantization of Nonlocal Field Theory

In this section we propose a procedure of canonical quantization of nonlocal fields which are described by the Lagrangian of type (2.3). According to the idea due to Efimov (1974) we introduce a certain regularization into the classical Lagrangian in such a way which permits us to carry out the usual canonical quantization. This quantization leads to the appearance of *additional ghost states with indefinite metrics.* The ghost states disappear when the regularization is removed, but a trace remains, namely the propagator of the scalar particle becomes nonlocal according to (2.5).

Distant analogy of the procedure proposed is the method of quantization of the electromagnetic field (see, for example, Bogolubov and Shirkov, 1980). The real physical photon can be in *two states with transverse* polarizations only. However, these states cannot lead to the correct propagator of the photon. In order to get the propagator of the virtual photon in the form

$$D_{\mu\nu}(k^2) = \delta_{\mu\nu}/(-k^2 - i\varepsilon), \tag{2.6}$$

it is necessary to introduce non-physical *scalar and longitudinal quanta* with indefinite metrics for consideration. The *Lorentz condition*

$$\partial_\mu A_\mu^{(-)}(x)|\text{any physical states}\rangle = 0$$

and the current conservation $\partial_\mu J_\mu(x) = 0$ guarantee that the scalar and longitudinal quanta *do not appear* in any interactions of physical transverse photons with charged particles, say, electrons. However, their role is that they contribute to the propagator (2.6), i.e., the virtual photon consists of transverse, longitudinal and scalar quanta.

2.2.1. *Formulation of the Quantization Problem*

Let us consider a one-component scalar field $\varphi(x)$. The Lagrangian density describing the classical field $\varphi(x)$ can be written in the form with accordance to our above proposals:

$$\mathscr{L}(x) = \tfrac{1}{2}\varphi(x)(\Box - m^2)\varphi(x) + gU(K(l^2\Box)\varphi(x)), \tag{2.7}$$

here g is the coupling constant, $U(z)$ is a function describing self-interaction of the scalar field $\varphi(x)$. We suppose that $U(z)$ is analytic in the neighborhood of the real axis in the complex $z = u + iv$-plane. In all other respects it is an arbitrary function. The classes of functions $U(z)$ for which the finite nonlocal S-matrix can be constructed, have been considered in detail by Alebastrov and Efimov (1972, 1973).

The operator $K(l^2\Box)$ in (2.7) is nonlocal and can be presented in the form of (1.11), so that the generalized function $K(x - y) = K(l^2\Box)\delta^{(4)}(x - y)$ belongs to one of the spaces of generalized functions, considered in Chapter 1. In what follows, for convenience, we consider the operator

$$V(-l^2\Box) = [K(l^2\Box)]^2 \tag{2.8}$$

and suppose that the function $V(z)$ satisfies the following

CONDITION 2.1

(1) $V(z)$ is an entire function of the finite order $\frac{1}{2} \leqslant \rho \leqslant \infty$, i.e., $\exists C > 0,\quad b > 0$

$$|V(z)| \leqslant C \exp\{\mathrm{b}\,|\,\mathrm{z}\,|^{\rho}\},$$

(2) $[V(z)]^* = V(z^*)$,

(3) $V(-m^2 l^2) = 1$,

(4) $V(x) \geqslant 0$ for real x,

(5) $$V(z) = \begin{cases} \mathrm{O}(1/|z|^2) & \mathrm{Re}\, z \to +\infty\ (k^2 \to -\infty) \\ \mathrm{O}(\exp(\mathrm{b}|z|^{\rho})) & \mathrm{Re}\, z \to -\infty\ (k^2 \to +\infty). \end{cases}$$

In the expansion

(6) $V(z) = \sum_{n=0}^{\infty} v_n[-z - m^2 l^2]^n$ the coefficients v_n satisfy

(7) $v_0 = 1,\quad v_n > 0,\quad \forall n,$

(8) $\exists C > 0,\quad A > 0$ that $v_n < CA^n/\Gamma(n/\rho + 1)$.

Thus we have described all the quantities in the Lagrangian density (2.7). Notice that the generalized function $K(x) = K(l^2\Box)\,\delta^{(4)}(x)$, where $K(l^2\Box)$ satisfies Condition 2.1, is defined on the space Z_α of test functions for $\alpha < 2\rho/(2\rho - 1)$.

The wave equation for the classical scalar field $\varphi(x)$ can be obtained according to the *principle of stationary action* and may be represented in the form

$$(\Box - m^2)\varphi(x) = -gK(l^2\Box)U'(K(l^2\Box)\varphi(x)).$$

Let us introduce the extended field (1.8)

$$\phi(x) = \varphi_R(x) = K(l^2\Box)\varphi(x)$$

Then the total Lagrangian density (2.7) for the field $\phi(x)$ is

$$\mathscr{L}(x) = \tfrac{1}{2}\phi(x)E(\Box)\phi(x) + gU(\phi(x)), \tag{2.9}$$

where

$$E(\Box) = (\Box - m^2)/V(-l^2\Box). \tag{2.10}$$

Formally the wave equation is written as

$$E(\Box)\phi(x) = -gU'(\phi(x)) \tag{2.11}$$

for the interaction field, and

$$E(\Box)\phi(x) = [(\Box - m^2)/V(-l^2\Box)]\phi(x) = 0 \tag{2.12}$$

for the free field.

The problem is how to understand these equations, how to study and solve them and how to perform the quantization of the field $\phi(x)$. We proceed in the following way in accordance with Efimov's method. Instead of equations (2.11) or (2.12) we consider a *regularized equation*

$$E^{\delta}(\Box)\phi^{\delta}(x) = -gU'(\phi^{\delta}(x)) \tag{2.13}$$

or, in the case $g = 0$,

$$E^{\delta}(\Box)\phi^{\delta}(x) = 0 \tag{2.14}$$

Here δ is a parameter of the regularization such that

$$\lim_{\delta \to 0} E^{\delta}(\Box) = E(\Box) = (\Box - m^2)/V(-l^2\Box). \tag{2.15}$$

Accordingly, instead of the Lagrangian (2.9) we obtain

$$\mathscr{L}(x) = \tfrac{1}{2}\phi^{\delta}(x)E^{\delta}(\Box)\phi^{\delta}(x) + gU(\phi^{\delta}(x)). \tag{2.16}$$

Our regularization is chosen in such a way that the function

$$E^{\delta}(k^2) = (k^2 - m^2)/V^{\delta}(-l^2k^2)$$

has zeros in some set of points:

$$E^{\delta}(k^2) \sim (k^2 - m^2) \prod_{j=1}^{\infty} [k^2 - m_j^2(\delta)] \tag{2.17}$$

such that $m_j^2(\delta) > 0$ ($j = 1, 2, \ldots$) and $m_j^2(\delta) \to \infty$ when $\delta \to 0$. Then the field $\phi^{\delta}(x)$ ($\delta > 0$) can be quantized by the methods with *indefinite metrics* (Blokhintsev, 1947; Pais and Uhlenbeck, 1950; and Nady, 1966).

The Hamiltonian H_0^{δ} and the vector space of states $\mathscr{H}^{\delta}$ with indefinite metrics can be constructed for the free system when $\delta > 0$. Further, the S^{δ}-matrix, and also different operators: the current

$$J^{\delta}(x) = i[\delta S^{\delta}/\delta\phi^{\delta}(x)]S^{\delta +},$$

the *Green functions*

$$G^{\delta}(x - y) = \langle 0 \mid T[\phi^{\delta}(x)\phi^{\delta}(y)S^{\delta}] \mid 0\rangle,$$

and the *Wightman functions*

$$W^{\delta}(x_1, \ldots, x_n) = \langle 0 \mid \phi^{\delta}(x_1) \cdots \phi^{\delta}(x_n) \mid 0\rangle,$$

etc., can be found for the interacting system.

By definition, we consider that when $\delta \to 0$ the limits of all these physical quantities are the quantum field solution of the initial system (2.9). The problem indicates such a regularization procedure which provides the existence of the limits of all operators and matrix elements for any physical quantities at $\delta \to 0$. It means that we have to obtain a self-consistent theory in the limit $\delta \to 0$. This section is devoted to this problem.

2.2.2. *Regularization Procedure*

The regularization is introduced in the following way. Instead of the function

$$V(-k^2l^2) = \sum_{n=0}^{\infty} v_n(k^2 - m^2)^n l^{2n}, \tag{2.18}$$

we introduce the regularized function

$$V^{\delta}(-k^2l^2) = \sum_{n=0}^{\infty} \frac{v_n(k^2 - m^2)^n l^{2n}}{\prod_{j=1}^{n+2} (1 - \delta/j^{\sigma} (k^2 - m^2)/m^2)}, \tag{2.19}$$

where $\sigma < 1/\rho \leqslant 2$. It is convenient to denote

$$z = k^2/m^2, \qquad \lambda = m^2 l^2, \qquad u_n = v_n \lambda^n.$$

Then

$$V^{\delta}(-z\lambda) = \sum_{n=0}^{\infty} \left\{ \frac{u_n(z-1)^n}{\prod\limits_{j=1}^{n+2} [1 - \delta/j^{\sigma}(z-1)]} \right\}$$

$$= \sum_{n=0}^{\infty} \left\{ \frac{(-1)^n u_n(z-1)^n[(n+2)!]^{\sigma}}{\delta^{n+2} \prod\limits_{j=1}^{n+2} [z - \mu_j(\delta)]} \right\}, \tag{2.20}$$

where $\mu_j(\delta) = 1 + j^{\sigma}/\delta$. Let us consider the function

$$E^{\delta}(z) = (z-1)/V^{\delta}(-z\lambda). \tag{2.21}$$

The function

$$1/E^{\delta}(z) = V^{\delta}(-z\lambda)/(z-1) \tag{2.22}$$

is a *meromorphic* analytic function in the complex z-plane and has the simple poles at the points

$$z = \mu_j(\delta) = 1 + j^{\sigma}/\delta \quad (j = 0, 1, 2, \ldots)$$

and decreases as [except the ray (z: $\arg z = 0$)]

$$\frac{1}{E^{\delta}(z)} \approx |z|^{-3} \sum_{n=0}^{\infty} (-1)^n \delta^{-n-2} u_n[(n+2)!]^{\sigma},$$

when $|z| \to \infty$. Here the series converges according to Condition 2.1. For the meromorphic function (2.22) the following representation is valid

$$\frac{1}{E^{\delta}(z)} = \sum_{j=0}^{\infty} \left\{ \frac{(-1)^j A_j(\delta)}{z - \mu_j(\delta)} \right\}, \tag{2.23}$$

where

$$A_j(\delta) = \sum_{n=\max\{0,j-2\}}^{\infty} u_n c_{n,j}(j^\sigma/\delta)^n, \quad A_0(\delta) = 1, \tag{2.24}$$

$$c_{n,j} = \frac{[(n+2)!]^\sigma}{\Sigma_{l=0}^{j-1}} \frac{1}{\Sigma_{k=j+1}^{n+2}}(k^\sigma - j^\sigma). \tag{2.25}$$

The numbers $A_j(\delta)$ satisfy the following conditions

$$\sum_{j=0}^{\infty} (-1)^j A_j(\delta)[\mu_j(\delta)]^N = 0, \quad \text{for } N = 0, 1, 2. \tag{2.26}$$

The inequality (Efimov, 1977a)

$$A_j(\delta) \leqslant \text{const}\cdot\exp\{B(j^\sigma/\delta)^\rho\}/\Gamma((1/\rho - \sigma)j)$$

is valid for the numbers $A_j(\delta)$ at the large j, where B is some number and $\sigma\rho < 1$.

Let us consider now the function

$$D^\delta(k^2) = -[E^\delta(k^2)]^{-1}$$

which has the properties:

(1) It is a meromorphic analytic function in the complex k^2-plane and has the simple poles at the points

$$k^2 = m_j^2(\delta) = m^2\cdot\mu_j(\delta) = m^2(1 + j^\sigma/\delta) \quad (j = 0, 1, 2, \ldots)$$

(2) The residues of $D^\delta(k^2)$ at these points are

$$\operatorname*{Res}_{k^2=m_j^2(\delta)} \mathrm{D}^\delta(k^2) = (-1)^{j+1} A_j(\delta)$$

(3) When $|k^2| \to \infty$ in all k^2-planes, except the ray $[m^2, \infty)$,

$$D^\delta(k^2) = V^\delta(-k^2 l^2)/(m^2 - k^2) = O(1/|k^2|^3).$$

(4) The function $D^\delta(k^2)$ may have zeros at the points $k^2 = a_r (r = 1, 2, \ldots)$

(5) $\lim_{\delta\to 0} D^\delta(k^2) = V(-k^2 l^2)/(m^2 - k^2)$.

2.2.3. *Quantization of the Regularized Equation*

Let us consider the classical system described by the Lagrangian density $\mathscr{L}^\delta(x)$ (2.16), where the regularized operator $E^\delta(\Box)$ satisfies the properties enumerated above. According to the principle of stationary action the wave equation for the system described by (2.16) has the form (2.13). It is the differential equation of the *infinite order*, i.e., it is an *integral equation.* In order to solve the

Cauchy problem we have to know the values of the function $\phi^\delta(x)$ and all its derivatives in the initial moment of time.

We analyse the solution of this equation following the scheme proposed by Pais and Uhlenbeck (1950). Let us introduce a system of fields

$$\phi_j^\delta(x) = [A_j(\delta)]^{1/2}\{E^\delta(\Box)/[\Box - m_j^2(\delta)]\}\phi^\delta(x), \tag{2.27}$$

where the coefficients $A_j(\delta)$ are introduced by formula (2.24), and

$$m_j^2(\delta) = m^2(1 + j^\sigma/\delta) \quad (j = 0, 1, 2, \ldots) \tag{2.28}$$

According to definition (2.27), the fields $\phi_j^\delta(x)$ are not independent for different j and they satisfy the relations

$$[A_j(\delta)]^{1/2}\left\{\frac{E^\delta(\Box)}{[\Box - m_j^2(\delta)]}\right\}\phi_i^\delta(x) = [A_i(\delta)]^{1/2}\left\{\frac{E^\delta(\Box)}{[\Box - m_i^2(\delta)]}\right\}\phi_j^\delta(x). \tag{2.29}$$

The field $\phi^\delta(x)$ can be expressed as

$$\phi^\delta(x) = \sum_{j=0}^{\infty}(-1)^j[A_j(\delta)]^{1/2}\phi_j^\delta(x). \tag{2.30}$$

In fact, on the one hand, the following chain of equalities

$$\begin{aligned}\phi^\delta(x) &= \sum_{j=0}^{\infty}(-1)^j A_j(\delta)\left\{\frac{E^\delta(\Box)}{\Box - m_j^2(\delta)}\right\}\phi^\delta(x)\\ &= [1/E^\delta(\Box)]E^\delta(\Box)\phi^\delta(x) = \phi^\delta(x)\end{aligned}$$

is valid. On the other hand, using (2.29) it is possible to obtain

$$\begin{aligned}\phi_j^\delta(x) &= [A_j(\delta)]^{1/2}\left\{\frac{E^\delta(\Box)}{\Box - m_j^2(\delta)}\right\}\sum_{i=0}^{\infty}(-1)^i[A_i(\delta)]^{1/2}\phi_i^\delta(x)\\ &= \sum_{i=0}^{\infty}(-1)^i[A_i(\delta)]^{1/2}[A_i(\delta)]^{1/2}\left\{\frac{E^\delta(\Box)}{\Box - m_i^2(\delta)}\right\}\phi_j^\delta(x)\\ &= \sum_{i=0}^{\infty}(-1)^i\left\{\frac{A_i(\delta)}{\Box - m_i^2(\delta)}\right\}E^\delta(\Box)\phi_j^\delta(x) = \phi_j^\delta(x).\end{aligned}$$

On the basis of the correlations (2.27), (2.29) and (2.30) the Lagrangian density can be expressed in terms of the fields $\phi_j^\delta(x)$

$$\begin{aligned}\mathscr{L}^\delta(x) = \frac{1}{2}\sum_{j=0}^{\infty}(-1)^j\phi_j^\delta(x)[\Box - m_j^2(\delta)]\phi_j^\delta(x) +\\ + gU\left(\sum_{j=0}^{\infty}(-1)^j[A_j(\delta)]^{1/2}\phi_j^\delta(x)\right).\end{aligned} \tag{2.31}$$

Assuming the fields $\phi_j^\delta(x)$ are independent variables and making use of the principle of stationary action, from equation (2.31) we obtain the following infinite system of equations

$$[\Box - m_j^2(\delta)]\phi_j^\delta(x) = -g[A_j(\delta)]^{1/2}U'\left(\sum_{i=0}^{\infty}(-1)^i[A_i(\delta)]^{1/2}\phi_i^\delta(x)\right)$$
$$(j = 0, 1, 2, \ldots). \tag{2.32}$$

The system of equations obtained here is equivalent to (2.13). Indeed, first, the relation (2.29) results indirectly from the system of equations (2.32), secondly, equation (2.13) is fulfilled if $\phi^\delta(x)$ is connected with the fields $\phi_j^\delta(x)$ by relation (2.30) and making use of (2.30), (2.29) and (2.32) it is easy to verify

$$E^\delta(\Box)\phi^\delta(x) = E^\delta(\Box)\sum_{j=0}^{\infty}(-1)^j[A_j(\delta)]^{1/2}\phi_j^\delta(x)$$

$$= \sum_{j=0}^{\infty}(-1)^j[A_j(\delta)]^{1/2}\left\{\frac{\Box - m_i^2(\delta)}{[A_i(\delta)]^{1/2}}\right\}\times$$

$$\times [A_i(\delta)]^{1/2}\frac{E^\delta(\Box)}{\Box - m_i^2(\delta)}\phi_j^\delta(x)$$

$$= \left\{\frac{\Box - m_i^2(\delta)}{[A_i(\delta)]^{1/2}}\right\}\sum_{j=0}^{\infty}(-1)^j A_j(\delta)\left\{\frac{E^\delta(\Box)}{\Box - m_j^2(\delta)}\right\}\phi_i^\delta(x)$$
$$= [\Box - m_i^2(\delta)][A_i(\delta)]^{-1/2}\phi_i^\delta(x) = -gU'(\phi^\delta(x)),$$

where $i = 0, 1, 2, \ldots$.

Thus the Lagrangian (2.31) and the system of equations (2.32) are completely equivalent to the Lagrangian (2.16) and equation (2.13). Therefore, one can assume that our initial system (2.16) is described by the Lagrangian (2.31) where the fields $\phi_j^\delta(x)$ are independent and satisfy the equation of motion (2.32).

The method used here is well known in the theory of differential equations. It is usually used when a differential equation of the highest order is replaced by a system of differential equations of the first order.

All the above-stated arguments concerned the classical field theory. The quantization of the system of the classical fields $\{\phi_j^\delta(x)\}$ can be performed according to the *canonical procedure* of *quantization*. Let us introduce a momentum field, conjugate to $\phi_j^\delta(\mathbf{x}, 0)$

$$\Pi_j^\delta(\mathbf{x},0) = [\delta/\delta\dot{\phi}_j^\delta(\mathbf{x},0)]\int d^3y\,\mathscr{L}^\delta(\mathbf{y},0) = (-1)^j\dot{\phi}_j^\delta(\mathbf{x},0). \tag{2.33}$$

The dot denotes the differential with respect to time

$$\dot{\phi}_j^\delta(\mathbf{x},0) = \frac{\partial}{\partial t}\phi_j^\delta(\mathbf{x},t)|_{t=0}.$$

We treat ϕ_j^δ and Π_j^δ as operators with the *commutation relations*:

$$[\phi_j^\delta(\mathbf{x},0), \phi_i^\delta(\mathbf{y},0)]_- = [\Pi_j^\delta(\mathbf{x},0), \Pi_i^\delta(\mathbf{y},0)]_- = 0,$$

$$[\phi_j^\delta(\mathbf{x},0), \Pi_i^\delta(\mathbf{y},0)]_- = i\,\delta_{ij}\,\delta^{(3)}(x-y) \tag{2.34}$$

or

$$[\phi_j^\delta(\mathbf{x},0), \dot{\phi}_i^\delta(\mathbf{y},0)]_- = i(-1)^j\,\delta_{ij}\,\delta^{(3)}(x-y). \tag{2.35}$$

It is seen that the *indefinite metrics* is to be employed to quantize our system in a regular manner (see Nady, 1966).

As we are unable to solve the system of equations (2.32) exactly, our problem is to construct the perturbation series for the S-matrix and we perform the quantization of the non-interacting system of fields $\{\phi_j^\delta(x)\}$, i.e., the case when $g = 0$. Instead of (2.32) we have

$$[\square - m_j^2(\delta)]\,\phi_j^\delta(x) = 0 \quad (j = 0, 1, 2, \ldots). \tag{2.36}$$

The solution of these equations can be written in the form

$$\phi_j^\delta(x) = (2\pi)^{-3/2}\int d^3k[2\omega_{j\mathbf{k}}^\delta]^{-1/2}(d_{j\mathbf{k}}\,e^{-ikx} + d_{j\mathbf{k}}^+\,e^{ikx}), \tag{2.37}$$

where

$$\omega_{j\mathbf{k}}^\delta = [\mathbf{k}^2 + m_j^2(\delta)]^{1/2} = [\mathbf{k}^2 + m^2(1 + j^\sigma/\delta)]^{1/2}.$$

The quantization conditions (2.34) and (2.35) lead to the operators $d_{j\mathbf{k}}$ and $d_{j\mathbf{k}}^+$ to satisfy

$$[d_{j\mathbf{k}}, d_{j'\mathbf{k}'}]_- = [d_{j\mathbf{k}}^+, d_{j'\mathbf{k}'}^+]_- = 0.$$

$$[d_{j\mathbf{k}}, d_{j'\mathbf{k}'}^+]_- = (-1)^j\,\delta_{jj'}\,\delta^{(3)}(k-k'). \tag{2.38}$$

The *Hamiltonian* of the non-interacting system can easily be obtained. Let us write it in the normal form

$$H_0^\delta = \sum_{j=0}^{\infty}(-1)^j\int d^3k\,\omega_{j\mathbf{k}}^\delta\,d_{j\mathbf{k}}^+\,d_{j\mathbf{k}}. \tag{2.39}$$

The system under consideration consists of quanta with the following mass spectra

$$m_j^2(\delta) = \begin{cases} m^2 & j = 0, \\ m^2(1 + j^\sigma/\delta) & j = 1, 2, \ldots \end{cases} \tag{2.40}$$

Let us denote

$$d_{0\mathbf{k}} = a_{\mathbf{k}}, \qquad d_{0\mathbf{k}}^+ = a_{\mathbf{k}}^+ \tag{2.41}$$

When $\delta \to 0$, the masses of quanta with $j = 1, 2, \ldots$ go to infinity, according to (2.40). These quanta are called *ghost states or ghosts.* The quanta with $j = 0$ have the finite mass m. We call them normal particles or scalar particles with mass m.

The *space of states* $\mathscr{H}^\delta$ is a vector space with *indefinite metrics*. It consists of:

(1) a *vacuum state* $|0\rangle$, that is unique, defined by the conditions

$$\mathrm{d}_{j\mathbf{k}}|0\rangle = 0 \quad \text{and normalized by } \langle 0|0\rangle = 1;$$

(2) one-particle states $|j, \mathbf{k}\rangle = \mathrm{d}^+_{j\mathbf{k}}|0\rangle$ which are normalized by

$$\langle j, \mathbf{k}|j', \mathbf{k}'\rangle = (-1)^j \, \delta_{jj'} \, \delta^{(3)}(k - k')$$

and are eigenstates of the Hamiltonian H_0^δ

$$H_0^\delta |j, \mathbf{k}\rangle = \omega^\delta_{j\mathbf{k}} |j, \mathbf{k}\rangle;$$

(3) many-particle states. If there are n particles with momenta $\mathbf{k}_1, \ldots, \mathbf{k}_n$ and among them there are $\nu_1, \nu_2, \ldots, \nu_\alpha$ $(n = \nu_1 + \nu_2 + \cdots + \nu_\alpha)$ identical particles, i.e., with the same index j, then the following is given:

$$|n\rangle = |j_1, \mathbf{k}_1; \ldots; j_n, \mathbf{k}_n\rangle = (\nu_1! \cdots \nu_\alpha!)^{-1/2} \, \mathrm{d}^+_{j_1\mathbf{k}_1} \cdots \mathrm{d}^+_{j_n\mathbf{k}_n}|0\rangle.$$

These states are also eigenstates of H_0^δ:

$$H_0^\delta |n\rangle = (\omega^\delta_{j_1\mathbf{k}_1} + \cdots + \omega^\delta_{j_n\mathbf{k}_n})|n\rangle.$$

All these states generate a complete system of eigenstates in the vector space $\mathscr{H}^\delta$, i.e.,

$$\circledast \underset{\mathrm{Df}}{=} |0\rangle\langle 0| + \sum_{n=1}^{\infty} \sum_{j_1, \ldots, j_n = 0}^{\infty} (-1)^{j_1 + \cdots + j_n} \int \mathrm{d}^3 k_1 \cdots \int \mathrm{d}^3 k_n \, |n\rangle\langle n| = 1.$$

What happens with the space $\mathscr{H}^\delta$ when $\delta \to 0$? At $\delta \to 0$ the masses of all ghosts increase according to (2.40). Therefore, if any physical state is characterized by a definite value of energy, then in the limit $\delta \to 0$ no physical states with arbitrary but finite energy can consist of ghost quanta. In this sense we have

$$\lim_{\delta \to 0} \mathscr{H}^\delta = \mathscr{H}, \tag{2.42}$$

where $\mathscr{H}$ is the Hilbert space which contains:

(1) a vacuum state $|0\rangle$, $a_\mathbf{k}|0\rangle = 0$,
(2) single- and many-particle states

$$|n\rangle = |\mathbf{k}_1, \ldots, \mathbf{k}_n\rangle = (n!)^{-1/2} a^+_{\mathbf{k}_1} \cdots a^+_{\mathbf{k}_n}|0\rangle.$$

All these states generate the complete system in $\mathscr{H}$:

$$\circledast \underset{\mathrm{Df}}{=} |0\rangle\langle 0| + \sum_{n=1}^{\infty} \int \mathrm{d}^3 k_1 \cdots \int \mathrm{d}^3 k_n |\mathbf{k}_1, \ldots, \mathbf{k}_n\rangle\langle \mathbf{k}_1, \ldots, \mathbf{k}_n| = 1.$$

2.2.4. *Green Functions of the Field* $\phi^\delta(x)$

First of all, let us consider the commutator

$$\Delta^\delta(x-y) = [\phi^\delta(x), \phi^\delta(y)]_- . \tag{2.43}$$

Substituting the representation (2.30) into (2.43) and using (2.37) we obtain

$$\Delta^\delta(x) = \sum_{j=0}^{\infty} (-1)^j A_j(\delta)\, \Delta_j(x). \tag{2.44}$$

Calculation of the explicit form of the different two-point Green functions carried out in any textbooks of field theory (see, for example Section 1.4.4 and Bogolubov and Shirkov, 1980). We use here their results. Thus we have

$$\Delta_j^\delta(x) = (2\pi)^{-3} \int d^4 k \varepsilon(k_0)\, \delta(k^2 - m_j^2(\delta))\, e^{-ikx}$$

$$= \frac{1}{2\pi i} \varepsilon(x_0)\, \delta(x^2) - \left[\frac{m_j(\delta)}{4\pi i}\right] (x^2)^{-1/2}\, \theta(x^2) J_1(m_j(\delta)(x^2)^{1/2}), \tag{2.45}$$

where

$$\varepsilon(t) = \begin{cases} 1 & t>0 \\ -1 & t<0 \end{cases}, \qquad \theta(t) = \begin{cases} 1 & t>0 \\ 0 & t<0 \end{cases}.$$

Because the series (2.44) converges absolutely we have

$$\Delta^\delta(x) = 0 \quad \text{when } x^2 < 0. \tag{2.46}$$

Thus the operator $\phi^\delta(x)$ satisfies the *local commutation relations*.

Now let us introduce functions $\Delta^\delta_{(\pm)}(x)$ according to

$$\Delta^\delta_{(-)}(x-y) = \Delta^\delta_{(+)}(y-x) = \langle 0|\phi^\delta(x)\phi^\delta(y)|0\rangle.$$

We have

$$\Delta^\delta_{(\pm)}(x) = \sum_{j=0}^{\infty} (-1)^j A_j(\delta)\, \Delta^\delta_{j(\pm)}(x), \tag{2.47}$$

where

$$\Delta^\delta_{j(\pm)}(x) = (2\pi)^{-3} \int d^4 k \theta(\mp k_0)\, \delta(k^2 - m_j^2(\delta))\, e^{-ikx}.$$

For $x^2 \to 0$, according to Bogolubov and Shirkov (1980), one can obtain

$$\Delta^\delta_{j(\pm)}(x) = -\frac{i}{4\pi} \varepsilon(\mp x_0)\, \delta(x^2) - \frac{1}{4x^2\pi^2} + \frac{m_j^2}{16\pi^2} \ln(m_j^2|x^2|/4) +$$

$$+ (im_j^2/16\pi)\varepsilon(\mp x_0)\theta(x^2) + O(x^2 \ln x^2).$$

Substituting this expansion into (2.47) we obtain

$$\Delta^{\delta}_{(\pm)}(x) = \frac{m^2}{16\pi^2}\sum_{j=0}^{\infty}(-1)^j A_j(\delta)\mu_j \ln \mu_j + O(x^2 \ln x^2).$$

Here we have used the correlations (2.26), hence it appears that the function $\Delta^{\delta}_{(\pm)}(x)$ is finite at $x = 0$ and

$$\Delta^{\delta}_{(\pm)}(0) = \frac{m^2}{16\pi^2}\sum_{j=0}^{\infty}(-1)^j A_j(\delta)\mu_j \ln \mu_j < \infty. \tag{2.48}$$

It means that the operator $\phi^{\delta}(x)$ is well-defined because of

$$\langle 0|\phi^{\delta}(x)\phi^{\delta}(x)|0\rangle = \Delta^{\delta}_{(-)}(0) < \infty.$$

Let us consider the *causal Green function*

$$\Delta^{\delta}_c(x-y) = \langle 0|\, T[\phi^{\delta}(x)\phi^{\delta}(y)]|0\rangle = \sum_{j=0}^{\infty}(-1)^j A_j(\delta)\,\Delta^{\delta}_{jc}(x-y), \tag{2.49}$$

where

$$\Delta^{\delta}_{jc}(x) = -i(2\pi)^{-4}\int \mathrm{d}^4 k\, \mathrm{e}^{-ikx}[m_j^2(\delta) - k^2 - i\varepsilon]^{-1}.$$

Otherwise

$$\Delta^{\delta}_c(x) = -i(2\pi)^{-4}\int \mathrm{d}^4 k\, \tilde{\Delta}^{\delta}_c(k^2)\, e^{-ikx}, \tag{2.50}$$

where

$$\tilde{\Delta}^{\delta}_c(k^2) = \sum_{j=0}^{\infty}(-1)^j A_j(\delta)[m_j^2(\delta) - k^2 - i\varepsilon]^{-1}$$

$$= (m^2 - k^2 - i\varepsilon)^{-1}\sum_{n=0}^{\infty}\left\{\frac{v_n l^{2n}(k^2 - m^2)^n}{\prod_{j=1}^{n+2}[1 - \delta j^{-\delta} m^{-2}(k^2 - m^2 + i\varepsilon)]}\right\}. \tag{2.51}$$

The function $\tilde{\Delta}^{\delta}_c(k^2)$ is analytic in the complex k^2-plane for $\mathrm{Im}\, k^2 \geqslant 0$ and has simple poles at the points $k^2 = m_j^2(\delta) - i\varepsilon$ $(j = 0, 1, \ldots)$.

The *retarded and advanced Green functions* can be defined in the following way

$$\begin{aligned}\Delta^{\delta}_{\text{ret}}(x) &= \theta(x_0)\,\Delta^{\delta}(x) = \Delta^{\delta}_c(x) + \Delta^{\delta}_{(+)}(x),\\ \Delta^{\delta}_{\text{adv}}(x) &= -\theta(-x_0)\,\Delta^{\delta}(x) = \Delta^{\delta}_c(x) - \Delta^{\delta}_{(-)}(x).\end{aligned} \tag{2.52}$$

They satisfy the conditions

$$\Delta^{\delta}_{\text{ret}}(x) = 0 \quad \text{for}\begin{cases} x^2 < 0\\ x^2 > 0, \quad x_0 < 0,\end{cases}$$

$$\Delta^{\delta}_{\text{adv}}(x) = 0 \quad \text{for} \begin{cases} x^2 < 0 \\ x^2 > 0, \quad x_0 > 0. \end{cases} \tag{2.53}$$

Thus we can see that all Green functions satisfy all requirements of the *local quantum field theory*. It means that the field $\phi^{\delta}(x)$ is *local*. The following correlations are valid for the regularized Green $\Delta^{\delta}_c(x)$ and $\Delta^{\delta}_{(\pm)}(x)$ functions:

$$\begin{aligned} \Delta^{\delta}_c(x) &= \theta(x_0)\,\Delta^{\delta}_{(-)}(x) + \theta(-x_0)\,\Delta^{\delta}_{(+)}(x), \\ \Delta^{\delta *}_c(x) &= \theta(x_0)\,\Delta^{\delta}_{(+)}(x) + \theta(-x_0)\,\Delta^{\delta}_{(-)}(x). \end{aligned} \tag{2.54}$$

An additional correlation should be mentioned because it is important when proving the *unitarity of the S-matrix regularized*

$$\langle 0|T\{\partial_\mu \phi^{\delta}(x)\,\partial_\nu \phi^{\delta}(y)\}|0\rangle = (\partial^2/\partial x_\mu\,\partial y_\nu)\,\langle\, 0|T[\phi^{\delta}(x)\phi^{\delta}(y)]|0\rangle. \tag{2.55}$$

In other words, the T-product in the Wick sense coincides with the T-product in the Dyson sense, i.e., symbolically $T_{\text{w}} = T_{\text{d}}$. Indeed, it is easy to obtain for the fields $\phi^{\delta}_j(x)$

$$\begin{aligned} &(\partial^2/\partial x_\mu\,\partial y_\nu)\,\langle 0|T\{\phi^{\delta}_j(x)\phi^{\delta}_i(y)\}|0\rangle \\ &\quad = \langle 0|T\left\{\frac{\partial}{\partial x_\mu}\phi^{\delta}_j(x)\frac{\partial}{\partial y_\nu}\phi^{\delta}_i(y)\right\}|0\rangle + i(-1)^j\,\delta_{ji}\,\delta_{\mu 0}\,\delta_{\nu 0}\,\delta^{(4)}(x-y). \end{aligned}$$

Therefore,

$$\begin{aligned} &\frac{\partial^2}{\partial x_\mu\,\partial y_\nu}\langle\, 0|T\{\phi^{\delta}(x)\phi^{\delta}(y)\}|0\rangle - \langle 0|T\left\{\frac{\partial}{\partial x_\mu}\phi^{\delta}(x)\frac{\partial}{\partial y_\nu}\phi^{\delta}(y)\right\}|0\rangle \\ &\quad = i\,\delta_{\mu 0}\,\delta_{\nu 0}\,\delta^{(4)}(x-y)\sum_{j=0}^{\infty}(-1)^j A_j(\delta) = 0 \end{aligned}$$

according to (2.26). Thus the following correlations take place

$$\begin{aligned} \partial_\mu\partial_\nu\,\Delta^{\delta}_c(x) &= \theta(x_0)\,\partial_\mu\partial_\nu\,\Delta^{\delta}_{(-)}(x) + \theta(-x_0)\,\partial_\mu\partial_\nu\,\Delta^{\delta}_{(+)}(x) \\ \partial_\mu\partial_\nu\,\Delta^{\delta}_{(-)}(x) &= \theta(x_0)\,\partial_\mu\partial_\nu\,\Delta^{\delta}_c(x) + \theta(-x_0)\,\partial_\mu\partial_\nu\,\Delta^{\delta *}_c(x) \end{aligned} \tag{2.56}$$

2.2.5. *The Interacting System Before Removal of the Regularization*

The interacting system is described by the Lagrangian (2.31). The total Hamiltonian of this system has the form

$$H^{\delta} = H^{\delta}_0 + H^{\delta}_{\text{in}}, \tag{2.57}$$

where H^{δ}_0 is given by the relation (2.39) and

$$H^{\delta}_{\text{in}} = -g\int d^3x\!: U(\phi^{\delta}(\mathbf{x},0))\!:. \tag{2.58}$$

In the *interaction picture*, we have

$$H^{\delta}_{\text{in}}(t) = \mathrm{e}^{-itH^{\delta}_{0}} H^{\delta}_{\text{in}} \mathrm{e}^{itH^{\delta}_{0}} = -g \int \mathrm{d}^3x \colon U(\phi^{\delta}(\mathbf{x}, t)) \colon. \tag{2.59}$$

Though the operator $\phi^{\delta}(x)$ is well-defined as we have shown in the previous section, the Hamiltonian $H^{\delta}_{\text{in}}(t)$ is not defined, since the theory is translationally invariant and, thus, there are difficulties connected with the Haag theorem (see Wightman, 1964). It is necessary to introduce an operation of 'switching on' and 'switching off' the interaction g:

$$g \to g(\mathbf{x}/L, t/L) = g(x/L).$$

Such regularization takes into account at the same time that, first, our system is situated in a box, hereby violating the Euclidean invariance and, secondly, the interaction is adiabatically switched on and switched off at infinity, i.e., when $t \to \pm\infty$. The large parameter L defines the intensity of switching on the interaction. The function $g(x)$ satisfies the conditions:

(1) $0 \leqslant g(x) < g$,
(2) $g(0) = g$
(3) $\mathrm{d}^n g(x)|_{x=0} = 0$, for $n = 1, 2, 3, 4$; where $\mathrm{d}^n = \partial_{\mu_1} \cdots \partial_{\mu_n}$
(4) $\int \mathrm{d}^4 x g(x) < \infty$,
(5) $g(x) \in Z_{\alpha}$, for $\alpha < 2\rho/(2\rho - 1)$,

where ρ is the order of growth of the nonlocal form factors.

Removal of the regularization corresponds to the limit $L \to \infty$, at which

$$g(x/L) \underset{L \to \infty}{\to} g(0) = g. \tag{2.60}$$

Thus the Hamiltonian (2.59) in the interaction picture takes the form

$$H^{(\delta, L)}_{\text{in}}(t) = -\int \mathrm{d}^3x\, g(\mathbf{x}/L, t/L) \colon U(\phi^{\delta}(x, t)) \colon. \tag{2.61}$$

We recall again that when δ is positive, then the Hamiltonian $H^{(\delta, L)}_{\text{in}}(t)$ is well-defined, so that it is not necessary to introduce any ultraviolet cutoff.

The regularized S-matrix can be written in the standard form

$$S^{\delta, L} = T\exp\left\{-i\int_{-\infty}^{\infty} \mathrm{d}t\, H^{(\delta, L)}_{\text{in}}(t)\right\} = T\exp\left\{i\int \mathrm{d}^4x g(x/L) \colon U(\phi^{\delta}(x)) \colon\right\}. \tag{2.62}$$

Here T has the meaning of strict ordering operator of the quantized fields $\phi^{\delta}(x)$ with respect to time. The chronological pairing or the Green function of the operators $\phi^{\delta}(x)$ is given by formulas (2.49)–(2.51). Since the operators $\phi^{\delta}(x)$ and $H^{(\delta, L)}_{\text{in}}(t)$ are well-defined and *local* then the $S^{\delta, L}$-matrix on the vector space

$\mathscr{H}^{\delta}$ is *unitary* and *microcausal*, i.e.,

$$S^{\delta,L}(S^{\delta,L})^{+} = S^{\delta,L} \circledast^{\delta} (S^{\delta,L})^{+} = 1,$$
$$\frac{\delta}{\delta g(x)} \left\{ \frac{\delta S^{\delta,L}}{\delta g(y)} (S^{\delta,L})^{+} \right\} = 0, \quad \text{when } x \lesssim y. \tag{2.63}$$

The *typical problem* of the nonlocal quantum field theory consists in proving of the existence of the following sequence of limits in each perturbation order

$$S^{L} = \lim_{\delta \to 0} S^{\delta,L}, \qquad S = \lim_{L \to \infty} S^{L}. \tag{2.64}$$

Here the first limit $\delta \to 0$ means removal of all 'ghost' states from the theory. The second limit $L \to \infty$ means passage to infinite volume and switching on the interaction over all four-dimensional space.

2.2.6. *The Green Functions in the Limit $\delta \to 0$*

The Green functions in the limit $\delta \to 0$ are *generalized functions* which are defined on a space of test functions $Z_{\alpha}(\alpha < 2\rho/(2\rho - 1))$. Therefore we have to consider *improper transitions* to the limit, i.e., investigate the limits

$$\lim_{\delta \to 0} \int d^4x G^{\delta}(x) f(x) = \lim_{\delta \to 0} \int d^4 k \tilde{G}^{\delta}(k) \tilde{f}(k), \tag{2.65}$$

where $G^{\delta}(x)$ is a Green function and $f(x) \in Z_{\alpha}$. First of all, let us consider the commutator $\Delta^{\delta}(x)$. We have

$$\lim_{\delta \to 0} \int d^4x\, \Delta^{\delta}(x) f(x) = \lim_{\delta \to 0} \int d^4 k \tilde{\Delta}^{\delta}(k) \tilde{f} = \lim_{\delta \to 0} (2\pi)^{-3} \int d^4 k \tilde{f}(k) \times$$
$$\times \sum_{j=0}^{\infty} (-1)^j A_j(\delta) \varepsilon(k_0)\, \delta(k^2 - m_j^2(\delta))$$
$$= (2\pi)^{-3} \int d^4 k \varepsilon(k_0)\, \delta(k^2 - m^2) \tilde{f}(k) + \lim_{\delta \to 0} Q^{\delta},$$

where

$$Q^{\delta} = (2\pi)^{-3} \int d^4 k \tilde{f}(k) \sum_{j=1}^{\infty} (-1)^j A_j(\delta) \varepsilon(k_0)\, \delta(k^2 - m_j^2(\delta))$$
$$= \sum_{j=1}^{\infty} (-1)^j A_j(\delta)(2\pi)^{-3} \iint d^3 k\, dk_0 \tilde{f}(\mathbf{k}, k_0) \varepsilon(k_0) \delta(k^2 - m_j^2(\delta)). \tag{2.66}$$

The following estimation can be obtained

$$I_j^\delta = \left| (2\pi)^{-3} \iint d^3k\, dk_0\, \tilde{f}(\mathbf{k}, k_0)\varepsilon(k_0)\,\delta(k^2 - m_j^2(\delta)) \right|$$

$$\lesssim (2\pi)^{-3}\frac{1}{2}\int d^3k \times \omega_{j\mathbf{k}}^{-\delta}\,[|\tilde{f}(\mathbf{k}, \omega_{j\mathbf{k}}^\delta)| + |\tilde{f}(\mathbf{k}, -\omega_{j\mathbf{k}}^\delta)|]$$

$$\lesssim \text{const} \int d^3k\omega_{j\mathbf{k}}^{-\delta} \exp\{-B[|\mathbf{k}|^\gamma + (\omega_{j\mathbf{k}}^\delta)^\gamma]\},$$

where $\gamma = \alpha/(\alpha - 1) > 2\rho \geqslant 1$. B is some positive number. Making use of the inequality

$$|\mathbf{k}|^\gamma + (\omega_{j\mathbf{k}}^\delta)^\gamma \geqslant h_1 m_j^\gamma + h_2|\mathbf{k}|^\gamma,$$

where $h_1 = 2^{-1+\gamma/2}$, $h_2 = 1 + 2^{-1+\gamma/2}$, further we obtain

$$I_j^\delta \leqslant \text{const}\cdot\exp(-Bh_1 m_j^\gamma) \int d^3k[\mathbf{k}^2 + m_j^2(\delta)]^{-1/2} \exp(-B\cdot h_2|\mathbf{k}|^\gamma)$$

$$\leqslant \text{const}\cdot\delta j^{-\sigma}\exp\{-B_1(j^\sigma/\delta)^{\gamma/2}\},$$

where $B_1 = Bh_1 m^\gamma$. Now Q^δ is evaluated to be

$$|Q^\delta| \leqslant \sum_{j=1}^{\infty} A_j(\delta)I_j^\delta \leqslant \text{const.} \sum_{j=1}^{\infty} j^{-\sigma}\,\delta \exp[B(j^\sigma/\delta)^\rho - B_1(j^\sigma/\delta)^{\gamma/2}] \times$$
$$\times\, \Gamma^{-1}((1/\rho - \sigma)j).$$

Because of $\rho < \frac{1}{2}\gamma$, we have

$$|Q^\delta| \leqslant \delta\cdot\text{const.} \sum_{j=1}^{\infty} j^{-\sigma}\Gamma^{-1}((1/\rho - \sigma)j) \leqslant \delta\cdot\text{const}$$

and finally

$$\lim_{\delta\to 0} |Q^\delta| = 0.$$

Thus in the limit $\delta \to 0$ the commutator $\Delta^\delta(x)$ changes into the commutator of the scalar field $\varphi(x)$

$$\lim_{\delta\to 0} \Delta^\delta(x) = \Delta(x) = (2\pi)^{-3} \int d^4k\varepsilon(k_0)\,\delta(k^2 - m^2)\,e^{-ikx}. \tag{2.67}$$

In the same manner it is easy to show that in the improper sense

$$\lim_{\delta\to 0} \Delta_{(\pm)}^\delta(x) = \Delta_{(\pm)}(x) = (2\pi)^{-3} \int d^4k\theta(\mp k_0)\,\delta(k^2 - m^2)\,e^{-ikx}. \tag{2.68}$$

The existence of these limits means that there exists a *weak limit* (Section 1.4)

$$\lim_{\delta \to 0} \phi^\delta(x) = \varphi(x), \tag{2.69}$$

where $\varphi(x)$ is the scalar field of mass m satisfying the free wave equation

$$(\Box - m^2)\varphi(x) = 0.$$

Consequently, all *ghost states disappear* in the limit $\delta \to 0$. This result confirms the statement that

$$\lim_{\delta \to 0} \mathscr{H}^\delta = \mathscr{H}$$

which was accomplished in Section 2.2.3.

Now we consider the causal function $\Delta_c^\delta(x)$. We have

$$\lim_{\delta \to 0} \int d^4x\, \Delta_c^\delta(x) f(x)$$

$$= \lim_{\delta \to 0} i^{-1}(2\pi)^{-4} \int d^4k \tilde{f}(k) \sum_{j=0}^{\infty} (-1)^j A_j(\delta)[m_j^2(\delta) - k^2 - i\varepsilon]^{-1}$$

$$= \lim_{\delta \to 0} i^{-1}(2\pi)^{-4} \int d^4k \frac{\tilde{f}(k)}{m^2 - k^2 - i\varepsilon} \sum_{n=0}^{\infty} \left\{ \frac{v_n l^{2n}(k^2 - m^2)^n}{\prod_{j=1}^{n+2}[1 - (k^2 - m^2 + i\varepsilon)m^{-2}\delta j^{-\sigma}]} \right\}$$

$$= i^{-1}(2\pi)^{-4} \int d^4k V(-k^2 l^2)\, \tilde{f}(k)[m^2 - k^2 - i\varepsilon]^{-1}$$

because of $f(x) \in Z_\alpha$.

Thus the *causal function* $\Delta_c^\delta(x)$ changes in the limit $\delta \to 0$ into a nonlocal propagator

$$\lim_{\delta \to 0} \tilde{\Delta}_c^\delta(k) = \tilde{D}_c(k^2) = V(-k^2 l^2)[m^2 - k^2 - i\varepsilon]^{-1}. \tag{2.70}$$

The function $\tilde{D}_c(k)$ has a single pole at $k^2 = m^2$. This pole corresponds to the scalar particle with mass m. The poles corresponding to all ghost states have disappeared. The form factor $V(-k^2 l^2)$ is an entire function and it does not correspond to any real state. This form factor describes the *nonlocal character of the interaction* of our scalar particles.

Thus in the limit $\delta \to 0$, our theory becomes nonlocal. In other words, the 'ghosts' (which disappear in the limit $\delta \to 0$) as a self-memory make the theory nonlocal. Consequently, the nonlocal character of the interaction of the classical field (2.7) is revealed in quantum field theory as a *residual effect* of the nonphysical ghost states (the origin of which is caused by the stochastic nature of space-time) when these ghosts are removed by the transition to the limit $\delta \to 0$.

2.3. The Physical Meaning of the Form Factors

We now give the physical meaning of the form-factor introduced in (1.9). On the basis of the simple model of quantum field theory, we show that the form-factors of the class investigated above give the *relativistic invariant description of extended objects* in QFT. Let there be an infinitely heavy source situated at the point $x = 0$. A field is generated by the radiation and absorption of scalar particles with mass m. We assume that the interaction Lagrangian density of this system has the form

$$\mathcal{L}_{\mathrm{in}}(x) = g\varphi(x)K(l^2\Box)J(x), \tag{2.71}$$

where $\varphi(x)$ is the field of the scalar particles with mass m and the function $J(x)$ describes the source. Further, let us assume the source as point-like (situated at the point $x = 0$) and the *switching on and switching off* as at the infinitely remote past and future, respectively. For definiteness, we read

$$J(x) \to J(x, T) = J(\mathbf{x}, x_0, T) = \exp(-x_0^2/T^2)\,\delta^{(3)}(x). \tag{2.72}$$

Here T is a large parameter characterizing the time interval during which the source is switched on. At the end of calculations, it is necessary to pass to the limit $T \to \infty$.

The function $K(l^2\Box)$ characterizes the region of the nonlocal interaction near the source. We assume that the function $\tilde{K}(k^2 l^2)$ satisfies all of Condition 2.1 of Section 2.2.1 for $\rho = \frac{1}{2}$ except for the normalization condition which we will discuss below. The following representation is valid for $\tilde{K}(k^2 l^2)$ when $\rho = \frac{1}{2}$

$$\tilde{K}(k^2 l^2) = \int_{b^2 \leqslant l^2} \mathrm{d}^4 b w(b^2) \exp(k_0 b_4 - i\mathbf{bk}). \tag{2.73}$$

When $k_0 = 0$, we set

$$\tilde{K}(-\mathbf{k}^2 l^2) = \int \mathrm{d}^3 b a(\mathbf{b}^2)\, \mathrm{e}^{i\mathbf{kb}}. \tag{2.74}$$

The connection between the functions $w(u)$ and $a(u)$ can be found by the relations

$$\tilde{K}(k^2 l^2) = \int \mathrm{d}^4 b w(b^2)\, \mathrm{e}^{ik_E b} = \int \mathrm{d}^3 r a(\mathbf{r}^2)\, \mathrm{e}^{i\mathbf{kr}},$$

where $k^2 = -k_E^2 = -\mathbf{k}^2, k_E = (ik_0, \mathbf{k})$, or

$$\tilde{K}(k^2 l^2) = 2\pi^2 \int_0^\infty \mathrm{d}t \sqrt{t}\, J_1[(t(-k^2))^{1/2}](-k^2)^{-1/2} w(t)$$

$$= 2\pi \int_0^\infty \mathrm{d}s a(s) \frac{\sin(s(-k^2))^{1/2}}{(-k^2)^{1/2}}.$$

From these correlations, it is easy to obtain

$$a(s) = \int_s^l \mathrm{d}t w(t)(t-s)^{-1/2},$$

$$w(t) = -(2\pi t)^{-1} \int_t^l \mathrm{d}s(s-t)^{-1/2}[a(s) + 2sa'(s)]. \tag{2.75}$$

The potential generated by the source at the point $\mathbf{x} = \mathbf{r}$ is equal to the Heisenberg field operator $\varphi(x)$ averaged over the vacuum:

$$W(x) = \langle 0| T[\varphi(x)S]|0\rangle. \tag{2.76}$$

In the given case, the vacuum state $|0\rangle$ containing no scalar particles is a state describing the source. We will calculate the potential generated by the field $\varphi(x)$ near the source.

The method of calculation is as follows: We construct the total S-matrix according to the methods developed above. It must satisfy the *stability condition* $\langle 0|S|0\rangle = 1$. However, in order to satisfy this condition it is necessary to add a counter-term to the Lagrangian (2.71). This counter-term ΔE is responsible for the renormalization of the energy of the basic state of the source, or the vacuum energy, and describes the interaction energy of the source with the field of the scalar particles. Since we assume that an operation of 'switching on' and 'switching off' the interaction is adiabatic, this counter-term must have the form

$$\Delta E(x_0) = \exp(-2x_0^2/T^2)\Delta E. \tag{2.77}$$

As shown above, we now introduce the regularization procedure associated with introducing the nonlocality into the interaction Lagrangian. According to the quantization procedure presented in the preceding section, we must quantize two fields $\varphi(x)$ and $\phi(x) = K(l^2 \Box)\varphi(x)$ is such a way that the quantized fields $\varphi^\delta(x)$ and $\phi^\delta(x)$ lead to the following *causal functions*:

$$\Delta_c(x-y) = \langle 0|T[\varphi^\delta(x)\varphi^\delta(y)]|0\rangle = i^{-1}(2\pi)^{-4}\int \mathrm{d}^4k \frac{\mathrm{e}^{-ik(x-y)}}{m^2 - k^2 - i\varepsilon},$$

$$\mathscr{D}_c^\delta(x-y) = \langle 0|T[\varphi^\delta(x)\phi^\delta(y)]|0\rangle = i^{-1}\int \frac{\mathrm{d}^4k}{(2\pi)^4}[\tilde{K}(k^2l^2)]^\delta \frac{\mathrm{e}^{-ik(x-y)}}{m^2 - k^2 - i\varepsilon},$$

$$D_c^\delta(x-y) = \langle 0|T[\phi^\delta(x)\phi^\delta(y)]|0\rangle = i^{-1}\int \frac{\mathrm{d}^4k}{(2\pi)^4}[\tilde{K}^2(k^2l^2)]^\delta \frac{\mathrm{e}^{-ik(x-y)}}{m^2 - k^2 - i\varepsilon},$$

this can be done as shown above.

In accordance with the above deduction for the S-matrix we obtain

$$S = \lim_{T\to\infty}\lim_{\delta\to 0} S^{\delta,T}, \tag{2.78}$$

where

$$S^{\delta,T} = T\exp\left\{ig\int d^4x\phi^\delta(x)J(x,T) - i\int_{-\infty}^{\infty} dx_0\,\Delta E(x_0)\right\}. \tag{2.79}$$

The averaged values of the field operators, i.e., the *Green functions*, are calculated by the formula

$$G_n(x_1,\ldots,x_n) = \langle 0|T[\varphi(x_1)\cdots\varphi(x_n)S]|0\rangle$$
$$= \lim_{T\to\infty}\lim_{\delta\to 0}\langle 0|T[\varphi^\delta(x_1)\cdots\varphi^\delta(x_n)S^{\delta,T}]|0\rangle. \tag{2.80}$$

Making use of the Wick theorem we obtain

$$S^{\delta,T} = S_r^{\delta,T}\langle 0|S^{\delta,T}|0\rangle,$$

where

$$S_r^{\delta,T} = :\exp\left\{ig\int d^4x\phi^\delta(x)J(x,T)\right\}: \tag{2.81}$$

$$\langle 0|S^{\delta,T}|0\rangle = \exp\left\{-\tfrac{1}{2}g^2\iint d^4x_1\,d^4x_2 J(x_1,T)D_c^\delta(x_1-x_2)J(x_2,T) - \right.$$
$$\left. - i\int_{-\infty}^{\infty} dx_0\,\Delta E(x_0)\right\}. \tag{2.82}$$

The problem is how to choose such ΔE that

$$\lim_{T\to\infty}\lim_{\delta\to 0}\langle 0|S^{\delta,T}|0\rangle = 1. \tag{2.83}$$

Let us consider expression (2.82). Using (2.72) and (2.77), and carrying out some simple calculations, we obtain

$$\lim_{\delta\to 0}\left(-\tfrac{1}{2}g^2\right)\iint d^4x_1\,d^4x_2 J(x_1,T)D_c^\delta(x_1-x_2)J(x_2,T)$$

$$= iT^2g^2\pi^{-3}2^{-5}\int_{-\infty}^{\infty} dk_0\exp\left(\frac{-T^2k_0^2}{2}\right)\int d^3k\frac{[\tilde{K}((k_0^2-\mathbf{k}^2)l^2)]^2}{m^2+\mathbf{k}^2-k_0^2-i\varepsilon}$$

$$= iT(2\pi)^{-5/2}\left(\tfrac{1}{4}g^2\right)\int d^3k[\tilde{K}(-\mathbf{k}^2l^2)]^2(m^2+\mathbf{k}^2)^{-1} + O(1/T),$$

$$-i\int_{-\infty}^{\infty} dx_0\,\Delta E(x_0) = -iT(2\pi)^{1/2}\cdot\frac{1}{2}\cdot\Delta E.$$

Thus the limits in (2.78) and (2.83) exist if

$$\Delta E = (2\pi)^{-3}(\tfrac{1}{2}g^2)\int d^3k[\tilde{K}(-\mathbf{k}^2l^2)]^2(m^2+\mathbf{k}^2)^{-1} < \infty. \tag{2.84}$$

The regularized S-matrix is given by expression (2.81).

Now we consider the potential $W(x)$. Substituting the $S_r^{\delta,T}$-matrix (2.81) into (2.76), we obtain

$$\begin{aligned} W(x) &= \lim_{T\to\infty}\lim_{\delta\to 0}\langle 0|T[\varphi^\delta(x)S_r^{\delta,T}]|0\rangle \\ &= \lim_{T\to\infty}\lim_{\delta\to 0} ig\int d^4y\mathcal{D}_c^\delta(x-y)J(y,T). \end{aligned} \tag{2.85}$$

Passing to the limits $\delta \to 0$ and $T \to \infty$, and making some simple calculations, we see that $W(x)$ does not depend on x_0 and takes the form

$$W(\mathbf{r}) = W(\mathbf{r},0) = (2\pi)^{-3}g\int d^3k\tilde{K}(-\mathbf{k}^2l^2)\,e^{i\mathbf{kr}}\,(m^2+\mathbf{k}^2)^{-1}. \tag{2.86}$$

Making use of representation (2.74), we have

$$W(\mathbf{r}) = \frac{g}{4\pi}\int d^3ba(\mathbf{b}^2)\exp\left(\frac{-m|\mathbf{r}-\mathbf{b}|}{|\mathbf{r}-\mathbf{b}|}\right)$$

$$= \begin{cases} \dfrac{g}{rm}\left\{e^{-mr}\displaystyle\int_0^r dbba(\mathbf{b}^2)\operatorname{sh} mb + \operatorname{sh}(mr)\int_r^l dbba(\mathbf{b}^2)e^{-mb}\right\}, & r<l \\ \dfrac{g}{4\pi}\tilde{K}(m^2l^2)\dfrac{e^{-mr}}{r}, & r>l \end{cases} \tag{2.87}$$

Here

$$\tilde{K}(m^2l^2) = \int d^3ra(\mathbf{r}^2)\frac{\operatorname{sh} m(\mathbf{r}^2)^{1/2}}{m(\mathbf{r}^2)^{1/2}}.$$

Thus the potential generated by the point-like source $J(x) = \delta^{(3)}(x)$ indeed represents the potential of a sphere of the radius l. Therefore the function

$$\mathcal{J}(x) = K(l^2\Box)J(x) = K(l^2\Box)\,\delta^{(3)}(x) = K(l^2\Delta)\,\delta^{(3)}(x)$$

describes the source which is the sphere of the radius l. Distribution of matter inside this sphere is given by the function $a(\mathbf{r}^2)$. The interaction potential generated by a single point-like element of matter is the Yukawa potential.

Notice that the potential $W(\mathbf{r})$ outside the sphere of the radius l coincides with a potential of a point-like particle, while the interaction force is given by the constant $g_{\text{eff}} = g\tilde{K}(m^2l^2)$. The normalization condition $\tilde{K}(m^2l^2) = 1$ means that the coupling constant g in the interaction Lagrangian is chosen in such a way that it just defines the potential of the source at large distances.

Investigations carried out up to now have shown that in the *statical limit* the relativistic invariant distributions (generalized functions) $K(x-x') = K(l^2\Box)\delta^{(4)}(x-x')$ used here give the correct description of extended.objects. As one might expect, these distributions yield a reasonable starting point for the relativistic invariant description of the nonlocal interactions.

Finally, we write the Coulomb and Yukawa potentials generated by the form factor (1.16) corresponding to the Gaussian measure $w_G(b_E^2/l^2)$ in the stochastic space-time $\mathbb{R}^4(x)$. In this case, the change of these potentials at small distances is given by the nonlocal propagators of the photon and scalar particle with mass m (i.e., π-meson) in the static limit:

$$\varphi_C(r) = e(2\pi)^{-3}\int d^3p\, e^{-i\mathbf{pr}} \exp(-\mathbf{p}^2 l^2/4)/\mathbf{p}^2 = (e/4\pi r)\phi(r/l), \tag{2.88}$$

and

$$\varphi_Y(r) = g(2\pi)^{-3}\int d^3p\, e^{-i\mathbf{pr}} \exp(-\mathbf{p}^2 l^2/4)(m^2+\mathbf{p}^2)^{-1}$$

$$= \frac{g}{8\pi r}\exp\left(\frac{m^2l^2}{4}\right)\left\{\phi\left(\frac{ml}{2}+\frac{\mathrm{r}}{l}\right) - \phi\left(\tfrac{1}{2}ml - \frac{r}{l}\right) - 2\,\mathrm{sh}\, rm\right\} \tag{2.89}$$

respectively. Here

$$\bar{\bar{\phi}}(x) = \mathrm{erf}(x) = \frac{2}{\sqrt{\pi}}\int_0^x dt\, \exp(-t^2).$$

Now we consider some peculiarity of the form factor $V(-p^2l^2)$ in the quantum electrodynamical case. In this case, the function (2.84) has the form

$$W(\mathbf{r}) = e^2(2\pi)^{-3}\int d^3k\, \frac{[\tilde{K}(-\mathbf{k}^2l^2)]^2}{\mathbf{k}^2}\, e^{i\mathbf{kr}}. \tag{2.90}$$

Substituting the expression (2.74) for $\tilde{K}(-\mathbf{k}^2l^2)$ into it, we get

$$W(\mathbf{r}) = \frac{e^2}{4\pi}\iint d^3r_1\, d^3r_2 \frac{a(\mathbf{r}_1^2)a((\mathbf{r}_2-\mathbf{r})^2)}{|\mathbf{r}_1-(\mathbf{r}_2-\mathbf{r})|}. \tag{2.91}$$

This formula is the electrostatical energy of two interacting spherically-symmetric charges whose centres are situated at distance r from each other. In this case a charge distribution is described by the function $a(\mathbf{r}^2)$.

The function $W(r)$ at $\mathbf{r}=0$ represents the so-called proper electrostatical energy of the electron in the classical field theory. If we assume that the electron is point-like [this means that it should be $a(\mathbf{r}^2) = \delta(\mathbf{r})$ in formula (2.91)] then we obtain the well-known classical electrodynamical result – the electron's proper energy increases as $W(0) \sim e^2/l$ for $l \to 0$, where l is the electron size.

However, the interaction energy or potential $W(r)$ of two electrons is finite at $\mathbf{r} = 0$ if the function $a(\mathbf{r}^2)$ is some smooth distribution. As shown by Efimov (1977a) the case of form-factors with the order of growth $\rho = \frac{1}{2}$ means that the distribution $a(\mathbf{r}^2)$ is bounded, i.e., all charge is distributed inside the sphere with the given radius l. If the order of growth is $\rho > \frac{1}{2}$ then it means that $a(\mathbf{r}^2)$ differs from zero in the whole space, but that it decreases rapidly as

$$a(\mathbf{r}^2) = \exp[-\text{const}\,(|\mathbf{r}|/l)^\gamma],$$

where

$$\gamma = 2\rho/(2\rho - 1) > 1.$$

An interesting extremal problem was done by Efimov (1977a) namely to find such form-factors $V(-p^2 l^2)$ for which the potential $W(\mathbf{r})$ of two rest electrons has the smallest possible value at zero, i.e., at $\mathbf{r} = 0$. Mathematically, the following problem has the solution: among entire functions $V(z)$ of the order of growth $\rho = \frac{1}{2}$ and of the type σ, satisfying conditions:

$V(0) = 1$, $V(x) = |K(x)|^2$ and on the real half-axis $0 \leqslant x < \infty$,

$$\int_0^\infty \frac{du}{\sqrt{u}} V(u) < \infty$$

there exists a function minimizing the functional

$$\mu[V] = \int_0^\infty du\, u^{-1/2} V(u) = \text{const}\, W(0). \tag{2.92}$$

Comparing (2.92) with (2.90) and (2.91) it is easy to see that the consideration of the functional represents proper energy of electron.

This problem pertains to the class of extremal problems in the theory of entire functions (for details see Ibragimov, 1962) and has a unique solution. To solve the given problem rewrite condition (2.92) in the form

$$\mu[V] = \int_0^\infty du\, u^{-1/2} V(u) = 2\int_0^\infty dt |K(t^2)|^2. \tag{2.93}$$

The function $K(t^2)$ is an entire function of the first order of growth and of the type $\frac{1}{2}\sigma$. Therefore, in accordance with Paley–Wiener's (1934) theorem, the following representation holds

$$K(t^2) = \frac{1}{\sqrt{2\pi}} \int_{-\sigma/2}^{\sigma/2} du\, e^{iut}\, g(u),$$

where $g(u) \in L_2(-\frac{1}{2}\sigma, \frac{1}{2}\sigma)$. Since K depends on t^2, then the function $g(u)$ is even and therefore

$$K(t^2) = \sqrt{\frac{2}{\pi}} \int_0^{\sigma/2} du\, g(u) \cos(ut).$$

Making use of Parseval's equality (Titchmarsh, 1939), from (2.93) we get

$$\mu[V] = \int_0^{\infty} dt |K(t^2)|^2 = \int_0^{\sigma/2} du |g(u)|^2. \tag{2.94}$$

Moreover, we have

$$K(0) = \sqrt{\frac{2}{\pi}} \int_0^{\sigma/2} du\, g(u) = 1. \tag{2.95}$$

Let there exist a system of orthonormalized polynomials $\{P_n(x)\}$ with weight 1 on the interval $[0, \frac{1}{2}\sigma]$, i.e.,

$$\int_0^{\sigma/2} du\, P_n(u) P_{n'}(u) = \delta_{nn'}, \qquad P_0(u) = (2/\sigma)^{1/2}.$$

Then the function $g(u)$ can be decomposed by this system

$$g(u) = \sum_{n=0}^{\infty} c_n P_n(u).$$

In this case formulas (2.94) and (2.95) acquire the form

$$\mu[V] = \int_0^{\sigma/2} du |g(u)|^2 = \sum_{n=0}^{\infty} |c_u|^2, \qquad \sqrt{\frac{2}{\pi}} \int_0^{\sigma/2} du\, g(u) = \sqrt{\frac{2}{\pi}} c_0 \sqrt{\frac{\sigma}{2}} = 1.$$

It is clear that

$$\min_{\{g\}} \mu[V] = \min_{\{c_n\}} \sum_{n=0}^{\infty} |c_n|^2 = |c_0|^2 = \frac{\pi}{\sigma}.$$

It means $g(u) = \sqrt{\pi/\sigma}$ and therefore

$$K(t^2) = \frac{2}{\sigma} \int_0^{\sigma/2} du \cos(ut) = \frac{\sin(\sigma t/2)}{(\sigma t/2)}.$$

Finally, we have

$$V(z) = [\sin(\sigma\sqrt{z}/2)/(\sigma\sqrt{z}/2)]^2. \tag{2.96}$$

Consider what charge distribution is described by the form-factor (2.96) obtained. For this, according to (2.74) we calculate the density of charge distribution $a(\mathbf{r}^2)$ corresponding to the function

$$\tilde{K}(\mathbf{k}^2 l^2) = \frac{\sin[(\sigma/\sqrt{2})\sqrt{\mathbf{k}^2 l^2}]}{(\sigma/2)\sqrt{\mathbf{k}^2 l^2}}.$$

After obvious calculation we have

$$a(\mathbf{r}^2) = \frac{1}{(2\pi)^3} \int \mathrm{d}^3 k \tilde{K}(\mathbf{k}^2 l^2) e^{i\mathbf{kr}} = \frac{1}{4\pi} \frac{\delta(|\mathbf{r}| - \sigma l/2)}{\mathbf{r}^2}.$$

This distribution describes a uniformly charged sphere with the radius $\sigma l/2$. Now we define the energy of two interacting electrons by formula (2.91). As a result, we get

$$W(\mathbf{r}) = \begin{cases} \dfrac{e^2}{4\pi} \dfrac{1}{\sigma l} \left(2 - \dfrac{|\mathbf{r}|}{\sigma l}\right) & r < \sigma l, \\ \dfrac{e^2}{4\pi} \dfrac{1}{|\mathbf{r}|} & r > \sigma l. \end{cases}$$

Turning to the following *Mellin representation*

$$V(k_E^2 l^2) = \frac{1}{2i} \int_{\substack{-\beta + i\infty \\ (0<\beta<1)}}^{-\beta - i\infty} \mathrm{d}\zeta \frac{v(\zeta)}{\sin \pi\zeta} (k_E^2 l^2)^{\zeta}, \tag{2.97}$$

we write finally the form of the form-factor (2.96):

$$V_s(-k^2 l^2) = \left[\frac{\sin\sqrt{-k^2 l^2}}{\sqrt{-k^2 l^2}}\right]^2, \qquad k^2 = k_0^2 - \mathbf{k}^2, \tag{2.98}$$

For the function $v(\zeta)$ in the representation (2.97) we obtain

$$v_s(\zeta) = 2^{1+2\zeta}/\Gamma(2\zeta + 3). \tag{2.99}$$

If we assume that the electron is a uniformly charged ball of radius l, then, it is easy to show that

$$V_b(-k^2 l^2) = \frac{9}{(-k^2 l^2)^2} \left[\frac{\sin(-k^2 l^2)^{1/2}}{(-k^2 l^2)^{1/2}} - \cos(-k^2 l^2)^{1/2}\right]^2$$

and $v_b(\zeta)$ is given by formula (1.15).

Thus, we see that in the quantum electrodynamics the form-factors with the order of growth $\rho = \frac{1}{2}$ have greatest interest, when the rest electron may be interpreted as a charged sphere (or ball) with radius l.

2.4. The Causality Condition and Unitarity of the S-Matrix in Nonlocal Quantum Field Theory

2.4.1. *Introduction*

It is well known that the principles of causality and unitarity of the S-matrix in quantum field theory are the basis of all approaches in the elementary particle theory which make claims to self-consistency and physical acceptability. Therefore, the *proof of the unitarity and causality condition is crucial* in constructing various models of QFT. These problems were considered in detail by Bogolubov and Efimov (and their coworkers) in both the local and nonlocal cases, respectively. In this section we formulate only briefly the causality condition and prove the unitarity of the non-local S-matrix in the functional form within the framework of the Efimov (1977a) approach.

First we turn to the mathematical formulation of the causality condition. In the local QFT the causality is manifested as a requirement imposed on the Heisenberg fields $\varphi(x)$: they must be *locally commutable*

$$[\varphi(x), \varphi(y)]_- = 0 \quad \text{for } x \sim y \tag{2.100}$$

or as the *microcausality condition* for the S-matrix

$$\frac{\delta}{\delta\varphi(x)}\left(\frac{\delta S}{\delta\varphi(y)}S^+\right) = 0 \quad \text{for } x \lesssim y. \tag{2.101}$$

Both the conditions are of a formal mathematical nature and they represent the postulates describing the mathematical structure of the fields $\varphi(x)$ in (2.100) and of the S-matrix in (2.101) rather than the requirements of physical causality. This is explained by the fact that the *concept of point-like nature of events suggested* in (2.100) and (2.101) is *incompatible* with the ideas of relativistic quantum mechanics

In the recent past, numerous attempts have been made to formulate the so-called physical condition of causality (Blokhintsev, 1973b; Wanders, 1959; Eden and Landshoff, 1965: Stapp, 1965, etc.), i.e., to find the minimal requirements on the amplitudes of physical processes which would guarantee the absence of any obviously noncausal phenomena in the macroworld. However, this problem remains as yet open and conditions (2.100) and (2.101) are actually the only working formulations of causality ensuring well-known results in different approaches to the local QFT.

It should be noted that the predictions of the local theory are in satisfactory agreement with the experiment up to the presently attainable distances (10^{-15}–10^{-16} cm). This appears to indicate that, first, the mathematical condition of causality guarantees without doubt the fulfilment of the physical condition provided one is able to give a reasonable formulation of the latter. Secondly, the concept of the point-like nature of events is a good approximation up to presently attainable energies.

However, the nonlocality which is caused by the stochastic nature of space-time as well as by the fact that the concept of strict localizability of events is only approximate, has stimulated many attempts to formulate a self-consistent theory at the *expense of locality*, and obviously of microcausality too. In doing so, we may consider the theory valid only in so far as the violation of microcausality is localized in sufficiently small space-time regions. The investigation of the causal S-matrix properties in theories of this kind is of predominant importance.

2.4.2. *The Causality Condition*

Our formulation of the causality condition is based on the Bogolubov and Shirkov (1980) approach. We introduce an operation of 'switching on' and 'switching off' the interaction. Instead of the coupling constant g, we introduce the function $g(x)$ on the interval $[0, g]$, which characterizes the intensity of 'switching on' the interaction (see Section 2.2.5). Let $g_1(x)$ be different from zero in some region $G_1 \subset \mathbb{R}^4$ and $g_2(x)$ – in $G_2 \subset \mathbb{R}^4$. Then the S-matrix of theory satisfies the *microcausality condition*, if

$$S[g_1 + g_2] = S[g_2]S[g_1] \tag{2.102}$$

for $G_2 \gtrsim G_1$, i.e., if and only if all the points of the region G_2 belong to the future cone (or to the space-like region) with respect to all points of the region G_1. The condition (2.102) can be written in the *differential form*

$$R(x, y) = \frac{\delta}{\delta g(x)}\left(\frac{\delta S}{\delta g(y)} S^+\right) = 0 \tag{2.103}$$

for $x \lesssim y$. This relation represents a formulation of causality in the differential form and it has been given by Bogolubov and Shirkov (1980).

We now verify the condition (2.103) for the S-matrix which is constructed by means of the interaction Lagrangian $\mathscr{L}_{\text{in}}(x)$ in the case of the local QFT. As usual, the S-matrix is presented as a functional series over the powers of the function $g(x)$:

$$S[g] = 1 + \sum_{n=1}^{\infty} \frac{i^n}{n!} \int d^4x_1 \cdots \int d^4x_n S_n(x_1, \ldots, x_n) g(x_1) \cdots g(x_n), \tag{2.104}$$

where the operator-valued expressions $S_n(x_1, \ldots, x_n)$ are given by

$$S_n(x_1, \ldots, x_n) = T[\mathscr{L}_{\text{in}}(x_1) \cdots \mathscr{L}_{\text{in}}(x_n)].$$

First of all, we write the expression (2.104) in some convenient form. The nth term of the series can be represented in the form

$$\frac{i^n}{n!} T\left\{\int d^4x_1 \mathscr{L}_{\text{in}}(x_1) g(x_1) \cdots \int d^4x_n \mathscr{L}_{\text{in}}(x_n) g(x_n)\right\}$$

$$= \frac{i^n}{n!} T\left\{\left(\int d^4x \mathscr{L}_{\text{in}}(x) g(x)\right)^n\right\}.$$

Thus, by introducing T-exponent we can formally summarize series (2.104)

$$S[g] = T\left\{1 + \sum_{n=1}^{\infty} \frac{i^n}{n!} \left[\int d^4x \mathscr{L}_{\text{in}}(x) g(x)\right]^n\right\}$$

$$= T \exp\left\{i \int d^4x \mathscr{L}_{\text{in}}(x) g(x)\right\}. \tag{2.105}$$

In order to check condition (2.103) we calculate the variational differential of $S[g]$ at the point y

$$-i\frac{\delta S[g]}{\delta g(y)} = T\left\{\mathscr{L}_{\text{in}}(y)\exp\left[i\int d^4x\mathscr{L}_{\text{in}}(x)g(x)\right]\right\},$$

taking into account here that $\delta g(x)/\delta g(y) = \delta^4(x-y)$ (for detail, see Chapter 5). Further, the four-dimensional space $\mathbb{R}^4$ is divided into two parts G_+ and G_- by the space-like surface $x_0 = \text{const.} = y_0$ with respect to which G_+ lies to 'the future' and G_- to 'the past'. Thus, we have

$$\begin{aligned}
-i\frac{\delta S[g]}{\delta g(y)} &= T\left\{\mathscr{L}_{\text{in}}(y)\exp\left[i\int_{G_+} d^4x\mathscr{L}_{\text{in}}(x)g(x)\right]\times\right.\\
&\qquad\left.\times\exp\left[i\int_{G_-} d^4z\mathscr{L}_{\text{in}}(z)g(z)\right]\right\}\\
&= T\left\{\mathscr{L}_{\text{in}}(y)\exp\left[i\int_{G_+} d^4x\mathscr{L}_{\text{in}}(x)g(x)\right]\right\}\times\\
&\qquad\times T\left\{\exp\left[i\int_{G_-} d^4z\mathscr{L}_{\text{in}}(z)g(z)\right]\right\}.
\end{aligned} \tag{2.106}$$

On the other hand, we reach by analogy

$$\begin{aligned}
S[g] &= T\left\{\exp\left[i\int_{G_+} d^4x\mathscr{L}_{\text{in}}(x)g(x) + i\int_{G_-} d^4z\mathscr{L}_{\text{in}}(z)g(z)\right]\right\}\\
&= T\left\{\exp\left[i\int_{G_+} d^4x\mathscr{L}_{\text{in}}(x)g(x)\right]\right\}T\left\{\exp\left[i\int_{G_-} d^4z\mathscr{L}_{\text{in}}(z)g(z)\right]\right\}
\end{aligned}$$

and also

$$\begin{aligned}
S^+[g] &= \left\{T\left[\exp\left(i\int_{G_-} d^4x\mathscr{L}_{\text{in}}(x)g(x)\right)\right]\right\}^+\times\\
&\qquad\times\left\{T\left[\exp\left(i\int_{G_+} d^4z\mathscr{L}_{\text{in}}(z)g(z)\right)\right]\right\}^+.
\end{aligned}$$

From this, taking into account the unitarity property for

$$\mathrm{T}\left\{\exp\left[i\int_{G_-} d^4x\mathscr{L}_{\text{in}}(x)g(x)\right]\right\}$$

and making use of (2.106), we get

$$-i\frac{\delta S[g]}{\delta g(y)}S^+[g] = T\left\{\mathscr{L}_{\text{in}}(y)\exp\left[i\int_{G_+} d^4z\mathscr{L}_{\text{in}}(z)g(z)\right]\right\}\times$$

$$\times \left\{ T\left[\exp\left(i \int_{G_+} \mathrm{d}^4 z \mathscr{L}_{\mathrm{in}}(z) g(z) \right) \right] \right\}^+ .$$

Therefore, the product

$$[\delta S[g]/\delta g(y)] S^+[g]$$

does not depend on the behavior of the function $g(x)$ in the region G_-, i.e., for $x_0 < y_0$. In accordance with the covariance principle of the relativistic QFT it takes place also in the case when $x \sim y$. We recall that in this region $(x \sim y)$ the commutator of the field operator $\varphi(x)$ (in particular, for the scalar theory):

$$[\varphi(x), \varphi(y)]_- = \Delta(x-y), \quad .$$

where

$$\begin{aligned}\Delta(x-y) &= \Delta_{(+)}(x-y) + \Delta_{(-)}(x-y) \\ &= (2\pi)^{-3} \int \mathrm{d}^4 k e^{-ik(x-y)} \varepsilon(k_0) \delta(k^2 - m^2) \\ &= \frac{1}{2\pi i} \varepsilon(x_0)\, \delta(\lambda) - \frac{m}{4\pi i \sqrt{\lambda}} \varepsilon(x_0) \theta(\lambda) J_1(m\sqrt{\lambda}), \\ &\quad \lambda = (x_0 - y_0)^2 - (\mathbf{x} - \mathbf{y})^2\end{aligned}$$

disappears, which ensures the independence of events separated by space-like intervals, i.e., the causality condition in space-time (or the *locality condition* of the theory). In the nonlocal case the proof of causality condition will be carried out in Section 2.4.5.

Now we consider how the causality condition is formulated depending on the choice of spaces of test functions. If $y = 0$ in (2.103), then it is verifiable that for

$$M(x) = \langle \alpha | R(x,0) | \beta \rangle,$$

(where $|\alpha\rangle$ and $|\beta\rangle$ are arbitrary physical states) the integral

$$\int \mathrm{d}^4 x M(x) f_G(x) = 0$$

for any functions $f_G(x)$ which differ from zero only in a space-time region $G \subset \mathbb{R}^4$ lying outside the future cone: $G \not\subset \bar{V}^+$, here

$$V^+ = \{x\colon x_0 > 0, x^2 > 0\}, \qquad \bar{V}^+ = \{x\colon x_0 \geqslant 0, x^2 \geqslant 0\}.$$

Thus we arrive at the definition.

Microcausality condition. If the space $\mathscr{U}$ of test functions satisfies the condition of strict locality, i.e., that in the space $\mathscr{U}$ there is found some subspace D of test functions such that its intersection is dense in D (for detail, see Jaffe, 1966) or in short that $D \cap \mathcal{U}$ is dense in D, it is said that the S-matrix satisfies the microcausality condition, if

$$\int d^4 x f(x) \langle \alpha | R(x,0) | \beta \rangle = 0 \tag{2.107}$$

for any $f(x) \in D(G)$ such that $G \cap \bar{V}^+ = 0$.

Thus, in order to verify the causality condition (2.107), it is enough to have in the space of test functions on which the theory under consideration is constructed, some subspaces of the *functions with finite supports*. Moreover, the fact of the presence of the functions with finite supports in the space of test functions was adopted as a principle of strict locality in QFT (see Jaffe, 1966, 1967; Jost, 1965).

The main trouble with formulating and verifying the causality condition in the (nonlocal) stochastic QFT presented above, is the fact that the *functions with finite supports are absent* in the space $\mathcal{U} \equiv Z$ of test functions. If the theory is nonlocal but causal, then its S-matrix should satisfy condition (2.102) only for $G_2 > G_1$. For $G_2 \sim G_1$ (i.e., they are space-like with respect to each other), the matrix elements

$$\langle \alpha | \{ S[g_1 + g_2] - S[g_2] S[g_1] \} | \beta \rangle \tag{2.108}$$

should decrease rapidly enough at the increasing distance between the regions G_1 and G_2.

Since in the (nonlocal) stochastic theory the space Z of test functions consists of analytical functions among which the functions with finite supports are absent, then it is necessary to use the following condition.

DEFINITION 2.1 (*macrocausality condition*). Let $\langle \alpha | R(x,0) | \beta \rangle \in Z'_\rho$ $(\rho > 1)$ and let $f(x) \in Z^\rho_\rho \subset Z_\rho$. Then the nonlocal S-matrix satisfies the *macrocausality condition*, if

$$\phi(x) = \int d^4 x' f(x - x') \langle \alpha | R(x', 0) | \beta \rangle \tag{2.109}$$

decreases when $x \to \infty$ outside the region $\bar{V}^+$ as

$$\phi(x) = O[\exp(-\|x\|^\rho)], \tag{2.110}$$

where $\|x\| = (x_0^2 + \mathbf{x}^2)^{1/2}$ is the Euclidean distance.

Now we discuss the macrocausality condition by means of the concept of the generalized functions. From the physical point of view, the functions of a test space describe a space-time distribution of some physical quantities. These test functions we call conditionally wave packets which are given in real space-time, and the functions describing them are assumed to fall off rapidly outside some bounded region in the space $\mathbb{R}^4$.

The *generalized function* $K(x - x')$ used as the form-factor has the meaning of spreading-out operator connected with the definite dynamics of the process under consideration. In this case we are interested in its space-time properties. Therefore it is necessary to choose some function $\varphi(x)$ supported by a bounded region $G \subset \mathbb{R}^4$, and to see the *support of the function* (see Section 1.4.3)

$$\phi(x) = \varphi_R(x) = (K, \varphi)(x) = \int d^4x' K(x - x')\varphi(x')$$

$$= \int d\mu(\zeta) \int_{b^2=1} d^4 b \varphi(x_0 + ib_4\zeta, \mathbf{x} + \mathbf{b}\zeta) \tag{2.111}$$

in $\mathbb{R}^4$. According to the basic idea of the nonlocal theory (see, for example, Block, 1952; Chretien and Peierls, 1953) the *macrocausality condition* is equivalent to the requirement that the 'effective wave packet' $\phi(x)$ will be located in some bounded region $G_\varphi \subset \mathbb{R}^4$ as soon as the 'initial wave packet' $\varphi(x)$ is located in a bounded region $G_\varphi \subset \mathbb{R}^4$. Since, in our spaces of test functions, finite functions are absent, it is necessary to choose such a 'wave packet' which is described by a function falling off rapidly enough outside some bounded region $G_\varphi \subset \mathbb{R}^4$. The fact that for $\mathbf{x}, t \to \infty$ the causal signals attenuate as $|t|^{-3/2}$ (the property of smooth solutions to the Klein–Gordon equation), it is a common assumption in physics that the function falls off rapidly enough if it falls off exponentially, i.e., it satisfies the limiting relation:

$$\lim_{\|x\| \to \infty} |\varphi(x)| \exp(a\|x\|^N) = 0. \tag{2.112}$$

for some $a > 0$ and $N > 0$. Then the *macrocausality condition* is equivalent to the requirement that 'effective wave packet' $\phi(x)$ would fall off rapidly enough if the function $\varphi(x)$ describing the 'initial wave packet' falls off rapidly enough outside some region G_φ.

Let us consider the functional (2.111) in the case when the functions of a test space belong to $Z_\alpha(1 < \alpha < 2/a, 0 < a < 2)$. Wave packets fall off rapidly when $\|x\| \to \infty$, and therefore it is natural to describe them by the functions of the space $Z_\alpha^\alpha \subset Z_\alpha$. If $\varphi(x) \in Z_\alpha^\alpha$, then

$$|\varphi(x)| \sim \exp\left\{-\text{const} \sum_{\mu=0}^{3} |x_\mu|^\alpha\right\},$$

when $|x_\mu| \to \infty$. Notice that no functions exist in the space Z_α which would fall off more rapidly. However, such order of decay is quite acceptable from the point of view of the physical requirements [see formula (2.112)].

Let $\varphi(x)$ in (2.111) be the wave packet which is described by a function of Z_α^α. Let us write the functional (2.111) in the momentum representation

$$\phi(x) = (K, \varphi)(x) = \int d^4 p \tilde{K}(p^2)\tilde{\varphi}(p)\, e^{ipx}.$$

Let the function $\tilde{\varphi}(p) \in Z_{\alpha'}^{\alpha'}$, where $\alpha' = \alpha/(\alpha - 1) > 2/(2 - a)$ and $\tilde{K}(p^2) \in Z_{2/(2-a)}$. Since $\alpha' > 2/(2 - a)$, the entire function $\tilde{K}(p^2)$ is the *multiplier* for the space $Z_{\alpha'}^{\alpha'}$, i.e.,

$$\tilde{\phi}(p) = \tilde{K}(p^2)\tilde{\varphi}(p) \in Z_{\alpha'}^{\alpha'}.$$

Therefore, $\phi(x) \in Z_{\alpha}^{\alpha}$ and it satisfies condition (2.112). Thus the 'effective wave packet' $\phi(x)$ falls off rapidly, and the same is true for the 'initial wave packet' $\varphi(x)$.

Thus, if the S-matrix satisfies one of the causality conditions formulated above, we say that causality is fulfilled in the theory under consideration. The causality condition in nonlocal QFT was studied by Efimov and Alebastrov (see Efimov, 1977a).

It is necessary to note that if the Schrödinger equation in the interaction picture

$$i\frac{\partial}{\partial t}\Psi(t) = H_{\text{in}}(t)\Psi(t) \tag{2.113}$$

were to have mathematical meaning, the causality condition would be automatically valid as a property of solutions to that equation. This problem will be investigated in Section 2.5.

2.4.3. *The Scheme of Proof of Unitarity of the S-Matrix in Perturbation Theory*

Now we turn to the formulation of the unitarity in the nonlocal QFT. If the Lagrangian of interacting quantum fields is given, the S-matrix is then sought in the form (2.104) of a formal expansion in powers of the function of 'switching on' the interaction $g(x)$, while the operator expression $S_n(x_1, \ldots, x_n)$ in (2.104) can be represented in a series of normal products of the quantized fields $\varphi(x)$:

$$S_n(x_1, \ldots, x_n) = \sum_{m_1, \ldots, m_n} K_{\ldots m_j \ldots}(x_1, \ldots, x_n) \colon \varphi^{m_1}(x_1) \cdots \varphi^{m_n}(x_n) \colon \tag{2.114}$$

where the coefficient functions $\{K_i\}$ are in turn expressed in terms of the products

$$K_{\ldots m_j \ldots}(x_1, \ldots, x_n) = \prod_{s,t} D_c(x_s - x_t) \tag{2.115}$$

of the nonlocal propagators

$$D_c(x - y) = i^{-1}(2\pi)^{-4} \int d^4k \frac{V(-k^2 l^2)}{m^2 - k^2 - i\varepsilon} e^{-ik(x-y)}.$$

In accordance with the above deduction (see Section 1.4 and Condition 2.1 of Section 2.2.1) the nonlocal propagator $D_c(x)$ exists as a generalized function on some space Z of test functions and their products – the coefficient functions $\{K_i\}$ in (2.114) are therefore not defined mathematically. Thus, the *formulation of a working method for the definition* of the products of type (2.115) is the *main problem* to be solved in the construction of the finite S-matrix in the nonlocal theory. As usual, the coefficient functions $\{K_i\}$ in (2.114) are constructed as a limit of the locally integrable functions $K^{\delta}_{\dots m_j \dots}(x_1, \dots, x_n)$ by introducing a *regularization procedure* given by the parameter δ, so there exists the limit

$$\lim_{\delta \to 0} K^{\delta}_{\dots m_j \dots}(x_1, \dots, x_n) = K_{\dots m_j \dots}(x_1, \dots, x_n)$$

in the improper sense or in the *weak limit* (see Section 1.4 and Bogolubov and Shirkov, 1980), or otherwise

$$S[g] = \lim_{\delta \to 0} S^{\delta}[g]. \tag{2.116}$$

It is unclear whether the S-matrix obtained in such a way satisfies the initial axioms and especially the unitarity relation:

$$S[g]S^{+}[g] = S^{+}[g]S[g] = 1. \tag{2.117}$$

In the local theory, the unitarity and causality conditions are directly used to formulate the subtractive method of regularization. This circumstance ensures the fulfilment of unitarity of the S-matrix in each order of perturbation theory, at least in the case of renormalizable interactions.

In the nonlocal theory the regularization procedure is formulated in such a way that the limit (2.116) exists and the $S[g]$-matrix is unitary in each order of the perturbation theory (Alebastrov and Efimov, 1973). We present here the outline of their proof of the unitarity of the S-matrix in the perturbation theory. Suppose that the S-matrix is known in the form of a functional expansion (2.104), where the operator $S_n(x_1, \dots, x_n)$ are given by expansion of type (2.114). If the S-matrix is finite and satisfies the axioms of QFT (see Bogolubov and Shirkov, 1980), then the coefficient functions $K_{\dots m_j \dots}(x_1, \dots, x_n)$ satisfy the following requirements:

(1) They are *translationally invariant*, i.e.,

$$K_{\dots m_j \dots}(x_1 + a, \dots, x_n + a) = K_{\dots m_j \dots}(x_1, \dots, x_n).$$

(2) They are integrable on some space of sufficiently smoothly varying test functions $\mathscr{U}$, i.e., for any $f(x_1, \dots, x_n) \in \mathscr{U}$ there exists an integral

$$\int d^4x_1 \cdots \int d^4x_n K_{\dots m_j \dots}(x_1, \dots, x_n) f(x_1, \dots, x_n) < \infty.$$

If the coefficient functions $K_{\dots m_j \dots}(x_1, \dots, x_n)$ are known, then the expansion for $S^{+}[g]$ is known as well. Following Bogolubov and Shirkov (1980), one can show that the coefficient functions obtained by *multiplying two operator* functions with different independent arguments

$$S_n(x_1,\ldots,x_n) \circledast S_m^+(y_1,\ldots,y_m) \tag{2.118}$$

are of the form

$$K_{\ldots m_j \ldots}(x_1,\ldots,x_n)K_{(-)}(x-y)K_{\ldots m_i \ldots}(y_1,\ldots,y_m), \tag{2.119}$$

where the functions

$$K_{(-)}(x-y) = \prod_{s,t} \Delta_{(-)}(x_s - y_t), \qquad \tilde{\Delta}_{(-)}(k) = 2\pi\theta(k_0)\delta(k^2 - m^2) \tag{2.120}$$

may be defined as generalized integrable functions on the space $\mathscr{U}$. The sign $\circledast$ denotes the transition to a normal product of field operators $\varphi(x)$ in the product (2.118), according to the Wick theorem.

So, if the S-matrix in the form of expansion (2.104) is known, then the product

$$S[g]S^+[g] \underset{\mathrm{Df}}{=} S[g] \circledast S^+[g]$$

is given in each order of the perturbation theory as a generalized operator-valued function on the space of the test functions $\mathscr{U}$.

The problem of the nonlocal theory is to prove that

$$S[g] \circledast S^+[g] = 1.$$

We shall start from the method defining a finite S-matrix through use of an *improper limiting transition*, and we construct our proof in the following manner. It is obvious that the S-matrix will be unitary, if there exists a regularization procedure which possesses the following features:

(i) The regularized functions $K^{\delta}_{\ldots m_j \ldots}(x_1,\ldots,x_n)$ are continuous and bounded, and

$$\lim_{\delta\to 0} K^{\delta}_{\ldots m_j \ldots}(x_1,\ldots,x_n) = K_{\ldots m_j \ldots}(x_1,\ldots,x_n),$$

i.e., the regularized $S^{\delta}[g]$-matrix is defined and there exists the improper limit

$$\lim_{\delta\to 0} S^{\delta}[g] = S[g].$$

(ii) The Green positive-frequency functions which determine the *multiplication* in (2.118) and (2.119) can also be regularized and there exists the limit

$$\lim_{\delta\to 0} K^{\delta}_{(-)}(x-y) = K_{(-)}(x-y),$$

or symbolically

$$\lim_{\delta\to 0} \circledast = \circledast .$$

(iii) In the relation

$$J[g] = \lim_{\delta_1 \to 0} \lim_{\delta_2 \to 0} \lim_{\delta_3 \to 0} S^{\delta_1}[g] \circledast^{\delta_2} S^{\delta_3 +}[g]$$

the limit is independent of the order of limiting transitions to the points $\delta_1 = \delta_2 = \delta_3 = 0$, i.e., the operator

$$J^{\delta_1, \delta_2, \delta_3}[g] = S^{\delta_1}[g] \circledast^{\delta_2} S^{\delta_3 +}[g]$$

is continuous at the point $\delta_1 = \delta_2 = \delta_3 = 0$.

(iv) Regularization is chosen so that

$$S^{\delta}[g] \circledast^{\delta} S^{\delta +}[g] = 1.$$

Then there is the following chain of equalities

$$J[g] = S[g] \circledast S^{+}[g] = \lim_{\delta_1 \to 0} \lim_{\delta_2 \to 0} \lim_{\delta_3 \to 0} S^{\delta_1}[g] \circledast^{\delta_2} S^{\delta_3 +}[g]$$

$$= \lim_{\delta \to 0} S^{\delta}[g] \circledast^{\delta} S^{\delta +}[g] = \lim_{\delta \to 0} 1 = 1.$$

Hence

$$S[g] S^{+}[g] = 1.$$

The main idea is to indicate a regularization procedure in the nonlocal QFT which would satisfy all the requirements listed above. This will prove the unitarity of the S-matrix.

We give here only a regularization procedure and prove the unitarity of the S-matrix in the functional form. Proof of the unitarity in the perturbation theory was given by Alebastrov and Efimov (1973).

2.4.4. *An Intermediate Regularization Scheme*

In Section 2.2.2 we introduced the regularization procedure which allowed the performance of the canonical quantization of the nonlocal interaction (2.7). However, for the construction of the perturbation theory by means of the propagators (2.5):

$$\tilde{D}_c(k^2) = V(-k^2 l^2)(m^2 - k^2 - i\varepsilon)^{-1},$$

and for the proof of the unitarity of the S-matrix, it is convenient to use a somewhat different regularization.

Let us consider the regularization method for the propagators (2.5). Let $V(-k^2 l^2)$ be the nonlocal form-factor which satisfies Condition 2.1 (or some conditions of this type), then the *Mellin representation*

$$V(-k^2 l^2) = \frac{1}{2i} \int_{-\beta + i\infty}^{-\beta - i\infty} d\zeta \, \frac{v(\zeta)}{\sin \pi\zeta} \, l^{2\zeta} (m^2 - k^2 - i\varepsilon)^{\zeta}, \tag{2.121}$$

holds in the region $m^2 > k^2$. The function $v(\zeta)$ possesses the following properties:

(1) It is regular in the half-plane $\operatorname{Re}\zeta \geqslant -\beta$, and in this region,

$$|v(\xi + i\eta)| \leqslant c(1 + |\eta|)^{-N} \exp(\pi|\eta|)/\Gamma(|\xi|/\rho + 1)$$

for $\forall N$ and some number $C > 0$.

(2) At the points $\zeta = -1, -2, \ldots, -[\beta]$, it has zero (at least of the first order).

(3) $v(0) = 1$.

(4) $v^*(\zeta) = v(\zeta)$.

In particular, for the following functions:

$$\begin{aligned} V'_G &= \exp[-l^2(m^2 - k^2)], \\ V_1 &= [\sin l(m^2 - k^2)^{1/2}/l(m^2 - k^2)^{1/2}]^4, \\ V_2 &= 2^{\mu}\cdot\Gamma(1 + \mu)J_{\mu}(l(m^2 - k^2)^{1/2})[l(m^2 - k^2)^{1/2}]^{-\mu}, \end{aligned} \tag{2.122}$$

we have

$$\begin{aligned} v'_G(\zeta) &= 1/\Gamma(1 + \zeta), \\ v_1(\zeta) &= (2^{4\zeta + 5} - 2^{2\zeta + 3})/\Gamma(2\zeta + 5), \\ v_2(\zeta) &= 2^{-2\mu}\Gamma(1 + \mu)/\Gamma(1 + \zeta)\Gamma(1 + \mu + \zeta). \end{aligned} \tag{2.123}$$

It is easy to see that these functions satisfy the above-mentioned conditions.

We emphasize once more that the representation (2.121) holds for $k^2 < m^2$ only. For the passage to the region $k^2 > m^2$, one has to go from the integration in the ζ-plane over the contour

$$L_0 = \{\zeta : \zeta = \xi + i\eta, \xi = -\beta, -\infty < \eta < \infty\}$$

to that over the contour (see Figure 2).

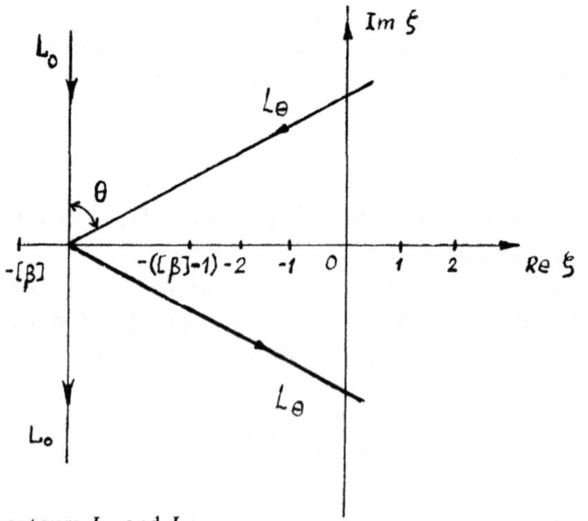

Fig. 2. Integration contours L_0 and L_θ.

$$L_\theta = \{\zeta : \zeta = -\beta + r\exp[\pm i(\tfrac{1}{2}\pi - \theta)], 0 \leqslant r < \infty\} \quad (0 \leqslant \theta \leqslant \tfrac{1}{2}\pi)$$

Let us introduce the regularized function

$$\tilde{D}_c^\delta(k^2) = \frac{1}{2i}\int_{L_\theta} d\zeta \frac{v(\zeta)}{\sin \pi\zeta} l^{2\zeta} \exp(\delta\zeta^2)(m^2 - k^2 - i\varepsilon)^{\zeta - 1},$$

where $0 \leqslant \theta \leqslant \frac{1}{4}\pi$. One can readily recognize that if $\delta > 0$, the function $\tilde{D}_c^\delta(k^2)$ is:

(a) regular and analytic in all the complex k^2-plane, except for the cut along the half-line $[m^2, +\infty)$,

(b) $\tilde{D}_c^\delta(k^2) = O(1/|k^2|^{1+\beta})$ for $|k^2| \to \infty$,

(c) $\lim_{\delta \to 0} \tilde{D}_c^\delta(k^2) = \tilde{D}_c(k^2) = V(-k^2 l^2)(m^2 - k^2 - i\varepsilon)^{-1}$.

For the function $\tilde{D}_c^\delta(k^2)$, there exists the Fourier transform

$$D_c^\delta(x) = i^{-1}(2\pi)^{-4}\int d^4k\, e^{-ikx}\, \tilde{D}_c^\delta(k^2)$$

$$= \frac{1}{2i}\int_{-\beta + i\infty}^{-\beta - i\infty} d\zeta \frac{v(\zeta)}{\sin \pi\zeta} l^{2\zeta} \exp(\delta\zeta^2) D_c(\zeta, m^2, x), \tag{2.124}$$

where

$$D_c(\zeta, m^2, x) = i^{-1}(2\pi)^{-4}\int d^4k\, e^{-ikx}\,(m^2 - k^2 - i\varepsilon)^{\zeta - 1}$$

$$= i2^\zeta e^{-i\pi\zeta}\frac{m^{2(1+\zeta)}}{8\pi\Gamma(1-\zeta)} H_{1+\zeta}^{(2)}(m\sqrt{x^2 - i\varepsilon})\,(m\sqrt{x^2 - i\varepsilon})^{-1-\zeta} \tag{2.125}$$

$H_n^{(2)}(x)$ is the Hankel function, for which the expansion

$$\left(\frac{2}{x}\right)^\nu H_\nu^{(2)}(x) = \frac{1}{i\sin\pi\nu}\left\{e^{i\pi\nu}\sum_{n=0}^{\infty}(-1)^n\frac{(x/2)^{2n}}{n!\,\Gamma(1+n+\nu)} - \right.$$

$$\left. - \left(\frac{2}{x}\right)^{2\nu}\sum_{n=0}^{\infty}(-1)^n \times \left(\frac{x}{2}\right)^{2n}\frac{1}{n!\,\Gamma(1+n-\nu)}\right\}$$

$$= \frac{1}{i\sin\pi\nu}\left\{\frac{e^{i\pi\nu}}{\Gamma(1+\nu)} - \frac{(2/x)^{2\nu}}{\Gamma(1-\nu)} + O(x^2)\right\}, \tag{2.126}$$

holds at the neighborhood of the point $x = 0$, when $\operatorname{Re}\nu < 0$. Using this representation, we obtain

$$D_c(\zeta, m^2, 0) = m^{2(1+\zeta)}(4\pi)^{-2}/\zeta(1+\zeta). \tag{2.127}$$

Therefore the function $D_c^\delta(x)$ is bounded at the point $x = 0$ because of

$$D_c^\delta(0) = \frac{(4\pi)^{-2}}{2i}\int_{-\beta+i\infty}^{-\beta-i\infty} d\zeta \frac{v(\zeta)}{\sin\pi\zeta}\frac{m^{2(1+\zeta)}}{\zeta(1+\zeta)} l^{2\zeta}\exp(\delta\zeta^2) < \infty. \tag{2.128}$$

From the representation (2.124) and (2.126), it follows that the function $D_c^\delta(x)$ is $[\beta - 1]$ times continuously differentiable with respect to x at the point $x^2 = 0$, where $[\beta]$ denotes the entire part of β.

Thus, if $\delta > 0$, then the function $D_c^\delta(x)$ is locally an integrable function and the products of the type $\Pi_{i,j} D_c^\delta(x_i - x_j)$ are also *locally integrable*. In the sense of generalized functions, $\lim_{\delta\to 0} D_c^\delta(x) = D_c(x)$, i.e.,

$$\lim_{\delta\to 0}\int d^4x D_c^\delta(x) f(x) = \int d^4x D_c(x) f(x) = \int d^4p \tilde{D}_c(p^2)\tilde{f}(p)$$

for any $f(x) \in Z_\alpha$ $[\alpha < 2\rho/(2\rho - 1)]$.

Now we regularize the function $\tilde{D}_{(\pm)}(k)$ corresponding to the propagator $\tilde{D}_c(k^2)$ (2.5). For the positive- and negative-frequence Green functions $\tilde{\Delta}_{(\pm)}(k)$ the following equalities hold:

$$\begin{aligned}\tilde{\Delta}_{(\pm)}(k) &= 2\pi\theta(\mp k_0)\,\delta(k^2 - m^2) = 2\pi\theta(\mp k_0)V(-k^2l^2)\,\delta(k^2 - m^2)\\ &= i^{-1}\left[\frac{V(-k^2l^2)}{m^2 - k^2 - i\varepsilon} - \frac{V(-k^2l^2)}{m^2 - k^2 \mp i\varepsilon k_0}\right]\\ &= \frac{1}{2i}\int_{L_\theta} d\zeta \frac{v(\zeta)}{\sin\pi\zeta} l^{2\zeta} \times\\ &\quad\times i^{-1}\{(m^2 - k^2 - i\varepsilon)^{\zeta-1} - (m^2 - k^2 \mp i\varepsilon k_0)^{\zeta-1}\},\end{aligned} \tag{2.129}$$

where $0 \leqslant \theta \leqslant \frac{1}{4}\pi$. Let us introduce the regularized functions

$$\tilde{D}_{(\mp)}^\delta(k) = \frac{1}{2i}\int_{-\beta+i\infty}^{-\beta-i\infty} d\zeta \frac{v(\zeta)}{\sin\pi\zeta} l^{2\zeta}\exp(\delta\zeta^2)\tilde{D}_{(\mp)}(\zeta, m^2, k), \tag{2.130}$$

where

$$\begin{aligned}\tilde{D}_{(\pm)}(\zeta, m^2, k) &= i^{-1}\{(m^2 - k^2 - i\varepsilon)^{\zeta-1} - (m^2 - k^2 \mp i\varepsilon k_0)^{\zeta-1}\}\\ &= 2\pi\theta(\mp k_0)\theta(k^2 - m^2)\pi^{-1}\sin(\pi\zeta)(k^2 - m^2)_+^{\zeta-1}.\end{aligned} \tag{2.131}$$

The function $(k^2 - m^2)_+^{\zeta-1}$ may be understood in the sense of the generalized functions (see Gel'fand and Shilov, 1964). Then it has poles of the first order at the point $\zeta = 0, -1, -2, \ldots$ of the complex ζ-plane. Besides, it is important that

$$\tilde{D}_{(\pm)}^\delta(k) \sim \theta(\mp k_0)\theta(k^2 - m^2).$$

In the x-space, we obtain

$$D^{\delta}_{(\pm)}(x) = i^{-1}(2\pi)^{-4}\int d^4k\, e^{-ikx}\,\tilde{D}^{\delta}_{(\pm)}(k)$$

$$= \frac{1}{2i}\int_{-\beta+i\infty}^{-\beta-i\infty} d\zeta\,\frac{v(\zeta)}{\sin\pi\zeta}\,l^{2\zeta}\exp(\delta\zeta^2)D_{(\pm)}(\zeta, m^2, x), \tag{2.132}$$

where

$$D_{(\pm)}(\zeta, m^2, x) = i^{-1}(2\pi)^{-4}\int d^4k\, e^{-ikx}\,\tilde{D}_{(\pm)}(\zeta, m^2, k)$$

$$= i\frac{2^{\zeta}m^{2(1+\zeta)}}{8\pi\Gamma(1-\zeta)}\left\{e^{-i\pi\zeta}\frac{H^{(2)}_{1+\zeta}(m\sqrt{x^2-i\varepsilon})}{(m\sqrt{x^2-i\varepsilon})^{1+\zeta}} + \right.$$

$$\left. + 2\theta(\pm x_0)\theta(x^2)J_{-(1+\zeta)}(m\sqrt{x^2})(m\sqrt{x^2})^{-1-\zeta}\right\}. \tag{2.133}$$

The first term of (2.133) coincides with the causal function $D_c(\zeta, m^2, x)$ considered in (2.124). This function is bounded at the point $x = 0$. The second term is also bounded at the point $x = 0$ because of the expansion for the Bessel function

$$\left(\frac{2}{x}\right)^{v} J_{-v}(x) = \left(\frac{x}{2}\right)^{-2v}\sum_{k=0}^{\infty}(-1)^k\frac{(x/2)^{2k}}{k!\,\Gamma(k+1-v)} = \left(\frac{x}{2}\right)^{-v}[1 + O(x^2)]. \tag{2.134}$$

Thus, the functions $D_{(\pm)}(\zeta, m^2, x)$ as well as $D_c(\zeta, m^2, x)$ are bounded at the point $x = 0$ and

$$D_{(\pm)}(\zeta, m^2, 0) = D_c(\zeta, m^2, 0) = (4\pi)^{-2}m^{2(1+\zeta)}/\zeta(1+\zeta).$$

Comparing the representations (2.132) with (2.124), it is easy to see that there exist relations between the functions $D^{\delta}_c(x)$ and $D^{\delta}_{(\pm)}(x)$:

$$D^{\delta}_c(x) = \theta(x_0)D^{\delta}_{(-)}(x) + \theta(-x_0)D^{\delta}_{(+)}(x),$$
$$D^{\delta}_{(-)}(x) = \theta(x_0)D^{\delta}_c(x) + \theta(-x_0)D^{\delta *}_c(x). \tag{2.135}$$

Moreover, the following relations

$$\partial^{v}D^{\delta}_c(x) = \theta(x_0)\partial^{v}D^{\delta}_{(-)}(x) + \theta(-x_0)\partial^{v}D^{\delta}_{(+)}(x) \tag{2.136}$$

hold, when $v \leqslant 2[\beta] + 2$ for $\beta > [\beta]$ and $v \leqslant 2\beta + 1$ for $\beta = [\beta]$.

In order to show equation (2.136), we consider the difference

$$R_v(x) = \partial_0^{v}D^{\delta}_c(x) - \{\theta(x_0)\,\partial_0^{v}D^{\delta}_{(-)}(x) + \theta(-x_0)\,\partial_0^{v}D^{\delta}_{(+)}(x)\}$$

$$= \frac{1}{2}\sum_{v_1=1}^{v}\frac{v!}{v_1!(v-v_1)!}\,\delta^{v_1-1}(x_0)\partial_0^{v-v_1}[D^{\delta}_{(-)}(x) - D^{\delta}_{(+)}(x)].$$

Here we have used the equalities (2.135) and the relation

$$\partial_0^{\nu_1}\theta(\pm x_0) = \pm\frac{1}{2}\delta^{\nu_1-1}(x_0).$$

When $x^2 \to 0$, we have, according to (2.133) and (2.134)

$$D_c^\delta(x) = D_{(-)}^\delta(x) - D_{(+)}^\delta(x) = \varepsilon(x_0)\theta(x^2)(x^2)^{\beta-1}O(1).$$

Therefore, the function

$$F^\delta(x_0) = \int d^3x D^\delta(x) f(\mathbf{x})$$

for any $f(\mathbf{x}) \in S$, behaves as

$$F^\delta(x_0) = \varepsilon(x_0)|x_0|^{2\beta+1}O(1).$$

when $x_0 \to 0$, i.e., it is $2[\beta]+1$ times continuously differentiable at the point $x_0 = 0$, and when $\beta > [\beta]$, all its differentials are equal to zero at the point $x_0 = 0$. If $\beta = [\beta]$, then only the 2β-differentials of the function $F^\delta(x_0)$ are equal to zero at $x_0 = 0$. This means that the functional equals

$$(R_\nu, f) = \int dx_0 \int d^3x R_\nu(x_0, \mathbf{x}) f(x_0, \mathbf{x}) = \sum_{\nu_1=1}^{\nu} (-1)^{\nu_1+1} \frac{\nu!}{\nu_1!(\nu-\nu_1)!} \times$$

$$\times \int dx_0 \delta(x_0) \partial_0^{\nu_1-1} \int d^3x f(x_0, \mathbf{x}) \partial_0^{\nu-\nu_1} D^\delta(x_0^2 - \mathbf{x}^2) = 0$$

for any $\nu \leqslant 2[\beta]+2$ for $\beta > [\beta]$ and $\nu \leqslant 2\beta+1$ for $\beta = [\beta]$ and any $f(x_0, \mathbf{x}) \in S$.

2.4.5. *Proof of the Unitarity of the S-Matrix in a Functional Form*

After the preliminary calculations carried out above, we verify the unitarity of the nonlocal *S*-matrix presented in the functional form [see formula (2.105) and Chapter 5]:

$$S_{t_2t_1}[\varphi] = T\exp\left\{i\int_{t_1}^{t_2} d^4x g(x)\mathscr{L}_{in}(x)\right\}, \tag{2.137}$$

where the function $\mathscr{L}_{in}(x) = U(\varphi(x))$ can be represented in the form:

$$U(\varphi) = \sum_{n=0}^{\infty} \frac{u_n}{n!}\varphi^n.$$

Here $\{u_0, u_1, \ldots, u_n, \ldots\}$ are some sequence of real numbers giving the form of the interaction. We assume that for the sequence of real numbers $\{u_n\}$, there exists in the complex ζ-plane a measure $\sigma(\zeta)$ such that

$$u_n = i^n \int \zeta^n \, \mathrm{d}\sigma(\zeta).$$

The function $U(\varphi)$ can then be written in the form

$$U(\varphi) = \int \mathrm{d}\sigma(\zeta) \, \mathrm{e}^{i\zeta\varphi}. \tag{2.138}$$

Notice that the interaction Lagrangian is usually chosen in the normal ordering form. To take this circumstance into account when the formal function transformation is performed, it is necessary to pass to the *usual product* of operators from thc *normal product.* This means that instead of (2.138), we must consider

$$U(\varphi(x)) = \int \mathrm{d}\sigma(\zeta) \exp\,[D_c(0)\zeta^2/2] \exp(\mathrm{i}\zeta\varphi(x)). \tag{2.139}$$

Denote

$$S[\varphi] = S_{\infty}{}'_{-\infty}[\varphi].$$

The T-ordering operation is given by the formula

$$T = \exp\left\{\frac{1}{2}\iint \mathrm{d}^4x_1 \, \mathrm{d}^4x_2 D_c(x_1 - x_2) \frac{\delta^2}{\delta\varphi(x_1)\,\delta\varphi(x_2)}\right\}.$$

The multiplying operator of two (normally-ordered) operator-valued functions is

$$NB[\varphi] \times NC[\varphi] = NE[\varphi],$$

where

$$E[\varphi] = B[\varphi] \circledast C[\varphi]$$

$$= \exp\left\{\iint \mathrm{d}^4x_1 \, \mathrm{d}^4x_2 \sum_{i,j=1} \Delta_{ij}^{(-)}(x_1 - x_2) \frac{\delta^2}{\delta\varphi_i'(x_1)\delta\varphi_j''(x_2)}\right\} \times$$

$$\times \; B[\varphi']C[\varphi''] \,|_{\varphi_j'(x) = \varphi_j''(x) = \varphi_j(x)}$$

The Fourier transform $\tilde{\Delta}_{ij}^{(-)}(k)$ of the function $\Delta_{ij}^{(-)}(x)$ is equal to

$$\tilde{\Delta}_{ij}^{(-)}(k) = 2\pi\theta(k_0)\,\delta(k^2 - m^2).$$

Further, we assume that the functions $D_c(x)$ and $D_{(-)}(x)$ are bounded and continuous, and $D_c(0) = D_c^*(0) = D_{(-)}(0) = D_{(+)}(0) < \infty$, according to our regularization scheme presented above. All these results are valid for the regularized or averaged functions $D_j^\delta(x)$ ($j = c, -, +$) at $\delta > 0$. For simplicity we omit the index δ for the D-functions in this section.

We show that the S-matrix (2.137) satisfies the following relations

$$S_{t_2t_1}[\varphi] \circledast S_{t_1t_0}[\varphi] = S_{t_2t_0}[\varphi],$$
$$S^+_{t_1t_0}[\varphi] \circledast S^+_{t_2t_1}[\varphi] = S^+_{t_2t_0}[\varphi], \tag{2.140}$$

$$T\{\Psi[\varphi(x)]S[\varphi]\} = S_{\infty,x_0}[\varphi] \circledast \Psi[\varphi(x)] \circledast S_{x_0,-\infty}[\varphi], \tag{2.141}$$

$$S_{t_2t_1}[\varphi] \circledast S^+_{t_2t_1}[\varphi] = 1, \tag{2.142}$$

$$\frac{\delta}{\delta\varphi(x)}\left(\frac{\delta}{\delta\varphi(y)} S[\varphi] \circledast S^+[\varphi]\right) = 0, \quad \text{for } x_0 < y_0. \tag{2.143}$$

We notice that relation (2.143), the so-called *causality condition* (see Section 2.4.2), is a consequence of the first three equalities. Indeed, let formulas (2.140) – (2.142) be valid then we have

$$\frac{\delta}{\delta\varphi(y)} S[\varphi] = ig(y)T\{U'[\varphi(y)]S[\varphi]\}$$
$$= ig(y)S_{\infty,y_0}[\varphi] \circledast U'[\varphi(y)] \circledast S_{y_0,-\infty}[\varphi].$$

Further, making use of expressions (2.140) and (2.142), we get

$$\frac{\delta}{\delta\varphi(y)} S[\varphi] \circledast S^+[\varphi] = ig(y)S_{\infty,y_0}[\varphi] \circledast U'[\varphi(y)] \circledast S_{y_0,-\infty}[\varphi] \times$$
$$\times \circledast S^+_{y_0,-\infty}[\varphi] \circledast S^+_{\infty,y_0}[\varphi]$$
$$= ig(y)S_{\infty,y_0}[\varphi] \circledast U'[\varphi(y)] \circledast S^+_{\infty,y_0}[\varphi].$$

Condition (2.143) is obviously fulfilled, since the obtained expression does not depend on the field $\varphi(x)$ at $x_0 < y_0$.

Now consider relations (2.140). Owing to our definitions, we have

$$S_{t_2t_1}[\varphi] \circledast S_{t_1t_0}[\varphi] = \exp\left\{\frac{1}{2}\iint d^4x_1\, d^4x_2\left[D_c(x_1 - x_2)\frac{\delta^2}{\delta\varphi_1(x_1)\,\delta\varphi_1(x_2)} + \right.\right.$$
$$+2D_{(-)}(x_1 - x_2)\frac{\delta^2}{\delta\varphi_1(x_1)\,\delta\varphi_2(x_2)} +$$
$$\left.\left. +D_c(x_1 - x_2)\frac{\delta^2}{\delta\varphi_2(x_1)\delta\varphi_2(x_2)}\right]\right\} \times$$
$$\times \exp\left\{i\int_{t_1}^{t_2} d^4x g(x)U(\varphi_1(x)) + \right.$$
$$\left. +i\int_{t_0}^{t_1} d^4x g(x)U(\varphi_2(x))\right\}\Bigg|_{\varphi_1=\varphi_2=\varphi}$$

Since the defining region of the function $\varphi_1(x)$ is the interval $[t_1, t_2]$ and for the function $\varphi_2(x)$ it is the interval $[t_0, t_1]$, then first, one can put $\varphi_1(x) = \varphi_2(x) = \varphi(x)$ and secondly, $D_{(-)}(x_1 - x_2) = D_c(x_1 - x_2)$ in the second term of the first exponential, because of $x_{10} > x_{20}$ for $x_{10} \in [t_1, t_2]$, $x_{20} \in [t_0, t_1]$. Therefore we have

$$S_{t_2t_1}[\varphi] \circledast S_{t_1t_0}[\varphi] = \exp\left\{\frac{1}{2}\int_{t_0}^{t_2} d^4x_1 \int_{t_0}^{t_2} d^4x_2 D_c(x_1 - x_2)\frac{\delta^2}{\delta\varphi(x_1)\,\delta\varphi(x_2)}\right\} \times$$

$$\times \exp\left\{i\int_{t_0}^{t_2} d^4x g(x) U(\varphi(x))\right\} = S_{t_2t_0}[\varphi].$$

Relation (2.141) is proven simply; to this end it is sufficient to use the definition of T and $\circledast$-operators.

Now we turn to relation (2.142) for the unitarity. We have

$$S_{t_2t_1}[\varphi] \circledast S^+_{t_2t_1}[\varphi] = T_1 \circledast {}_{12}T_2^+ E_{t_2t_1}[\varphi_1, \varphi_2]|_{\varphi_1=\varphi_2=\varphi}, \tag{2.144}$$

where

$$T_i = T[\delta/\delta\varphi_i(x)] \quad (i = 1, 2),$$

$$\mathrm{E}_{t_2t_1}[\varphi_1, \varphi_2] = \exp\left\{i\int_{t_1}^{t_2} d^4x g(x)[U(\varphi_1(x)) - U(\varphi_2(x))]\right\}. \tag{2.145}$$

We make use of the equality:

$$\exp\left\{i\int_{t_1}^{t_2} dx_0 f(x_0)\right\} = 1 - \int_{t_1}^{t_2} dx_0 \frac{d}{dx_0} \exp\left[i\int_{x_0}^{t_2} dx'_0 f(x'_0)\right]$$

$$= 1 + i\int_{t_1}^{t_2} dx_0 f(x_0) \exp\left[i\int_{x_0}^{t_2} dx'_0 f(x'_0)\right]$$

for

$$f(x_0) = \int d^3x g(x_0, \mathbf{x})\{U(\varphi_1(x_0, \mathbf{x})) - U(\varphi_2(x_0, \mathbf{x}))\}.$$

Substituting this equality into (2.145) and using representation (2.139) for the interaction Lagrangian, we obtain

$$S_{t_2t_1}[\varphi] \circledast S^+_{t_2t_1}[\varphi] = 1 + i\int_{t_1}^{t_2} d^4x \int d\sigma(\zeta) T_1 \circledast {}_{12}T_2^+ \{\exp[i\zeta\varphi_1(x)] -$$

$$- \exp[i\zeta\varphi_2(x)]\}\exp[\zeta^2 D_c(0)/2] \times$$

$$\times E_{t_2x_0}[\varphi_1, \varphi_2]|_{\varphi_1=\varphi_2=\varphi}. \tag{2.146}$$

One can easily obtain the following relations:

$$T_1 \exp[i\zeta\varphi_1(x)]T_1^{-1} = \exp\Big\{-\zeta^2\tfrac{1}{2}D_c(0) +$$

$$+ i\zeta\varphi_1(x) + i\zeta\int d^4x' D_c(x-x')\frac{\delta}{\delta\varphi_1(x')}\Big\}, \quad T_2 \exp[i\zeta\varphi_2(x)]T_2^{-1}$$

$$= \exp\Big\{-\tfrac{1}{2}\zeta^2 D_c(0) + i\zeta\varphi_2(x) +$$

$$+ i\zeta\int d^4x' D_c^*(x-x')\frac{\delta}{\delta\varphi_2(x')}\Big\}, \quad \circledast_{12}\exp[i\zeta\varphi_1(x)]\circledast_{12}^{-1}$$

$$= \exp\Big\{i\zeta\varphi_1(x) + i\zeta\int d^4x' D_{(-)}(x-x')\frac{\delta}{\delta\varphi_2(x')}\Big\}, \circledast_{12}\exp[i\zeta\varphi_2(x)]\circledast_{12}^{-1}$$

$$= \exp\Big\{i\zeta\varphi_2(x) + i\zeta\int d^4x' D_{(+)}(x-x')\frac{\delta}{\delta\varphi_1(x')}\Big\}, \quad T_1 \exp[i\zeta\varphi_2(x)]T_1^{-1}$$

$$= \exp\{i\zeta\varphi_2(x)\},$$

$$T_2^+ \exp[i\zeta\varphi_1(x)]\,(T_2^+)^{-1} = \exp[i\zeta\varphi_1(x)].$$

By means of these relations, one can write formula (2.146) in the form

$$S_{t_2t_1}[\varphi]\circledast S^+_{t_2t_1}[\varphi] = 1 + i\int_{t_1}^{t_2} d^4x \int d\sigma(\zeta)\Big\{\exp[i\zeta\varphi_1(x)]\exp\Big[i\zeta\int d^4x' \times$$

$$\times\Big(D_c(x-x')\frac{\delta}{\delta\varphi_1(x')} + D_{(-)}(x-x')\frac{\delta}{\delta\varphi_2(x')}\Big)\Big] -$$

$$- \exp[i\zeta\varphi_2(x)] \times \tag{2.147}$$

$$\times \exp\Big[i\zeta\int d^4x'\Big(D_{(+)}(x-x')\frac{\delta}{\delta\varphi_1(x')} +$$

$$+ D_c^*(x-x')\frac{\delta}{\delta\varphi_2(x')}\Big)\Big]\Big\}E_{t_2t_0}[\varphi_1,\varphi_2]|_{\varphi_1=\varphi_2=\varphi}.$$

Now we note that the argument $x_0 - x_0' < 0$ for all the functions $D_c(x)$, $D_c^*(x)$, $D_{(-)}(x)$ and $D_{(+)}(x)$ in the curley brackets, since x_0' belongs to the interval $[x_0, t_2]$. Therefore, according to (2.135), we have

$$D_{(-)}(x-x') = D_c^*(x-x'), \qquad D_{(+)}(x-x') = D_c(x-x').$$

Formula (2.147) takes the form

$$S_{t_2t_1}[\varphi] \circledast S^+_{t_2t_1}[\varphi] = 1 + i\int_{t_1}^{t_2} d^4x \int d\sigma(\zeta)\{\exp[i\zeta\varphi_1(x)] - \exp[i\zeta\varphi_2(x)]\} \times$$

$$\times \exp\left\{i\zeta \int d^4x' \left[D_c(x-x')\frac{\delta}{\delta\varphi_1(x')} + \right.\right.$$

$$\left.\left. + D_c^*(x-x')\frac{\delta}{\delta\varphi_2(x')}\right]\right\} \times$$

$$\times S_{t_2x_0}[\varphi_1] \circledast S^+_{t_2x_0}[\varphi_2]|_{\varphi_1=\varphi_2=\varphi}.$$

On the right-hand side of this expression one can put $\varphi_1 = \varphi_2 = \varphi$, since all the operators depending on the variational derivatives stand to the right with respect to the square brackets. Hence, the whole expression is equal to zero, and finally, we have

$$S_{t_2t_1}[\varphi] \circledast S^+_{t_2t_1}[\varphi] = 1. \tag{2.148}$$

Notice again that the presence of the derivatives in the interaction Lagrangain does not change the correctness of the procedures undertaken. The fulfilment of relations (2.136) guarantees that in this case the unitarity is fulfilled.

2.5. The Schrödinger Equation in Quantum Field Theory with Nonlocal Interactions

2.5.1. *Introduction*

We have already noted above that our scheme of construction of QFT in the stochastic space-time ($\mathbb{R}^4(\hat{x})$ with the measure $w(b_E^2/l^2)$ lead to the quantum field theory with nonlocal interactions. In this case, both the fields averaged in $\mathbb{R}^4(\hat{x})$ and the interaction Lagrangian constructed by means of these fields became nonlocal quantities. Now our main proposal is that the Schrödinger equation, describing the dynamics of physical systems, must be modified according to the idea of nonlocality in nature resulting from the stochasticity of space-time at small distances. The present section is devoted to the investigation of this problem.

Thus, as shown above, the construction of a self-consistent theory of nonlocal interactions of quantized fields became possible owing to the following two ideas: First, the form-factors must be entire analytic functions in the momentum space, and must decrease rapidly enough in the Euclidean space. Second, the form-factor must be *quantized*, i.e., it is necessary to introduce supplementary degrees of freedom, which determine the regularization, in order to facilitate the transition to the Euclidean metric, and the reestablishment of the form-factor in the limit of the cancelled regularization. The development of these ideas allowed the construction of the finite and unitary S-matrix for arbitrary interaction Lagrangians in each perturbation order (see preceding sections and Efimov, 1977a).

What is incomplete in this construction? The following problem arose from investigations into causality conditions. The coefficient functions of the S-matrix in the configuration space turn out to be analytical functions. Analytic methods used in the study of local properties of analytical functionals make it impossible, in principle, to determine a space localization of studied functionals within an accuracy of a certain distance given by the nonlocality (Fainberg and Soloviev, 1978; Soloviev, 1980). Obviously, the use of non-analytical methods is needed. However, these methods are not developed. The results obtained by the projecting sequences of functions (Efimov, 1977a) give rise to doubts because, as is shown in the paper of Fainberg and Soloviev (1977), there are examples of explicitly nonlocal functionals which are, as local ones, characterized by the projecting sequences of functions. Therefore, the existence of the microcausality condition, understood as a strict equality of the corresponding functional outside the light cone, remains an open problem in the theory of nonlocal interactions.

On the other hand, *causality* is nothing other than the correctness of the *Cauchy problem* of Schrödinger quantum-field equation (or the equation of Tomanaga and Schwinger). However, the utilization of the regularization procedure in the construction of the S-matrix both in the local and nonlocal theories is reduced in that the S-matrix is not a solution of the corresponding equation, and is determined by a series of limits. It seems therefore that the natural properties of the Schrödinger equation solutions defined as unitarity and causality are to be proved separately.

Difficulties in nonlocal theory arise usually when a nonlocal form-factor is introduced into the interaction Lagrangian, but the Schrödinger field equation (or Tomonaga–Schwinger equation) remains local

$$i\frac{\delta}{\delta\sigma(x)}\Psi[\sigma(x)] = \mathscr{H}_{\rm in}(x)\Psi[\sigma(x)], \tag{2.149}$$

where

$$\mathscr{H}_{\rm in}(x) = g\left[\int d^4x' K(x-x')\varphi(x')\right]^4.$$

The integrability conditions are violated within this approach and numerous other difficulties arise (see Markov, 1958, for example).

We assume that while introducing nonlocality in the interaction Hamiltonian the equation of Tomonaga and Schwinger must be treated also *nonlocally but with retardation.* In this way, in our opinion, one can get rid of difficulties caused by the integrability conditions in the nonlocal theory.

In this section we shall show that the S-matrix describing nonlocal interactions of quantized fields solves the *Cauchy problem* of the *evolution equation* (or Schrödinger equation in the interaction picture at imaginary time, i.e., in the Euclidean metric) *with retardation.* In this way supplementary degrees of freedom with respect to the *Fock space* of physical particles need not be introduced.

Strictly speaking, the formulation of such an equation with the correctly stated Cauchy problem at *imaginary time* does not answer directly the question about the causality condition of the S-matrix in Minkowski space. However, a *simple analytical connection* between the S-matrices both in Euclidean and Minkowski spaces means without doubt that *causality of the evolution equation must ensure the absence of any physically observable noncausal phenomena.*

2.5.2. *The Field Operator at Imaginary Time*

We shall consider the theory of a one component scalar field $\varphi(x)$ describing particles with mass m. The field operator $\varphi(x)$ may be written in a standard way (see Bogolubov and Shirkov, 1980, for example)

$$\varphi(x) = \varphi(\mathbf{x}, t) = (2\pi)^{-3/2} \int \frac{d^3 k}{\sqrt{2\omega}} (a_{\mathbf{k}}\, e^{-i\omega t + i\mathbf{k}\mathbf{x}} + a_{\mathbf{k}}^{+}\, e^{i\omega t - i\mathbf{k}\mathbf{x}}), \tag{2.150}$$

where $\omega = (m^2 + \mathbf{k}^2)^{1/2}$. *Creation* $a_{\mathbf{k}}^{+}$ and *annihilation* $a_{\mathbf{k}}$ Boson operators satisfy ordinary *commutation rules*:

$$[a_{\mathbf{k}}, a_{\mathbf{k}'}]_{-} = [a_{\mathbf{k}}^{+}, a_{\mathbf{k}'}^{+}]_{-} = 0, \qquad [a_{\mathbf{k}}, a_{\mathbf{k}'}^{+}]_{-} = \delta^{(3)}(k - k'). \tag{2.151}$$

We assume that there exists a single vacuum state $\Psi_0 = |0\rangle$ which obeys the conditions:

$$\langle 0|0\rangle = 1, \qquad a_{\mathbf{k}}|0\rangle = 0 \quad \forall \mathbf{k}. \tag{2.152}$$

State vectors of scalar particles are represented by rays in the Fock space which is, as usual, constructed over the basis

$$\Psi^{(n)}_{\mathbf{k}_1 \cdots \mathbf{k}_n} = \frac{1}{\sqrt{n!}} a_{\mathbf{k}_1}^{+} \cdots a_{\mathbf{k}_n}^{+} |0\rangle, \tag{2.153}$$

where $n = 0, 1, 2, \ldots$

Now we pass to *imaginary time* $t \to -i\tau$, or to the Euclidean metric. In the constructive quantum field theory the physical Hilbert space F of a free field in *Minkowski space* is considered as a subspace of the Hilbert space N of a free Euclidean field (see Simon, 1974). Especially, in order to obtain Euclidean Green functions and to construct the scattering matrix (Petrina *et al.*, 1979), in creation and annihilation operators $a_{\mathbf{k}}^{+}$ and $a_{\mathbf{k}}$ the supplementary degree of freedom $a_{\mathbf{k}}^{+} \to a_{\mathbf{k},\varepsilon}^{+}$ ($a_{\mathbf{k}} \to a_{\mathbf{k},\varepsilon}$) connected with imaginary time is introduced so that commutation rules are of the form

$$[a_{\mathbf{k},\varepsilon}, a^+_{\mathbf{k}',\varepsilon'}]_- = \delta^{(3)}(k - k')\,\delta(\varepsilon - \varepsilon') = \delta^{(4)}(k_E - k'_E), \tag{2.154}$$

where $k_E = (\varepsilon, \mathbf{k})$.

Here we shall not enlarge the number of degrees of freedom of a scalar field and we shall construct a space of our Euclidean states over the same basis (2.153).

So, we introduce the free field $\phi(x_E)$, where $x_E = (\tau, \mathbf{x})$ in the Euclidean space, i.e., at imaginary time by the replacement $t \to -i\tau$ in expression (2.150) for the field operator $\varphi(\mathbf{x}, t)$:

$$\phi(x_E) = \phi(\tau, \mathbf{x}) = \varphi(-i\tau, \mathbf{x}) = (2\pi)^{-3/2} \int \frac{d^3 k}{\sqrt{2\omega}} \times$$

$$\times\, (a_k e^{-\omega\tau - i\mathbf{kx}} + a_{\mathbf{k}}^+ e^{\omega\tau + i\mathbf{kx}}). \tag{2.155}$$

Let us introduce the T-product operation, i.e., the imaginary time τ ordering operation. In representation (2.155) the parameter τ is introduced explicitly, so the T-product operation is defined straightforwardly:

$$T[\phi(x_{1E})\cdots\phi(x_{nE})] = \phi(x_{nE})\cdots\phi(x_{1E}) \quad (\tau_1 \leqslant \tau_2 \leqslant \tau_3 \leqslant \cdots \leqslant \tau_n). \tag{2.156}$$

Further, we shall define the two-point Euclidean Green function which will be called causal

$$\Delta_c(x_{1E} - x_{2E}) = \langle 0 | T\{\phi(x_{1E})\phi(x_{2E})\} | 0 \rangle$$

$$= (2\pi)^{-3} \int \frac{d^3 k}{2\omega} \exp[-\omega|\tau_1 - \tau_2| + i\mathbf{k}(\mathbf{x}_1 - \mathbf{x}_2)]$$

$$= (2\pi)^{-4} \int d^4 k_E \exp[ik_E(x_{1E} - x_{2E})](m^2 + k_E^2)^{-1}$$

$$= m(2\pi)^{-2} K_1(m[(x_{1E} - x_{2E})^2]^{1/2})[(x_{1E} - x_{2E})^2]^{-1/2}, \tag{2.157}$$

where $k_E = (\varepsilon, \mathbf{k})$, $k_E^2 = \varepsilon^2 + \mathbf{k}^2$ and $K_1(z)$ is the Mac'Donald function. The obtained function $\Delta_c(x_E)$ represents an analytical continuation to $t \to -i\tau$ of the causal Green function in Minkowski space:

$$\Delta_c(x - y) = i^{-1}(2\pi)^{-4} \int d^4 k \frac{e^{-ik(x-y)}}{m^2 - k^2 - i\varepsilon}.$$

Now let us consider the commutator of the fields $\phi(x_E)$ and the function

$\Delta_{(-)}(x - y)$:

$$[\phi(x_{1E}), \phi(x_{2E})]_- = (2\pi)^{-3} \int \frac{d^3 k}{2\omega} e^{ik(x_1 - x_2)} 2\,\mathrm{sh}[\omega(\tau_1 - \tau_2)],$$

$$\Delta_{(-)}(x_{1E} - x_{2E}) = \langle 0 | \phi(x_{1E}) \phi(x_{2E}) | 0 \rangle \tag{2.158}$$

$$= (2\pi)^{-3} \int \frac{d^3 k}{2\omega} e^{ik(x_1 - x_2) - \omega(\tau_1 - \tau_2)}.$$

We see easily that integrals on the right-hand side of formulas (2.158) do not exist under an arbitrary choice of the difference $(\tau_1 - \tau_2)$. Therefore, only the T-product of operators $\phi(x_E)$ has a reasonable mathematical meaning.

Let us introduce operator $R[\phi]$ of the following type:

$$R[\phi] = \sum_{n=0} \frac{1}{n!} \int d^4 x_{1E} \cdots \int d^4 x_{nE} R_n(x_{1E}, \ldots, x_{nE}) T[\phi(x_{1E}) \cdots \phi(x_{nE}]. \tag{2.159}$$

Operators $R[\phi]$ are defined by a set of functions $\{R_n(x_{1E}, \ldots, x_{nE})\}$ the properties of which will be elaborated below.

Let us define the operation of conjugation

$$[\phi(x_E)]^* = [\varphi^+(i\tau, x)]^+ = \varphi(-i\tau, x) = \phi(x_E). \tag{2.160}$$

Then for the operator R in (2.159) we obtain the following expression

$$R^*[\phi] = \sum_{n=0} \frac{1}{n!} \int d^4 x_{1E} \cdots \int d^4 x_{nE} R_n^*(x_{1E}, \ldots, x_{nE}) \times$$

$$\times T[\phi(x_{1E}) \cdots \phi(x_{nE})]. \tag{2.161}$$

Now we determine the operation of *multiplication* of two operators $R_1[\phi]$ and $R_2[\phi]$ of type (2.159) by definition:

$$R_1[\phi] R_2[\phi] \underset{\mathrm{Df}}{=} T\{R_1[\phi] R_2[\phi]\} = \sum_{n_1, n_2} \frac{1}{n_1! n_2!} \int d^4 x_{1E} \cdots \int d^4 x_{n_1 E} \times$$

$$\times \int d^4 y_{1E} \cdots \int d^4 y_{n_2 E} R_{n_1}(x_{1E}, \ldots, x_{n_1 E}) R_{n_2}(y_{1E}, \ldots, y_{n_2 E}) \times$$

$$\times T[\phi(x_{1E}) \cdots \phi(x_{n_1 E}) \phi(y_{1E}) \cdots \phi(y_{n_2 E})]. \tag{2.162}$$

2.5.3. *The State Space at Imaginary Time*

Let us define the state space N of the system as a set of vectors of the type

$$\Psi = R[\phi] | 0 \rangle \tag{2.163}$$

where R has the form (2.159). It will be assumed that the state (2.163) is given if all the functions $R_n(x_{1E}, \ldots, x_{nE})$ determining the operator $R[\phi]$ in (2.159) are known.

Notice that any state in the Fock space F must be represented as shown in (2.163) in the case of the corresponding choice of the set of functions $\{R_n(x_{1E}, \ldots, x_{nE})\}$ in the operator $R[\phi]$ since there exists a simple linear connection between sets of basic vectors:

$$\left\{\begin{array}{r} |0\rangle \\ a^+_{\mathbf{k}_1}|0\rangle \\ a^+_{\mathbf{k}_1} a^+_{\mathbf{k}_2}|0\rangle \\ a^+_{k_1} \quad \cdots \quad a^+_{\mathbf{k}_3}|0\rangle \\ a_{\mathbf{k}_1} \quad \cdots \quad a^+_{\mathbf{k}_n}|0\rangle \end{array}\right\} \rightleftarrows \left\{\begin{array}{r} |0\rangle \\ \phi(x_{1E})|0\rangle \\ T[\phi(x_{1E})\phi(x_{2E})]|0\rangle \\ T[\phi(x_{1E}) \quad \cdots \quad \phi(x_{3E})]|0\rangle \\ T[\phi(x_{1E}) \quad \cdots \quad \phi(x_{nE})]|0\rangle \end{array}\right\}. \tag{2.164}$$

The appropriate formula of this correspondence can be easily obtained, if necessary (Appendix A).

The most important here is the following: the *Fock space F* consists of vectors of the type

$$\Psi_F = \sum_n \frac{1}{n!} \int d^3 k_1 \cdots \int d^3 k_n f_n(\mathbf{k}_1, \ldots, \mathbf{k}_n) \frac{1}{\sqrt{n!}} a^+_{\mathbf{k}_1} \cdots a^+_{\mathbf{k}_n}|O\rangle, \tag{2.165}$$

but the space N of vectors

$$\Psi = \sum_n \frac{1}{n!} \int d^4 x_{1E} \cdots \int d^4 x_{nE} R_n(x_{1E}, \ldots, x_{nE}) T[\phi(x_{1E}) \cdots \phi(x_{nE})]|0\rangle. \tag{2.166}$$

Obviously, from representation (2.165) and (2.166) it follows that the space F is a subspace of N, since some set of mutually different vectors [different sets of functions $R_n(\cdots)$] from N corresponds to each vector from F.

In the scattering problem the initial state of the type (2.165) is given at some $\tau_0 = -T$, where $T \to \infty$. In this case such a state can be written in the form

$$\Psi(\tau_0) = \sum_n \frac{1}{n!} \int d^3 x_1 \cdots \times$$

$$\times \int d^3 x_n r_n(\mathbf{x}_1, \ldots, \mathbf{x}_n) T[\phi(\tau_0, \mathbf{x}_1) \cdots \phi(\tau_0, \mathbf{x}_n)]|0\rangle, \tag{2.167}$$

where $r_n(\mathbf{x}_1, \ldots, \mathbf{x}_n)$ is connected with $f_n(\mathbf{k}_1, \ldots, \mathbf{k}_n)$ in (2.165). The state (2.167) can also be rewritten in the form (2.166), where

$$R_n(x_{1E}, \ldots, x_{nE}) = r_n(\mathbf{x}_1, \ldots, \mathbf{x}_n) \prod_{j=1}^{n} \delta(\tau_j - \tau_0).$$

The norm of state vectors from N is given by definition:

$$\begin{aligned}\|\Psi\|^2 &= (\Psi, \Psi) = \langle 0|R^*[\phi]R[\phi]|0\rangle \\ &= \sum_{n_1, n_2} \frac{1}{n_1! n_2!} \int d^4 x_{1E} \cdots \int d^4 x_{n_1 E} \times \\ &\times \int d^4 y_{1E} \cdots \int d^4 y_{n_2 E} R^*_{n_1}(x_{1E}, \ldots, x_{n_1 E}) R_{n_2}(y_{1E}, \ldots, y_{n_2 E}) \times \\ &\times \langle 0|T[\phi(x_{1E}) \cdots \phi(x_{n_1 E}) \phi(y_{1E}) \cdots \phi(y_{n_2 E})]|0\rangle. \end{aligned} \tag{2.168}$$

It is known [see formula (7.38) of Chapter 7 and Simon, 1974, and Efimov, 1979] that for a free Euclidean field there exists such a *Gaussian positive measure* $d\mu_f$ that

$$\int d\mu_f f(x_{1E}) \cdots f(x_{nE}) = \langle 0|T[\phi(x_{1E}) \cdots \phi(x_{nE})]|0\rangle. \tag{2.169}$$

Then the Euclidean norm (2.168) is

$$\|\Psi\|^2 = \int d\mu_f |R[f]|^2 < \infty, \tag{2.170}$$

where the functional $R[f]$ is given by

$$R[f] = \sum_n \frac{1}{n!} \int d^4 x_{1E} \cdots \int d^4 x_{nE} R_n(x_{1E}, \ldots, x_{nE}) f(x_{1E}) \cdots f(x_{nE}).$$

Therefore, functions $R_n(x_{1E}, \ldots, x_{nE})$ must be such that the norm (2.170) is finite.

So, we have constructed the state space N, which includes all vectors of type (2.163) for which the norm (2.168) [or (2.170)] is finite. Furthermore, the operators $R[\phi]$ (2.159) have been introduced for which the operations of conjugation (2.161) and multiplication (2.162) are defined.

2.5.4. *The Interaction Hamiltonian and the Evolution Equation*

The dynamics of a quantum field system is described by the Schrödinger equation. In the *interaction picture* it has the form

$$i \frac{\partial}{\partial t} \Psi(t) = H_{\text{in}}(t) \Psi(t) \tag{2.171}$$

with some initial condition

$$\Psi(t_0) = \Psi_0. \tag{2.172}$$

Passing to imaginary time $t \to -i\tau$ we obtain

$$\frac{\partial}{\partial \tau}\Psi(\tau) = -H_{\text{in}}(\tau)\Psi(\tau), \qquad \Psi(\tau_0) = \Psi_0. \tag{2.173}$$

It is customary to call equations (2.173) *evolution equations.*

In quantum field theory the local interaction Hamiltonian are usually of the form of polynomials in the field operators. For example, for the self-interacting scalar field we have

$$H_{\text{in}}(t) = g\int \mathrm{d}^3x\varphi^4(t, \mathbf{x}), \tag{2.174}$$

or in the imaginary time formulation

$$H_{\text{in}}(\tau) = g\int \mathrm{d}^3x\phi^4(\tau, \mathbf{x}). \tag{2.175}$$

If we look for a solution of equation (2.173) with the Hamiltonian (2.175) in the form (2.163), then we have the standard problem of the local quantum field theory with all its difficulties.

Now we consider the quantum field theory with the nonlocal interaction. We introduce the extended-nonlocal field (see Chapter 1)

$$\phi_n(\mathrm{x}) = \phi_n(\tau, \mathbf{x}) = \int \mathrm{d}^4y_{\mathrm{E}}\, w(y_{\mathrm{E}}^2)\phi(\tau + y_4, \mathbf{x} + \mathbf{y}), \tag{2.176}$$

where $y_{\mathrm{E}} = (y_4, \mathbf{y})$, $y_{\mathrm{E}}^2 = y_4^2 + \mathbf{y}^2$ and $w(y_{\mathrm{E}}^2)$ is a real function-*measure*, the properties of which have been discussed in Chapter 1.

Let us calculate the *causal Green function* of the extended field $\phi_n(x_{\mathrm{E}})$ in the Euclidean space. According to formula (1.13) we have

$$D_c(x_{1\mathrm{E}} - x_{2\mathrm{E}}) = (2\pi)^{-4}\int \mathrm{d}^4p_{\mathrm{E}}\, \mathrm{e}^{-ip_{\mathrm{E}}(x_{1\mathrm{E}} - x_{2\mathrm{E}})}[\tilde{K}(-p_{\mathrm{E}}^2)]^2(m^2 + p_{\mathrm{E}}^2)^{-1}, \tag{2.177}$$

where

$$\tilde{K}(-p_{\mathrm{E}}^2) = \int \mathrm{d}^4y_{\mathrm{E}}\, w(y_{\mathrm{E}}^2)\exp(-ip_{\mathrm{E}}y_{\mathrm{E}}). \tag{2.178}$$

Here for the sake of simplicity we omit the index l for the functions $K(\cdot)$ and $w(\cdot)$. As usually, we shall assume that the function $w(y_{\mathrm{E}}^2)$ is chosen so that the $\tilde{K}(-p_{\mathrm{E}}^2)$ is an entire analytic function in the complex p^2-plane and that it decreases so rapidly as $p_{\mathrm{E}}^2 \to \infty$ that

$$D_c(0) = (2\pi)^{-4}\int \mathrm{d}^4p_{\mathrm{E}}[\tilde{K}(-p_{\mathrm{E}}^2)]^2(m^2 + p_{\mathrm{E}}^2)^{-1} < \infty. \tag{2.179}$$

Instead of the interaction (2.175) we now write

$$H_{\text{in}}[\tau,\phi] = g\int \mathrm{d}^3x\, T[\phi_n^4(\tau,\mathbf{x})]$$

$$= g\int \mathrm{d}^3x \prod_{j=1}^{4}\int \mathrm{d}^4 y_{j\mathrm{E}}\, w(y_{j\mathrm{E}}^2)\, T\left\{\prod_{n=1}^{4}\phi(\tau+\tau_n,\mathbf{x}+\mathbf{y}_n)\right\}. \quad (2.180)$$

Therefore, the interaction Hamiltonian (2.180) belongs to the class of operators $R[\phi]$ (2.159). Notice that the interaction Hamiltonian may be chosen in the normal form:

$$H_{\text{in}}[\tau,\phi] = g\int \mathrm{d}^3x\, {:}T[\phi_n^4(\tau,\mathbf{x})]{:}$$

$$= g\int \mathrm{d}^3x\, T[\phi_n^4(\tau,\mathbf{x}) - 6\phi_n^2(\tau,\mathbf{x})D_c(0) + 3D_c^2(0)], \quad (2.181)$$

where $D_c(0)$ is given by (2.179). In this case

$$\langle 0|{:}H_{\text{in}}[\tau,\phi]{:}|0\rangle = 0.$$

Using the representation of the state vector $\Psi(\tau)$ (2.163) we shall write the evolution equation (2.173) for the nonlocal interaction (2.180) or (2.181) in the following form:

$$\frac{\partial}{\partial\tau}R[\tau,\phi]|0\rangle = -T\{H_{\text{in}}[\tau,\phi]R[\tau,\phi]\}|0\rangle. \quad (2.182)$$

The operation of multiplication on the right-hand side of (2.182) is defined by (2.162). The initial condition for equation (2.182) is

$$\Psi(\tau_0) = R[\tau_0,\phi]|0\rangle = R_0[\phi]|0\rangle. \quad (2.183)$$

So, the obtained evolution equation is *retarded*, since the 'time' τ in the nonlocal Hamiltonian (2.180) or (2.181) may *precede* the times in field operators $R[\tau,\phi]$. However, the T-ordering operation in equation (2.182) arranges the times appropriately.

Equation (2.182) with the initial condition (2.183) may be rewritten as

$$\frac{\partial}{\partial\tau}R[\tau,\phi] = -T\{H_{\text{in}}[\tau,\phi]R[\tau,\phi]\}, \quad (2.184)$$

$$R[\tau_0,\phi] = R_0[\phi]. \quad (2.185)$$

Notice that if the solution (2.184) is written in the form $R(\tau,\tau_0)$, it does not satisfy the condition

$$R(\tau,\tau_0) = R(\tau,\tau_1)R(\tau_1,\tau_0) \quad (\tau_0<\tau_1<\tau),$$

where multiplication is understood in the ordinary sense. This is an immediate consequence of the *nonlocality* of the theory.

Nevertheless, the Cauchy problem for equation (2.184) can be formulated. If we introduce the space cut-offs $g \to g(\mathbf{x})$ in order to eliminate the difficulties caused by the *Haag theorem*, the Hamiltonian $H_{\text{in}}[\tau, \phi]$ represents an operator in the considered space N. We have

$$H_{\text{in}}[\tau, \phi] = \int d^3x g(\mathbf{x}) : T\{\phi_n^4(\tau, \mathbf{x})\}. \tag{2.186}$$

Considering the norm of the state

$$\Psi = H_{\text{in}}[\tau, \phi]|0\rangle$$

we obtain

$$||\Psi||^2 = \langle 0|T\{H_{\text{in}}[\tau, \phi]H_{\text{in}}[\tau, \phi]\}|0\rangle = \int d^3x_1 \int d^3x_2 g(\mathbf{x}_1)g(\mathbf{x}_2) \times$$

$$\times\, 6D_c^2(x_1 - x_2) \leqslant 6D_c^2(0)\left[\int d^3xg(\mathbf{x})\right]^2 < \infty.$$

Therefore, the ultraviolet catastrophe is absent in the considered nonlocal theory.

Let us turn now to the solution of equation (2.184) with the initial condition (2.185). The solution is given by

$$R[\tau, \phi] = T\left\{\exp\left[-\int_{\tau_0}^{\tau} d\tau' H_{\text{in}}[\tau', \phi]\right]R_0[\phi]\right\}. \tag{2.187}$$

For the state vector we obtain

$$\Psi(\tau) = R[\tau, \phi]|0\rangle. \tag{2.188}$$

The norm of state $\Psi(\tau)$ (2.188) equals

$$||\Psi(\tau)||^2 = \int d\mu_f \exp\left\{-2\int_{\tau_0}^{\tau} d\tau' H_{\text{in}}[\tau', f]\right\}|R_0[f]|^2. \tag{2.189}$$

If the Hamiltonian $H_{\text{in}}[\tau, f]$ in (2.189) is chosen in the form (2.186) then according to (2.181) we have

$$H_{\text{in}}[\tau, f] \geqslant -6D_c^2(0) \int d^3xg(\mathbf{x}).$$

For the norm (2.189) we have

$$||\Psi(\tau)|| \leqslant ||\Psi_0|| \exp\left\{6(\tau - \tau_0)D_c^2(0) \int d^3xg(\mathbf{x})\right\}. \tag{2.190}$$

Thus, we see that the solution exists in the state space N. The uniqueness of the obtained solution follows from (2.187), which produces the S-matrix that coincides completely with the S-matrix constructed and investigated in the preceding sections.

2.5.5. *Appendix A*

Before obtaining formulas (2.164) [or (2.165) and (2.166)] we give some definitions.

DEFINITION. Functions $R_n(x_1, \ldots, x_n)$ $(n = 1, 2, \ldots)$ belong to a class $G(n)$ if their Fourier transforms

$\tilde{R}_n(k_1, \ldots, k_n) = F_n(\mathbf{k}_1, \ldots, \mathbf{k}_n)g_n(k_{10}, \ldots, k_{n0})$ satisfy the following conditions (E):

(1) $F_n(\mathbf{k}_1, \ldots, \mathbf{k}_n) = f_n(\mathbf{k}_1, \ldots, \mathbf{k}_n) \prod_{i=1}^{n} \sqrt{2\omega_i}, \quad \omega_i = (m^2 + \mathbf{k}_i^2)^{1/2},$

(2) $g_n(\omega_1, \ldots, \omega_n) = 1,$

(3) $g_n^* (k_{10}, \ldots, k_{n0})g(k_{10}, \ldots, k_{n0}) = 1,$

(4) $R_n(x_{10}, \ldots, x_{n0}) = \int \mathrm{d}k_{10} \ldots \int \mathrm{d}k_{n0} \exp\left\{i \sum_{j=1}^{n} k_{j0}x_{j0}\right\} g_n(k_{10}, \ldots, k_{n0})$

is a regular function with respect to all variables x_{i0}.

If Ψ is the state vector of the single particle system, then among the components $R_n(x_1, \ldots, x_n)$ in the *Fock space* only the amplitude $R_1(x)$ differs from zero. Thus

$$\Psi_1 = R_1[\varphi]\Psi_0 = \int \mathrm{d}^4x R_1(x) T\varphi(x)|0\rangle$$

and

$$\Psi_1^+ \underset{\mathrm{Df}}{=} \int \mathrm{d}^4x R_1^*(x)\langle 0| T\varphi(x).$$

Let $R_1(x) \in G(1)$, for example, be the function

$$\tilde{R}_1(\mathbf{k}, k_0) = \sqrt{2\omega_{\mathbf{k}}}\, f_1(\mathbf{k}) \exp[i\beta(k_0^2 - \omega_{\mathbf{k}}^2)], \tag{A.1}$$

$\beta > 0$, $f_1(\mathbf{k}) \in L_2(\mathbb{R}^3)$, satisfies all conditions (E). Taking into account that $\varphi(x) = \varphi^+(x) + \varphi^-(x)$, where

$$\varphi^\pm(x) = \int_{k_0 > 0} \mathrm{d}^4k \exp(\pm ikx)\, \delta(k^2 - m^2)\tilde{\varphi}(\pm k),$$

$$a^+(k) = \tilde{\varphi}(k)|_{k_0 > 0} \quad \text{and} \quad a(k) = \tilde{\varphi}(-k)|_{k_0 > 0},$$

and according to (A.1) we have

$$\Psi_1 = \int d^4x R_1(x)\varphi(x)|0\rangle = \int d^3k f_1(\mathbf{k}) a^*(\mathbf{k})|0\rangle$$

here

$$a^*(\mathbf{k}) = a^+(\mathbf{k}, k_0)(2\omega_{\mathbf{k}})^{-1/2}|_{k_0 = \omega_{\mathbf{k}}}.$$

The norm of the vector Ψ_1 is given by

$$||\Psi_1||^2 = \Psi_1^+ \circledast \Psi_1 = \iint d^4x\, d^4y R_1^*(x) R_1(y)\langle 0|T[\varphi(x)\varphi(y)]|0\rangle.$$

In the momentum representation this formula takes the form

$$||\Psi_1||^2 = 2\int d^3k |f_1(\mathbf{k})|^2\, \omega_{\mathbf{k}} i^{-1} \int_{-\infty}^{\infty} dk_0 \frac{g^*(k_0)g(k_0)}{m^2 + \mathbf{k}^2 - k_0^2 - i\varepsilon}. \tag{A.2}$$

According to conditions (E) the last integral in (A.2) has the form

$$i^{-1}\int_{-\infty}^{\infty} dk_0 \frac{g^*(k_0)g(k_0)}{m^2 + \mathbf{k}^2 - k_0^2 - i\varepsilon} = i^{-1}\int dk_0(\omega_{\mathbf{k}}^2 - k_0^2 - i\varepsilon)^{-1} = \frac{\pi}{\omega_{\mathbf{k}}}$$

and therefore

$$||\Psi_1||^2 = (2\pi)\int d^3k |f_1(\mathbf{k})|^2.$$

Similarly, the two-particle state is given by

$$\Psi_2 = \frac{1}{\sqrt{2!}}\left\{\iint d^4x_1\, d^4x_2 R_2(x_1, x_2) T[\varphi(x_1)\varphi(x_2)] - C_2\right\}|0\rangle \equiv R_2|0\rangle,$$

where the Fourier transform of the function $R_2(x_1, x_2) \subset G(2)$ has the form

$$\tilde{R}_2(k_1, k_2) = f_2(\mathbf{k}_1, \mathbf{k}_2)(2\omega_{\mathbf{k}_1})^{-1/2}(2\omega_{\mathbf{k}_2})^{-1/2} g(k_{10})g(k_{20})$$

and the constant C_2 is chosen as follows:

$$C_2 = i^{-1}\iint d^4x_1\, d^4x_2 R_2(x_1, x_2)\,\Delta_c(x_1 - x_2) = \int d^3p f_2(\mathbf{p}, -\mathbf{p})\omega_{\mathbf{p}} \cdot Q,$$

here

$$Q = i^{-1}\int_{-\infty}^{\infty} dp_0 g(p_0) g(-p_0)(\omega_p^2 - p_0^2 - i\varepsilon)^{-1}.$$

The choice of the function $g(p_0)$ being such that the value $|Q| < \infty$, in particular, for $g(k_0) = \exp[-it(k_0 - \omega)]$ (t is some constant parameter), one can easily verify that

$$Q = \exp(2it\, \omega_{\mathbf{p}})(\pi/\omega_{\mathbf{p}}).$$

Then

$$\Psi_2 = \frac{1}{\sqrt{2!}} \iint d^3k_1 \, d^3k_2 \, f_2(\mathbf{k}_1, \mathbf{k}_2) a^*(\mathbf{k}_1) a^*(\mathbf{k}_2) |0\rangle$$

and the norm of the vector Ψ_2 is defined as

$$||\Psi_2||^2 = \langle 0 | R_2^+ \circledast R_2 | 0 \rangle, \tag{A.3}$$

where

$$R_2^+ = \left\{ \iint d^4x_1 \, d^4x_2 R_2^*(x_1, x_2) T[\varphi(x_1)\varphi(x_2)] - C_2^+ \right\} 2^{-1/2}$$

and

$$C_2^+ = i^{-1} \iint d^4x_1 \, d^4x_2 R_2^*(x_1, x_2) \, \Delta_c(x_1 - x_2).$$

The symbol $\circledast$ denotes the following operation

$$T[\varphi(x_1) \cdots \varphi(x_j)] \circledast T[\varphi(y_1) \cdots \varphi(y_i)] \equiv T[\varphi(x_1) \cdots \varphi(x_j)\varphi(y_1) \cdots \varphi(y_i)]$$

Then, the norm of the vector Ψ_2 defined by the equality (A.3) becomes the following

$$||\Psi_2||^2 = i^{-2} \iint d^4x_1 \, d^4x_2 \iint d^4x_3 \, d^4x_4 R_2^*(x_1, x_2) R_2(x_3, x_4) \times$$

$$\times \Delta_c(x_1 - x_3) \, \Delta_c(x_2 - x_4) = (2\pi)^{-2} \iint d^3k_1 \, d^3k_2 \, |f(\mathbf{k}_1, \mathbf{k}_2)|^2.$$

Similar calculations can be carried out for an arbitrary N-particle ($N = 3, 4, \ldots$) state vector of the system. Thus, any normalized state of the system in the Fock space F can be represented in the form (2.163) by means of the operator $R[\phi]$ if its coefficient functions belong to the class $G(n)$.

Chapter 3

Electromagnetic Interactions in Stochastic Space-Time

3.1. Introduction

Quantum electrodynamics (QED) is a beautiful theory describing the dynamics of photon and fermion interactions (electrons, muons etc.) with enormous accuracy in a very wide range of physical quantities (describing for example phenomena occurring in a region extending from very large length scales (5×10^{10} cm) to those of the microworld (10^{-15}–10^{-16} cm)). Up to now no effect has been found experimentally which could not be confined within the local QED.

The tremendous quantitative success of QED yields a fundamental criterion for the construction of new physical models based on different assumptions and hypothesis concerning the space-time structure at small distances (particularly, a possible violation of locality in QED and the existence of fundamental length).

Tests of locality are usually performed by the modification of the particle propagators and vertex functions. In particular, tests of QED are carried out in terms of the modified electron and photon propagators (see Kraus, 1975; Magg *et al.*, 1972; and Ringhofer and Salecker, 1980). On the other hand, it is well-known that the introduction of nonlocality into the theory leads to a change of particle propagators (Chapters 1 and 2). As shown above, however, this change is not arbitrary and is determined by fundamental principles of QFT: Lorentz covariance, finiteness, unitarity, causality and gauge invariance. Roughly speaking, the aim of the nonlocal theory is to find restrictions on the choice of a form for nonlocal particle propagators (or the so-called form-factors of the theory).

Thus, an analysis of experimental data for testing locality must be performed within a theory which satisfies the basic principles of QFT. The problem of constructing the nonlocal theory for the scalar field satisfying the above-mentioned basic principles has been solved successfully (see Chapter 2 and Efimov, 1977a).

According to the method presented in Chapter 1, both neutral and charged fields are extended in our scheme over the large scale of space-time. The propagators of these extended-nonlocal fields are given by formulas of the type (1.13). However the change of charged particle propagators (in particular, in the momentum representation)

$$S(\hat{p}) \Rightarrow S_R(\hat{p}) = V(-p^2 l^2) S(\hat{p})$$

essentially complicates the proof of the gauge invariance of the theory.

This chapter is devoted to the construction of gauge-invariant QED of particles with spin 0, $\frac{1}{2}$ and 1 in the stochastic space-time $\mathbb{R}^4(\hat{\mathbf{x}})$ considered above. We introduce here a definite algebraical prescription which ensures the gauge invariance of QED obtained in the perturbation theory. Further, the electromagnetic processes are investigated and contributions to the anomalous magnetic moments of leptons and to the Lamb shift due to nonlocality (or stochasticity) are estimated. Here the considerations are mainly concerned with the low-energy processes. Notice that at very high energies testing of locality of the theory may be difficult because of *interference effects between the electromagnetic and weak interactions*. For example, in the standard model of electroweak interactions, the testing of QED will be disturbed by interference from the weak effects due to Z^0 bosons.

3.2. Gauge Invariance of the Theory and Generalization of Kroll's Procedure

The concept of the gauge invariance in QED is crucial, within which the interaction of fermion fields with photons is formed. Moreover, modern theory (including unified theories of all type forces) of elementary particle interactions is constructed in analogy with QED. A particularly striking example is the electroweak theory of electromagnetic and weak interactions due to A. Salam (1968, 1980); S. Weinberg (1967, 1980); and Sh. Glashow (1961, 1980).

We start this section with a short discussion concerning the method of introducing the interaction Lagrangian between charged particles and photons. As an example, consider the field of electrons $\psi(x)$. The free Lagrangian of this field has the standard form

$$\mathscr{L}_0(x) = \bar{\psi}(i\hat{\partial} - m)\psi, \tag{3.1}$$

where

$$\hat{\partial} = \gamma_\nu \partial_\nu, \qquad \partial_\nu = \partial/\partial x_\nu = (\partial/\partial x_0, \partial/\partial \mathbf{x}),$$

$\gamma_\nu = (\gamma_0, \gamma)$ are the Dirac γ-matrices and m is the mass of the electron. Lagrangian $\mathscr{L}_0(x)$ is invariant with respect to the *global gauge transformation*

$$\psi(x) \to \psi'(x) = e^{i\lambda}\psi(x). \tag{3.2}$$

Here λ is an arbitrary real constant. It is obvious that the Lagrangian (3.1) is not invariant with respect to the *local gauge transformation*

$$\psi(x) \to \psi'(x) = U(x)\psi(x), \tag{3.3}$$

where $U(x) = \exp[i\lambda(x)]$ and $\lambda(x)$ is an arbitrary real function of x. This is due to the fact that the derivative $\partial_\nu\psi(x)$ is not transformed under (3.3) as the field $\psi(x)$ itself. Indeed, we have

$$\partial_\nu\psi'(x) = U(x)(\partial_\nu + i\,\partial_\nu\lambda(x))\psi(x).$$

It is well known that the local gauge invariance (3.3) can be maintained, provided the *interaction* of the field $\psi(x)$ with the electromagnetic field A_ν is introduced. Consider the quantity $(\partial_\nu - ieA_\nu)\psi$ (e is the electron charge). We have

$$(\partial_\nu - ieA_\nu(x))\psi(x) = U^{-1}(x)(\partial_\nu - ieA'_\nu(x))\psi'(x), \tag{3.4}$$

where

$$A'_\nu(x) = A_\nu(x) - \frac{1}{e}\partial_\nu\lambda(x). \tag{3.5}$$

From (3.4) it is obvious that the Lagrangian which follows from (3.1) by the substitution

$$\partial_\nu\psi \to (\partial_\nu - ieA_\nu)\psi \tag{3.6}$$

is invariant with respect to the gauge transformations (3.3) and (3.5). To construct the complete Lagrangian of the system under consideration, we have to add the gauge invariant Lagrangian of the electromagnetic field. The tensor of the electromagnetic field is given as

$$F_{\mu\nu} = \partial_\mu A_\nu - \partial_\nu A_\mu. \tag{3.7}$$

Clearly

$$F'_{\mu\nu} = F_{\mu\nu}.$$

Consequently, the gauge invariant Lagrangian of the fields of electrons and photons takes the form

$$\mathcal{L}_{\text{tot}} = i\bar{\psi}\gamma_\nu(\partial_\nu - ieA_\nu)\psi - m\bar{\psi}\psi - \tfrac{1}{4}F_{\mu\nu}F_{\mu\nu}. \tag{3.8}$$

The substitution of the derivative $\partial_\nu\psi$ by the covariant derivative $(\partial_\nu - ieA_\nu)\psi$ in the free Lagrangian of the field $\psi(x)$ leads to the following interaction Lagrangian of electrons and photons:

$$\mathcal{L}_{\text{in}}(x) = ej_\nu A_\nu, \tag{3.9}$$

where $j_\nu = \bar{\psi}\gamma_\nu\psi$ is the electromagnetic current. Thus, the substitution (3.6) fixes uniquely the form of the interaction Lagrangian. Such an interaction is called a *minimal electromagnetic interaction.* In this connection, we notice that the principle of gauge invariance alone does not fix the interaction Lagrangian uniquely. For example, the addition of the Pauli term $\mu \cdot \bar{\psi}\sigma_{\mu\nu}\psi F_{\mu\nu}$ to the Lagrangian (3.8) does not mar the gauge invariance of the theory. Here $\sigma_{\mu\nu} = \frac{1}{2}(1/i)(\gamma_\mu \gamma_\nu - \gamma_\nu \gamma_\mu)$ and μ is the anomalous magnetic moment of the electron (see Section 3.4).

All available experimental data confirm that the Lagrangian (3.9) is, indeed, the true Lagrangian which governs the interactions of electrons and photons in the local theory. It is also well known that electrodynamics with its minimal interaction (3.9) is the renormalizable theory.

The application of the above-mentioned idea to the non-Abelian *Yang–Mills theory* based on the gauge SU(2) × U(1) invariance has been made in the construction of the unified theory of electromagnetic and weak interactions. The exposition of this theory transcends the bounds of this book and its fundamentals may be found in any monographs and textbooks devoted to recent developments in quantum field theory (see, for example, Lai, ed., 1983; Chaichian, and Nelipa, 1984).

In the language of the perturbation theory (or the Feynman diagrammatical techniques) the *gauge invariance* of QED means that *every matrix elements of the S-matrix* defining the concrete electromagnetic processes have a definite structure, and algebraical relations exist between them. For example, in the momentum representation, the so-called vacuum polarization diagram in the second order of the perturbation theory, the following form (see below) exists:

$$\tilde{\Pi}_{\mu\nu}(q^2) = (g_{\mu\nu}q^2 - q_\mu q_\nu)\tilde{\Pi}(q^2) \tag{3.10}$$

as well as the following relation between the vertex function $\tilde{\Gamma}_\mu(p, q)$ and the self-energy of the electron $\tilde{\Sigma}(p)$:

$$\frac{\partial \tilde{\Sigma}(p)}{\partial p_\mu} = -\tilde{\Gamma}_\mu(p, q)|_{q=0}. \tag{3.11}$$

The latter relation is called the *Ward–Takahashi identity*. The explicit forms of these functions may be written (for detail, see Bogolubov and Shirkov, 1980):

$$\tilde{\Sigma}(p) = \frac{-ie^2}{(2\pi)^4}\int d^4k\, \Delta(k^2)\gamma_\mu S(\hat{p} - \hat{k})\gamma_\mu \tag{3.12}$$

and

$$\tilde{\Gamma}_\mu(p, q) = \frac{ie^2}{(2\pi)^4}\int d^4k\, \Delta((p - k)^2)\gamma_\nu S(\hat{q} + \hat{k})\gamma_\mu S(\hat{k})\gamma_\nu, \tag{3.13}$$

and the identity (3.11) proved where $S(\hat{p}) = (m - \hat{p})^{-1}$, $\Delta(k^2) = (-k^2 - i\varepsilon)^{-1}$. For the proof of relation (3.11) consider the identity

$$\frac{\partial S(\hat{p})}{\partial p_\mu} = S(\hat{p})\gamma_\mu S(\hat{p}), \tag{3.14}$$

where the vertex γ_μ is given by

$$\gamma_\mu = -\frac{\partial}{\partial p_\mu} S^{-1}(\hat{p}). \tag{3.15}$$

Further, differentiating $\tilde{\Sigma}(p)$ over p_μ and making use of the identity (3.14), we have

$$\frac{\partial \tilde{\Sigma}(p)}{\partial p_\mu} = \frac{-ie^2}{(2\pi)^4} \int d^4k\, \Delta(k^2)\gamma_\nu S(\hat{p} - \hat{k})\gamma_\mu S(\hat{p} - \hat{k})\gamma_\nu. \tag{3.16}$$

Choosing other momentum variables in (3.13) and assuming $q = 0$, $p' = p + q = p$, we get

$$\tilde{\Gamma}_\mu(p, 0) = \frac{ie^2}{(2\pi)^4} \int d^4k\, \Delta(k^2)\gamma_\nu S(\hat{p} - \hat{k})\gamma_\mu S(\hat{p} - \hat{k})\gamma_\nu.$$

Comparing this expression with (3.16) we obtain the Ward–Takahashi identity (3.11). The relations of the type (see Section 4.2):

$$q_\mu \tilde{\Gamma}_\mu(p, q)|_{p'^2 = p^2 = m^2} = 0$$

$$\cdots$$

follows from the following identity

$$q_\mu\{S(\hat{p}_1)\gamma_\mu S(\hat{p}_2)\} = S(\hat{p}_1) - S(\hat{p}_2)$$

if $q = p_1 - p_2$.

It is important to notice that the vertex γ_μ in any Feynman diagram is connected with the propagator $S(\hat{p})$ of the charged particle by relation (3.15). It implies that if the particle propagator $S(\hat{p})$ is changed, it should be to modify the vertex $\gamma_\mu \to u_\mu(\cdot)$ at the same time in accordance with formula (3.15).

Now we discuss the gauge invariance of QED for the case of the nonlocal theory. Thus, the hypothesis of the stochastic space-time leads to a change of propagators of both neutral and charged particles. It is well known that any modification of the local causal Green functions of charged particles results in a violation of some algebraical relations, as mentioned above, and the fulfilment of which grants the gauge invariance of the theory. There are numerous papers (Kroll, 1966; Kraus, 1975; Magg *et al.*, 1972; and Ringhofer and Salecker, 1980) devoted to this problem. Among these, Kroll's work plays an important role. We generalize here his procedure, the essence of which consists of the following (Dineykhan and Namsrai, 1977):

(1) To satisfy the conditions of gauge invariance for the modified theory (with the changing propagators of the charged particles) one must change the form of the *one-photon vertex* (for example, in the case of spinor QED):

$$\gamma_\mu \to u_\mu(q, k) = -d_\mu(k)S_R^{-1}(\hat{q}) \tag{3.17}$$

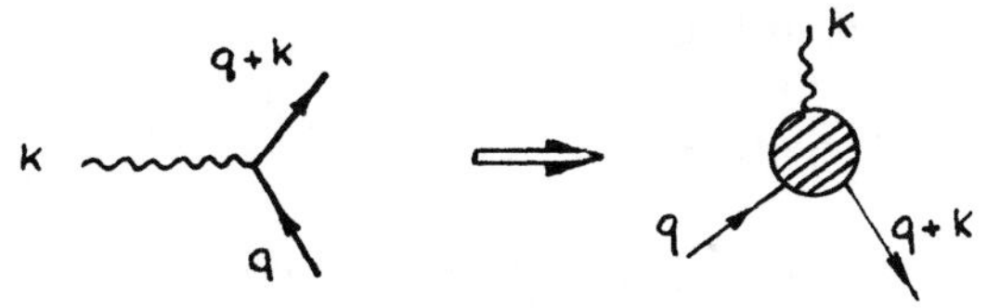

Fig. 3. The illustration of the change of the one-photon vertex in stochastic (nonlocal) QED.

(Figure 3) due to the *Ward–Takahashi identity*

$$k_\mu \tilde{\Gamma}_\mu(p, q) = S_R(\hat{p}) - S_R(\hat{q}), \tag{3.18}$$

where

$$\tilde{\Gamma}_\mu(p, q) = S_R(\hat{p}) u_\mu(k, q) S_R(\hat{q}) \quad (p = k + q). \tag{3.19}$$

Here $\mathrm{d}_\mu(k)$ is some operator whose action on the entire functions is determined below.

(2) Any theory with the modified propagators and the vertex functions contains the minimal number of the many-photon vertices $e^n u_n$ satisfying the condition

$$u_n(q; k_1, \ldots, k_n) = -\mathrm{d}(k_n) u_{n-1}(q; k_1, \ldots, k_{n-1}) \tag{3.20}$$

with $u_0 = S_R^{-1}$. If S_R^{-1} and u_1 are polynomial functions, the minimal number of necessary multi-photon vertices equals $(n_0 - 1)$, where n_0 is a number for which the relation (3.20) turns to zero. Otherwise, S_R^{-1} and u_1 would be nonpolynomial functions, the number n_0 being infinite.

(3) The interactions of n-photons with the open charged propagators is given by the following formula (see Figure 4)

$$S_R(\hat{q}_n) \tilde{\Gamma}_n(q; k_1, \ldots, k_n) S_R(\hat{q}) = (-1)^n \mathrm{d}(k_1) \cdots \mathrm{d}(k_n) S_R(\hat{q}), \tag{3.21}$$

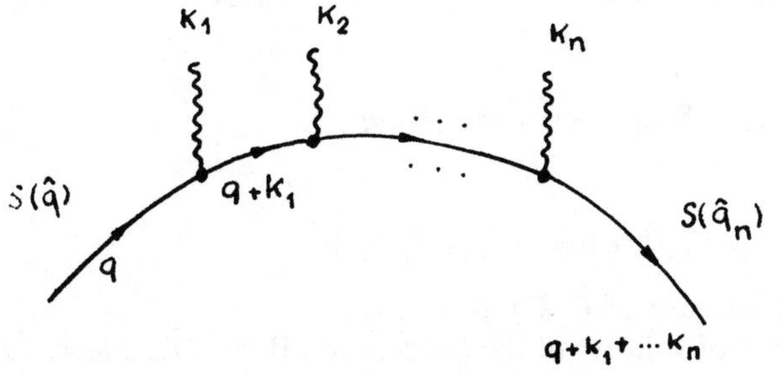

Fig. 4. Interactions of n-photons with open charged propagators.

where $\tilde{\Gamma}_n$ and u_n are related by

$$\tilde{\Gamma}_n(q; k_1, \ldots, k_n) = u_n(q; k_1, \ldots, k_n) + \mathscr{P}_n \sum_{j=1}^{n-1} \frac{u_j(q_j; k_{j+1}, \ldots, k_n)}{j!(n-j)!} \times$$

$$\times S_R(\hat{q}_j)\tilde{\Gamma}_{n-j}(q; k_1, \ldots, k_j),$$

$$u_1 = \tilde{\Gamma}_1, \quad q_j = k_j + q_{j-1}, \quad q_1 = k_1 + q, \quad q_n = \sum_{i=1}^{n} k_i.$$

Here the symbol $\mathscr{P}_n$ means summation over all permutations of the variables $k_1, \ldots, k_n$.

(4) Now we shall determine the *d-operation for entire functions*: Acting on the identity

$$\hat{q}V(-q^2l^2) = V(-q^2l^2)\hat{q}$$

by the d-operation and making use of its definition for polynomial function (Kroll, 1966), we obtain

$$\mathrm{d}_\mu(k)V(-q^2l^2) = [V(-(q+k)^2l^2) - V(-q^2l^2)]\frac{\hat{k}\gamma_\mu}{k^2}. \tag{3.22}$$

(5) The d-operation for the inverse of entire functions. Acting on the identity

$$V^{-1}(-q^2l^2)\cdot V(-q^2l^2) = 1.$$

by the d-operation we get

$$\mathrm{d}_\mu(k)V^{-1}(-q^2l^2) = -V^{-1}(-(q+k)^2l^2)[\mathrm{d}_\mu(k)V(-q^2l^2)]V^{-1}(-q^2l^2). \tag{3.23}$$

(6) Calculation of $\mathrm{d}_\mu(k)S_R(\hat{q})$ where $S_R(\hat{q})$ is the modified propagator of charged particles. From the identity

$$\mathrm{d}_\mu(k)\,[S_R(\hat{q})S_R^{-1}(\hat{q})] = 0,$$

it follows that

$$\mathrm{d}_\mu(k)S_R(\hat{q}) = -S_R\,(\hat{q} + \hat{k})[\mathrm{d}_\mu(k)S_R{}^{-1}(\hat{q})]S_R(\hat{q}) \tag{3.24}$$

or

$$\mathrm{d}_\mu(k)S_R(\hat{q}) = S_R(\hat{q} + \hat{k})\tilde{\Gamma}_{1\mu}(k, q)S_R(\hat{q}),$$

where

$$\tilde{\Gamma}_{1\mu}(k, q) = u_{1\mu}(k, q) = -\mathrm{d}_\mu(k)S_R^{-1}(\hat{q}).$$

It is a particular case of (3.21) at $n = 1$.

(7) The proof of validity of the *generalized Ward–Takahashi identity:*

$$(p_\mu - q_\mu)\tilde{\Gamma}_\mu(p, q) = S_R(\hat{p}) - S_R(\hat{q}), \tag{3.25}$$

where

$$\tilde{\Gamma}_\mu(p, q) = S_R(\hat{p}) u_\mu(k, q) S_R(\hat{q}), \quad k = p - q.$$

Taking into account the relation

$$u_\mu(k, q) = \gamma_\mu V^{-1}(-q^2 l^2) + (m - \hat{q} - \hat{k}) V^{-1}(-p^2 l^2) \times \\ \times [\mathrm{d}_\mu(k) V(-q^2 l^2)] V^{-1}(-q^2 l^2) \tag{3.26}$$

and equation (3.22) we obtain after some calculations identity (3.25).

(8) *Charged closed loop*: first of all notice that due to Kroll (1966) the closed loop in the local theory is given by

$$\tilde{\Pi}_n(k_1, \ldots, k_n) = \frac{1}{n} \int \mathrm{d}^4 q \, \mathrm{Sp}\{\tilde{\Gamma}_n(q; k_1, \ldots, k_n) S_R(\hat{q})\},$$

where

$$S_R(\hat{q}_n) \tilde{\Gamma}_n(q; k_1, \ldots, k_n) S_R(\hat{q}) = (-1)^n \mathrm{d}(k_1)\, \mathrm{d}(k_2) \cdots \mathrm{d}(k_n) S_R(\hat{q}).$$

A generalization of this equation to the modified theory with entire form factors represents no difficulty and the charged loop is determined by the following expression

$$\tilde{\Pi}_n^R(k_1, \ldots, k_n) = \frac{1}{n} \int \mathrm{d}^4 q \, \mathrm{Sp}\{\tilde{\Gamma}_n^R(q; k_1, \ldots, k_n) S_R(\hat{q})\}, \tag{3.27}$$

where

$$\tilde{\Gamma}_n^R(q; k_1, \ldots, k_n) = V(-q_n^2 l^2) \tilde{\Gamma}_n(q; k_1, \ldots, k_n),$$

$$S_R(\hat{q}_n) \tilde{\Gamma}_n(q; k_1, \ldots, k_n) S_R(\hat{q}) = \mathscr{P}_n \sum_{j=1}^{n} \frac{1}{j!(n-j)!} S_R(\hat{q}_n) \times \\ \times u_j(q_j; k_{j+1}, \ldots, k_n) S_R(\hat{q}_j) \times \tilde{\Gamma}_{n-j}(q; k_1, \ldots, k_j) S_R(\hat{q}) \\ = (-1)^n \, \mathrm{d}(k_1) \cdots \mathrm{d}(k_n) S_R(\hat{q}),$$

with $\Gamma_0 = S_R^{-1}(q)$. Tensor indexes are omitted here.

Thus we have generalized Kroll's prescription for our case and obtained the necessary algebraic relations which provide gauge invariance for the S-matrix in any order of the perturbation series. Investigations of gauge invariance of the S-matrix for concrete interactions will be given below.

3.3. The Interaction Lagrangian and the Construction of the S-Matrix

According to the above deduction (Chapter 1), we must construct all physical quantities (for example, interaction Lagrangian, causal Green function, etc.) by means of the nonlocal fields $\varphi_R(x)$ which are associated with the fields $\varphi(\hat{x})$ by formula (1.8). The *causal Green function* of the fields $\varphi_R(x)$ is determined by expression (1.13):

$$D_c^R(x-y) = \langle 0|T\{\varphi_R(x)\varphi_R(y)\}|0\rangle_{R^4(\hat{x})}$$

in the physical configuration space, i.e., in the space of a large scale, where

$$D_c^R(x) = i^{-1}(2\pi)^{-4}\int d^4q\, e^{iqx}\, V(-q^2l^2)\,\tilde{\Delta}_c(q) \tag{3.28}$$

is the Efimov nonlocal Green function if $V(-z) = |K(z)|^2$ is an entire function and $\tilde{\Delta}_c(q)$ is the Fourier transform of the local Green function (see Chapter 2).

The Lagrangian of a system of fields is constructed in terms of the averaged fields $\varphi_R(x) = \langle\varphi(\hat{x})\rangle_{R^4(\hat{x})}$ in the Minkowski space. Thus the initial Lagrangian describing the electromagnetic and four-fermion weak interactions of leptons is chosen in the form

$$\begin{aligned}
\mathscr{L}(x) &= \mathscr{L}_0(x) + \mathscr{L}_{em}(x) + \mathscr{L}_w(x),\\
\mathscr{L}_0(x) &= -\tfrac{1}{2}\!:[\partial_\nu A_\mu(x)][\partial_\nu A_\mu(x)]\!: + \sum_j :\bar{\psi}_j(x)(i\hat{\partial} - m_j)\psi_j(x):,\\
\mathscr{L}_{em}(x) &= e\!:\bar{\psi}_R(x)\hat{A}_R(x)\psi_R(x)\!:,\\
\mathscr{L}_w(x) &= \frac{G}{\sqrt{2}}\!:[\bar{\psi}_R(x)0_\mu\nu_R(x)][\bar{\nu}_R(x)0_\mu\psi_R(x)]\!:,
\end{aligned} \tag{3.29}$$

where $A_\nu^R(x)$ and $\psi_j^R(x)$, $\nu_R(x)$ are the nonlocal fields of photon and leptons. The summation in the expression of $\mathscr{L}_0(x)$ extends over all considered fermion fields $(j = e, \mu, \nu_e, \nu_\mu)$.

Formally, the S-matrix can be written in the form of the T-products:

$$\begin{aligned}
S &= 1 + i\sum_{n=1}^{\infty}\frac{1}{n!}S_n,\\
S_n &= i^{n-1}\int d^4x_1\cdots\int d^4x_n T_d\left\{\prod_{j=1}^{n}[\mathscr{L}_{em}(x_j) + \mathscr{L}_w(x_j)]\right\}.
\end{aligned} \tag{3.30}$$

Here the symbol T_d means the so-called Wick T-product or T^*-operation (see Chapter 2, and Bogolubov and Shirkov, 1980 and Efimov, 1977a) and the lower case d corresponds to the algebraic prescription determined in Section 3.2.

In order to construct the perturbation series for the S-matrix (3.30) by prescription of the usual local theory, it is necessary to change (in the Feynman diagrams):

$$\frac{m+\hat{k}}{m^2-k^2-i\varepsilon} \to \frac{m+\hat{k}}{m^2-k^2-i\varepsilon}V_m(-k^2l^2), \qquad \frac{\hat{k}}{-k^2-i\varepsilon} \to \frac{\hat{k}}{-k^2-i\varepsilon}V_0(-k^2l^2),$$

$$g_{\mu\nu}(-k^2-i\varepsilon)^{-1} \to g_{\mu\nu}V_0(-k^2l^2)(-k^2-i\varepsilon)^{-1},$$

and at the same time to insert the modified vertex (3.17) into the vertices at the *external photon lines*. The calculation of the matrix elements for the charged lepton loops will be undertaken using formulas (3.27) and (3.21).

3.4. Construction of a Perturbation Series for the S-Matrix in Quantum Electrodynamics

The construction of a perturbation series for the S-matrix is possible only within the framework of an intermediate regularization procedure. We shall here use the *regularization procedure* presented in Section 2.4.4. The regularization introduced there makes it possible to pass to the Euclidean metric in any diagrams of the perturbation theory. We recall that the form factors $V(-q^2 l^2)$ decrease only in the Euclidean direction, i.e., when $q^2 \to -\infty$. Therefore we shall investigate the Feynman diagrams in the Euclidean momentum space. At the end of all calculations it is necessary to remove this intermediate regularization, i.e., to pass to the limit $\delta \to 0$. Of course, final results obtained in such a way do not depend on a concrete form of regularization.

Let us calculate the matrix elements for the S-matrix corresponding to the following primitive diagrams (see Figure 5) which are divergent in the usual local quantum electrodynamics (Akhiezer and Berestetskii, 1965, 1981).

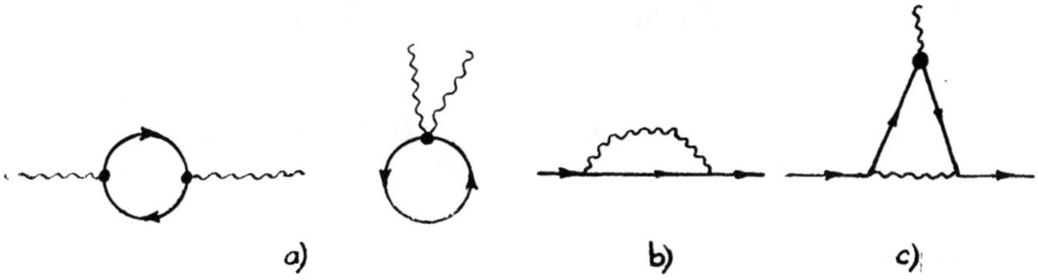

Fig. 5. The primitive Feynman diagrams in stochastic (nonlocal) QED.

3.4.1. *The Diagrams of Vacuum Polarization*

In the gauge invariant stochastic theory the vacuum polarization in the second order of the perturbation theory is determined by diagrams sketched in Figure 5a. In the momentum representation the term of the S-matrix which corresponds to these diagrams is given by an expression of type (3.27):

$$\Pi^R_{\mu\nu}(k_1, k_2) = \lim_{\delta \to 0} \frac{ie^2}{2(2\pi)^4} \int d^4q V^\delta_m(-q^2 l^2)\, \mathrm{Sp}\{\tilde{\Gamma}^\delta_{\mu\nu}(q; k_1, k_2) S^\delta_R(\hat{q})\}, \quad (3.31)$$

where $k_1 + k_2 = 0$,

$$S_R(\hat{q}_2)\tilde{\Gamma}_{\mu\nu}(q; k_1, k_2) S_R(\hat{q}) = (-1)^2\, d_\mu(k_1) d_\nu(k_2) S_R(\hat{q}) \quad (q_2 = q + k_1 + k_2 = q).$$

Equation (3.31) is simplified by the d-operation determined in Section 3.2. The regularization procedure δ guarantees the possibility of passing to the Euclidean metric. Thus making use of the definition of the d-operation for the entire functions and taking the trace of γ-matrices:

$$\mathrm{Sp}\,\gamma_\mu\gamma_\nu = 4g_{\mu\nu},$$

$$\mathrm{Sp}\,\gamma_\mu\gamma_\nu\gamma_\delta\gamma_\rho = 4g_{\mu\nu}g_{\delta\rho} + 4g_{\mu\rho}g_{\nu\delta} - 4g_{\mu\delta}g_{\nu\rho},$$

we obtain

$$\tilde{\Pi}^R_{\mu\nu}(k) = \lim_{\delta\to 0}\frac{ie^2}{(2\pi)^4}\int d^4q\,\mathrm{Sp}\left\{\gamma_\mu\frac{V^\delta(-q^2l^2)}{m-\hat{q}_1}\gamma_\nu\frac{1}{m-\hat{q}} + \gamma_\mu\frac{1}{m-\hat{q}_1}\times\right.$$

$$\left.\times\,[d_\nu(k)V^\delta(-q^2l^2)] + \tfrac{1}{2}\,d_\nu(k)\,d_\mu(-k)V^\delta(-q^2l^2)\right\}$$

$$= \tilde{\Pi}^{(1)}_{\mu\nu}(k) + \tilde{\Pi}^{(2)}_{\mu\nu}(k) + \tilde{\Pi}^{(3)}_{\mu\nu}(k),$$

where $q_1 = q + k$,

$$\tilde{\Pi}^{(1)}_{\mu\nu}(k) = 4\rho(1-\zeta)\int_0^1 dx(1-x)^{-\zeta}\int d^4q\,\mathscr{L}^{\zeta-2}\times$$

$$\times\,[2q_\mu q_\nu + 2k_\mu q_\nu + g_{\mu\nu}(m^2 - q^2 - (kq))],$$

$$\tilde{\Pi}^{(2)}_{\mu\nu}(k) = \frac{4\rho}{k^2}\int d^4q[q_\mu k_\nu - q_\nu k_\mu + g_{\mu\nu}(k^2 + (kq))]\times$$

$$\times\left\{(m^2 - q_1^2 - i\varepsilon)^{\zeta-1} + \zeta\int_0^1 dx\,(1-x)^{-1-\zeta}\mathscr{L}^{\zeta-1}\right\},$$

$$\tilde{\Pi}^{(3)}_{\mu\nu}(k) = \frac{\rho}{k^4}(2k_\mu k_\nu - k^2 g_{\mu\nu})\times$$

$$\times\int d^4q[(m^2 - q^2 - i\varepsilon)^\zeta - (m^2 - q_1^2 - i\varepsilon)^\zeta].$$

Here we have used the representation for the form-factor (Chapter 2)

$$V^\delta_m(-q^2l^2) = \frac{1}{2i}\int_{-\beta+i\infty}^{-\beta-i\infty} d\zeta\,\frac{v(\zeta)}{\sin\pi\zeta}\,l^{2\zeta}\exp(\delta\zeta^2)(m^2 - q^2 - i\varepsilon)^\zeta \tag{3.32}$$

and the notation

$$\mathscr{L} = m^2 - q^2 - (kq)2x - k^2x,\qquad \rho = \frac{ie^2}{(2\pi)^4}\frac{1}{2i}\int_{\substack{-\beta+i\infty\\(0<\beta<1)}}^{-\beta-i\infty} d\zeta\frac{v(\zeta)}{\sin\pi\zeta}\,l^{2\zeta}.$$

Integrating over d^4q and going to the Euclidean metric we obtain in the limit $\delta \to 0$

$$\tilde{\Pi}^R_{\mu\nu}(k) = \frac{e^2}{2\pi^2}(k_\mu k_\nu - k^2 g_{\mu\nu})\frac{1}{2i}\int_{\substack{-\beta+i\infty \\ (0<\beta<1)}}^{-\beta-i\infty} d\zeta \frac{v(\zeta)}{\sin \pi\zeta} \times$$

$$\times (m^2 l^2)^\zeta \int_0^1 dx x(1-x)^{1-\zeta} \frac{\Gamma(-\zeta)}{\Gamma(1-\zeta)} \mathscr{L}_0^\zeta,$$

where

$$\mathscr{L}_0 = 1 - \frac{k^2}{m^2}(1-x)x.$$

Assuming $m^2 l^2 \ll 1$ we obtain

$$\tilde{\Pi}^R_{\mu\nu}(k) = \frac{e^2}{2\pi^2}(k_\mu k_\nu - k^2 g_{\mu\nu}) \int_0^1 dx(1-x)x \times$$

$$\times \left\{ \ln \frac{\mathscr{L}_0(x)}{x(1-x)} - \frac{5}{6} + v'(0) + \ln m^2 l^2 + O(m^2 l^2) \right\}.$$

We see that after the charge renormalization the value obtained for the vacuum polarization coincides with the renormalized expression in the usual local theory (see, for example, Bogolubov and Shirkov, 1980).

3.4.2. *The Diagram of Self-Energy* (Figure 5b)

The corresponding term in the S-matrix can be written in the form

$$-i{:}\, \bar{\psi}(x)\Sigma_R(x-y)\psi(y){:},$$

where

$$\Sigma_R(x-y) = (2\pi)^{-2} \int d^4p\, e^{ip(x-y)}\, \tilde{\Sigma}_R(p).$$

Here

$$\tilde{\Sigma}_R(p) = \lim_{\delta \to 0} \frac{-ie^2}{(2\pi)^4} \int d^4k \frac{V_0^\delta(-k^2 l^2)}{-k^2 - i\varepsilon} \gamma_\mu \frac{m + \hat{p} - \hat{k}}{m^2 - (p-k)^2 - i\varepsilon} \gamma_\mu V_m^\delta(-(p-k)^2 l^2).$$

We can use the representation (3.32) for the form factors $V_m^\delta(z)$ and $V_0^\delta(z)$ and the *general Feynman parametric formula*

$$b_1^{-n_1} b_2^{-n_2} \cdots b_j^{-n_j} = \frac{\Gamma(n_1 + \cdots + n_j)}{\Gamma(n_1)\cdots\Gamma(n_j)} \int_0^1 dx_1 \cdots \int_0^1 dx_j\, \delta\left(1 - \sum_{i=1}^{j} x_i\right) \times$$

$$\times x_1^{n_1-1} \cdots x_j^{n_j-1} \left[\sum_{i=1}^{j} x_i b_i\right]^{-n_1-\cdots-n_j} \tag{3.33}$$

Then, after some calculations we arrive at:

$$\tilde{\Sigma}_R(p) = \frac{e^2}{8\pi^2}\frac{1}{2i}\int\limits_{\substack{-\beta+i\infty \\ (0<\beta<1)}}^{-\beta-i\infty} d\zeta \frac{v(\zeta)}{\sin \pi\zeta}(m^2 l^2)^{\zeta}\frac{1}{2i}\int\limits_{\substack{-\gamma+i\infty \\ (0<\gamma<1)}}^{-\gamma-i\infty} d\eta \frac{v(\eta)}{\sin \pi\eta}(m^2 l^2)^{\eta} \times$$

$$\times \frac{\Gamma(-\zeta-\eta)}{\Gamma(1-\zeta)\Gamma(1-\eta)} \times$$

$$\times \int_0^1 dx(1-x)^{\zeta}x^{-\zeta}\left(1-\frac{p^2}{m^2}x\right)^{\zeta+\eta}(2m-\hat{p}x). \tag{3.34}$$

Assuming the value of $m^2 l^2$ to be small, one can obtain for the self-energy the following expression:

$$\tilde{\Sigma}_R(p) = \frac{e^2}{8\pi^2}\int_0^1 dx(2m-\hat{p}x)\ln\left[\frac{m^2}{m^2-p^2x}\right] + \frac{e^2 m}{16\pi^2}\left\{\left[3\ln(m^{-2}l^{-2}) - \right.\right.$$

$$\left.\left. - 3v'(0) - 1 + \frac{\pi}{2i}\int\limits_{\substack{-\beta+i\infty \\ (0<\beta<1)}}^{-\beta-i\infty} d\zeta \frac{v(\zeta)v(-\zeta)}{\sin^2 \pi\zeta}(3-\zeta)\right] + O(m^2 l^2)\right\} + \frac{e^2}{16\pi^2}(m-\hat{p}) \times$$

$$\times\left\{\left[\ln(m^{-2}l^{-2}) - v'(0) + 1 + \frac{\pi}{2i}\int_{-\beta+i\infty}^{-\beta-i\infty} d\zeta \frac{v(\zeta)v(-\zeta)}{\sin^2 \pi\zeta}(1-\zeta)\right] + O(m^2 l^2)\right\}. \tag{3.35}$$

The calculation of integrals of the type

$$\frac{\pi}{2i}\int_{-\beta+i\infty}^{-\beta-i\infty} d\zeta \frac{v(\zeta)v(-\zeta)}{\sin^2 \pi\zeta}(\ldots)$$

is simple for a concrete choice of the function $v(\zeta)$. For example, in the case of the form factor V_s (2.98) describing the extended electron as the uniformly charged sphere with radius l (Section 2.3) the first integral in (3.35) is of the form

$$\frac{\pi}{2i}\int_{-\beta+i\infty}^{-\beta-i\infty} d\zeta \frac{v_s(\zeta)v_s(-\zeta)}{\sin^2 \pi\zeta}(3-\zeta) = -\frac{23}{9} - \sum_{n=2}^{\infty}\frac{(n-3)}{n(4n^2-1)(n^2-1)} \approx -\frac{229}{90},$$

where $v_s(\zeta)$ is given by formula (2.99).

Thus the value obtained for the self-energy differs slightly from the value calculated in Efimov's (1972) nonlocal theory.

3.4.3. *The Vertex Diagram and the Corrections to the Anomalous Magnetic Moment (AMM) of Leptons and to the Lamb Shift*

Let us consider the vertex diagram shown in Figure 5c. In the momentum representation it has the following form

$$\tilde{\Gamma}^R_\mu(p,q) = i^{-1}e^2(2\pi)^{-4}\int d^4k D_c(-(p-k)^2 l^2)\gamma_\nu\, d_\mu(q) S_R(\hat{k})\gamma_\nu$$

where

$$\mathrm{d}_\mu(q)S_R(\hat{k}) = (m - \hat{k} - \hat{q})^{-1}\gamma_\mu V_m(-k^2 l^2)(m - \hat{k})^{-1} + \\ + (m - \hat{k} - \hat{q})^{-1}\{V_m(-(k+q)^2 l^2) - V_m(-k^2 l^2)\}\hat{q}\gamma_\mu q^{-2}.$$

The symbol δ for the intermediate regularization procedure is omitted here and below. By using the identity

$$a^n - b^n = n\int_b^a \mathrm{d}x\, x^{n-1} = n(a-b)\int_0^1 \mathrm{d}x[(a-b)x + b]^{n-1}, \tag{3.36}$$

the difference of the form factor values can be transformed to the form

$$V_m(-(k+q)^2 l^2) - V_m(-k^2 l^2) = \\ -[q^2 + 2(kq)]\frac{1}{2i}\int_{-\beta+i\infty}^{-\beta-i\infty} \mathrm{d}\zeta\frac{v(\zeta)}{\sin\pi\zeta}\cdot\zeta\cdot l^{2\zeta}\int_0^1 \mathrm{d}x\,\times \\ \times[m^2 - k^2 - 2x(kq) - q^2 x]^{\zeta-1}$$

which is convenient for calculations.

Carrying out the necessary estimates we obtain the usual form for the matrix element of the vertex functions between two free single-lepton states:

$$\tilde{\Gamma}^R_\mu(p,q) = \bar{u}_j(p')\left\{\gamma_\mu F_1(q^2) + \frac{i}{2m_j}\sigma_{\mu\nu}q^\nu F_2(q^2)\right\}u_j(p), \tag{3.37}$$

where

$$p' = q + p, \qquad \sigma_{\mu\nu} = \frac{1}{2i}(\gamma_\mu\gamma_\nu - \gamma_\nu\gamma_\mu),$$

$$F_1(q^2) = f_1(q^2) + f_2(q^2), \qquad F_2(q^2) = g_1(q^2) + g_2(q^2).$$

Here

$$f_1(q^2) = N(\zeta,\eta)\left\{\left[-2 + 8\beta - 2\beta^2 - 2\frac{q^2}{m_j^2}(1-\gamma)(1-\alpha)\right]\Gamma(1-\eta-\zeta)\,\times \right. \\ \left. \times\,\mathscr{L}_1^{-1+\eta+\zeta} - 2\Gamma(-\eta-\zeta)\mathscr{L}_1^{\eta+\zeta}\right\},$$

$$g_1(q^2) = 4N(\zeta,\eta)\beta(1-\beta)\Gamma(1-\eta-\zeta)\mathscr{L}_1^{-1+\eta+\zeta},$$

$$f_2(q^2) = N(\zeta,\eta)\eta\int_0^1 \mathrm{d}t\left\{\left[2\frac{q^2}{m_j^2} - 2\frac{q^2}{m_j^2}(\alpha + t\gamma)\right](1 - \beta - 2\alpha - 2t\gamma)\,\times \right. \\ \left. \times\,\Gamma(1-\eta-\zeta)\mathscr{L}_2^{-1+\eta+\zeta} - 2\Gamma(-\eta-\zeta)\mathscr{L}_2^{\zeta+\eta}\right\},$$

$$g_2(q^2) = 4N(\zeta, \eta)\eta \int_0^1 dt\beta(1 - \beta - 2\alpha - 2t\gamma)\Gamma(1 - \eta - \zeta)\mathscr{L}_2^{-1+\eta+\zeta},$$

$$N(\zeta, \eta) = \frac{e^2}{(2\pi)^4}\frac{\pi^2}{2i}\int_{-\delta+i\infty}^{-\delta-i\infty} d\zeta \frac{v(\zeta)}{\sin \pi\zeta}(m_j^2 l^2)^\zeta \frac{1}{2i}\int_{-\rho+i\infty}^{-\rho-i\infty} d\eta \frac{v(\eta)}{\sin \pi\eta} \times$$

$$\times (m_j^2 l^2)^\eta \frac{1}{\Gamma(1-\zeta)\Gamma(1-\eta)} \times$$

$$\times \iiint_0^1 d\alpha\, d\beta\, d\gamma \cdot \beta^{-\zeta}\gamma^{-\eta}\, \delta(1 - \alpha - \beta - \gamma),$$

$$\mathscr{L}_1(q^2) = (1-\beta)^2 - (q^2/m_j^2)\alpha\gamma, \quad m_j = (m_e, m_\mu);$$

$$\mathscr{L}_2(q^2) = \mathscr{L}_1(q^2) - (q^2/m_j^2)t\gamma(1 - t\gamma) + 2(q^2/m_j^2)t\alpha\gamma + (q^2/m_j^2)t\gamma\beta.$$

The first term of (3.37) in the limit $q^2 \to 0$ and with the assumption $m_j^2 l^2 \ll 1$ is

$$F_1(q^2) = -\frac{\alpha}{4\pi}\left\{\ln(m_j^{-2} l^{-2}) - 2\ln\left(\frac{m_j^2}{m_\gamma^2}\right) - v'(0) + \frac{9}{4} - m_j^2 l^2 v'(1) \times\right.$$

$$\left.\times\left[-1\tfrac{1}{3}v'(0) + 4\frac{v'(1)}{v(1)} - \frac{8}{3}\ln(m^{-2}l^{-2}) - \frac{20}{9}\right]\right\} -$$

$$-\frac{\alpha}{2\pi}\frac{q^2}{m_j^2}\left\{\frac{2}{3}\left[\ln\left(\frac{m_j}{m_\gamma}\right) - \frac{5}{8}\right] +\right.$$

$$\left. + v(1)m_j^2 l^2\left[\tfrac{1}{3}v'(0) + \frac{2}{3}\frac{v'(1)}{v(1)} + \ln(m_j^2 l^2) - \frac{3}{4}\right]\right\},$$

$$\alpha = e^2/4\pi$$

(m_γ is photon mass) and contains the terms corresponding to the charge renormalization of the leptons. The second ferm of (3.37) at $q^2 = 0$ contributes to the AMM of the leptons by

$$a_j = F_2(0) = \frac{4}{(2i)^2}\frac{\alpha}{2\pi}\int_{-\beta+i\infty}^{-\beta-i\infty} d\zeta \int_{-\gamma+i\infty}^{-\gamma-i\infty} d\eta \frac{v(\zeta)v(\eta)}{\sin \pi\zeta \sin \pi\eta}(m_j^2 l^2)^{\eta+\zeta} \times$$

$$(0 < \gamma, \beta < 1)$$

$$\times (1 - \eta)(1 - \zeta) \times$$

$$\times \frac{\Gamma(1 - \eta - \zeta)\Gamma(1 + 2\zeta + \eta)}{\Gamma(3 - \eta)\Gamma(3 + \eta + \zeta)}.$$

Assuming $m_j^2 l^2 \ll 1 (m_j = m_e, m_\mu)$ we obtain

$$a_j = \frac{\alpha}{2\pi}\left\{1 + m_j^2 l^2\left[-\frac{2}{3}\frac{\pi}{2i}\int_{-\beta+i\infty}^{-\beta-i\infty} d\zeta \frac{v(\zeta)v(1-\zeta)}{\sin^2 \pi\zeta}\zeta(1-\zeta) - \tfrac{2}{3}v(1)\right]\right\}$$

$$\approx \frac{\alpha}{2\pi}[1 - \tfrac{2}{3}m_j^2 l^2 v(1)]. \tag{3.38}$$

The present experimental values of lepton AMM (Van Dyck *et al.*, 1979, 1977 and Bailey *et al.*, 1979)

$$a_{\text{exp}}(e^-) = (1159652200 \pm 40)10^{-12}, \qquad a_{\text{exp}}(\mu) = (1165924 \pm 8.5)10^{-9}$$

are reliably confirmed by quantum electrodynamics (Calmet *et al.*, 1977 and Kinoshita, 1979). It is natural to suppose that the contributions calculated here should be of an order not greater than the experimental errors. This makes it possible to establish the following restrictions on the parameter l:

$$l \lesssim \begin{cases} 1.1 \times 10^{-14}\,\text{cm} & \text{for } V = V_1 \\ 2.0 \times 10^{-14}\,\text{cm} & \text{for } V = V_b \end{cases} \quad (e^-),$$

$$l \lesssim \begin{cases} 0.76 \times 10^{-15}\,\text{cm} & \text{for } V = V_1 \\ 1.40 \times 10^{-15}\,\text{cm} & \text{for } V = V_b \end{cases} \quad (\mu),$$

Here $V = V_1$ is given by

$$V_1 = \sin^4(l(m^2 - q^2)^{1/2})[l(m^2 - q^2)^{1/2}]^{-4}$$

$$= \frac{1}{2i}\int_{\substack{-\beta+i\infty \\ (0<\beta<1)}}^{-\beta-i\infty} d\zeta \frac{v_1(\zeta)}{\sin \pi\zeta} l^{2\zeta}(m^2 - q^2 - i\varepsilon)^\zeta,$$

$$v_1(\zeta) = 2^{3+2\zeta}(2^{2+2\zeta} - 1)/\Gamma(5 + 2\zeta) \tag{3.39}$$

and $V = V_b$ equals (1.14).

Similarly, for the level $n = 2$ of the hydrogen atom, the calculated shift $2S_{1/2} - 2P_{1/2}$ due to (3.37) is (see Efimov, 1977a and Faustov, 1972, for detail)

$$\Delta E_l(2S_{1/2} - 2P_{1/2}) = \alpha^2 Ry\{m^2 F_2'(0) - \tfrac{1}{2}F_2(0)\}$$

$$= -\frac{\alpha^3}{2\pi} Ry \cdot v(1) m_e^2 l^2 \left[\frac{1}{3}v'(0) + \frac{2}{3}\frac{v'(1)}{v(1)} + \ln(m_e^2 l^2) - \frac{13}{12}\right].$$

For the function $v_b(\zeta)$ determined by (1.15) this expression acquires the form

$$\Delta E_l(2S_{1/2} - 2P_{1/2}) = \frac{\alpha^3}{2\pi} Ry \frac{m_e^2 l^2}{5}[\ln(m_e^{-2} l^{-2}) + 4.15], \tag{3.40}$$

where

$$\alpha^3 Ry = \tfrac{1}{2} m_e \alpha^5 = 1.25 \times 10^3 \text{ MHz/sec}.$$

The observed shift of 1057.912 ± 0.011 MHz is well explained by QED (Brodsky and Drell, 1970, see also Scadron, 1980). Therefore $|\Delta E_l(2S_{1/2} - 2P_{1/2})| \lesssim 0.011$ MHz and substituting formula (3.40) into this inequality we get $l \lesssim 1.9 \times 10^{-13}$ cm.

In the conclusion of this subsection, we note that the *S-matrix* obtained is *gauge invariant*. Indeed, in the stochastic electrodynamics under consideration, the Ward identity

$$\frac{\partial}{\partial p_\mu} \tilde{\Sigma}_R(p) = -\tilde{\Gamma}^R_\mu(p, 0) \tag{3.41}$$

is valid, since this identity is a direct consequence of the identities (3.24) and (3.17). Since we must not subtract any infinite counterterms, no dangerous terms which can break the Ward identity (3.41) when formula (3.24) is valid will appear in the perturbation theory. The diagram of the vacuum polarization is gauge invariant due to our choice of the gauge invariant regularization procedure of Kroll (1966).

3.5. The Electrodynamics of Particles with Spins 0 and 1

3.5.1. *Introduction*

Now let us construct within the framework of our approach the gauge invariant quantum electrodynamics of particles with spins 0 and 1 on the basis of the first-order Duffin–Kemmer equation (see Efimov and Namsrai, 1975; Duffin, 1938; Kemmer, 1939; Kinoshita, 1950; Kinoshita and Namby, 1950).

We write the Duffin–Kemmer equation

$$(i\hat{\partial} - m)\psi(x) = 0, \quad (\hat{\partial} = \beta_\mu \partial_\mu), \tag{3.42}$$

with

$$\beta_\mu \beta_\lambda \beta_\nu + \beta_\nu \beta_\lambda \beta_\mu = \delta_{\mu\lambda} \beta_\nu + \delta_{\nu\lambda} \beta_\mu. \tag{3.43}$$

The field $\psi(x)$ consists of three irreducible fields $\psi = \{\psi_0, \psi^{(0)}, \psi^{(1)}\}$. The first of which is trivial $\psi_0 = 0$, while the other two are spin 0 and spin 1 fields, respectively. From (3.43), one can obtain

$$[\beta_i, 2\beta_0^2 - 1]_+ = 0; \qquad [\beta_i, 2\beta_0^2 - 1]_- = 0, \quad (i = 1, 2, 3).$$

In accordance with this, the adjoint field $\bar{\psi}(x) = \psi^+(x)\eta$ with $\eta = 2\beta_0^2 - 1$ obeys the equation $\bar{\psi}(x)(i\hat{\partial} + m) = 0$. The following operators

$$\Lambda(\partial) = i\hat{\partial} - m \quad \text{and} \quad \mathrm{d}(\partial) = i\hat{\partial} - \tfrac{1}{2}(\Box - m^2) + \frac{1}{m}(i\hat{\partial})^2$$

satisfy the relation

$$\Lambda(\partial)\, \mathrm{d}(\partial) = \mathrm{d}(\partial)\Lambda(\partial) = \Box - m^2, \tag{3.44}$$

from which we deduce

$$(\Box - m^2)\psi(x) = 0.$$

The main problem of the theory is to find such matrices β_μ which would grant that equation (3.42) describes a field $\psi(x)$ with definite spin s, namely $s = 0, 1$ and does not contain any unnecessary components. By studying two spinor basic representations we obtain here equations for a particle with spins 0 and 1. The wave functions in this representation can be written as $\Psi_{\alpha\beta}$, where α and β are the Dirac spinor indices. The matrices γ_μ connected with these indices denote $\gamma_\mu^{(1)}$ and $\gamma_\mu^{(2)}$, respectively. Under the Lorentz transformations

$$x'_\mu = x_\mu + \delta\varepsilon_\mu + \delta\omega_{\mu\nu}x_\nu, \qquad \Psi'(x') = [I + \tfrac{1}{2}S_{\mu\nu}\,\delta\omega_{\mu\nu}]\Psi(x)$$

the spin matrix $S_{\mu\nu}$ reads

$$S_{\mu\nu} = S_{\mu\nu}^{(1)} + S_{\mu\nu}^{(2)} = [\beta_\mu, \beta_\nu], \tag{3.45}$$

where $\delta\varepsilon_\mu$ is a translation by an infinitesimal small real vector, $\delta\omega_{\mu\nu} = -\delta\omega_{\nu\mu}$ is the antisymmetric matrix for an infinitesimal Lorentz transformation, and

$$S_{\mu\nu}^{(j)} = \tfrac{1}{4}[\gamma_\mu^{(j)}, \gamma_\nu^{(j)}], \qquad \beta_\mu = \tfrac{1}{2}[\gamma_\mu^{(1)} + \gamma_\mu^{(2)}], \quad (j = 1, 2).$$

It is easily proven that the matrices β_μ satisfy the Duffin–Kemmer commutation relations (3.43). As usual, the generators for infinitesimal translations and for 'rotations' are formed by the Hermitian operators $p_\mu = i\partial_\mu$ and the matrices $M_{\mu\nu} = S_{\mu\nu} + (x_\mu p_\nu - x_\nu p_\mu)$, respectively.

The scalar operators

$$p^2 = p_\mu p_\mu, \qquad W = W_\mu W_\mu = \tfrac{1}{2}S_{\mu\nu}S_{\mu\nu}p^2 - S_{\mu\sigma}S_{\nu\sigma}p_\mu p_\nu, \tag{3.46}$$

commute with all the generators $M_{\mu\nu}$ and p_μ. They are therefore invariants of the group and they are multiples of the identity in any irreducible representation of the inhomogeneous Lorentz group, and their eigenvalues can be used for the classification of irreducible representations.

The eigenvalues of the operator p^2 equal m^2, where m is the mass of a particle. The operator W is an invariant of the Lorentz group and therefore it is convenient to consider it in the rest of the frame, where $p = (m, 0)$. Then

$$W = m^2\mathbf{S}^2 = m^2(S_1^2 + S_2^2 + S_3^2).$$

where

$$S_i = \tfrac{1}{2}\varepsilon_{ijk}S_{jk}\ (i, j, k = 1, 2, 3) \quad \text{and} \quad [S_i, S_j] = i\varepsilon_{ijk}S_k.$$

The operators S_j obey the angular momentum commutation rules. The eigenvalues of S^2 are therefore $s(s+1)$, where s can take integer and half-integer values and is called the spin of the field under consideration.

Thus, the eigenvalues of the operator W acquire the form $W = m^2 s(s+1)$. We can therefore label each irreducible representation of the inhomogeneous Lorentz group with a pair of indices (m, s). The wave functions for fixed m and s are the eigenfunctions of the operators p^2 and W:

$$(p^2 - m^2)\Psi = 0, \qquad W\Psi = s(s+1)p^2\Psi. \tag{3.47}$$

In order to obtain the wave function with a definite value of spins s it is necessary to construct a projecting operator choosing the spin representation s. Such *projecting operator* is easily constructed:

$$\hat{P}(s) = \prod_{s' \neq s} \frac{W - s'(s'+1)p^2}{s(s+1)p^2 - s'(s'+1)p^2}. \tag{3.48}$$

In our basic representation there are only two spinor states: $s = 0$ and $s = 1$. Both are irreducible. According to definition (3.48) we get

$$\hat{P}(0) = 1 - W/2p^2, \tag{3.49}$$

$$\hat{P}(1) = W/2p^2. \tag{3.50}$$

A wave function of the basic representation satisfies the so-called *Bargmann–Wigner* (1948) equation:

$$(\hat{p}^{(i)} - m)\Psi = 0 \quad (i = 1, 2). \tag{3.51}$$

Then the function

$$\Psi^{(s)} = \hat{P}(s)\Psi \tag{3.52}$$

belongs to the irreducible representation (m, s).

In accordance with (3.52) the general solution of equation (3.51) can be written in the form

$$\Psi^{(s)} = \prod_{j=1}^{2} \left(\frac{\hat{p}^{(j)} + m}{2m}\right) \hat{P}(s)\Psi, \tag{3.53}$$

where

$$\left(\frac{\hat{p}^{(j)} + m}{2m}\right)^2 = \frac{\hat{p}^{(j)} + m}{2m} \quad \text{for } p^2 = m^2.$$

The problem now is to find an equation of type (3.42). For which the obtained solution (3.53) is satisfactory. The operator $\Lambda(\partial)$ (3.44) must be constructed from the invariant operators with respect to the transformations of the inhomogeneous Lorentz group, and must be linear with respect to the mass parameter m outside the mass shell. For this purpose we transform the solution (3.53), and introduce the projecting operator

$$\chi(p) = \tfrac{1}{2}[1 + \hat{p}^{(2)}\hat{p}^{(1)}/p^2] \tag{3.54}$$

so that

$$[\chi(p)]^2 = \chi(p).$$

This operator on the mass shell $p^2 = m^2$ satisfies the condition

$$\chi(p)(\hat{p}^{(1)} + m) = \left(\frac{\hat{p}^{(2)} + m}{2m}\right)(\hat{p}^{(1)} + m).$$

Then, making use of these, relations (3.53) can be rewritten in the form

$$\Psi^{(s)} = \chi(p)\frac{\hat{p}^{(1)} + m}{2m}\hat{P}(s)\Psi. \tag{3.55}$$

Relations (3.53) and (3.55) give a general expression for the wave functions from the irreducible representation (m, s). Thus, the Bargmann–Wigner equation (3.51) is equivalent to the following two system equations:

$$(\hat{p}^{(1)} - m)\Psi_i^{(s)} = 0, \qquad [1 - \chi(p)]\Psi_i^{(s)} = 0, \quad (i = 1,2). \tag{3.56}$$

In turn, this system is equivalent to one equation

$$\lambda(a, p)\Psi_i^{(s)} = 0. \tag{3.57}$$

where

$$\lambda(a, p) = (\hat{p}^{(1)} - m)\chi(p) + a[1 - \chi(p)]$$

and a is an arbitrary parameter. In order to exclude the singular terms with $1/p^2$ we choose $a = -m$ and obtain

$$\lambda(m, p) = \beta_\mu p_\mu - m = \hat{p} - m.$$

Wave functions with definite spin values are the solutions of the following system equations:

$$\lambda(a, p)\hat{P}(s)\Psi = 0, \qquad \hat{P}(s')\Psi = 0, \quad (s' \neq s).$$

This system is equivalent to the equation

$$\Lambda^{(s)}(p)\Psi^{(s)} = 0,$$

where

$$\Lambda^{(0)}(p) = (\hat{p} - m)(1 - W/2p^2) + a_0 W/2p^2, \tag{3.58}$$

$$\Lambda^{(1)}(p) = (\hat{p} - m)W/2p^2 + a_1(1 - W/2p^2). \tag{3.59}$$

Here the parameters a_0 and a_1 are arbitrary. The singular terms are removed if it is assumed $a_0 = a_1 = -m$. Making use of the commutation relations (3.43) we obtain the following expression for the invariant W:

$$W = B^2 p^2 - 3Bp^2 - 2B\hat{p}^2 + 4\hat{p}^2, \quad B = \beta_\nu \beta_\nu.$$

After simple transformations we have

$$\Lambda^{(s)}(p) = \beta_\mu^{(s)} p_\mu - m = \hat{p}^{(s)} - m, \tag{3.60}$$

where

$$\beta_\mu^{(0)} = (1 - P)\beta_\mu, \qquad \beta_\mu^{(1)} = P\beta_\mu, \qquad P = -\tfrac{1}{2}(B - 4)(B - 1). \tag{3.61}$$

Using relations (3.43) and (3.45) one can show that the matrix P satisfies the equalities:

$$P\beta_\mu - \beta_\mu P = 0, \qquad P^2\beta_\mu = P\beta_\mu, \qquad P(1 - P)\beta_\nu = 0.$$

Therefore, the matrices $\beta_\nu^{(0)}$ and $\beta_\nu^{(1)}$ satisfy also the Duffin–Kemmer commutation relations (3.43). Moreover, $\beta_\nu^{(0)}\beta_\nu^{(1)} = 0$.

Traces of the matrices $\beta_\nu^{(0)}$ and $\beta_\nu^{(1)}$ are given by formulas following from the representations (3.45) and (3.61) and from the properties of the γ-matrix traces. Thus, we have for $s = 0$

$$\mathrm{Sp}\,\underbrace{\{\beta_\mu\beta_\nu\beta_\lambda\beta_\sigma\cdots\beta_\rho\beta_\tau\beta_\chi\}}_{n} = \begin{cases} g_{\mu\nu}g_{\lambda\sigma}\cdots g_{\tau\chi} + g_{\nu\lambda}g_\sigma\cdots g_{\rho\tau}g_{\chi\mu}, & \text{if } n \text{ is even} \\ 0, \quad \text{if } n \text{ is odd} \end{cases}$$

and for $s = 1$

$$\begin{aligned} \mathrm{Sp}\,\beta_\nu\beta_\mu &= 6g_{\mu\nu}, \\ \mathrm{Sp}\,\beta_\nu\beta_\mu\beta_\lambda\beta_\rho &= 3(g_{\mu\nu}g_{\lambda\rho} + g_{\mu\lambda}g_{\nu\rho}), \\ &\cdots \end{aligned}$$

In our case, the divisor $d^{(s)}(p)$ is given by

$$d^{(s)}(p) = m + \hat{p}^{(s)} - \frac{1}{m}[p^2 - (\hat{p}^{(s)})^2] \quad (s = 0, 1).$$

Now to separate fields corresponding to spin 0 and 1 from the 16-component wave function $\Psi_{\alpha\beta}$, where α and β are spinor indices taking the values 0, 1, 2, 3. The field Ψ in its basic representation is decomposed into the linear independent 16-matrices:

$$\begin{aligned} \Psi_{\alpha\beta} = \tilde{C}_{\alpha\beta}\Psi + (\gamma_\mu\tilde{C})_{\alpha\beta}\Psi_\mu + (\gamma_\mu\gamma_5\tilde{C})_{\alpha\beta}\Psi_{\mu 5} + \tfrac{1}{4}i(\sigma_{\mu\nu}\tilde{C})_{\alpha\beta}\varepsilon_{\mu\nu\rho\sigma}\Psi_{\rho\sigma} + \\ + (\gamma_5\tilde{C})_{\alpha\beta}\Psi_5 = \sum_J (Q^J\tilde{C})_{\alpha\beta}\Psi_J, \end{aligned} \tag{3.62}$$

where definition of Q^J follows from the formula (3.62) itself. Here $\tilde{C} = \gamma_5 C$ and C is a charge conjugate matrix, so that $\tilde{C}$ obeys the condition $\tilde{C}\gamma_\mu\tilde{C}^{-1} = \gamma_\mu^T$.

Making use of formulas (3.45) and (3.61) we arrive at the following representation for the operators P and $\hat{p}$:

$$\begin{aligned} P_{\alpha\alpha',\beta\beta'} &= [1 + \tfrac{1}{4}(\gamma_\rho^{(1)}\gamma_\rho^{(2)}) - \tfrac{1}{8}(\gamma_\rho^{(1)}\gamma_\rho^{(2)})^2]_{\alpha\alpha',\beta\beta'} \\ &= g_{\alpha\alpha'}g_{\beta'\beta} + \tfrac{1}{4}(\gamma_\rho)_{\alpha\alpha'}(\tilde{C}\gamma_\rho\tilde{C}^{-1})_{\beta'\beta} - \tfrac{1}{8}(\gamma_\rho\gamma_\sigma)_{\alpha\alpha'}(\tilde{C}\gamma_\sigma\gamma_\rho\tilde{C}^{-1})_{\beta'\beta}, \quad (3.63) \\ \hat{p}_{\alpha\alpha',\beta\beta'} &= \tfrac{1}{2}p_\rho[\gamma_\rho^{(1)} + \gamma_\rho^{(2)}]_{\alpha\alpha',\beta\beta'} \\ &= \tfrac{1}{2}p_\rho\{(\gamma_\rho)_{\alpha\alpha'}g_{\beta'\beta} + g_{\alpha\alpha'}(\tilde{C}\gamma_\rho\tilde{C}^{-1})_{\beta'\beta}\} \end{aligned}$$

The operation of P and $\hat{p}$ upon the matrix $(Q^J\tilde{C})_{\alpha\beta}$ in the decomposition (3.62) gives

$$P(Q^J\tilde{C}) = \sum_{\alpha'\beta'} P_{\alpha\alpha',\beta\beta'}(Q^J\tilde{C})_{\alpha'\beta'} = \{[Q^J + \tfrac{1}{4}\gamma_\rho Q^J\gamma_\rho - \tfrac{1}{8}\gamma_\rho\gamma_\sigma Q^J\gamma_\sigma\gamma_\rho]\tilde{C}\}_{\alpha\beta} \quad (3.64)$$

and

$$\hat{p}(Q^J\tilde{C}) = \sum_{\alpha'\beta'} \hat{p}_{\alpha\alpha',\beta\beta'}(Q^J\tilde{C})_{\alpha'\beta'} = \tfrac{1}{2}p_\rho\{[\gamma_\rho Q^J + Q^J\gamma_\rho]\tilde{C}\}_{\alpha\beta}. \quad (3.65)$$

After simple calculations using only the properties of the Dirac γ-matrices one can reduce equation (3.42) with the operator $\Lambda^{(0)}$ (3.60) to the following system equations:

$$m\Psi = p_\mu\Psi_\mu, \qquad m\Psi_\mu = p_\mu\Psi, \quad (3.66)$$

from this it follows $(p^2 - m^2)\Psi = 0$. Thus, we have obtained the usual *Klein–Gordon equation* for spinless particles. The system equation (3.66) contains five components so that for the matrices $\beta_\mu^{(0)}$ one can find a representation in the form of 5×5 matrices.

In the case of $s = 1$ we obtain

$$p_\mu\Psi_{\nu5} - p_\nu\Psi_{\mu5} - m\Psi_{\mu\nu} = 0, \qquad p_\alpha\Psi_{\alpha\mu} - m\Psi_{\mu5} = 0, \quad (3.67)$$

from this it follows

$$(p^2 - m^2)\Psi_{\mu5} - p_\mu p_\nu\Psi_{\nu5} = 0, \qquad \Psi_{\mu\nu} = \frac{1}{m}(p_\mu\Psi_{\nu5} - p_\nu\Psi_{\mu5}). \quad (3.68)$$

This is the *Proc equation* for spin $s = 1$. Equations (3.67) contain 10 components and one can therefore find a representation in the form of 10×10 matrices for $\beta_\mu^{(1)}$-matrices.

For the function Ψ_5 in the decomposition (3.62) we get $\Psi_5 = 0$. Thus, the basic representation containing 16-components have decomposed into three irreducible representations: the 10-component for spin 1, the 5-component for spin 0, and the one-component for the trivial case $\Psi_5 = 0$.

It should be noted that linear equations for the definite spin automatically permit the necessary supplementary conditions to be taken into account with the exception of unnecessary spinor components.

The investigation of the perturbation theory for the S-matrix constructed on the basis of Duffin–Kemmer equations will be formally identical to the QED of leptons. Therefore, we should use the methods and procedures develop in the construction of the gauge-invariant spinor electrodynamics in terms of the stochastic space-time concept.

Hence, as was shown above, the averaged field $\psi_R(x)$ of a boson field in the space $\mathbb{R}^4(\hat{x})$ leads to the change of the free particle propagator (in momentum representation):

$$T(\hat{p}) = \frac{\hat{p}(\hat{p}+m) - p^2 + m^2}{m(m^2 - p^2 - i\varepsilon)} \to V_m(-p^2 l^2) T(\hat{p}) \equiv T_R(\hat{p}),$$

where $\hat{p} = p_\mu \beta_\mu$ and β_μ are the four 16-rank matrices which are split into 5- and 10-rank matrices for particles with spins 0 and 1, respectively. In this case it is also necessary to change the form of the one-photon vertex:

$$\beta_\mu \to u_\mu(q,k) = -\mathrm{d}_\mu(k) T_R^{-1}(\hat{q}).$$

Here the following identities hold:

$$\mathrm{d}_\mu(k) T_R(\hat{q}) = T_R(\hat{q} + \hat{k}) u_\mu(q,k) T_R(\hat{q}), \tag{3.69}$$

$$(p_\mu - q_\mu)\Gamma_\mu^R(p,q) = T_R(\hat{p}) - T_R(\hat{q}), \tag{3.70}$$

where

$$\Gamma_\mu^R(p,q) = T_R(\hat{p}) u_\mu(k,q) T_R(\hat{q}), \quad k = p - q;$$

$$u_\mu(k,q) = \beta_\mu V^{-1}(-q^2 l^2) + (m - \hat{q} - \hat{k}) V^{-1}(-p^2 l^2)[\mathrm{d}_\mu(k) V(-q^2 l^2)] V^{-1}(-q^2 l^2)$$

and

$$\mathrm{d}_\mu(k) V(-q^2 l^2) = \{V(-p^2 l^2) - V(-q^2 l^2)\} k^{-2}(k_\mu - 2\beta_\mu \hat{k} + 2\hat{k}\beta_\mu). \tag{3.71}$$

Let us now examine the perturbation series for the S-matrix.

3.5.2. *The Diagrams of the Vacuum Polarization of Boson Fields*

Expression (3.31) for boson fields becomes

$$\tilde{\Pi}_{\mu\nu}^R(k_1,k_2) = \lim_{\delta \to 0} \frac{ie^2}{2(2\pi)^4} \int \mathrm{d}^4 q V_m(-q_2^2 l^2)\, \mathrm{Sp}\{\Gamma_{\mu\nu}^\delta(q;k_1,k_2) T_R(\hat{q})\},$$

where

$$k_1 + k_2 = 0, \qquad q_2 = q + k_1 + k_2 = q,$$

$$T_R(\hat{q}_2)\Gamma_{\mu\nu}(q,k_1,k_2) T_R(\hat{q}) = (-1)^2\, \mathrm{d}_\mu(k_1)\, \mathrm{d}_\nu(k_2) T_R(\hat{q}).$$

Making use of the definition of the d-operation (3.71) for an entire function we obtain

$$\tilde{\Pi}^R_{\mu\nu}(k) = \lim_{\delta \to 0} \frac{ie^2}{(2\pi)^4} \int d^4q \, \mathrm{Sp}\left\{ \beta_\mu \frac{V_m(-q^2l^2)}{m - \hat{q}_1} \beta_\nu \frac{1}{m - \hat{q}} + \beta_\mu \frac{1}{m - \hat{q}_1} \times \right.$$

$$\left. \times \, [d_\nu(k) V_m(-q^2l^2)] + \tfrac{1}{2} d_\mu(k) \, d_\nu(-k) V_m(-q^2l^2) \right\}$$

$$= \tilde{\Pi}^{(1)}_{\mu\nu}(k) + \tilde{\Pi}^{(2)}_{\mu\nu}(k) + \tilde{\Pi}^{(3)}_{\mu\nu}(k),$$

where

$$q_1 = q + k, \qquad (m - \hat{p})^{-1} = m^{-1}[\hat{p}(\hat{p} + m) - p^2 + m^2](m^2 - p^2 - i\varepsilon)^{-1},$$

$$\tilde{\Pi}^{(1)}_{\mu\nu}(k) = \frac{-1}{4} \rho_\zeta 2^\zeta \frac{\Gamma(2 - \zeta)}{\Gamma(1 - \zeta)} \int_{-1}^{1} dy (1 - y)^\zeta (1 - y^2)^{-\zeta} \int d^4 q_{\mathrm{E}} \, \times$$

$$\times \{(k_\mu k_\nu - k^2 g_{\mu\nu})[\sigma_1 y^2 + (\tfrac{1}{2}\sigma_1 - 1)(q_{\mathrm{E}}^2/m^2)] +$$

$$+ \, g_{\mu\nu} \sigma_1 [q_{\mathrm{E}}^2 + 2m^2 - \tfrac{1}{2}k^2(1 - y^2)]\} \times$$

$$\times \, [q_{\mathrm{E}}^2 + m^2 - \tfrac{1}{4}k^2(1 - y^2)]^{\zeta - 2},$$

$$\tilde{\Pi}^{(2)}_{\mu\nu}(k) = \tfrac{1}{2}\sigma_1 g_{\mu\nu} \rho_\zeta 2^\zeta \frac{\Gamma(1 - \zeta)}{\Gamma(-\zeta)} \int_{-1}^{1} dy (1 - y)^\zeta \, \times$$

$$\times \, (1 - y^2)^{-\zeta} \int d^4 q_{\mathrm{E}} [q_{\mathrm{E}}^2 + m^2 - \tfrac{1}{4}k^2(1 - y^2)]^{\zeta - 1},$$

$$\tilde{\Pi}^{(3)}_{\mu\nu}(k) = \frac{\rho_\zeta}{k^4} [4\sigma_1 (k_\mu k_\nu - k^2 g_{\mu\nu}) + \sigma_2 k_\mu k_\nu] \, \times$$

$$\times \int d^4 q_{\mathrm{E}} \{(m^2 + q_{\mathrm{E}}^2)^\zeta - (m^2 + q_{1\mathrm{E}}^2)^\zeta\},$$

$$\sigma_1 = \begin{cases} 2 \\ 6 \end{cases} \quad \text{and} \quad \sigma_2 = \begin{cases} 5 & \text{for spin } s = 0 \\ 10 & \text{for spin } s = 1 \end{cases}$$

Here

$$\rho_\zeta = e^2 (2\pi)^{-4} \frac{1}{2i} \int_{\substack{-\gamma + i\infty \\ (1 < \gamma < 2)}}^{-\gamma - i\infty} d\zeta \frac{v(\zeta)}{\sin \pi\zeta} l^{2\zeta}.$$

Further, making use of representation (3.32) and of the definition for the trace of β-matrices (Harish – Chandra, 1946, see also previous section), and performing necessary calculations we get

$$\tilde{\Pi}^{(1)}_{\mu\nu}(k) = \frac{\alpha}{4\pi}\frac{1}{2i}\int_{-\gamma+i\infty}^{-\gamma-i\infty} d\zeta \frac{v(\zeta)}{\sin \pi\zeta}(m^2 l^2)^{\zeta} \int_0^1 dx(1-x)^{-\zeta}\Big\{(g_{\mu\nu}k^2 - k_\mu k_\nu) \times$$
$$\times (1-2x)^2 \frac{\Gamma(-\zeta)}{\Gamma(1-\zeta)}\mathscr{L}_0^{\zeta} - 2m^2 g_{\mu\nu}\frac{\Gamma(-1-\zeta)}{\Gamma(-\zeta)}\mathscr{L}_0^{1+\zeta}\Big\},$$

$$\tilde{\Pi}^{(2)}_{\mu\nu}(k) = \frac{\alpha}{2\pi} g_{\mu\nu}\frac{m^2}{2i}\int_{-\gamma+i\infty}^{-\gamma-i\infty} d\zeta \frac{v(\zeta)}{\sin \pi\zeta}(m^2 l^2)^{\zeta}\int_0^1 dx(1-x)^{-\zeta}\mathscr{L}_0^{1+\zeta}\frac{\Gamma(-1-\zeta),}{\Gamma(-\zeta),}$$

$$\tilde{\Pi}^{(3)}_{\mu\nu}(k) = 0, \qquad \mathscr{L}_0(x) = 1 - (k^2/m^2)x(1-x),$$

for the scalar boson, and

$$\tilde{\Pi}^{(1)}_{\mu\nu}(k) = \frac{\alpha}{4\pi}\frac{1}{2i}\int_{-\gamma+i\infty}^{-\gamma-i\infty} d\zeta \frac{v(\zeta)}{\sin \pi\zeta}(m^2 l^2)^{\zeta}\int_0^1 dx(1-x)^{-\zeta} \times$$
$$\times \Big\{3\Big[(g_{\mu\nu}k^2 - k_\mu k_\nu)(1-2x)^2 \times \frac{\Gamma(-\zeta)}{\Gamma(1-\zeta)}\mathscr{L}_0^{\zeta} -$$
$$- 6m^2 g_{\mu\nu}\frac{\Gamma(-1-\zeta)}{\Gamma(-\zeta)}\mathscr{L}_0^{1+\zeta}\Big] +$$
$$+ 2(g_{\mu\nu}k^2 - k_\nu k_\mu)\frac{\Gamma(-1-\zeta)}{\Gamma(1-\zeta)}\mathscr{L}_0^{1+\zeta}\Big\},$$

$$\tilde{\Pi}^{(2)}_{\mu\nu}(k) = 6m^2 g_{\mu\nu}\frac{\alpha}{4\pi}\frac{1}{2i}\int_{-\gamma+i\infty}^{-\gamma-i\infty} d\zeta \frac{v(\zeta)}{\sin \pi\zeta}(m^2 l^2)^{\zeta}\frac{\Gamma(-1-\zeta)}{\Gamma(-\zeta)} \times$$
$$\times \int_0^1 dx(1-x)^{-\zeta}\mathscr{L}_0^{1+\zeta},$$

$$\tilde{\Pi}^{(3)}_{\mu\nu}(k) = 0,$$

for the vector boson, respectively. Here the variables x and y are connected by the formula $y = 1 - 2x$.

From the above formulas it is easy to see that in both the cases of scalar and vector particles the expression $\tilde{\Pi}^{(2)}_{\mu\nu}(k)$ and the second term of $\tilde{\Pi}^{(1)}_{\mu\nu}(k)$ cancel mutually.

Let us write the other terms:

$$\tilde{\Pi}^{(s)}_{\mu\nu}(k) = \frac{\alpha}{4\pi}(g_{\mu\nu}k^2 - k_\mu k_\nu)\frac{1}{2i}\int_{-\gamma+i\infty}^{-\gamma-i\infty} d\zeta \frac{v(\zeta)}{\sin \pi\zeta}(m^2 l^2)^{\zeta}\frac{\Gamma(-\zeta)}{\Gamma(1-\zeta)} \times$$
$$\times \int_0^1 dx(1-x)^{-\zeta} \times (1-2x)^2\mathscr{L}_0^{\zeta}, \quad (0 < \gamma < 1)$$

for the scalar boson, and

$$\tilde{\Pi}^{(v)}_{\mu\nu}(k) = \frac{\alpha}{4\pi}(g_{\mu\nu}k^2 - k_\mu k_\nu)\frac{1}{2i}\int_{\substack{-\gamma+i\infty \\ (1<\gamma<2)}}^{-\gamma-i\infty} d\zeta\, \frac{v(\zeta)}{\sin \pi\zeta}(m^2 l^2)^\zeta \times$$

$$\times \int_0^1 dx(1-x)^{-\zeta} \times \left[3(1-2x)^2 \frac{\Gamma(-\zeta)}{\Gamma(1-\zeta)}\mathscr{L}_0^\zeta + \right.$$

$$\left. \times\, 2\frac{\Gamma(-1-\zeta)}{\Gamma(1-\zeta)}\mathscr{L}_0^{1+\zeta}\right]$$

for the vector boson.

Assuming the value $m_j^2 l^2$ ($m_j = m_s, m_v$) to be small we obtain finally

$$\tilde{\Pi}^{(s)}_{\mu\nu}(k) = \frac{\alpha}{24\pi}(g_{\mu\nu}k^2 - k_\mu k_\nu)\left\{\frac{k^2}{m^2}\int_0^1 dy y^4[1 - \tfrac{1}{4}(k^2/m^2)(1-y^2)]^{-1} + \right.$$

$$\left. + 2(v'(0) + \ln(m^2 l^2)) - \tfrac{8}{3}\right\},$$

$$\tilde{\Pi}^{(v)}_{\mu\nu}(k) = \frac{\alpha}{4\pi}(g_{\mu\nu}k^2 - k_\mu k_\nu)\left\{\int_0^1 dx[2 - 2x(1-x)(k^2/m^2) - 3(1-2x)^2] \times \right.$$

$$\times \ln[1 - x(1-x)(k^2/m^2)] + (1 - \tfrac{1}{3}(k^2/m^2)) \times$$

$$\left. \times (v'(0) + \ln m^2 l^2) + \frac{\delta}{m^2 l^2} + \frac{k^2}{18m^2} - \frac{4}{3}\right\},$$

$$\delta = v'(-1).$$

It is necessary to perform the charge renormalization of the scalar boson and the charge and mass renormalization for the vector boson, and after these procedures we obtain the usual quantum-electrodynamical quantities for the vacuum polarization of these particles:

$$\tilde{\Pi}^{(s)}_{\mu\nu}(k) = (g_{\mu\nu}k^2 - k_\mu k_\nu)\{\Pi^{(s)}(k^2) - \Pi^{(s)}(0)\} = \frac{\alpha}{24\pi}(k^2/m^2)(g_{\mu\nu}k^2 - k_\mu k_\nu) \times$$

$$\times \int_0^1 dy y^4[1 - \tfrac{1}{4}(k^2/m^2)(1-y^2)]^{-1},$$

$$\tilde{\Pi}^{(v)}_{\mu\nu}(k) = (g_{\mu\nu}k^2 - k_\mu k_\nu)\left\{\Pi^{(v)}(k^2) - \Pi^{(v)}(0) - \frac{d}{dk^2}\Pi^{(v)}(k^2)\bigg|_{k^2=0} k^2\right\}$$

$$= \frac{\alpha}{4\pi}(g_{\mu\nu}k^2 - k_\mu k_\nu) \times \int_0^1 dx\{[2 - 2x(1-x)(k^2/m^2) - 3(1-2x)^2] \times$$

$$\times \ln[1 - x(1-x)(k^2/m^2)] - 3(1-2x)^2 x(1-x)(k^2/m^2) +$$

$$+ 2x(1-x)(k^2/m^2)\}.$$

3.5.3. *The Self-Energy of Bosons*

Let us now consider the self-energy diagram, shown in Figure 5b (here the solid line corresponds to the boson field). In our scheme the term of the S-matrix corresponding to this diagram has the form

$$-i\colon \bar{\psi}(x)\Sigma_R^b(x-y)\psi(y)\colon,$$

$$\Sigma_R^b(x-y) = (2\pi)^{-2}\int \mathrm{d}^4p\, e^{ip(x-y)}\,\tilde{\Sigma}_R^b(p),$$

where

$$\tilde{\Sigma}_R^b(p) = \lim_{\delta\to 0}\frac{-ie^2}{(2\pi)^4}\int \mathrm{d}^4k\,\frac{V_0^\delta(-k^2l^2)}{-k^2-i\varepsilon}\beta_\mu\,\frac{m^2-(p-k)^2+(\hat{p}-\hat{k})(\hat{p}-\hat{k}+m)}{m(m^2-(p-k)^2-i\varepsilon)}\times$$
$$\times\,\beta_\mu\cdot V_m^\delta(-(p-k)^2l^2).$$

Passing to the Euclidean metric in this expression, and substituting representation (3.32) in a way similar to that used in the previous section, it is easy to obtain the following formula:

$$\tilde{\Sigma}_R^b(p) = \frac{\alpha}{4\pi}m\left\{\left[B+\frac{\hat{p}}{m}+\frac{\hat{p}^2}{m^2}(1-B)\right]L_{0,0}-2(1-B)\frac{\hat{p}^2}{m^2}L_{1,0}-\right.$$
$$\left.-\tfrac{1}{2}B(1-B)L_{1,1}+(1-B)\frac{\hat{p}^2}{m^2}L_{2,0}-\frac{\hat{p}}{m}L_{1,0}\right\},$$

where

$$B=\beta_\mu\beta_\mu,\qquad \hat{p}=\beta_\mu p_\mu=\beta_0p_0-\boldsymbol{\beta}\mathbf{p},$$

$$L_{a,b}=L_{a,b}\left(m^2l^2,\frac{p^2}{m^2}\right)=w\,\frac{\Gamma(-b-\zeta-\eta)}{\Gamma(1-\zeta)}\times$$
$$\times\int_0^1 \mathrm{d}u(1-u)^{\zeta+a}u^{-\zeta}\left(1-\frac{p^2}{m^2}u\right)^{\zeta+\eta+b},$$

Here

$$w=\frac{1}{2i}\int_{-\beta_{a+b}+i\infty}^{-\beta_{a+b}-i\infty}\mathrm{d}\zeta\,\frac{v(\zeta)}{\sin\pi\zeta}(m^2l^2)^{\zeta}\,\frac{1}{2i}\int_{-\gamma_{a+b}+i\infty}^{-\gamma_{a+b}-i\infty}\mathrm{d}\eta\,\frac{v(\eta)}{\sin\pi\eta}(m^2l^2)^{\eta},$$

$$\tfrac{1}{2}(a+b)<\gamma_{a+b}+\beta_{a+b}<\tfrac{1}{2}(a+b+1),\quad a,b=0,1,2.$$

We calculate the electromagnetic correction to the boson mass in the two limits $m^2l^2 \ll 1$ and $m^2l^2 \gg 1$. We have: (i) for $m^2l^2 \ll 1$

$$\tilde{\Sigma}_R^b(m) = -\frac{3\alpha}{48\pi}B(B-1)\frac{\delta m}{m^2l^2} = \begin{cases}-(3\alpha/4\pi)(m/m^2l^2)\delta & \text{for } s=0,\\ (\alpha/4\pi)(m/m^2l^2)\delta & \text{for } s=1,\end{cases} \tag{3.72}$$

where

$$\delta = \frac{\pi}{2i}\int_{-\beta+i\infty}^{-\beta-i\infty} d\zeta \frac{v(\zeta)v(-1-\zeta)}{\sin^2 \pi\zeta}[1 + O(\ln m^2 l^2)], \quad 0 < \beta < 1$$

and (ii) $m^2l^2 \gg 1$, i.e., in the classical limit

$$\tilde{\Sigma}_R^b(m) = \tfrac{1}{2}\alpha\frac{v(-\frac{1}{2})}{ml} m[1 + O(1/m^2 l^2)] \quad \text{for } s = 0, 1.$$

The latter means that contributions to the mass of the particles with spin 0 and 1 coincide with the classical limit. In general, as shown by Efimov (1977a) in this limit contribution to the mass of charged particles with spin does not depend on their spin value, although in the quantum field limit, i.e., $ml \ll 1$ it has a larger singularity with respect to the growth of spin.

It is interesting to calculate the mass difference $\Delta m(\pi^+ - \pi^0)$ of π^+- and π^0-mesons. Since π^0-meson is electrical neutral, the value $\Delta m(\pi^+ - \pi^0)$ is caused by the proper electromagnetic energy of the π^+-meson only. It is roughly assumed that π^+-meson represents a uniformly charged ball, radius l of which determined by the relation

$$l^2 = \tfrac{5}{3}\langle r^2\rangle_{\text{exp}} \sim \frac{5}{3}\frac{0.19}{m_\pi^2} = \frac{0.32}{m_\pi^2},$$

where

$$\langle r^2\rangle_{\text{exp}} \sim (0.6 \times 10^{-13}\,\text{cm})^2 = 0.19/m_\pi^2.$$

For the charged ball the function $v_b(\zeta)$ is given by formula (1.15) and therefore

$$-\frac{\pi}{2\pi i}\int_{-\beta+i\infty}^{-\beta-i\infty} d\eta \frac{v_b(\eta)v_b(-1-\eta)}{\sin^2 \pi\eta} \approx \frac{3.41}{\pi} = 1.086.$$

Substituting this value into (3.72) we get

$$\Delta m(\pi^+ - \pi^0) = \alpha m \frac{3}{4}\frac{1.086}{0.32} = 2.6\,\text{MeV},$$

which gives a rough agreement with the experimental value of $\Delta m_{\text{exp}}(\pi^+ - \pi^0) = 4.6\,\text{MeV}$. It is clear that by an appropriate choice of a charge distribution for π^+-meson it is easy to obtain a complete coincidence with the experimental value of the mass difference $\Delta m_{\text{exp}}(\pi^+ - \pi^0)$.

The vertex diagram and the static characteristics for the bosons were investigated by Dineykhan *et al.* (1977) within the framework of the stochastic space-time, and by Efimov and Namsrai (1975) in the nonlocal theory. The vertex function $\tilde{\Gamma}_\mu(p, q)$ corresponding to the diagram shown in Figure 5c in the boson field case takes the form

$$\tilde{\Gamma}_\mu(p, q) = \frac{ie^2}{(2\pi)^4}\int d^4k \tilde{D}_R((p-k)^2)\beta_\nu T_R(\hat{k} + \hat{q})u_\mu(k, q)T_R(\hat{k})\beta_\nu.$$

After simple but long calculations, the function $e\bar{\psi}(\beta_\mu + \tilde{\Gamma}_\mu(p, q))\psi$ is reduced to the following form

$$e\bar{\psi}(\beta_\mu + \tilde{\Gamma}_\mu(p, q))\psi = e\bar{\psi}\beta_\mu\left(1 + \frac{1}{2m}F_0(q^2)\right)\psi + \\ + M_{\mu\nu}q_\nu F_1(q^2) + Q_{\mu\nu}q_\nu F_2(q^2), \tag{3.73}$$

where $M_{\mu\nu} = (e/2m)\,\bar{\psi}s_{\mu\nu}(2B-4)\psi$ and $Q_{\mu\nu} = -(e/2m)\,\bar{\psi}s_{\mu\nu}(2B-5)\psi$, $(s_{\mu\nu} = \beta_\mu\beta_\nu - \beta_\nu\beta_\mu)$ are antisymmetric tensors which define, as we see below, the dipole and quadripole moments of the boson with spin $s = 1$. Functions $F_i(q^2)$ $(i = 0, 1, 2)$ depend on the invariant variable $q^2 = (p' - p)^2$ and contribute to the physical quantities of the boson field. Here we shall not present them explicitly but recommend that the reader calculate them as an exercise.

Now we briefly discuss the static electromagnetic characteristics of charged bosons. Taking into account the identity

$$e\bar{\psi}\beta_\mu\psi\cdot A_\mu = e\bar{\psi}\left\{\frac{1}{2m}(p'_\mu + p_\mu) - \frac{1}{2m}s_{\mu\nu}q_\nu\right\}\psi A_\mu \\ = \{e(P_\mu/2m)\bar{\psi}\psi - M_{\mu\nu}q_\nu - Q_{\mu\nu}q_\nu\}A_\mu,$$

expression (3.73) can be rewritten in the form

$$e\bar{\psi}(\beta_\mu + \tilde{\Gamma}_\mu)\psi A_\mu \\ = \{e_r(P_\mu/2m)\bar{\psi}\psi - (1 - F_1(0))M_{\mu\nu}q_\nu - (1 - F_2(0))Q_{\mu\nu}q_\nu\}A_\mu, \tag{3.74}$$

where A_μ is an external electromagnetic potential. In order to consider the static characteristics of charged bosons, we pass to the static limit in expression (3.74). First, let us consider the case of scalar boson, i.e., $s = 0$. In this case, through relations (3.62) and (3.66), we get

$$N_{\mu\nu} = \bar{\psi}s_{\mu\nu}\psi = \bar{\psi}s_{\mu\nu}B\psi = \frac{1}{m}\left(\frac{\partial\phi^*}{\partial x_\mu}\frac{\partial\phi}{\partial x_\nu} - \frac{\partial\phi^*}{\partial x_\nu}\frac{\partial\phi}{\partial x_\mu}\right),$$

where ϕ is the field of scalar particles which denotes Ψ in (3.62) and (3.66).

In the static limit we have $N_{\mu\nu}F_{\mu\nu} \to \mathbf{v}(\partial\mathbf{E}/\partial t) \to 0$ for $\mathbf{v} \to 0$, where $F_{\mu\nu} = \partial_\mu A_\nu - \partial_\nu A_\mu$, $\mathbf{v}$ is the velocity of a scalar particle and $\mathbf{E}$ is the electric field. Thus, we see that the magnetic moment for scalar bosons is absent.

The vector $\phi_\mu(x)$, denoted through $\Psi_{\mu 5}$ in (3.62) and (3.67), has the components

$$\phi_\mu(x) = (2\pi)^{-3/2}\int\frac{d^3k}{\sqrt{2\omega_k}}(a_{\mu k}\,e^{-ikx} + b^+_{\mu k}\,e^{ikx}),$$

where $a_{\mu k}$ and $b^+_{\mu k}$ correspond to particle and antiparticle, respectively. If $b_{\mu k} = 0$ it means that antiparticle is absent in our consideration. In the static limit the vector $a_{\mu\mathbf{k}}$ has three components:

$$a_{\mu\mathbf{k}} = \begin{cases} a_0 = (|\mathbf{k}|/m)a_3 \to 0, \\ \mathbf{a} = \mathbf{e}_1 a_1 + \mathbf{e}_2 a_2 + (\mathbf{k}/|\mathbf{k}|)(\omega_{\mathbf{k}}/m)a_3, \end{cases}$$

where $\mathbf{a} = (a_1, a_2, a_3)$ is the polarization vector of the rest vector particle. The vector spin $\mathbf{S}$ and the tensor T_{ij} of the quadripole moment are given by

$$\mathbf{S} = i[\mathbf{a}^+ \times \mathbf{a}] \quad \text{and} \quad T_{ij} = a_i^+ a_j + a_j^+ a_i - \tfrac{2}{3}\delta_{ij}(\mathbf{a}^+ \mathbf{a}),$$

respectively. In the vector boson case we have

$$\bar{\psi} s_{\mu\nu}(B - 2)\psi = -m(\phi_\mu^* \phi_\nu - \phi_\nu^* \phi_\mu)$$

and

$$\bar{\psi} s_{\mu\nu}(2B - 5)\psi = -\frac{1}{m}(G_{\mu\rho}^* G_{\rho\nu} - G_{\nu\rho}^* G_{\rho\mu}) - m(\phi_\mu^* \phi_\nu - \phi_\nu^* \phi_\mu),$$

where

$$G_{\mu\nu} = \partial_\mu \phi_\nu - \partial_\nu \phi_\mu.$$

In the static limit we obtain

$$\tfrac{1}{2} M_{\mu\nu} F_{\mu\nu} = -\frac{e}{2m}\mathbf{SH} \quad \text{and} \quad Q_{\mu\nu} F_{\mu\nu} = \frac{e}{2m^2}\left[\frac{i}{4} T_{ij}(\partial_i E_j + \partial_j E_i)\right],$$

where $\mathbf{H}$ is the magnetic field. Thus, the tensors $M_{\mu\nu}$ and $Q_{\mu\nu}$ define the dipole and quadripole moments of the vector boson, respectively. Contributions to these quantities are given by $\delta\mu_1 = -F_1(0)$ and $\delta\mu_2 = -F_2(0)$, calculations of which have already been presented to the reader.

In conclusion it is easily verified that in the case of the electrodynamics of particles with spins 0 and 1, the gauge invariance is held automatically by the construction.

Chapter 4

Four-Fermion Weak Interactions in Stochastic Space-Time

4.1. Introduction

The four-fermion theory (Fermi theory) plays, and it seems, will play a fundamental role in the development of the theory of weak interactions. The four-fermion V-A interaction describes in a unified way some of the weak decays of leptons and fermions. Earlier success of this theory in the explanation of muon and β-decays has given the certain hope that within the framework of this theory weak processes might be described at least in the low-energy domain.

However, in four-fermion theory one meets the well-known difficulty caused by the ultraviolet divergences and the renormalization problem. For this reason, calculations of higher order corrections in the perturbation series in coupling constant G_F are difficult. Notice that a similar situation occurs in the theory with intermediate vector bosons.

Ways of eliminating these difficulties were proposed in different models which can be classified in two groups. One of them is connected with some schemes and approaches aimed at constructing a new theory of weak interactions within the framework of *gauge theory* [especially, the unified theory of weak and electromagnetic interactions, i.e., the standard model of *electroweak interactions* due to Weinberg, 1967; Salam, 1968, and Glashow, 1961, 1980]. The other approaches assume a modification of the usual theory of weak interactions based on the analysis of fundamental principles (causality condition, locality and properties of geometry on small scale, etc.) of modern local quantum field theory at small distances (see for example, Alebastrov *et al.*, 1973; Kadyshevsky, 1980; Efimov and Seltzer, 1971 and Arbuzov, 1975).

The description of weak, electromagnetic, and also strong interactions within the quark model framework based on gauge theories with a spontaneously broken symmetry, obviously represents a qualitatively new step in understanding elementary particle phenomena and their internal structure.

May be that usefulness of the old (free of the above-mentioned difficulties) and of the new weak interaction theories is revealed mainly in two limiting cases. Indeed, when the energy is not high enough for the production of intermediate particles (for example, $W^{\pm}$, Z^0, Higgs bosons, etc.) which are necessary in gauge theories, i.e., if the energy is small with respect to some limiting value E_l, the weak processes must be described by the four-fermion theory. Contrarily, when $E \gtrsim E_l$ the gauge theory will play an important role in weak processes. Here E_l is the value of energy at which new particles $W^{\pm}$, Z^0 etc. will be produced (see Banner *et al.*, 1983). Of course, there has to exist a reasonable correspondence of both the theories at $E \sim E_l$.

In the language of distance it means that starting from some small scale $l_w \sim 1/E_l$ the growth of weak-processes cross-section must be compensated by corrections given by the intermediate bosons. We shall assume here that this length of weak interactions is of an order of $l_w \sim (\hbar/m_W c) \sim 10^{-16}$ cm, and that the unitary limit is reached at energies $E_l \sim m_W \sim 100$ GeV depending on the choice of the form-factor of the theory. Notice that it is quite possible that at these energies the unification of electromagnetic and weak interactions, described by the standard model of Weinberg–Salam–Glashow, is achieved.

It seems that in view of this aim, the investigation of four-fermion theory within the framework of the second approach is undoubtedly interesting, and can give new information about weak interactions. Moreover, recently great attention has been paid to the low-energy weak interactions, connected with the problem of neutrino oscillations and its consequences, and of the proton instability in the grand unified theories (see, for example, Fiorini, 1979).

This chapter is devoted to the construction of a gauge-invariant theory of weak and electromagnetic interactions of leptons in the stochastic space-time $\mathbb{R}^4(\hat{x})$. In this case the investigation of the terms of the S-matrix is carried out by the methods elaborated in the previous chapter.

Notice that in the stochastic (nonlocal) theory the concept of stochasticity (locality breaking) is characterized not only by the length $l_w \sim 1/E_l$, but also by a form of the *measure* $w(b_E^2/l^2)$ (1.2) (or the shape of a form-factor or of potential) at small distances. It seems to us that in real physical processes apart from the value of elementary length an important role may be played by the form-factor of the theory. This occurred especially in the study of the decay $K_L^0 \to \mu^+\mu^-$ and of the mass difference $\Delta m(K_L^0 - K_S^0)$ in our scheme.

4.2. Gauge Invariance for the S-Matrix in the Stochastic-Nonlocal Theory of Weak Interactions

The expansion of the S-matrix in powers of the normally ordered operators of the electromagnetic field $A_\mu(x)$ and the lepton fields $\psi(x)$ has the form

$$S = \sum_{n,m,l} \frac{1}{n!\,m!\,l!} \int d^4k_1 \cdots \int d^4k_n \int d^4p_1 \cdots \int d^4p_m \int d^4q_1 \cdots \int d^4q_l \times$$
$$\times F_{\mu_1,\ldots,\mu_n}(k_1,\ldots,k_n; p_1,\ldots,p_m; q_1,\ldots,q_l) {:} A_{\mu_1}(k_1) \cdots A_{\mu_n}(k_n) \times$$
$$\times \psi_j(p_1) \cdots \psi_j(p_m) \bar{\psi}_j(q_1) \cdots \bar{\psi}_j(q_l){:}; \quad (j = e, \mu, \nu_e, \nu_\mu). \tag{4.1}$$

The requirement of *gauge invariance* means that the coefficient functions $F_{\mu_1,\ldots,\mu_n}(\cdots)$ in expansion (4.1) satisfy the following conditions:

$$\begin{aligned} &k_{\mu_j} F_{\mu_1,\ldots,\mu_j,\ldots,\mu_n}(\cdots) = 0, \\ &k_{\mu_i} k_{\mu_j} F_{\mu_1,\ldots,\mu_i,\ldots,\mu_j,\ldots,\mu_n}(\cdots) = 0, \\ &\cdot \quad \cdot \quad \cdot \end{aligned} \tag{4.2}$$

Let us remark that each of the conditions (4.2) are fulfilled when other momenta in the function $F_{\mu_1,\ldots,\mu_n}(\cdots)$ are on the mass shell. The series of the perturbation theory contains three types of diagrams: diagrams with purely weak vertices, with weak and electromagnetic, and with purely electromagnetic vertices. Investigation of the last type of diagrams will not be carried out here because they represent gauge-invariant quantum electrodynamics constructed by us in Chapter 3. Proof of the fulfilment of condition (4.2) for the diagrams with mixed weak and electromagnetic vertices is simple. Indeed, in the considered four-fermion theory of weak interactions the Ward identity

$$(\partial/\partial p_\mu)\tilde{\Sigma}_R(p) = -\tilde{\Gamma}^R_\mu(p, 0),$$

is valid since it is a consequence of identities (3.24) and (3.17) at $k = 0$. By definition:

$$\mathrm{d}_\mu(k)|_{k\to 0} F(q) = (\partial/\partial q_\mu)F(q).$$

Here $\tilde{\Sigma}_R(p)$ and $\tilde{\Gamma}^R_\mu(p, q)$ correspond to the diagrams of the *self-energy* (Figure 6a) and the vertex (Figure 6b), respectively.

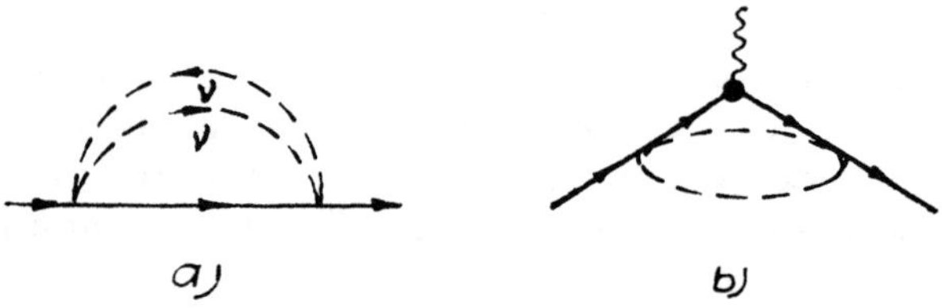

Fig. 6. The Feynman diagrams of the self-energy and of the vertex functions in four-fermion weak interactions.

Proof of the gauge invariance shown in (4.2) in the series of perturbation theory is quite simple and is based on the identity (3.18). The diagrams of *closed loops* constructed by propagators of charged leptons are gauge invariant due to the d-operation.

Now let us proceed to investigate these diagrams (see, for example, the diagrams shown in Figure 7) in the series of perturbation theory. First consider the diagram represented in Figure 7a. In the momentum representa-

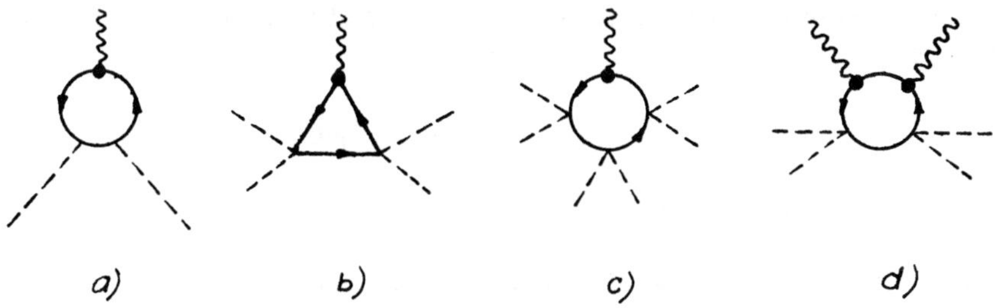

Fig. 7. The diagrams of some closed loops constructed by propagators of charged leptons in stochastic (nonlocal) theory.

tion the term of the S-matrix corresponding to this diagram is given by the following expression

$$F^{(2)}_{\mu\alpha}(k) = \frac{ieG}{\sqrt{2}\,(2\pi)^4} \int \mathrm{d}^4 p \,\mathrm{Sp}\, f^{(2)}_{\mu\alpha}(p, k), \tag{4.3}$$

where

$$f^{(2)}_{\mu\alpha} = \mathrm{d}_\mu(k) S_R(\hat{p}) 0_\alpha, \quad (G = G_F).$$

Taking into account formulas (3.18) and (3.24) we obtain

$$k_\mu \,\mathrm{d}_\mu(k) S_R(\hat{p}) = S_R(\hat{p} + \hat{k}) - S_R(\hat{p}). \tag{4.4}$$

From this it is easily seen that

$$k_\mu F^{(2)}_{\mu\alpha}(k) = 0.$$

By using the main property (4.4) for the d-operation we can easily prove the fulfilment of the gauge invariance condition for the other matrix elements of closed loops, shown in Figure 7. The matrix element corresponding to Figure 7b is given by

$$f^{(2)}_{\mu\alpha\beta}(p, k) = \mathrm{d}_\mu(k)[S_R(\hat{p} + \hat{q}) 0_\alpha S_R(\hat{q}) 0_\beta], \tag{4.5}$$

where p, k are the external momenta, q is the internal momentum over which one carries out integration. Making use of identity (4.4) we obtain

$$\begin{aligned} k_\mu f^{(3)}_{\mu\alpha\beta} &= [S_R(\hat{p} + \hat{q} + \hat{k}) - S_R(\hat{p} + \hat{q})] 0_\alpha S_R(\hat{q}) 0_\beta + \\ &\quad + S_R(\hat{p} + \hat{k} + \hat{q}) 0_\alpha [S_R(\hat{q} + \hat{k}) - S_R(\hat{q})] 0_\beta \\ &= -S_R(\hat{p} + \hat{q}) 0_\alpha S_R(\hat{q}) 0_\beta + S_R(\hat{p} + \hat{k} + \hat{q}) 0_\alpha S_R(\hat{q} + \hat{k}) 0_\beta. \end{aligned} \tag{4.6}$$

Elementary integration of the expression of type (4.3) with (4.6) over q gives the following identity:

$$k_\mu F^{(3)}_{\mu\alpha\beta}(p, k) = 0.$$

Now we consider the fourth-order diagrams (Figure 7c). In this case the terms of types $f^{(2)}_{\mu\alpha}$ and $f^{(3)}_{\mu\alpha\beta}$ have the form

$$f^{(4)}_{\mu\alpha\beta\gamma}(k, q_1, q_2) = \mathrm{d}_\mu(k)[S_R(\hat{q})0_\alpha S_R(\hat{q} + \hat{q}_1)0_\beta S_R(\hat{q} + \hat{q}_1 - \hat{q}_2)0_\gamma].$$

From this we easily obtain

$$\begin{aligned} k_\mu f^{(4)}_{\mu\alpha\beta\gamma} &= [S_R(\hat{q} + \hat{k}) - S_R(\hat{q})]0_\alpha S_R(\hat{q} + \hat{q}_1)0_\beta S_R(\hat{q} + \hat{q}_1 - \hat{q}_2)0_\gamma + \\ &\quad + S_R(\hat{q} + \hat{k})0_\alpha \times [S_R(\hat{q} + \hat{q}_1 + \hat{k}) - S_R(\hat{q} + \hat{q}_1)]0_\beta S_R(\hat{q} + \hat{q}_1 - \hat{q}_2)0_\gamma + \\ &\quad + S_R(\hat{q} + \hat{k})0_\alpha S_R(\hat{q} + \hat{q}_1 + \hat{k})0_\beta \times [S_R(\hat{q} + \hat{q}_1 - \hat{q}_2 + \hat{k}) - \\ &\quad - S_R(\hat{q} + \hat{q}_1 - \hat{q}_2)]0\gamma \\ &= -S_R(\hat{q})0_\alpha S_R(\hat{q} + \hat{q}_1)0_\beta S_R(\hat{q} + \hat{q}_1 - \hat{q}_2)0_\gamma + \\ &\quad + S_R(\hat{q} + \hat{k})0_\alpha S_R(\hat{q} + \hat{q}_1 + \hat{k})0_\beta S_R(\hat{q} + \hat{q}_1 - \hat{q}_2 + \hat{k}))0_\gamma. \end{aligned}$$

In the second term of this expression, the substitution $q + k \rightarrow q$ and integration over $\mathrm{d}^4 q$ give

$$k_\mu F^{(4)}_{\mu\alpha\beta\gamma} = 0.$$

Finally, let us consider one more diagram, shown in Figure 7d. The term $f^{(2,2)}_{\mu\nu\alpha\beta}(k_1, k_2, p)$ acquires the form

$$f^{(2,2)}_{\mu\nu\alpha\beta} = \mathrm{d}_\nu(k_2)\{\mathrm{d}_\mu(k_1)[0_\alpha S_R(\hat{p} + \hat{q})0_\beta S_R(\hat{q})]\}. \tag{4.7}$$

From this we see that due to equality (4.6) we obtain the following identity

$$k_{1\mu} k_{2\nu} F^{(2,2)}_{\mu\nu\alpha\beta} = 0.$$

Thus, the algebraic relations we have found show that within our model the terms of the S-matrix satisfy the condition of gauge invariance in form (4.2) in each perturbation order.

4.3. Calculation of the 'Weak' Corrections to the Anomalous Magnetic Moment (AMM) of Leptons

In testing the locality of quantum theory, the calculation of corrections due to the weak interactions to the quantum electrodynamical processes is always very interesting. In the case of actual discrepancy between the local quantum electrodynamics and an experiment, which is to be expected at very high energies, the 'weak' corrections would contain, however, just the breakdown of QED. From this the conclusion may be drawn that local QED can be violated, and weak and electromagnetic forces can be equal in magnitude. It is quite possible that in this domain of energies the process of unification of weak and electromagnetic interactions starts.

The present section is devoted to the calculation of corrections to the AMM of the leptons, based on the assumption that the parameter $l \to l_w \sim 10^{-16}$ cm, which characterizes the domain of unification of weak and electromagnetic interactions.

In the lowest order of the coupling constant G, the corrections to the anomaly of leptons are given by two types of diagrams (Figure 8) corresponding to both the diagonal and nondiagonal weak interactions, i.e., to two types of terms of the interaction Lagrangian $\mathscr{L}_w$ in (3.29). Within the framework of the nonlocal theory the diagrams of such a type was discussed in detail by Efimov *et al.* (1973). Therefore, we shall not calculate it in detail, but give only the main result. Hence, in the stochastic theory the terms corresponding to these diagrams, after a Fierz reordering, may be rewritten in the form

$$\tilde{\Gamma}^{(j)}_\mu(p, q) = \lim_{\delta \to 0} (2\pi)^{-4} \frac{G^2}{2i} \int d^4 k N^\delta_{\alpha\beta}(k) 0_\alpha S^\delta_R(\hat{p} + \hat{k}) u^\delta_\mu(p, k) \times$$
$$\times S^\delta_R(\hat{p}' + \hat{k}) 0_\beta \ (j = e, \mu), \tag{4.8}$$

where the symbol δ means the intermediate regularization procedure used above. Here p and p' are the external momenta of the leptons, $p = p' + q$ and

$$S_R(\hat{p} + \hat{k}) u_\mu(p, k) S_R(\hat{p}' + k) = (m_j - \hat{p} - \hat{k})^{-1} \gamma_\mu S_R(\hat{p}' + \hat{k}) + (m_j - \hat{p} - \hat{k})^{-1} \frac{\hat{q}\gamma_\mu}{q^2} \times$$
$$\times [V_{m_j}(-l^2(p' + k)^2) - V_{m_j}(-l^2(p + k)^2)]. \tag{4.9}$$

The function $N_{\alpha\beta}(k)$ corresponds to the neutrino loop and is given by (see below Section 4.4.2):

$$N_{\alpha\beta}(k) = \frac{1}{2\pi^2} \frac{1}{2i} \int\limits_{\substack{-\delta + i\infty \\ (0 < \delta < 1)}}^{-\delta - i\infty} d\zeta \frac{v(\zeta)}{\sin \pi\zeta} (-l^2 k^2)^\zeta \frac{1}{2i} \int\limits_{\substack{-\rho + i\infty \\ (0 < \rho < 1)}}^{-\rho - i\infty} d\eta \frac{v(\eta)}{\sin \pi\eta} \times$$
$$\times (-l^2 k^2)^\eta \frac{\Gamma(2 + \zeta)\Gamma(2 + \eta)}{\Gamma(1 - \zeta)\Gamma(1 - \eta)} \times \frac{1}{\Gamma(4 + \zeta + \eta)} \times$$
$$\times [-2k_\alpha k_\beta \Gamma(-\zeta - \eta) - g_{\alpha\beta} k^2 (2 + \zeta + \eta)\Gamma(-1 - \eta - \zeta)]. \tag{4.10}$$

Here $Re(-\eta - \zeta) = \delta + \rho < 0$ and $Re(-1 - \eta - \zeta) = \delta + \zeta - 1 < 0$ in first and second terms of (4.10), respectively. Substituting (4.9) and (4.10) into (4.8), integration over d^4k and assuming $m_j^2 l^2 \ll 1$, we get the terms giving the contribution of weak interactions to the AMM of the leptons:

$$\tilde{\Gamma}^{(j)}_\mu(q) = \tilde{\Gamma}^{(j)}_\mu(p, q)|_{p^2 = p'^2 = m_j^2} = \frac{i}{2m_j} \sigma_{\mu\nu} q^\nu F(q^2)$$

and

$$a^w_j = F(0) = (2\pi)^{-4} l^{-2} G^2 m_j^2 B.$$

Here the quantity B is given by

$$B = \frac{1}{12}\frac{\pi}{2i}\int_{-\delta+i\infty}^{-\delta-i\infty} d\zeta\,\frac{v(\zeta)}{\sin\pi\zeta}\,\frac{1}{2i}\int_{-\rho+i\infty}^{-\rho-i\infty} d\eta\,\frac{v(\eta)}{\sin\pi\eta}\times$$

$$\times\frac{\Gamma(2+\zeta)\Gamma(2+\eta)\Gamma(-\eta-\zeta)v(-1-\eta-\zeta)}{\Gamma(1-\zeta)\Gamma(1-\eta)\Gamma(4+\eta+\zeta)\sin\pi(\eta+\zeta)}\times$$

$$\times\frac{1}{1+\eta+\zeta}[48+54(\zeta+\eta)+9(\eta+\zeta)^2-(\eta+\zeta)^3]. \quad (4.11)$$

An integral of the type of (4.11) will be investigated in Appendix B. Shifting in turn the contour of integration to the right we can reduce this integral to the double series. The result of numerical calculations gives $B = 2.4$ for the form-factor v_b (1.15).

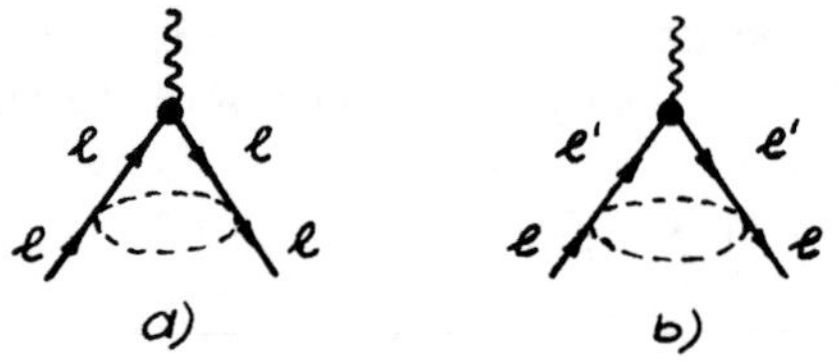

Fig. 8. The type of diagrams which give contributions to the anomalous magnetic moment of the leptons ($l = e, \mu$).

Thus, supposing $l \to l_w \sim 10^{-16}$ cm, it is easy to show that the obtained 'weak' contribution a_j^w is much less than the experimental errors for the value AMM $a_{\text{exp}}(e^-)$ and $a_{\text{exp}}(\mu)$ of the leptons.

4.4. Some Consequences of Neutrino Oscillations in Stochastic-Nonlocal Theory

4.4.1. *Introduction*

Many papers have appeared recently in which the problem of neutrino oscillations are considered within different approaches (see, for example, Bilenky and Pontecorvo, 1980a, b; Bilenky *et al.*, 1980; Vuilleumier, 1979; Barger *et al.*, 1980b; for earlier works see the reviews of Bilenky and Pontecorvo, 1978; Wachsmuth, 1979; Morrison, 1980; and for a more popular presentation see Thomsen, 1980; Sutton, 1980; De Rujula and Glashow, 1980). The possibility of neutrino oscillations was first considered by Pontecorvo (1967): he assumed that the oscillation may appear, if in addition to the usual weak interaction, there is another interaction which does not conserve the lepton number. Such a picture is similar to the *oscillation* in the system of neutral kaons. Pontecorvo showed that massive neutrinos might change their identities during the evolution of time. A particle which is born as an electron neutrino in a beta decay may periodically behave as if it were a muon neutrino or a tau neutrino.

Many recent papers (see, for example, Cheng and Ling-Fong Li, 1980; Kang *et al.*, 1980; Yanagida and Yoshimure, 1980; Neutrino 1982 conference and references therein) devoted to the problem of neutrino oscillations, deal with the *unified theory of weak and electromagnetic interactions*. Thus, in addition to the standard hypothesis of lepton–quark analogy, in these works a new conjecture is proposed that leptons, like quarks, are *mixed*. In such a theory the oscillations $\nu_e \rightleftarrows \nu_\mu \rightleftarrows \nu_\tau$ appear.

Possible indications for neutrino oscillations have been obtained in the beam-dump experiments at CERN (De Rujuila *et al.*, 1980) and in the experiments with reactor antineutrinos (Barger *et al.*, 1980a; and Reines *et al.*, 1980). A number of experiments searching for neutrino oscillations stimulated recent interest in the question, since they appear to give some indications for nonzero neutrino masses. A direct experiment of Lubimov and co-workers (see Kozik *et al.*, 1980) on measuring the $\tilde{\nu}_e$-mass from 3H-decay gives the mass of the electron neutrino as between 14 and 46 eV, and most probably to be 36 ± 10 eV at 95% confidence level. The results of Reines, Pasierb and Sobel (1980) gave no direct implication for the neutrino mass, although they implied that the difference in mass between the two basic states lies in the region of 1 eV. It is worth noting that in the recent developments of grand unified models, the possibility of the appearance of neutrino masses is also discussed (Barbieri *et al.*, 1980a, b; Gell-Mann *et al.*, 1979; Georgi and Nanopoulos, 1979; and Witten, 1980).

Thus, we believe that a neutrino oscillation mechanism does exist in nature and we will expect precise experiments on the properties of neutrinos.

Our aim here is modest, and consists only in considering this problem from the viewpoint of the stochasticity of space-time. We believe that due to the neutrino oscillation mechanism there exist mixed states of neutrinos which give nonorthogonality between neutrinos, say, ν_e and ν_μ, $\langle \nu_e | \nu_\mu \rangle \neq 0$ (ν_τ oscillations are not considered here). In our case it is postulated that the difference in behavior between ν_e and ν_μ is caused by the internal properties of these particles and depends on the nonlocal effects of weak interactions. For example,

$$\nu_\mu = \nu_e + f\nu = \nu_e - (1 - \phi)\nu, \tag{4.12}$$

where f(or ϕ) is some parameter which may be connected to a mixing angle and a value of mass difference of neutrino states, and ν is a massive neutrino (basic state) which possesses the properties of both the ν_e and ν_μ neutrinos. In the representation (4.12) the *transition propagator* between ν_e and ν_μ has the form

$$\langle 0|T[\psi_{\nu_e}(x)\bar{\psi}_{\nu_\mu}(y)]|0\rangle = \phi D_{\nu_e\nu}(x-y), \tag{4.13}$$

where we put $D_{\nu_e\nu_e}(x) = D_{\nu_e\nu}(x)$ since the ν neutrino possesses the property of ν_e.

If the neutrino mixing is assumed, the *exotic decays* $\mu \to e\gamma$, $\mu \to 3e$, $K^\pm \to \mu e\pi^\pm$, $K_L^0 \to \mu e$, etc., are principle, possible. Notice that these decays are forbidden by the usual theory. These processes appear in the higher orders of the perturbation theory. The present section is devoted to the investigation of these decays within the framework of our approach. It is shown here that if the parameter l for the weak interactions is of an order of $l \to l_w \sim 10^{-16}$ cm and if neutrino mixing takes place, then the probability for these decays is close to the experimental upper bounds (see, for example, Bricman *et al.*, 1980; and Fiorini, 1979). In this section, for example, we will consider the decays $\mu \to 3e$ and $K_L^0 \to \mu e$ in detail.

4.4.2. *The $\mu \to 3e$ Decay*

Within the framework of our model of weak interactions the probability decay is determined by the diagram shown in Figure 9a. Before we proceed to estimating corrections from these diagrams to the probability of decays $\mu \to 3e$ and $K_L^0 \to \mu e$, let us consider diagrams of the type of neutrino–neutrino, neutrino–lepton and lepton–lepton loops (Figure 10a, b, c). Here we study one of these diagrams, say, the neutrino–lepton loop corresponding to the diagram shown in Figure 10b. Its matrix element has the form

$$\Pi_{\alpha\beta}(p) = \frac{1}{(2\pi)^4 i}\int \mathrm{d}^4k\, \mathrm{Sp}[\hat{k}0_\alpha(m+\hat{k}+\hat{p})0_\beta]\frac{V_0(-k^2l^2)}{-k^2-i\varepsilon}\frac{V_m(-(k+p)^2l^2)}{m^2-(k+p)^2-i\varepsilon}.$$

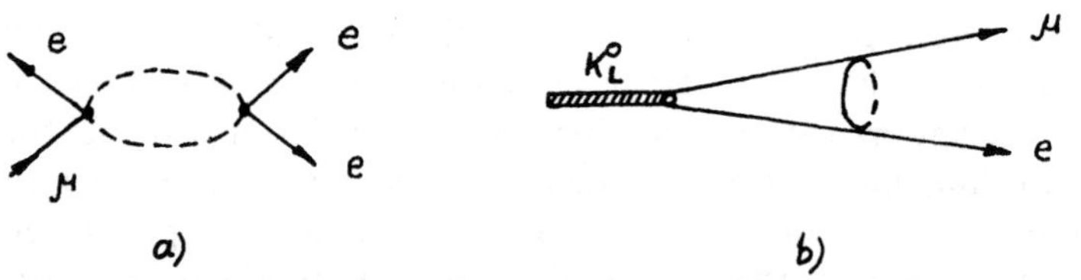

Fig. 9. The Feynman diagrams determining the probability of decays $\mu \to 3e$ and $K_L^0 \to \mu e$.

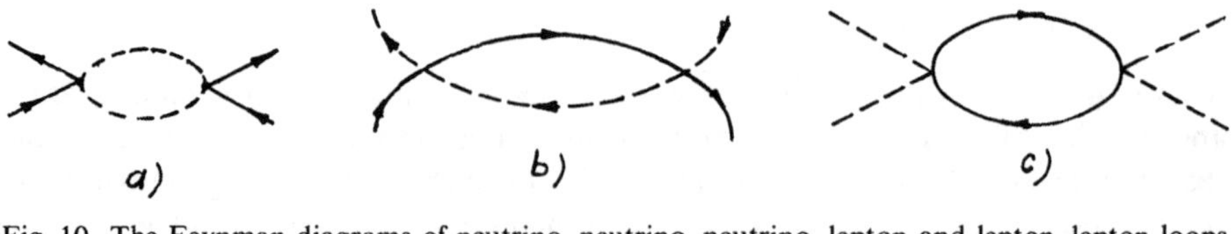

Fig. 10. The Feynman diagrams of neutrino–neutrino, neutrino–lepton and lepton–lepton loops.

After the standard calculations we have

$$\Pi_{\alpha\beta}(p) = \frac{1}{2\pi^2}\frac{1}{2i}\int_{\substack{-\delta+i\infty\\(0<\delta<1)}}^{-\delta-i\infty} \mathrm{d}\zeta \frac{v(\zeta)}{\sin \pi\zeta}(m^2 l^2)^{\zeta}\frac{1}{2i}\int_{\substack{-\gamma+i\infty\\(0<\gamma<1)}}^{-\gamma-i\infty} \mathrm{d}\eta \frac{v(\eta)}{\sin \pi\eta} \times$$

$$\times (m^2 l^2)^{\eta}\frac{1}{\Gamma(1-\zeta)\Gamma(1-\eta)} \times \int_0^1 \mathrm{d}x\left(\frac{1-x}{x}\right)^{\zeta}\left(1-\frac{p^2}{m^2}\right)^{\eta+\zeta} \times$$

$$\times [(-2p_\alpha p_\beta + g_{\alpha\beta}p^2)x(1-x)\Gamma(-\eta-\zeta) +$$

$$+ g_{\alpha\beta}\Gamma(-1-\eta-\zeta)(m^2 - p^2 x)(1-x)]. \tag{4.14}$$

We notice that the neutrino–neutrino loop (4.10) in Section 4.3 is obtained by substituting $m = 0$ in expression (4.14).

So, by definition (4.13) the matrix element corresponding to the $\mu \to 3e$ decay has the form

$$M(\mu \to 3e) = \phi(G/\sqrt{2})^2(\bar{e}0_\alpha \mu)N_{\alpha\beta}(q)(\bar{e}0_\beta e),$$

where $N_{\alpha\beta}(q)$ is the neutrino–neutrino loop determined by (4.10). Then the square of the matrix element equals

$$\sum_{\text{spin}} |M|^2 = 4(\pi l)^{-4} G^4\phi^2 c^2[(k_1 k_2)(k_0 k_3) + (k_0 k_2)(k_1 k_3)], \tag{4.15}$$

where

$$c = -\frac{\pi}{2i}\int_{\substack{-\beta+i\infty\\(0<\beta<1)}}^{-\beta-i\infty} \mathrm{d}\zeta \frac{v(\zeta)v(-1-\zeta)}{\sin^2 \pi\zeta}, \quad (k_0 = k_1 + k_2 + k_3). \tag{4.16}$$

In expression (4.15) for the square of the matrix element we take into account main terms of the order of (G^2/l^2). After integration over phase space of the electrons and averaging over the muon spin we have

$$W(\mu \to 3e) = \frac{1}{192}\frac{G^4\phi^2 c^2}{4l^4\pi^7}m_\mu^5.$$

The probability branching ratio of $\mu \to 3e$ to $\mu \to e\nu\tilde{\nu}$ constitutes

$$B(\mu \to 3e) = W(\mu \to 3e)/W(\mu \to e\nu\tilde{\nu}) = G^2\phi^2 c^2(\pi l)^{-4}/4, \tag{4.17}$$

where

$$c = \begin{cases} 2.3 & \text{for } v = v_b, \\ 1.4 & \text{for } v = v_s, \\ 0.65 & \text{for } v = v_1. \end{cases}$$

Here v_b, v_s, and v_1 are determined by formulas (1.15), (2.99) and (3.39), respectively, and $|\phi| \lesssim 0.055$ (see Lee *et al.*, 1977). Assuming $\phi \sim 10^{-2}$ and

$l \to l_w \sim 2 \times 10^{-16}$ cm we get

$$B(\mu \to 3e) \lesssim \begin{cases} 1.35 \times 10^{-8} & \text{for } v_b, \\ 4.9 \times 10^{-9} & \text{for } v_s, \\ 1.0 \times 10^{-9} & \text{for } v_1. \end{cases} \tag{4.18}$$

4.4.3. *The $K_L^0 \to \mu e$ Decay*

In the second order in G the decay $K_L^0 \to \mu e$ is described by the diagram in Figure 9b. The term of the S-matrix corresponding to this diagram can be rewritten in the form

$$i\sqrt{2}\, f_{KN\Lambda}(G/\sqrt{2})^2\, \bar{\mu}(p_-)\Gamma(p_-, q)e(p_+)\varphi_K \cos\theta_c \sin\theta_c, \tag{4.19}$$

where

$$\Gamma(p_-, q) = \frac{\phi}{(2\pi)^4 i}\int \mathrm{d}^4k\, \Pi^{\nu}_{\alpha\beta}(k) 0_\alpha \frac{(m_N + \hat{k} + \hat{p}_+)V_m(-(k+p_+)^2 l^2)}{m_N^2 - (k+p_+)^2 - i\varepsilon}\gamma_5 \times$$

$$\times \frac{(m_N + \hat{k} - \hat{p}_-)V_m(-(k-p_-)^2 l^2)}{m_N^2 - (k-p_-)^2 - i\varepsilon} 0_\beta, \quad (m_N \sim m_\Lambda).$$

Here $\Pi^{\nu}_{\alpha\beta}(k)$ corresponds to the neutrino–nucleon loop determined by (4.14) at $m_L \to m_N$. The hadron current is chosen in the Cabibbo form. After tedious but elementary calculations we obtain

$$\Gamma(p_-, q) = -m_N \hat{q}(1-\gamma_5)\frac{1}{16\pi^4 l^2} A,$$

where

$$A = \frac{\pi}{2i}\int_{-\gamma+i\infty}^{-\gamma-i\infty} \mathrm{d}y \frac{v(y)}{\sin \pi y}\frac{1}{2i}\int_{-\beta+i\infty}^{-\beta-i\infty} \mathrm{d}\zeta \frac{v(\zeta)}{\sin\pi\zeta}\frac{1}{2i}\int_{-\alpha+i\infty}^{-\alpha-i\infty} \mathrm{d}\eta \frac{v(\eta)}{\sin\pi\eta} \times$$

$$\times \frac{v(-1-\eta-\zeta-y)\Gamma(2+\zeta)\Gamma(2+\eta)\Gamma(-1-\eta-\zeta)}{\sin\pi(y+\eta+\zeta)\Gamma(1-\eta)\Gamma(1-\zeta)\Gamma(3+\eta+\zeta)}, \quad (0 < \alpha, \beta, \gamma < 1). \tag{4.20}$$

As above, after displacement of the contours of integration to the right we get (see, Appendix B)

$$A = 2\frac{\pi}{2i}\int_{-\beta+i\infty}^{-\beta-i\infty} \mathrm{d}\zeta \frac{v(\zeta)}{\sin^2\pi\zeta}\cdot\frac{1}{\zeta(1+\zeta)(2+\zeta)} \times$$

$$\times \sum_{n=0}^{\infty} [v(n)v(-1-n-\zeta) - v(n-\zeta)v(-1-n)]. \tag{4.21}$$

Now the matrix elements (4.19) acquire the form

$$M = i\sqrt{2}\, f_{KN\Lambda}(G/\sqrt{2})^2 \cos\theta_c \sin\theta_c \frac{m_N \phi}{16\pi^4 l^2} A\cdot\varphi_K[\bar{\mu}(p_-)\hat{q}e(p_+)],$$

and its square is

$$|M|^2 = 2m_K^2(1 - m_\mu^2/m_K^2)A^2 g^2,$$

where

$$g = f_{KN\Lambda}\frac{G^2}{2}\cos\theta_c \sin\theta_c \frac{m_\mu m_N \cdot \phi}{16\pi^4 l^2}.$$

Integration of the decay probability over the phase space of two leptons gives

$$W(K_L^0 \to \mu e) = \frac{m_K}{8}[1 - (m_\mu^2/m_K^2)]^2 \times$$
$$\times A^2 \cos^2\theta_c \sin^2\theta_c \cdot \phi^2 G^4 f_{KN\Lambda}^2 m_\mu^2 m_N^2 \pi^{-8} l^{-4}/512\pi \quad (4.22)$$

where $\sin\theta_c = 0.23$, $f_{KN\Lambda}^2/4\pi = 10.0 \pm 2.8$ (see, for example, Ebel, 1970; Nagels *et al.*, 1976), $\phi \sim 10^{-2}$, and $l \to l_w \sim 2\times 10^{-16}$ cm;

$$A = \begin{cases} -1.8 & \text{for } v = v_b, \\ -2.7\times 10^{-1} & \text{for } v = v_s. \end{cases} \quad (4.23)$$

Thus, the branching ratio of this exotic decay in our model is

$$B = \frac{W(K_L^0 \to \mu e)}{W(K_L^0 \to \text{all})} \lesssim \begin{cases} 1.2\times 10^{-8} & \text{for } v = v_b, \\ 1.4\times 10^{-10} & \text{for } v = v_s. \end{cases}$$

We see that the exotic decays $\mu \to 3e$ and $K_L^0 \to \mu e$ depend on the *form-factors* of the nonlocal theory. Our results, together with the calculations by

Table 4.1.

Processes	Our calculations		Results of Cheng and Li	Experimental upper limits
$\mu \to 3e$	1.3×10^{-8} 4.9×10^{-9} 1.0×10^{-9}	$v = v_b$ $v = v_s$ $v = v_1$	10^{-12}	1.9×10^{-9}
$K_L^0 \to \mu e$	1.2×10^{-8} 1.4×10^{-10}	$v = v_b$ $v = v_s$	10^{-10}	2×10^{-9}
$K_L^0 \to \mu^+\mu^-$	1.2×10^{-6}	$v = v_s$	—	$(9.1 \pm 1.9)\times 10^{-9}$
$\Delta m(K_L^0 - K_S^0)$	$5\times 10^{11}\,\hbar\,\text{sec}^{-1}$ $2\times 10^{11}\,\hbar\,\text{sec}^{-1}$	$v = v_s$ $v = v_1$	—	$0.5\times 10^{10}\,\hbar\,\text{sec}^{-1}$

Cheng and Ling-Fong Li (1977) and the experimental upper bounds [Bricman *et al.*, 1980 (Particle Data Group)], are summarized in Table 4.1. These numerical calculations are presented for a purely illustrative purpose. These are important in a sense that they permit one to estimate a parameter of mixing at a given value of elementary length $l \to l_w \sim 10^{-16}$ cm. The introduced parameter ϕ is connected to the mixing angle and the mass difference of neutrinos ν_1 and ν_2 (or heavy leptons N_1 and N_2) of other models of weak interactions by the formula

$$\phi \sim \sin\varphi \cdot \cos\varphi \cdot \Delta m_i, \quad (i = \nu_i, N_i).$$

4.5. Neutrino Electromagnetic Properties in the Stochastic-Nonlocal Theory of Weak Interactions

In recent years great attention has been paid to the physical properties of neutrinos. This is connected with the problem of neutrino oscillations and its mass, and astrophysical consequences (Faber and Gallagher, 1979; Schramm, 1979; Schramm and Steigman, 1980; this problem has been discussed by many authors at the Neutrino 1979 and 1982 Conferences), and also with the rapid development of neutrino experiments carried out at CERN, Caltech-FINAL, Serpukhov, etc. (see, for example, Busser, 1980; Winter, 1979; Baltay, 1979; Arbuzov, 1975).

As usual, the neutrino is considered as a weak-interacting particle with zero mass and without an electric charge. Therefore among the electromagnetic characteristics of neutrinos the only nonvanishing quantity is its *charged radius* r_ν. Possible experiments on measuring the charged radius of the neutrinos have been pointed out by Andryushin *et al.* (1971).

However, it seems that the neutrino mass is not zero (see previous section). Then the neutrino may possess the *magnetic moment* a_ν. Recently, the magnetic moment of a massive neutrino has been discussed by Fujikawa and Shrock (1980).

Starting with the analysis of experimental data on inclusive reactions $\nu(\tilde{\nu}) + N \to \nu(\tilde{\nu}) +$ anything and $\nu_\mu - e$ elastic scattering, Bardin and Mogilevski (1974); Kim *et al.*, (1974) have investigated the electromagnetic properties of neutrinos and calculated the correction to the cross-sections of these reactions due to Feynman diagrams involved in one-photon exchange, and obtained the restriction on the charged radius and magnetic moment of the neutrino.

The contribution due to the one-photon exchange calculated by these authors is called electromagnetic, although, as is known, the charged radius and magnetic moment of the neutrino must appear due to the effects of the weak interactions. The calculation of these quantities in the usual theory of weak interactions meets with difficulties because of divergences in the S-matrix elements.

The present section is devoted to the calculation of contributions to r_ν and a_ν within the framework of our approach formulated in this book. In Feynman diagrams of the order eG^2 giving the corrections to r_ν and a_ν there are *closed loops* constructed by propagators of the charged leptons and neutrinos for calculations of which it is necessary to apply the method of the stochastic theory.

In the stochastic-nonlocal theory of weak interactions the electromagnetic interaction of the neutrino, say, the muon one, in the lowest order of G is given by the following Feynman diagrams (Figure 11). It should be noted that in the stochastic-nonlocal four-fermion weak interaction theory the Fierz reordering is not yet studied so that in this section we consider all possible diagrams which give contributions to the electromagnetic form-factors of neutrino.

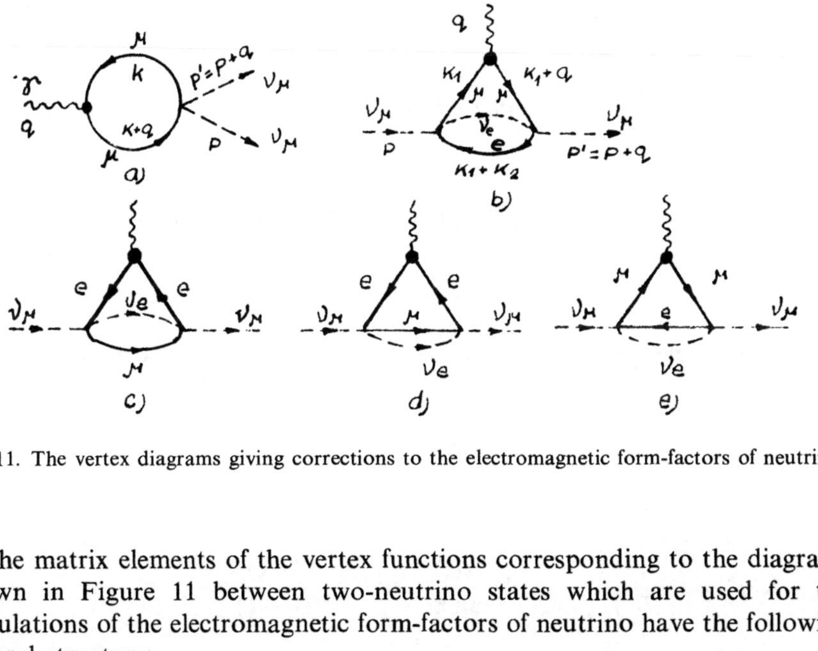

Fig. 11. The vertex diagrams giving corrections to the electromagnetic form-factors of neutrino.

The matrix elements of the vertex functions corresponding to the diagrams shown in Figure 11 between two-neutrino states which are used for the calculations of the electromagnetic form-factors of neutrino have the following general structure:

$$M = ie\bar{u}_\nu(p')\{\gamma_\mu F_1(q^2) + i\sigma_{\mu\lambda}q^\lambda F_2(q^2)\}u_\nu(p), \tag{4.24}$$

where

$$F_1(q^2) = \tfrac{1}{6}r_\nu^2 q^2, \qquad F_2(0) = \frac{1}{2m_e}a_\nu, \qquad q = p' - p.$$

Here r_ν^2 is the mean-square charged radius of the neutrino and a_ν is its magnetic moment in the units of electron Bohr magnetons.

In order to calculate the contributions from the diagrams shown in Figure

11 to r_ν and a_ν we shall study them separately. First, let us consider the lepton loop (Figure 11a). The term for the S-matrix corresponding to this diagram has the form

$$M_1 = ie\frac{G}{\sqrt{2}}\bar{u}_\nu(p')0_\alpha u_\nu(p)K_{\alpha\beta}(q)A^\beta(q), \tag{4.25}$$

where the $K_{\alpha\beta}(q)$ in the stochastic theory is given by

$$K_{\alpha\beta}(q) = \frac{1}{(2\pi)^4 i}\int d^4k\ \mathrm{Sp}[0_\alpha S_R(\hat{k}+\hat{q})u_\beta(q,k)S_R(\hat{k})],$$

where

$$u_\beta(q,k) = -\dot{d}_\beta(q)S_R^{-1}(\hat{k}).$$

On the other hand, taking into account the identity (3.24) for the d-operation we have

$$K_{\alpha\beta}(q) = \frac{1}{(2\pi)^4 i}\int d^4k\ \mathrm{Sp}[0_\alpha \dot{d}_\beta(q)S_R(\hat{k})].$$

By definition

$$d_\beta(q)S_R(\hat{k}) = (m-\hat{k}-\hat{q})^{-1}\gamma_\beta V_m(-k^2l^2)(m-\hat{k})^{-1} + (m-\hat{k}-\hat{q})^{-1}\times$$

$$\times [V_m(-(k+q)^2l^2) - V_m(-k^2l^2)]\frac{q\gamma_\beta}{q^2}.$$

After some elementary calculations we obtain the *gauge-invariant expression* for $K_{\alpha\beta}(q)$:

$$K_{\alpha\beta}(q) = \frac{1}{2\pi^2}(q_\alpha q_\beta - g_{\alpha\beta}q^2)\tilde{K}(q^2),$$

where

$$\tilde{K}(q^2) = -\frac{1}{2i}\int_{\substack{-\alpha+i\infty \\ (0<\alpha<1)}}^{-\alpha-i\infty} d\zeta\,\frac{v(\zeta)}{\sin\pi\zeta}(m^2l^2)^\zeta\frac{\Gamma(-\zeta)}{\Gamma(1-\zeta)}\times$$

$$\times\int_0^1 dx\ x(1-x)^{1-\zeta}\left[1-\frac{q^2}{m^2}x(1-x)\right]^\zeta.$$

Making use of the identity

$$\Gamma(-\zeta) = -\pi/\sin\pi\zeta\Gamma(1+\zeta),$$

we have

$$\tilde{K}(q^2) = \frac{\pi}{2i}\int_{-\alpha+i\infty}^{-\alpha-i\infty} d\zeta\,\frac{v(\zeta)}{\sin^2\pi\zeta}\frac{1}{\Gamma(1-\zeta)\Gamma(1+\zeta)}\int_0^1 dx\ x(1-x)^{1-\zeta}\times$$

$$\times \left[1 - \frac{q^2}{m^2} x(1-x)\right]^{\zeta}$$

$$= \int_0^1 \mathrm{d}x\, x(1-x)\left\{v'(0) + \ln m^2 l^2 + \ln \frac{m^2 - q^2 x(1-x)}{m^2(1-x)}\right\}.$$

Here

$$\int_0^1 \mathrm{d}x\, x(1-x)\ln(1-x) = \frac{\Gamma(2)\Gamma(2)}{\Gamma(4)}[\psi(2) - \psi(4)] = -\tfrac{5}{36},$$

$$\psi(1+n) = \psi(n) + \frac{1}{n}, \qquad \psi(1) = -C,$$

$C \simeq 0.577$ is the Eulerian constant.

The first diagram considered gives a contribution to the mean-square radius of the neutrino only:

$$r_{1\nu}^2 = 6\frac{\partial}{\partial q^2} F_1^{(1)}(q^2)|_{q^2=0} = \frac{G}{\sqrt{2}}\frac{1}{2\pi^2}[\tfrac{5}{6} - v'(0) - \ln m^2 l^2]. \tag{4.26}$$

Assuming, as above, $l \to l_w \sim 2 \times 10^{-16}$ cm we obtain

$$r_1^2 = \begin{cases} 1 \times 10^{-33}\,\mathrm{cm}^2 & \text{for } v = v_b, \\ 0.9 \times 10^{-33}\,\mathrm{cm}^2 & \text{for } v = v_s, \\ 1.1 \times 10^{-33}\,\mathrm{cm}^2 & \text{for } v = v_1. \end{cases}$$

We see that the contribution from the diagram (Figure 11a) to the mean-square radius of the neutrino is of the order of 10^{-33} cm^2.

Now we turn to discussion of the vertex diagrams. There are only eight diagrams for each neutrino $\nu_i (i = e, \mu)$. Among them are diagrams of different structure, e.g., those shown in Figure 11b–e we shall begin with the calculation of these diagrams. The terms for the S-matrix corresponding to these diagrams are:

$$\Gamma_{1\rho}(p,q) = N_\alpha[d_\rho(-q)S_R^{(\mu)}(\hat{k}_1 + \hat{p})]0_\beta\, \mathrm{Sp}\{0_\alpha S_R^{(e)}(\hat{k}_1 + \hat{k}_2)0_\beta S_R^{(\nu)}(\hat{k}_2)\},$$

$$\Gamma_{2\rho}(p,q) = N_\alpha S_R^{(\mu)}(\hat{k}_1 + \hat{p})0_\beta\, \mathrm{Sp}\{0_\alpha[d_\rho(-q)S_R^{(e)}(\hat{k}_1 + \hat{k}_2)]0_\beta S_R^{(\nu)}(\hat{k}_2 - \hat{q})\},$$

$$\Gamma_{3\rho}(p,q) = N_\alpha S_R^{(\nu)}(\hat{k}_2 - \hat{q})0_\beta\, \mathrm{Sp}\{0_\alpha[d_\rho(-q)S_R^{(e)}(\hat{k}_1 + \hat{k}_2)]0_\beta S_R^{(\mu)}(\hat{k}_1 + \hat{p})\},$$

$$\Gamma_{4\rho}(p,q) = N_\alpha S_R^{(\nu)}(\hat{k}_2)0_\beta\, \mathrm{Sp}\{0_\alpha S_R^{(e)}(\hat{k}_1 + \hat{k}_2)0_\beta[d_\rho(-q)S_R^{(\mu)}(\hat{k}_1 + \hat{p})]\},$$

$$N_\alpha = i^{-1}(2\pi)^{-4}\int \mathrm{d}^4k_1\, i^{-1}(2\pi)^{-4}\int \mathrm{d}^4k_2 \cdot 0_\alpha, \tag{4.27}$$

respectively.

Let us consider the first expression of (4.27). We are interested only in those terms which give contributions to r_ν and a_ν in the limit $q^2 \to 0$. The structure of type (4.24) is obtained in the usual way. After calculations we have

$$\Gamma_{1\rho}(p,q) = \frac{1}{16\pi^4 l^2}\frac{1}{3}[\gamma_\rho q^2 F_1^{(1)} + i\sigma_{\rho\alpha} q^\alpha m_\nu F_2^{(1)}], \tag{4.28}$$

where

$$F_1^{(1)} = \rho(\zeta,\eta)\{2(3+2\eta+2\zeta)\Gamma(-1-\eta-\zeta) + \Gamma(-\eta-\zeta)[\tfrac{9}{2} - (1+\eta+\zeta)]\},$$

$$F_2^{(1)} = \rho(\zeta,\eta)\Gamma(-\eta-\zeta)[-\frac{1+\eta+\zeta}{2}(1+2\eta+2\zeta) - 13 - 6\eta - 6\zeta].$$

Here

$$\rho(\zeta,\eta) = \frac{\pi}{2i}\int_{-\beta+i\infty}^{-\beta-i\infty} d\zeta \frac{v(\zeta)}{\sin\pi\zeta} \times$$

$$\times \frac{1}{2i}\int_{-\gamma+i\infty}^{-\gamma-i\infty} d\eta \frac{v(\eta)}{\sin\pi\eta}\frac{v(-1-\eta-\zeta)\Gamma(2+\eta)\Gamma(2+\zeta)}{\sin\pi(\eta+\zeta)\Gamma(1-\eta)\Gamma(1-\zeta)\Gamma(4+\eta+\zeta)}.$$

Similar calculations give the following structure for $\Gamma_{4\rho}(p,q)$ and $\Gamma_{2\rho}(p,q)$:

$$\Gamma_{4\rho}(p,q) = \frac{1}{16 l^2 \pi^4}\frac{1}{6}\{\gamma_\rho q^2 F_1^{(4)} + i\sigma_{\rho\alpha} m_\nu q^\alpha F_2^{(4)}\}, \tag{4.29}$$

where

$$F_1^{(4)} = \rho(\zeta,\eta)\{[-2+2(1+\eta+\zeta)]\Gamma(-\eta-\zeta) - 10\Gamma(-1-\eta-\zeta)\},$$
$$F^{(4)} = \rho(\zeta,\eta)[-10 - 17(1+\eta+\zeta) + 4(1+\eta+\zeta)^2]\Gamma(-\eta-\zeta),$$

and

$$\Gamma_{2\rho}(p,q) = \frac{1}{16\pi^4 l^2}\{\gamma_\rho q^2 F_1^{(2)} + i\sigma_{\rho\alpha} q^\alpha m_\nu F_2^{(2)}\}. \tag{4.30}$$

Here

$$F_1^{(2)} = \rho(\zeta,\eta)\Gamma(1-\eta-\zeta)\Bigg\{\frac{3+x}{(1-\eta)(1+\eta)}\left[\frac{x-1}{6} - \frac{2+x}{24}(3+x) - \frac{5}{2}\frac{2+x}{\eta}\right] +$$

$$+ \frac{1}{2-\eta}\times\left[\frac{2+\zeta}{1+\eta}\left(-\frac{7}{6} - \frac{5}{6}x + \frac{5}{2}\frac{3+x}{\eta}\right) - \frac{1}{12}(1-x)(4+x) + \right.$$

$$\left. + \frac{1}{3}\frac{3+x}{1+\eta}(1-4x)\right] + \frac{1}{3}\frac{1-x}{3-\eta}\left(1 - \frac{2+\zeta}{1+\eta}\right) + \frac{1}{(1-\eta)(2-\eta)}\left[\tfrac{1}{3}(1-x) - \right.$$

$$\left. - \frac{5}{2}\frac{2+x}{1+\eta}(3+x)\right] + \frac{1}{(2-\eta)(3-\eta)}\times\left[\frac{3+x}{1+\eta}(-\tfrac{41}{6} - \tfrac{25}{6}x) + \frac{2+\zeta}{1+\eta'}\times\right.$$

$$\times\left(-1-\tfrac{3}{2}x+\frac{5}{2}\frac{3+x}{\eta}\right)+(2+x)(\tfrac{11}{6}+\tfrac{5}{3}x)+\tfrac{1}{12}(2-x)(1-x)\Bigg]+$$

$$+\frac{\eta}{(1-\eta)(2-\eta)}\left[-\frac{1}{2}\frac{3+x}{1+\eta}(1-x)+(2+x)(\tfrac{17}{24}+\tfrac{7}{24}x)+\frac{1}{2}\frac{2+\zeta}{1+\eta}\right]+$$

$$+\frac{2}{3}\frac{1}{(1-\eta)(2-\eta)(3-\eta)}\frac{2+\zeta}{1+\eta}(2+x)+\frac{2+x}{3}\frac{3+x}{1+\eta}+$$

$$+\frac{\eta}{(2-\eta)(3-\eta)}\left[(2+x)\left(\frac{1}{6}-\frac{x}{4}+\frac{5}{6}\frac{2+\zeta}{1+\eta}\right)+\frac{3+x}{1+\eta}(-\tfrac{1}{2}+x)\right]\Bigg\},$$

$$F_2^{(2)}=\frac{\rho(\zeta,\eta)\cdot\Gamma(1-\eta-\zeta)}{6x(1-\eta^2)(2-\eta)}[20x(5x+6+x^2)+9\eta(1-\eta)+$$
$$+2\eta x^2(1-x)+18x\eta(1-\eta)+$$
$$+\eta x^2(\zeta-\eta)-8x\eta^2(1+x)-4\eta^2x^2(1+x)-$$
$$-7x\eta^2(x+\eta)-\eta^2x^2(3+2\zeta)],\qquad (x=\eta+\zeta).$$

It is easily seen that by substitution of the variables of integration, the third term can be transformed to the second term of expression (4.27). Thus, in our case of $q^2\to 0$ we get

$$\Gamma_{3\rho}(p,q)=\Gamma_{2\rho}(p,q).$$

Now we proceed to the numerical calculations for the concrete form of the function $v(\zeta)$, say, for v_b which is determined by (1.15). The results of a displacement of the contours of integration in (4.28)–(4.30) give

$$\begin{aligned} F_1^{(1)}&=-0.18, & F_2^{(1)}&=2.56,\\ F_1^{(4)}&=-0.28, & F_2^{(4)}&=6.7,\\ F_1^{(2)}&=-0.05, & F_2^{(2)}&=1.07.\end{aligned}\tag{4.31}$$

Finally, we shall consider expressions (4.28)–(4.30) together with (4.31). Then

$$M_\rho=\delta(\Gamma_{1\rho}+2\Gamma_{2\rho}+\Gamma_{4\rho})=\delta[\gamma_\rho F_1(q^2)+i\sigma_{\rho\alpha}q^\alpha m_v F_2(q^2)],\tag{4.32}$$

where

$$F_2(0)=\frac{1}{16\pi^4l^2}(f_1+2f_2+f_4),\quad \delta=ieG^2/2;$$

$$F_1(q^2)=(q^2/16\pi^4l^2)(R_1+2R_2+R_4).$$

Here

$$f_1=\tfrac{1}{3}F_2^{(1)}(0)=2.56/3,\qquad f_2=F_2^{(2)}=1.07,\qquad f_4=\tfrac{1}{6}F_2^{(4)}(0)=1.1,$$

$$R_1=\tfrac{1}{3}(-0.18),\qquad R_4=\tfrac{1}{6}(-0.28),\qquad R_2=-0.05.$$

Therefore

$$r_{2\nu}^2 = \langle r_{2\nu}^2 \rangle = 6q^{-2} G^2 F_1(q^2),$$

$$a_{\nu_j} = 4m_{\nu_j} m_e (G^2/2) F_2(0), \quad \nu_j = \nu_e, \nu_\mu$$

in the units of the electron Bohr magneton.

Recent experiments and data analysis [see, for example, Bardin and Mogilevski, 1974; Kim *et al.*, 1974; Daum *et al.*, 1978; Bricman *et al.*, 1980 (Particle Data Group)] established the following restrictions for a_ν, $\langle r_\nu^2 \rangle$, m_{ν_μ}, and m_{ν_e}

$$|a_\nu| \lesssim 10^{-7} \qquad m_{\nu_\mu} \lesssim 0.57 \text{ MeV},$$

$$\langle r_\nu^2 \rangle \lesssim 10^{-31} \text{ cm}^2 \qquad m_{\nu_e} \lesssim 6 \times 10^{-5} \text{ MeV}.$$

Thus, our result gives the following restrictions:

$$|a_{\nu_\mu}| \lesssim 2 \times 10^{-15}, \qquad |a_{\nu_e}| \lesssim 1.75 \times 10^{-19},$$

and

$$\langle r_\nu^2 \rangle \lesssim 1 \times 10^{-33} \text{ cm}^2,$$

on the assumption $l \to l_w \sim 2 \times 10^{-16}$ cm. If $(14 \leqslant m_{\nu_e} \leqslant 46)$ eV due to Kozik *et al.*, (1980), then

$$4 \times 10^{-20} \lesssim |a_{\nu_e}| \lesssim 1.34 \times 10^{-19}.$$

We notice in our model that the vertex diagrams, shown in Figure 11b–e give small contributions to $\langle r_\nu^2 \rangle$ of an order $(10^{-37}–10^{-38})\,\text{cm}^2$ with respect to diagram 11a.

4.6. Studies of the Decay $K_L^0 \to \mu^+ \mu^-$ and K_L^0-and K_S^0-Meson Mass Difference

4.6.1. *Introduction*

Some time ago the rare decay $K_L^0 \to \mu^+ \mu^-$ was observed [see Bricman *et al.*, 1980 (Particle Data Group)], whose branching ratio coincides, in order of magnitude, with the unitary limit (Sehgal, 1969; Quigg and Jackson, 1968)

$$B_u(K_L^0 \to \mu^+ \mu^-) = W(K_L^0 \to \mu^+ \mu^-)/W(K_L^0 \to \text{all}) \gtrsim 6 \times 10^{-9}.$$

This process occurs essentially due to electromagnetic interactions (through two-photon exchange). It is interesting to note that the well-known so-called *GIM mechanism* (the hypothesis of the existence of the charm quark) originated from the investigation of this process within the standard model of electroweak interactions (Glashow *et al.*, 1970).

The calculation of 'weak' corrections to the $B(K_L^0 \to \mu^+ \mu^-)$ and the mass difference of K_L^0- and K_S^0-mesons within the usual nonrenormalizable theory (the four-fermion theory and the theory with intermediate bosons) of weak interactions gives a very small value for the cut-off momentum Λ of an order of a few GeV. This contradicts the value of the natural cut-off $\Lambda \sim 10^2 \div 10^3$ GeV for the growth of weak interactions (see, for example, Ioffe, 1973).

In this section we want to calculate matrix elements of the order $O(G^2)$ in decay $K_L^0 \to \mu^+\mu^-$ and mass difference $\Delta m(K_L^0 - K_S^0)$ within the framework of the stochastic-nonlocal theory of weak interactions without introducing the fourth quark. It is quite possible that suppression of these matrix elements may be achieved by an appropriate choice of form-factors of the theory. However, as shown below, change of these quantities is slow for the form-factors v_s, v_b and v_1 under consideration.

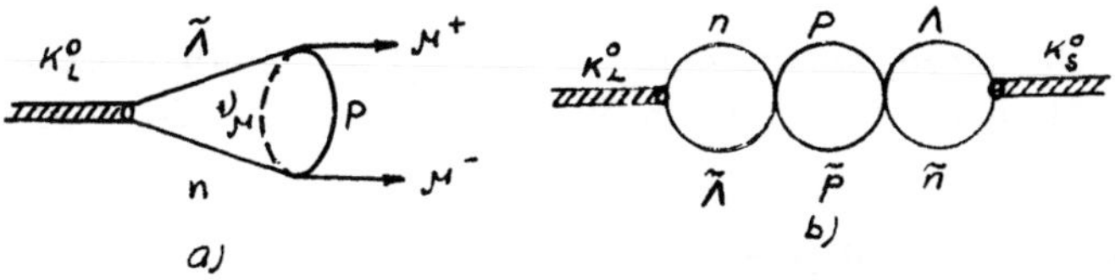

Fig. 12. The Feynman diagrams which give contributions to the $K_L^0 \to \mu^+\mu^-$ and the $\Delta m(K_L^0 - K_S^0)$.

In our model contributions of nonlocal interactions to the $K_L^0 \to \mu^+\mu^-$ and the $\Delta m(K_L^0 - K_S^0)$ arise from the diagrams, shown in Figure 12.

4.6.2. *The $K_L^0 \to \mu^+\mu^-$ Decay*

$K_L^0 \to \mu^+\mu^-$ decay in the second order in G is described by the diagram shown in Figure 12a. The corresponding term has the form

$$M(K_L^0 \to \mu^+\mu^-) = i\sqrt{2} f_{KN\Lambda} (G/\sqrt{2})^2 \bar{\mu}(p_-)\Gamma(p_-, q)\mu(p_+)\varphi_K \cos\theta_c \sin\theta_c$$

where $\Gamma(p_-, q)$ was calculated in Section 4.4.3. Thus, the branching ratio of this decay is

$$B(K_L^0 \to \mu^+\mu^-) = W(K_L^0 \to \mu^+\mu^-)/W(K_L^0 \to \text{all}) = 1.7A^2 \times 10^{-5}, \quad (4.33)$$

where

$$W(K_L^0 \to \mu^+\mu^-) = \frac{m_K}{8}\left(1 - 2\frac{m_\mu^2}{m_K^2}\right)^2 A^2 \cos^2\theta_c \sin^2\theta_c \frac{G^4}{l^4} \frac{m_N^2 m_\mu^2}{128\pi^8} \frac{f_{KN\Lambda}^2}{4\pi}$$

and A is given by (4.20) or (4.21). Substitution of numerical value (4.23) for A into (4.33) leads to the contribution

$$B(K_L^0 \to \mu^+\mu^-) = 1.25 \times 10^{-6} \quad \text{for } v = v_s. \quad (4.34)$$

4.6.3. *The Mass Difference of K_L^0- and K_S^0-Mesons*

Let us now find the energy operator for the transition K_L^0 into K_S^0. A typical diagram of the order G^2 giving the contribution to the $\Delta m(K_L^0 - K_S^0)$ is shown in Figure 12b. The expression corresponding to this diagram is

$$\tilde{\Sigma}(p) = G^2 \sin^2 \theta_c \cos^2 \theta_c \Pi_\alpha(p)\Pi_{\alpha\beta}(p)\Pi_\beta(p), \tag{4.35}$$

where

$$\Pi_\alpha(p) = f_{KN\Lambda} \frac{1}{(2\pi)^4 i} \int d^4 k \, \mathrm{Sp}[S_R(\hat{k})0_\alpha S_R(\hat{k} + \hat{p})\gamma_5]$$

and

$$\Pi_{\alpha\beta}(p) = i^{-1}(2\pi)^{-4} \int d^4 k \, \mathrm{Sp}[S_R(\hat{k})0_\alpha S_R(\hat{k} + \hat{p})0_\beta],$$

which is determined by the expression of (4.14). In the case $m_K^2 l^2 \ll 1$ the expression obtained for $\Pi_{\alpha\beta}(p)$ on the mass shell of the K-meson acquires the form

$$\Pi_{\alpha\beta}(p) = g_{\alpha\beta}(m_K^2/m_K^2 l^2)(1/4\pi^2)c, \tag{4.36}$$

where c is given by (4.16).

The calculation of functions $\Pi_j(p)$ ($j = \alpha, \beta$) is similar. As a result

$$\Pi_j(p) = \frac{f_{KN\Lambda} m_N p_j}{4\pi^2} \frac{1}{2i} \int_{-\delta + i\infty}^{-\delta - i\infty} d\zeta \frac{v(\zeta)}{\sin \pi\zeta} l^{2\zeta} \frac{1}{2i} \int_{-\gamma + i\infty}^{-\gamma - i\infty} d\eta \frac{v(\eta)}{\sin \pi\eta} l^{2\eta} \times$$

$$\times \frac{\Gamma(-\eta - \zeta)}{\Gamma(1 - \eta)\Gamma(1 - \zeta)} \int_0^1 dx (1 - x)^{-\zeta} x^{-\eta} [m_N^2 - p^2 x(1 - x)]^{\eta + \zeta}.$$

The contour integration gives

$$\Pi_j(p) = -f_{KN\Lambda} \frac{m_N p_j}{4\pi^2} \left\{ v'(0) + 1 + \ln(m_N^2 l^2) + \right.$$

$$\left. + \int_0^1 dx \ln\left[1 - \frac{p^2}{m_N^2} x(1 - x)\right] - \frac{\pi}{2i} \int_{-\delta + i\infty}^{-\delta - i\infty} d\zeta \frac{v(\zeta)v(-\zeta)}{\sin^2 \pi\zeta} \right\}. \tag{4.37}$$

Here it is necessary to carry out renormalization in the strong coupling constant $f_{KN\Lambda}$. After such renormalization expression (4.37) takes the form

$$\Pi_j(p) = -(m_N p_j/4\pi^2) f_{KN\Lambda} \cdot \rho,$$
$$\rho = \int_0^1 dx \ln[1 - (p^2/m_N^2)(1 - x)x]. \tag{4.38}$$

Substituting (4.38) and (4.36) into (4.35) we obtain the following expression for

$\Delta m(K_L^0 - K_S^0)$:

$$\Delta m(K_L^0 - K_S^0) = (2\pi)^{-6} f_{KN\Lambda}^2 G^2 \sin^2\theta_c \cos^2\theta_c m_N^2 \frac{c \cdot \rho^2}{l^2} m_K, \tag{4.39}$$

or

$$\Delta m(K_L^0 - K_S^0) = \begin{cases} 1 \times 10^{12}\, \hbar\, \sec^{-1} & \text{for } v = v_b, \\ 5 \times 10^{11}\, \hbar\, \sec^{-1} & \text{for } v = v_s, \\ 1.6 \times 10^{11}\, \hbar\, \sec^{-1} & \text{for } v = v_1. \end{cases} \tag{4.40}$$

Table 4.1 presents the contributions calculated within our nonlocal theory and experimental data on $K_L^0 \to \mu^+\mu^-$ and the mass difference $\Delta m(K_L^0 - K_S^0)$ for different form-factors of the theory. From Table 4.1 we observe some dependence of physical quantities under consideration from the form-factors of the theory.

4.7. Appendix B. Calculation of the Contour Integral

In this appendix we give the method of calculation of the contour integral, for example, the integral A given by formula (4.20):

$$A = \frac{\pi}{2i}\int_{-\gamma+i\infty}^{-\gamma-i\infty} dy \frac{v(y)}{\sin \pi y}\frac{1}{2i}\int_{-\beta+i\infty}^{-\beta-i\infty} d\zeta \frac{v(\zeta)}{\sin \pi\zeta}\frac{1}{2i}\int_{-\alpha+i\infty}^{-\alpha-i\infty} d\eta \frac{v(\eta)}{\sin \pi\eta} \times$$

$$\times \frac{v(-1-\eta-y-\zeta)\Gamma(2+\zeta)\Gamma(2+\eta)\Gamma(-1-\eta-\zeta)}{\sin \pi(y+\eta+\zeta)\Gamma(1-\eta)\Gamma(1-\zeta)\Gamma(3+\eta+\zeta)} \quad (0 < \alpha, \beta, \gamma < 1). \tag{B.1}$$

First we displace the contour γ to the right. Then the poles will appear at points $y = 0, 1, 2, \ldots$ and $y = n - \eta - \zeta$ $(n = 0, 1, 2, \ldots)$. In the first case $(y = 0, 1, 2, \ldots)$ it is necessary to displace one of the other contours, e.g., β-contour. The calculation of the residues at points $\zeta = 0, 1, 2, \ldots$ and $\zeta = N - \eta$ $(N = 0, 1, 2, \ldots)$ give

$$A = 2\frac{\pi}{2i}\int_{-\alpha+i\infty}^{-\alpha-i\infty} d\eta \frac{v(\eta)}{\sin^2 \pi\eta}\frac{1}{\eta(1+\eta)(2+\eta)}\sum_{k=0}^{\infty} v(k)v(-1-k-\eta). \tag{B.2}$$

From similar calculations of residues at points $\zeta = 0, 1, 2, \ldots$ and $\zeta = N - \eta$ in the second case $(y = n - \eta - \zeta)$ we get

$$A = -2\frac{\pi}{2i}\int_{-\alpha+i\infty}^{-\alpha-i\infty} d\eta \frac{v(\eta)}{\sin^2 \pi\eta}\frac{1}{\eta(1+\eta)(2+\eta)}\sum_{k=0}^{\infty} v(k-\eta)v(-1-k). \tag{B.3}$$

After uniting expressions (B.2) and (B.3),

$$A = 2\frac{\pi}{2i}\int_{-\alpha+i\infty}^{-\alpha-i\infty} \mathrm{d}\eta \frac{v(\eta)}{\sin^2 \pi\eta} \frac{1}{\eta(1+\eta)(2+\eta)} \sum_{k=0}^{\infty} [v(k)v(-1-k-\eta) -$$

$$- v(k-\eta)v(-1-k)]. \tag{B.4}$$

The last integral (B.4) for the concrete form of form-factors of the theory is calculated easily. For example, let

$$v = v_s = 2^{1+2\zeta}/\Gamma(3+2\zeta).$$

Then

$$\sum_{k=0}^{\infty} [v(k)v(-1-k-\eta) - v(k-\eta)v(-1-k)]$$

$$= 2^{-2\eta}\left[\frac{1}{\Gamma(3)\Gamma(1-2\eta)} + \frac{1}{\Gamma(5)\Gamma(-1-2\eta)} + \right.$$

$$\left. + \frac{1}{\Gamma(7)\Gamma(-3-2\eta)} + \cdots - \frac{1}{\Gamma(3-2\eta)}\right]$$

and

$$A = 4\frac{\pi}{2i}\int_{-\alpha+i\infty}^{-\alpha-i\infty} \mathrm{d}\eta \frac{1}{\sin^2 \pi\eta \cdot \Gamma(3+2\eta)}\left\{-\frac{1+2\eta}{\Gamma(3-2\eta)(1+\eta)(2+\eta)} + \right.$$

$$\left. + \frac{1+2\eta}{12(1+\eta)(2+\eta)\Gamma(1-2\eta)}[1 + \tfrac{1}{15}(3+2\eta)(1+\eta)]\right\}.$$

After calculation of the residues at points $\eta = 0, 1, 2, \ldots$ we obtain

$$A = -\frac{13}{45} + 4\left\{\frac{1}{12}\sum_{k=1}^{\infty} \frac{1}{4k(k+1)^2(2+k)}[1 + \tfrac{1}{15}(3+2k)(1+k)] - \right.$$

$$\left. - \frac{1}{8}\sum_{k=2}^{\infty} \frac{1}{k(1-k^2)(1-2k)(1+k)(2+k)}\right\} \approx -0.27.$$

Chapter 5

Functional Integral Techniques in Quantum Field Theory

In recent years, the investigation and application of stochastic processes and quantized fields has concentrated on the different functionals of these processes and fields. Such objects play, and it seems, will continue to play, a fundamental role in the development of mathematical and physical theories. Substantial success in the applications of *functional methods* has been obtained in *quantum field theory*. In this chapter the main attention is focussed on this problem, which was considered in more detail by Simon (1979); Glimm and Jaffe (1981); especially in the cases of gauge fields by Nash (1978); Faddeev and Slavnov (1980); and of stochastic fields by Langouche *et al.* (1979).

The study of the functionals of stochastic processes forms a new field uniting the probability theory with mathematical analysis, that may conditionally be called probability analysis. In Section 6.12 (Chapter 6) we discuss briefly some problems related to the functionals of stochastic processes, and present an elementary stochastic calculus. Recently there has been an increasing interest in functional integral formulations of diffusion processes (see Leschke and Schmutz, 1977 and references therein) and in the powerful synthesis of functional path integration, Euclidean methods and stochastic methods (see Guerra, 1981).

Before passing to consideration of the functional methods in quantum field theory, we expound some mathematical problems of functional analysis. At present, these problems are currently under intensive investigation, but a strict mathematical foundation (as in the case of ordinary analysis) is not yet available, so further careful investigation is required.

5.1. Mathematical Preliminaries

The mathematical methods which provide the foundation for functional analysis are the idea of differentiation with respect to functions [or (quantized) fields, or random processes] rather than with respect to real or complex numbers; and integration with respect to a measure defined on a space of functions (or paths and fields) rather than with respect to the Riemann or

Lebesque measure used for real and complex numbers. These methods are called functional differentiation and functional integration.

The rule of definition of *functional differentiation and functional integration* depends on the properties of functions (or fields and processes) over which the functional operations are carried out. An essential difference arises when functionals depend on *Grassmann variables.* Therefore, in concrete applications of functional methods one has often to choose a suitable method of defining the functional operations for the object under consideration. We first give an elementary definition of functional differentiation and the basic idea of functional integration starting from physical intuition. Next we give a more formal definition of the functional operations appropriate to quantum theory. This analysis leads to finite- and infinite-Grassmann algebras. Application of the method of functional analysis in the S-matrix theory will be considered in Section 5.5.

We begin with *functional (or variational) differentiation.* We have to go beyond the ordinary space of functions $f(x)$ and to include generalized functions [in particular the Dirac delta function $\delta(x - x')$], in order to define the functional derivative. The *Dirac $\delta(x)$ function* plays an important role in functional calculus (see Section 1.4). The *variational derivative* of a function $f(x)$ with respect to itself is defined in terms of the delta function

$$\delta f(x)/\delta f(y) = \delta(x - y). \tag{5.1}$$

Then we extend this definition in the case of a general functional. Let $F(f(x))$ be a functional of $f(x)$, for example, we might have

$$F(f) = \int f(x)\,\mathrm{d}x \quad \text{or} \quad F(f(x)) = f(x_0), \quad x_0 \text{ fixed.}$$

DEFINITION 5.1. The formal definition of the *functional derivative* (or variational derivative) of $F(f)$ with respect to $f(y)$ is written $\delta F/\delta f(y)$, where

$$\frac{\delta F}{\delta f(y)} = \lim_{\varepsilon \to 0} \varepsilon^{-1}\{F(f(x) + \varepsilon\cdot\delta(x - y)) - F(f(x))\},$$

here ε is a real positive number. If $F(f(x)) = f(x_0)$, then

$$\frac{\delta f(x_0)}{\delta f(y)} = \lim_{\varepsilon \to 0} \varepsilon^{-1}[f(x_0) + \varepsilon\cdot\delta(x_0 - y) - f(x_0)] = \delta(x_0 - y)$$

which corresponds to definition (5.1). When $F(f) = \int \mathrm{d}x f(x)$,

$$\frac{\delta F(f)}{\delta f(y)} = \lim_{\varepsilon \to 0} \varepsilon^{-1}\left\{\int \mathrm{d}x[f(x) + \varepsilon\cdot\delta(x - y) - f(x)]\right\} = \int \mathrm{d}x\,\delta(x - y) = 1.$$

However, this result is easily obtained if we make use of (5.1) to write

$$\frac{\delta F(f)}{\delta f(y)} = \int \mathrm{d}x \frac{\delta f(x)}{\delta f(y)} = \int \mathrm{d}x\,\delta(x - y) = 1.$$

We may consider another example which is provided by the kernel term occurring in a typical linear integral equation, i.e.,

$$F_x(f) = \int \mathrm{d}x' K(x, x') f(x'),$$

where x is a parameter and not an independent variable. Then by (5.1) we have $\delta F_x(f)/\delta f(y) = K(x, y)$. In the case of the functional derivative some rules of ordinary differentiation are still valid. Let $E = E(F)$ and $F(f(x))$, then the chain rule which we are looking for reads

$$\frac{\delta E}{\delta f(y)} = \int \mathrm{d}x' \left[\frac{\delta E}{\delta F(x')}\right]\left[\frac{\delta F(x')}{\delta f(y)}\right]. \tag{5.2}$$

For example, let $E_1(F) = F$ and $F(x) = F(f) = \iint \mathrm{d}x_1\, \mathrm{d}x_2 f(x_1) f(x_2)$. Making use of rule (5.2), we find that $\delta E(x)/\delta F(x') = \delta(x - x')$ and

$$\frac{\delta F(f)}{\delta f(y)} = 2 \int \mathrm{d}z f(z); \tag{5.3}$$

so the chain rule tells us that

$$\frac{\delta E(x)}{\delta f(y)} = 2 \int \mathrm{d}x'\, \delta(x - x') \int \mathrm{d}z f(z) = 2 \int \mathrm{d}z f(z).$$

It is convenient to compute $\delta E_1/\delta f(y)$ directly by using equality (5.1) and the fact that

$$E_1(f) = \iint \mathrm{d}x_1\, \mathrm{d}x_2 f(x_1) f(x_2).$$

Thus we have just shown that

$$\delta F/\delta f(y) = 1 \quad \text{and} \quad \delta F^2/\delta f(y) = 2F$$

in the case of the functional $F = \int \mathrm{d}x f(x)$. It is easy to see that $\delta F^n/\delta f(y) = nF^{n-1}$.

In practical applications (particularly in the S-matrix representation) one encounters functionals which can be represented in the following way:

$$P(f) = K_0(x) + \int \mathrm{d}x_1 K_1(x, x_1) f(x_1) + \iint \mathrm{d}x_1\, \mathrm{d}x_2 K_2(x, x_1, x_2) f(x_1) f(x_2) + \cdots$$

Such a functional $P(f)$ is called the *functional power series* in f. According to the above, we have

$$P(0) = K_0(x), \qquad \delta P(f)/\delta f(y) = K_1(x, y) \quad \text{at } f = 0,$$

and

$$\delta^2 P(f)/\delta f(y_1)\delta f(y_2) = K_2(x, y_1, y_2) + K(x, y_2, y_1) \quad \text{at } f = 0. \tag{5.4}$$

We see that if $K_2(x, y_1, y_2)$ is symmetric in y_1 and y_2, then equation (5.4) has the desired connection between the coefficient function and the functional derivative. This means that the general statement is that

$$K_n(x, y_1, \ldots, y_n) = \frac{1}{n!} \frac{\delta^n P(f)}{\delta f(y_1) \cdots \delta f(y_n)}, \quad \text{at } f = 0, \tag{5.5}$$

provided $K_n(x, y_1, \ldots, y_n)$ is symmetric under permutation of the variables $y_1, \ldots, y_n$. In the case of quantum field theory we shall consider such functional power series below.

DEFINITION 5.2. Now we define an alternative method for defining the functional derivative $\delta F/\delta f(y)$ by which the *functional Taylor series* may be easily obtained formally. The definition for the *derivative* is

$$\int dy f'(y) \frac{\delta F(f)}{\delta f'(y)} = \lim_{\lambda \to 0} \lambda^{-1}[F(f + \lambda f') - F(f)].$$

This definition gives the same result as the previous one. On the other hand, this definition resembles the definition of the variational derivative of a functional $F(f)$, where f belongs to some real Hilbert space $\mathscr{H}$. In this case, we shall say that the functional $F(f)$ is called differentiation at a point $f \in \mathscr{H}$, if for any $h \in \mathscr{H}$ there exists an element $\delta F/\delta f \in \mathscr{H}$ such that

$$|F(f + h) - F(f) - (h, \delta F/\delta f)| = o(||h||), \tag{5.6}$$

$\delta F/\delta f$ is called the *variational derivative of the functional* F at the point f. Here

$$||h|| = \left(\int |h|^2 \, dx\right)^{1/2},$$

is the *norm of the function h*, and the *scalar product* (h, f) is given by the standard formula

$$(h, f) = \int h(x)\bar{f}(x) \, dx.$$

We now suppose that the ordinary function $g(\lambda)$ of a real variable λ defined by $g(\lambda) = F(f + \lambda f')$, has a Taylor series:

$$g(\lambda) = \sum_{n=0}^{\infty} \frac{1}{n!} g^{(n)}(0)\lambda^n, \tag{5.7}$$

then

$$g^{(1)}(0) = \lim_{\lambda \to 0} \lambda^{-1}[F(f + \lambda f') - F(f)] = \int dy f'(y) \frac{\delta F(f)}{\delta f'(y)}.$$

By induction, it is easy to prove that

$$g^{(n)}(0) = \int dx_1 \cdots \int dx_n f'(x_1)\cdots f'(x_n)\frac{\delta^n F}{\delta f'(x_1)\cdots \delta f'(x_n)}.$$

Therefore the power series (5.7) acquires the form

$$F(f + \lambda f') = \sum_{n=0}^{\infty} \int dx_1 \cdots \int dx_n f'(x_1)\cdots f'(x_n)\left[\frac{\delta^n F(f)}{\delta f'(x_1)\cdots \delta f'(x_n)}\right]\frac{\lambda^n}{n!}.$$

Now suppose that this power series converges when $\lambda = 1$ and then set $\lambda = 1$ and $f = 0$, and delete the prime on the function $f'(x)$. The result is the desired functional Taylor series

$$F(f) = \sum_{n=0}^{\infty} \frac{1}{n!}\int dx_1 \cdots \int dx_n f(x_1)\cdots f(x_n)\frac{\delta^n F(0)}{\delta f(x_1)\cdots \delta f(x_n)}. \tag{5.8}$$

Thus, $F(f)$ has the functional Taylor series (5.8) when $g(\lambda) = F(f + \lambda f')$ has an ordinary Taylor series convergent at the point $\lambda = 1$.

5.2. Historical Background of Path Integrals

Functional integrals in the sense of integrals over a space of functions were first considerea by mathematician N. Wiener (1923). The first appearance of the method of functional integration in quantum theory was due to Feynman (see Feynman and Hibbs, 1965). In quantum theory, the term *path integral* is often used interchangeably with the *term functional integral.* This is because the integrals considered by Feynman can be described as integrals over a space of functions each of which describes the path of a particle from point *a* to point *b*. However, the functional integrals constructed by Wiener and those considered in quantum theory are not quite the same.

In order to compare the two kinds of integrals consider first those due to Wiener. They have their origin in the theory of *Brownian motion.* That is to say, a particle diffuses at random so that its *probability distribution* is given by the *diffusion equation* (see Part II)

$$\partial P/\partial t = v\cdot\partial^2 P/\partial x^2, \tag{5.9}$$

with

$$P = (4\pi v t)^{-1/2}\exp(-x^2/4vt).$$

We would like to construct an averaging process of the positions of the particle. Let the position of the particle at time t be $x(t)\in \mathbb{R}^1$. Divide the time interval $[0, t]$ into n equal parts by the points $t_1, \ldots, t_{n-1}, t_n$ ($t_n = t$). Let the position of the particle $x(t_i)$ at time t_i be in the interval $[a_i, b_i]$ for $i = 1, 2, \ldots, n$. Then the probability of this distribution of positions is given by the n-fold multiple integral (for algebraic convenience we have here set $\nu = \frac{1}{2}$)

$$[2^n \pi^n t_1(t_2 - t_1) \cdots (t_n - t_{n-1})]^{-1/2} \int_{a_1}^{b_1} \mathrm{d}x_1 \int_{a_2}^{b_2} \mathrm{d}x_2 \cdots \int_{a_n}^{b_n} \mathrm{d}x_n \times$$

$$\times \exp\left\{\frac{-x_1^2}{2t_1} - \frac{(x_2 - x_1)^2}{2(t_2 - t_1)} - \cdots - \frac{(x_n - x_{n-1})^2}{2(t_n - t_{n-1})}\right\}, \tag{5.10}$$

here $a_1 < b_1,\ a_2 < b_2, \ldots, a_n < b_n;\ 0 < t_1 < t_2 < \cdots < t_n,\ n \geqslant 1$. This definition is correct because of

$$(2\pi t)^{-1/2} \exp\left[\frac{-(b-a)^2}{2t}\right] = \int_{\mathbb{R}^1} \mathrm{d}c[2\pi(t-s)]^{-1/2}(2\pi s)^{-1/2} \times$$

$$\times \exp\left\{\frac{-(b-c)^2}{2(t-s)} - \frac{(c-a)^2}{2c}\right\}, \quad (t > s).$$

The family of the variables $\{x(t): t > 0\}$ is *Gaussian*, and

$$\mathbb{E}\{x(t_1)x(t_2)\} = \min(t_1, t_2),$$

where the symbol $\mathbb{E}$ denotes the (mathematical) *expectation value* (see Chapter 6).

Now we pass to the limit, where $\Delta t = (t/n) \to 0$, and we insert in the integrand of (5.10) a function $F(x_1, \ldots, x_n)$. Then we write

$$\mathbb{E}_W\{F(x(t))\} = \int \mathrm{d}\mu_W F(x(t))$$

$$= \lim_{n \to \infty} (2\pi \cdot \Delta t)^{-n/2} \int \mathrm{d}x_1 \cdots \int \mathrm{d}x_n F(x_1, \ldots, x_n) \times$$

$$\times \exp\left\{\frac{-x_1^2}{2\Delta t} - \sum_{i=1}^{n-1} \frac{(x_{i+1} - x_i)^2}{2\Delta t}\right\}. \tag{5.11}$$

The symbol $\mathrm{d}\mu_W$ stands for *Wiener measure*. We rewrite the right-hand side of (5.11) as

$$\frac{1}{A} \int \cdots \int F(x(t)) \exp\left\{-\frac{1}{2} \int_0^1 \mathrm{d}t' \left[\frac{\mathrm{d}x(t')}{\mathrm{d}t'}\right]^2\right\} \prod_0^t \mathrm{d}x(t). \tag{5.12}$$

This enables us to give a symbolic representation of the *Wiener measure*

$$\mathrm{d}\mu_W = \frac{1}{A} \exp\left\{-\frac{1}{2} \int_0^1 \mathrm{d}t' \left[\frac{\mathrm{d}x(t')}{\mathrm{d}t'}\right]^2\right\} \prod_0^t \mathrm{d}x(t). \tag{5.13}$$

The factor $1/A$ is included to normalize the measure so that $\int \mathrm{d}\mu_W = 1$. Notice that the integral (5.11) of $F(x(t))$ can be defined only in terms of this limiting procedure defined on $F(x_1, \ldots, x_n)$ if $F(x_1, \ldots, x_n)$ is well-defined itself.

For example, we now consider the *Wiener integral* $\mathbb{E}_W\{\exp[-\int_0^1 dt' U(x(t'))]\}$. Let $U(x)$ be a continuous function defined on the interval $(-\infty, \infty)$ and $U(x) \geqslant 0$. Then the question arises: does the integral exist?

Since the functional under the sign of integral is bounded it is sufficient to show that it is measurable. Since

$$\int_0^t dt' U(x(t')) = \lim_{n\to\infty} \Delta t \sum_{k=1}^n U(x(k\,\Delta t)), \quad \Delta t = t/n \tag{5.14}$$

[we recall that $x(t) \in \mathbb{R}^1$, and that we have assumed continuity of $U(x)$], and the functional $\Delta t\, \Sigma_{k=1}^n U(x(k\,\Delta t))$ is obviously measurable, then the measurability of the functional $\int_0^t dt' U(x(t'))$ and hence of the expression $\exp\{-\int_0^t dt' U(x(t'))\}$ is proved.

From equality (5.14) and according to the theorem about the convergence of a bounded sequence of functions (see, for example, Titchmarsh, 1939) it follows that

$$\mathbb{E}\left\{\exp\left[-\int_0^t dt' U(x(t'))\right]\right\} = \lim_{n\to\infty} \mathbb{E}\left\{\exp\left[-\Delta t \sum_{k=1}^n U(x(k\,\Delta t))\right]\right\},$$

and here the existence of the limit in the right-hand side is proved at the same time.

On the other hand, from (5.11) it is easy to see that

$$\mathbb{E}\left\{\exp\left[-\Delta t \sum_{k=1}^n U(x(k\,\Delta t))\right]\right\} = \int_{-\infty}^{\infty} dx_1 \cdots \int_{-\infty}^{\infty} dx_n \exp\left[-\Delta t \sum_{k=1}^n U(x_k)\right] \times$$

$$\times\, p(0, x_1; \Delta t)\, p(x_1, x_2; \Delta t) \cdots p(x_{n-1}, x_n; \Delta t),$$

where

$$p(x_i, x_j; \Delta t) = (2\pi\,\Delta t)^{-1/2} \exp\{-(x_i - x_j)^2/2\,\Delta t\}, \quad (\Delta t = t/n).$$

From this we obtain the following result: the limit

$$\lim_{n\to\infty} \int_{-\infty}^{\infty} dx_1 \cdots \int_{-\infty}^{\infty} dx_n \exp\left\{-\Delta t \sum_{k=1}^n U(x_k)\right\} \times$$

$$\times\, p(0, x_1; \Delta t) p(x_1, x_2; \Delta t) \cdots p(x_{n-1}, x_n; \Delta t) \tag{5.15}$$

exists and is equal to

$$\mathbb{E}\left\{\exp\left[-\int_0^1 dt' U(x(t'))\right]\right\}. \tag{5.16}$$

Existence of the limit (5.15) follows here simply from the measurability of the functional.

One can try to define the integral (5.16) as a limit of integrals (5.15); this procedure was indeed used by Feynman. The defects of such an approach are obvious from the mathematical point of view. In the case of quantum field theory, some definition of integrals of type (5.15) will given in Section 5.5.2.

Now we want to describe the *integrals introduced by Feynman* (1948). Feynman wrote the integral (5.15) in the form ($x_0 = 0$)

$$(2\pi\cdot\Delta t)^{-n/2}\int_{-\infty}^{\infty}\mathrm{d}x_1\cdots\int_{-\infty}^{\infty}\mathrm{d}x_n\exp\left\{-\Delta t\left[\sum_{k=1}^{n}U(x_k)+\frac{1}{2}\sum_{k=1}^{n}\left(\frac{x_k-x_{k-1}}{\Delta t}\right)^2\right]\right\}.$$

From which it follows that in the exponential there stands nothing but the integral sum for the integral $-\int_0^t \mathrm{d}t'[\frac{1}{2}(\mathrm{d}x/\mathrm{d}t')^2 + U(x(t'))]$. Instead of (5.16) Feynman writes

$$\int\exp\left\{-\int_0^t\left[\frac{1}{2}\left(\frac{\mathrm{d}x}{\mathrm{d}t'}\right)^2+U(x(t'))\right]\mathrm{d}t'\right\}\mathrm{d}\,(\text{paths}). \tag{5.17}$$

This symbolic notation is physically more appealing since $\frac{1}{2}(\mathrm{d}x/\mathrm{d}t)^2 + U(x)$, is the Hamilton function of a particle with mass $m = 1$, moving in the potential field $U(x)$.

Indeed, Feynman (1948) in his approach to nonrelativistic quantum mechanics arrived at the integral (see Section 5.5 also)

$$\int\exp\left\{i\hbar^{-1}\int_0^t\left[\frac{1}{2}\left(\frac{\mathrm{d}x}{\mathrm{d}t'}\right)^2-U(x(t'))\right]\mathrm{d}t'\right\}\mathrm{d}\,(\text{paths}), \tag{5.18}$$

where $\hbar$ is Planck's constant divided by 2π, and

$$\int_0^t\left[\frac{1}{2}\left(\frac{\mathrm{d}x}{\mathrm{d}t'}\right)^2-U(x(t'))\right]\mathrm{d}t'$$

is the *classical action* along the path $x(t)$. In other words, the integral can be written in the form

$$\int\exp\left[-\frac{i}{\hbar}\int_0^t\mathrm{d}t'U(x(t'))\right]\exp\left\{\frac{i}{2\hbar}\int_0^t\mathrm{d}t'\left(\frac{\mathrm{d}x(t')}{\mathrm{d}t'}\right)^2\right\}\prod\mathrm{d}x(t). \tag{5.19}$$

Equation (5.19) would appear to be the integral over Wiener measure of the quantity $\exp[-i\hbar^{-1}\int_0^t \mathrm{d}t' U(x(t'))]$. However, actually this is not so, because of the presence of the factor $i/\hbar$ in the exponential. The factor $\hbar$ is not important; its presence is simply the matter of units. But the factor i is of *fundamental importance*. This factor corresponds to a diffusion with *diffusion coefficient* ν which is *purely imaginary*. That is to say, the diffusion equation (5.9) is replaced by

$$\frac{i}{\hbar}\frac{\partial\psi}{\partial t}=\nu\frac{\partial^2\psi}{\partial x^2},\quad (U(x)=0).$$

This equation is recognized straight away as the *Schrödinger equation*. Because of $i = \sqrt{-1}$ occurring in the exponential, one cannot strictly apply the

Feynman path integral. Another approach to Feynman type integrals will be presented by us in Section 9.2 of the second part of the book. Recently, very delicate and useful new definitions for the Feynman path integrals have been given by DeWitt-Morette (1972); Albeverio and Høegh-Krohn (1976), and DeWitt-Morette *et al.* (1979) (see also Exner, 1984).

Before using the functional integral techniques in field theory, we consider analysis on a finite- and infinite-dimensional Grassmann algebra.

5.3. Analysis on a Finite-Dimensional Grassmann Algebra

5.3.1. *Definition*

A *Grassmann algebra* with n generators is an algebra with *generators* C_i, $n = 1, \ldots, n$ obeying the following relation:

$$C_i C_k + C_k C_i \equiv \{C_i, C_k\} = 0. \tag{5.20}$$

In particular, $C_i^2 = 0$. We denote the Grassmann algebra with n generators as G_n. It follows from (5.20) that G_n as a linear space has the dimension 2^n. As a basis in G_n, it is convenient to consider the following *monomials*:

$$1; \quad C_1, \ldots, C_n; \quad C_1 C_2, \ldots, C_{n-1} C_n; \quad , \ldots, C_1 C_2 \cdots C_n. \tag{5.21}$$

The monomial $C_{i_1} \cdots C_{i_p}$ will be called monomial of the order p.

Each element $f(C)$ of the algebra G_n can be represented in the form of a linear combination of monomials

$$f(C) = f_0 + \sum_k f_1(k) C_k + \sum_{k_i} f_2(k_1, k_2) C_{k_1} C_{k_2} + \cdots + \\ + \sum_{k_i} f_n(k_1, \ldots, k_n) C_{k_1} \cdots C_{k_n}. \tag{5.22}$$

The element of the type $\Sigma_{k_i} f_p(k_1, \ldots, k_p) C_{k_1} \cdots C_{k_p}$ will be called *homogeneous* of the order p. It is assumed that the coefficients $f_n(k_1, \ldots, k_n)$ are *skew-symmetric*, i.e., changing sign with respect to rearrangement of any pairs of the arguments $k_1, \ldots, k_n$. This property of the coefficient functions makes the expansion (5.22) *synonymous*. It is obvious that each element of G_n may be synonymously represented in the form $f = f' + f''$, where $f' \in G'_n$ and $f'' \in G''_n$. Elements f'' and f' of sets G''_n and G'_n consist of linear combinations of monomials of even

$$f'' = f_0 + \sum_{k_i} f_2(k_1, k_2) C_{k_1} C_{k_2} + \cdots$$

and of odd

$$f' = \sum_k f_1(k) C_k + \sum_{k_i} f_3(k_1, k_2, k_3) C_{k_1} C_{k_2} C_{k_3} + \cdots$$

orders, respectively.

5.3.2. *Derivatives*

Now we proceed to define *derivatives of elements of Grassmann algebras*. Unlike the case of ordinary functions, it turns out that, since the elements of a Grassmann algebra do not commute, there are two sorts of derivative. They are called the left and right derivative and their difference becomes important only when we are going to differentiate the products of generators. The derivatives are linear and so we have to specify only the derivatives of generators and their products. The left derivative of a product $C_{i_1} \cdots C_{i_m}$ of generators is defined by

$$\frac{\partial}{\partial C_j} C_{i_1} \cdots C_{i_m} = \delta_{ji_1} C_{i_2} \cdots C_{i_m} - \delta_{ji_2} C_{i_1} C_{i_3} \cdots C_{i_m} + \\ + (-1)^{m-1} \delta_{ji_m} C_{i_1} \cdots C_{i_{m-1}} \tag{5.23a}$$

and the right derivative is given by

$$C_{i_1} \cdots C_{i_m} \frac{\partial}{\partial C_j} = C_{i_1} C_{i_2} \cdots C_{i_{m-1}} \delta_{i_m j} - C_{i_1} \cdots C_{i_{m-2}} C_{i_m} \delta_{i_{m-1} j} + \cdots + \\ + (-1)^{m-1} C_{i_2} C_{i_3} \cdots C_{i_m} \delta_{i_1 j}. \tag{5.23b}$$

We can express these definitions verbally by saying: in order to compute the left (right) derivative of product $\{C_{i_s}\}$, $s = 1, \ldots, m$ with respect to C_j we have to commute each C_{i_s} to the furthest left (furthest right) place in the product $C_{i_1} \cdots C_{i_m}$ and then to replace simply C_{i_s} by $\delta_{i_s j}$ with the appropriate power of -1. If the monomial $C_{i_1} \cdots C_{i_m}$ does not contain C_j, both derivatives are zero.

The rules of differentiation for more general functions follow immediately from the definition of derivatives of monomials. We illustrate this by simple examples:

(1) Let

$$C_k = \sum_p a_{kp} y_p, \qquad f(C) = f(C(y)).$$

Then

$$\frac{\partial}{\partial y_p} f(C(y)) = \sum_k \left[-\frac{\partial}{\partial C_k} f(C) \right]_{C = C(y)} a_{kp},$$

$$f(C(y)) \frac{\partial}{\partial y_p} = \sum_k \left[f(C) \frac{\partial}{\partial C_k} \right]_{C = C(y)} a_{kp}.$$

(2) Let t be a real parameter and

$$C_k(t) = \sum_p a_{kp}(t) y_p.$$

Then

$$\frac{\mathrm{d}}{\mathrm{d}t}f(C(t))=\sum_k\frac{\mathrm{d}C_k}{\mathrm{d}t}\frac{\partial}{\partial C_k}f=\sum_k\left(f\frac{\partial}{\partial C_k}\right)\frac{\mathrm{d}C_k}{\mathrm{d}t}.$$

Notice also the formula differentiating products. Let $f_1\in G_n''$ be an event element, and f_2 be an arbitrary element of G_n, then

$$\frac{\partial}{\partial C_p}(f_1f_2)=\left(\frac{\partial}{\partial C_p}f_1\right)f_2+f_1\left(\frac{\partial}{\partial C_p}f_2\right),$$

$$(f_1f_2)\frac{\partial}{\partial C_p}=f_1\left(f_2\frac{\partial}{\partial C_p}\right)+\left(f_1\frac{\partial}{\partial C_p}\right)f_2.$$

If $f_1\in G_n'$ is an odd element and $f_2\in G_n$ is an arbitrary element, the formulas acquire the form

$$\frac{\partial}{\partial C_p}(f_1f_2)=\left(\frac{\partial}{\partial C_p}f_1\right)f_2-f_1\left(\frac{\partial}{\partial C_p}f_2\right),$$

$$(f_2f_1)\frac{\partial}{\partial C_p}=f_2\left(f_1\frac{\partial}{\partial C_p}\right)-\left(f_2\frac{\partial}{\partial C_p}\right)f_1.$$

Let us notice finally the following properties of repeated derivatives:

$$\frac{\partial}{\partial C_1}\left(\frac{\partial}{\partial C_2}f\right)=-\frac{\partial}{\partial C_2}\left(\frac{\partial}{\partial C_1}f\right),\quad\left(f\frac{\partial}{\partial C_1}\right)\frac{\partial}{\partial C_2}=-\left(f\frac{\partial}{\partial C_2}\right)\frac{\partial}{\partial C_1},$$

$$\left(\frac{\partial}{\partial C_1}f\right)\frac{\partial}{\partial C_2}=\frac{\partial}{\partial C_1}\left(f\frac{\partial}{\partial C_2}\right).$$

5.3.3. *Integration over a Grassmann Algebra (Finite-Dimensional Case)*

We shall present the definitions and show how they give rise to a satisfactory integration calculus. A volume element is written as $\mathrm{d}C_1\cdots\mathrm{d}C_i$, $1\leqslant i\leqslant n$, where C_i and $\mathrm{d}C_i$ must satisfy

$$\{\mathrm{d}C_i,\mathrm{d}C_k\}=\{C_k,\mathrm{d}C_i\}=0. \tag{5.24}$$

The (formal) integral is then defined by setting

$$\int\mathrm{d}C_i=0,\qquad\int C_i\,\mathrm{d}C_i=1. \tag{5.25}$$

Multiple integrals will be understood as repeated ones. Thus, formulas (5.24) and (5.25) define the integral $\int f(C)\,\mathrm{d}C_n\cdots\mathrm{d}C_1$ on all monomials.

DEFINITION 5.3. The integrals defined by formulas (5.24) and (5.25) are called the *integrals over a Grassmann algebra with generators* $C_1,\ldots,C_n$. It follows from the definition of the integral that for any element $f(C)$ the formula

$$\int f(C)\, \mathrm{d}C_m \cdots \mathrm{d}C_1 = m!\, f_m(1, \ldots, m), \quad (1 \leqslant m \leqslant n) \tag{5.26}$$

holds, where $f_m(1, \ldots, m)$ is assumed to be antisymmetric under interchange of the indices $1, 2, \ldots, m$. As an illustration, take $m = 3$. Then the first formula of (5.24) tells that

$$\int f(C)\, \mathrm{d}C_3\, \mathrm{d}C_2\, \mathrm{d}C_1 = f_3(m_1, m_2, m_3) \int C_{m_1} C_{m_2} C_{m_3}\, \mathrm{d}C_3 \mathrm{d}C_2 \mathrm{d}C_1,$$

but

$$\begin{aligned} &f_3(m_1, m_2, m_3) C_{m_1} C_{m_2} C_{m_3} \\ &\quad = f_3(1, 2, 3) C_1 C_2 C_3 + f_3(1, 3, 2) C_1 C_3 C_2 + f_3(2, 3, 1) C_2 C_3 C_1 + \\ &\qquad + f_3(2, 1, 3) C_2 C_1 C_3 + f_3(3, 1, 2) C_3 C_1 C_2 + f_3(3, 2, 1) C_3 C_2 C_1 \\ &\quad = 6 f_3(1, 2, 3) C_1 C_2 C_3. \end{aligned}$$

Hence

$$\int f(C)\, \mathrm{d}C_3\, \mathrm{d}C_2\, \mathrm{d}C_1 = 3!\, f_3(1, 2, 3) \int C_1 C_2 C_3\, \mathrm{d}C_3\, \mathrm{d}C_2\, \mathrm{d}C_1 = 3!\, f_3(1, 2, 3).$$

These examples should be sufficient to make the handling of integrals a quite straightforward matter.

As in ordinary analysis, in the case of the Grassmann algebra there are formulas of *partial integration*

$$\int f(C) \left(\frac{\partial}{\partial C_p} g(C) \right) \mathrm{d}C_n \cdots \mathrm{d}C_1 = \int \left[f(C) \frac{\partial}{\partial C_p} \right] g(C)\, \mathrm{d}C_n \cdots \mathrm{d}C_1,$$

$$\int f(C) \left[g(C) \frac{\partial}{\partial C_p} \right] \mathrm{d}C_n \cdots \mathrm{d}C_1 = (-1)^{n+1} \int \left[\frac{\partial}{\partial C_p} f(C) \right] g(C)\, \mathrm{d}C_n \cdots \mathrm{d}C_1.$$

For proof, it is sufficient to consider f and g as arbitrary monomials.

A useful formula is valid for the *change of variables in integrals over the Grassmann algebra.* We restrict ourselves to the simple case when the change is a linear one.

$$C_i = \sum_k a_{ik} e_k, \qquad \mathrm{d}C_i = \sum_k \tilde{a}_{ik}\, \mathrm{d}e_k,$$

where $\|\tilde{a}_{ik}\|$ is an inverse matrix to $\|a_{ik}\|$. The following formula is valid for such a change of variables:

$$\int f(C)\, \mathrm{d}C_n \cdots \mathrm{d}C_1 = \det \|\tilde{a}_{ik}\| \int f(C(e))\, \mathrm{d}e_n \cdots \mathrm{d}e_1. \tag{5.27}$$

Notice that in distinction to the usual formula for the change of variables, the independent variables and differentials transform by means of *mutually inverse matrices.* Now we shall make this more transparent. Let $f(x_1, \ldots, x_n)$ be a function of n real variables. Define the integral I of f by

$$I = \int f(\mathbf{x})\, \mathrm{d}x_1 \cdots \mathrm{d}x_n.$$

If $a = a_{ik}$ is a $n \times n$ matrix, we can also write, $\mathbf{x} = a\mathbf{y}$, $x_i = \| a_{ik} \| y_k$,

$$I = \det\| a_{ik} \| \int f(a\mathbf{y})\, \mathrm{d}y_1 \cdots \mathrm{d}y_n, \quad \det a \neq 0).$$

However, consider the integral over the Grassmann algebra

$$I = \int g(C)\, \mathrm{d}C_n \cdots \mathrm{d}C_1,$$

and write the linear change of variables in the Grassmann algebra as $\mathbf{C} = A\mathbf{e}$, then we find that

$$I = (\det A)^{-1} \int g(A\mathbf{e})\, \mathrm{d}e_n \cdots \mathrm{d}e_1.$$

The 'Jacobian' is the inverse of what we are used too. The reason for this is easy to find. We have $C_1 \cdots C_n = (\det A) e_1 \cdots e_n$ but $\mathrm{d}C_n \cdots \mathrm{d}C_1 = (\det A)^{-1}\, \mathrm{d}e_n \cdots \mathrm{d}e_1$ so that $\int C_i \mathrm{d}C_i = 1$ may be maintained. As an example, we calculate the '*Gaussian integral*'

$$I = \int \exp\left\{ \sum_{i,k} a_{ik} C_i C_k \right\} \mathrm{d}C_n \cdots \mathrm{d}C_1, \quad a_{ik} = -a_{ki}.$$

Let first $\| a_{ik} \|$ be a real matrix. Applying a proper orthogonal transformation $\| s_{ik} \|$, the matrix $\| a_{ik} \|$ can be brought into the form

$$\begin{pmatrix} 0 & \lambda_1 & 0 & 0 & \vdots & \ldots \\ -\lambda_1 & 0 & 0 & 0 & \vdots & \ldots \\ 0 & 0 & 0 & \lambda_2 & 0 & \ldots \\ 0 & 0 & -\lambda_2 & 0 & 0 & \ldots \\ . & . & . & . & . & . \end{pmatrix}$$

Perform the change of variables $C_i = \Sigma_k s_{ik} e_k$ in the integral I. Owing to formula (5.27), we get

$$I = \int \exp\{2[\lambda_1 e_1 e_2 + \lambda_2 e_3 e_4 + \cdots + \lambda_{n/2} e_{n-1} e_n]\}\, \mathrm{d}e_n \cdots \mathrm{d}e_1$$

for even n and

$$I = \int \exp\{2[\lambda_1 e_1 e_2 + \cdots + \lambda_{(n-1)/2} e_{n-2} e_{n-1}]\} \, de_n \cdots de_1$$

for odd n, respectively ($\det\|s_{ik}\| = 1$ since $\|s_{ik}\|$ is a proper orthogonal matrix). Using formula (5.27), we find that for even n

$$I = 2^{n/2} \lambda_1 \cdots \lambda_n = [\det(\|2a_{ik}\|)]^{1/2},$$

and for odd n, $I = 0$. Since for odd n, $\det\|a_{ik}\| = 0$, both these cases fall under the following rule

$$\int \exp\left[\sum_{i,k} a_{ik} C_i C_k\right] dC_n \cdots dC_1 = (\det\|2a_{ik}\|)^{1/2}. \tag{5.28}$$

Formula (5.28) is valid not only for real matrices but also for complex ones.

DEFINITION 5.4. An algebra K_n with generators, $k_1, \ldots, k_n$ obeying the following relations:

$$k_i k_j + k_j k_i \equiv \{k_i, k_j\} = 0$$

for $i \neq j$ is called *Clifford or spinor algebra* with n generators. A Clifford algebra K_{2n} with a doubled number of generators is closely associated with the Grassmann algebra G_n.

We recall that Bose systems with n degrees of freedom can be described by means of the Hilbert space of analytic functions $f(z_1, \ldots, z_n) = f(\mathbf{z})$ of variables $z_1, \ldots, z_n$ with the scalar product

$$(f_1 f_2) = (2\pi i)^{-n} \iint dz_1 \, d\bar{z}_1 \cdots \iint dz_n \, d\bar{z}_n \exp\left(-\sum_k z_k \bar{z}_k\right) f_1(z) \overline{f_2(z)}.$$

Coordinate and momentum operators are realized as follows

$$Q_k f = \frac{1}{\sqrt{2}}\left(\hat{z}_k + \frac{\partial}{\partial z_k}\right) f, \qquad P_k f = \frac{1}{\sqrt{2}\, i}\left(\hat{z}_k - \frac{\partial}{\partial z_k}\right) f.$$

It is easy to see that these operators satisfy the anti-commutation relations

$$\{P_j, Q_i\} = 0, \qquad \{P_i, P_j\} = \{Q_i, Q_j\} = \delta_{ij}.$$

Thus, the operators P_j, Q_j turn out to be *generators of the Clifford algebra* K_{2n}. At this stage, the operators $\hat{z}_k$ and $\partial/\partial z_k$ can be identified with the operators in G_n, that is those of the left multiplication with the generator $\hat{z}_k$ and the left derivative $\partial/\partial z_k$ with respect to z_k in G_n.

In conclusion, let us write the linear functionals and linear operators in G_n in the following form

$$F(f) = \int f(C) F_{\text{right}}(C) \, dC_n \cdots dC_1 \quad \text{or} \quad F(f) = \int F_{\text{left}}(C) f(C) \, dC_n \cdots dC_1 \tag{5.29}$$

and

$$(Tf)(C) = \int f(e_1, \ldots, e_n) K_{\text{right}}(e_1, \ldots, e_n | C_1, \ldots, C_n)\, de_n \cdots de_1$$

or

$$(Tf)(C) = \int K_{\text{left}}(C_1, \ldots, C_n | e_1, \ldots, e_n) f(e_1, \ldots, e_n) de_n \cdots de_1 \tag{5.30}$$

respectively. Here F_{right} and F_{left} are fixed elements of G_n, and K_{right} and K_{left} are elements of G_{2n}; C_i, e_k are generators in G_{2n} [see Berezin, 1966 for the proof of formulas (5.29) and (5.30)].

5.4. Grassmann Algebra with an Infinite Number of Generators

5.4.1. *Definition*

For use in field theories with fermions, we require an infinite-dimensional algebra G. In quantum theory, a Grassmann algebra equipped with a scalar product and with an *involution* are of particular importance. A strict mathematical definition of the algebra G and an analysis of it were given by Berezin (1966) using the notion of a direct sum of a *denumerable number of topological linear spaces.* Concrete applications of the analysis on an algebra of anticommuting variables will be considered in Section 5.5.

DEFINITION 5.5. As a *Grassmann algebra* G, we shall understand a *direct sum* of spaces E^n possessing the following properties:

(1) E^0 is one-dimensional space with fixed basis element f_0.
(2) For any $f \in E^p$, $g \in E^q$, a product $fg \in E^{p+q}$ is defined; and
  (a) if $f = \alpha f_0 \in E^0$, α is a complex number, then $fg = \alpha g$,
  (b) $fg = (-1)^{pq} gf$,
  (c) if $f_1, \ldots, f_n$ and $g_1, \ldots, g_m$ are linear independent elements of E^p and E^q, respectively, then $f_i g_k (i = 1, \ldots, n;\ k = 1, \ldots, m)$ are linear independent elements of E^{p+q},
  (d) finite linear combinations of elements of the type fg, $f \in E^p$, $g \in E^q$ generate a dense set in E^{p+q}.
(3) The space G is closed with respect to a *multiplication* which is defined in the following way if

$$f = f_0 + f_1 + \cdots + f_n + \cdots, \quad f_n \in E^n,$$
$$g = g_0 + g_1 + \cdots + g_n + \cdots, \quad g_n \in E^n,$$

then

$$fg = h = h_0 + h_1 + \cdots + h_n + \cdots$$

where

$$h_n = \sum_{k=0}^{n} f_k g_{n-k}.$$

Every element $f_n \in E^n$ is represented in the form

$$f_n = \int C(x_1)\cdots C(x_n)\varphi(x_1,\ldots,x_n)\,dx_1\cdots dx_n, \tag{5.31}$$

where $\varphi(x_1,\ldots,x_n)$ is, generally speaking, a generalized function. Notice that in the *infinite-dimensional case* we have labelled the anti-commuting variables (or generators) by a continuous index, as $C(x)$.

According to the definition of the Grassmann algebra G, each element belonging to G may be represented by the form

$$f = f_0 + f_1 + \cdots + f_n + \cdots, \quad f_n \in E^n.$$

Using the generic representation (5.31) for f_n, we get the following formula for any element $f \in G$:

$$f = f(C) = C_0 + \int dx C(x)\,\varphi(x) + \iint dx_1\,dx_2\,C(x_1)C(x_2)\varphi(x_1, x_2) + \cdots$$

$$= \sum_{n=0}^{\infty} \int dx_1 \cdots \int dx_n C(x_1)\cdots C(x_n)\varphi(x_1,\ldots,x_n), \tag{5.32}$$

where $\varphi(x_1,\ldots,x_n)$ are *skew-symmetric* (*generalized*) *functions* of n variables. The elements $C(x)$ we will call *generators* of the algebra G. They play a role analogous to the generators in the finite-dimensional algebras.

Notice that formula (5.32) recalls a representation of an analytical functional. In this connection, the elements $f = f(C)$ of the Grassmann algebra G with scalar product (f, f) can be considered as functionals of anti-commuting variables (or generators).

We now consider two elements f_1 and f'_1 belonging to E^1, i.e.,

$$f_1 = \int C(x)\varphi_1(x)\,dx \quad \text{and} \quad f'_1 = \int C(x)\varphi'(x)\,dx,$$

and write their product

$$f_1 f'_1 = \int dx C(x)\varphi_1(x) \int dy C(y)\varphi'(y).$$

Defining a product $C(x)C(y)$ of generalized generators $C(x)$ and $C(x)$, we get

$$f_1 f'_1 = \int dx C(x)\varphi_1(x) \int dy C(y)\varphi'(y) = \iint dx\,dy C(x)C(y)\varphi_1(x)\varphi'(y).$$

According to the definition of the Grassmann algebra

$$f_1 f'_1 = -f'_1 f_1, \qquad f_1(f_2 f_3) = (f_1 f_2) f_3.$$

It is evident that these equalities are equivalent to the relations

$$C(x)C(y) = -C(y)C(x), \qquad C(x)[C(y)C(z)] = [C(x)C(y)]C(z). \tag{5.33}$$

5.4.2. *Grassmann Algebra with Involution*

Let G be a Grassmann algebra with scalar product in which mutually homomorphous $f \longleftrightarrow f^*$ mapping onto itself is defined and which obeys the following conditions:

(a) $(f^*)^* = f$,

(b) $(f_1 f_2)^* = f_2^* f_1^*$,

(c) $(\beta f)^* = \bar{\beta} f^*$ (β is a complex number).

(d) If for f, g, a scalar product is defined, then it is also defined for f^*, g^* and they are related by $(f^*, g^*) = (g, f)$.
(e) The space E^1 is decomposable in direct sum $E^1 = F + F^*$. The mapping f^* satisfying these conditions is called *involution* in G.

Let us consider an orthonormalized basis $C(x)$ in the space F with the scalar product defined as $(C(x), C(y)) = \delta(x - y)$ where $\delta(x - y)$ is the delta function $\int \mathrm{d}y \varphi(y)\, \delta(x - y) = \varphi(x)$. In the space F^*, we introduce a conjugate basis $C^*(x)$, and extend the definition of the involution to linear functionals:

$$f^* = \int C^*(x)\bar{\varphi}(x)\,\mathrm{d}x \quad \text{for } f = \int C(x)\varphi(x)\,\mathrm{d}x.$$

The union of these bases is obviously the basis in E^1. As any basis in E^1, $\{C, C^*\}$ is a system of generators of the algebra G. By means of the generators $C(x)$, $C^*(x)$, all elements of G can be written in the form:

$$f = \sum_{m,n} \int \varphi_{mn}(x_1, \ldots, x_m \,|\, y_1, \ldots, y_n) C^*(x_1) \cdots C^*(x_m) \times$$
$$\times\, C(y_1) \cdots C(y_n)\, \mathrm{d}^m x\, \mathrm{d}^n y. \tag{5.34}$$

It is evident that the individual monomials in this sum are elements of the spaces E^{mn}. It is easy to see that if f has the form (5.34), then f^* has the following form

$$f^* = \sum_{m,n} \int \bar{\varphi}_{mn}(x_1, \ldots, x_m \,|\, y_1, \ldots, y_n) \times$$
$$\times\, C^*(y_n) \cdots C(y_1) C(x_m) \cdots C(x_1)\, \mathrm{d}^m x\, \mathrm{d}^n y.$$

We assume that the functions φ_{mn} are *skew-symmetric* separately with respect to $x_1, \ldots, x_m$ and $y_1, \ldots, y_n$.

5.4.3. *Functional (or Variational) Derivatives*

Let G be a Grassmann algebra with scalar product, and $C(x)$ is the system of generators of G. We define the left and right variational derivatives of the element $f \in G$ with respect to $C(x)$. We shall denote the left and right variational derivatives of the element f by $\delta f/\delta C(x)$ and $f\,\delta/\delta C(x)$, respectively. Both the derivatives are linear operators in G, hence it is sufficient to consider their action on the monomials $f_n \in E^n$. First define the derivatives of a product of generators:

$$\begin{aligned}\frac{\delta}{\delta C(x)} C(x_1)\cdots C(x_n) &= \delta(x - x_1)C(x_2)\cdots C(x_n) - \\ &\quad - \delta(x - x_2)C(x_1)C(x_3)\cdots C(x_n) + \\ &\quad + \cdots + (-1)^{n-1}\,\delta(x - x_n)C(x_1)\cdots C(x_{n-1}),\end{aligned} \tag{5.35}$$

$$\begin{aligned}C(x_1)\cdots C(x_n)\frac{\delta}{\delta C(x)} &= \delta(x - x_n)C(x_1)\cdots C(x_{n-1}) - \delta(x - x_{n-1}) \times \\ &\quad \times C(x_1)\cdots C(x_{n-2})C(x_n) + \\ &\quad + \cdots + (-1)^{n-1}\,\delta(x - x_1)C(x_2)\cdots C(x_n).\end{aligned}$$

The derivatives of products of the generators have an obvious meaning as generalized functions; differentiating formally under the integral sign and using formulas (5.35), we find that if

$$f = \int dx_1 \cdots \int dx_n\, \varphi(x_1, \ldots, x_n)C(x_1)\cdots C(x_n),$$

then

$$\begin{aligned}\frac{\delta}{\delta C(x)} f &= n\int dx_1 \cdots \int dx_{n-1}\, \varphi(x, x_1, \ldots, x_{n-1})C(x_1)\cdots C(x_{n-1}), \\ f\frac{\delta}{\delta C(x)} &= n\int dx_1 \cdots \int dx_{n-1}\, \varphi(x_1, \ldots, x_{n-1}, x)C(x_1)\cdots C(x_{n-1}).\end{aligned} \tag{5.36}$$

Here one has to take into account that the function $\varphi(x_1, \ldots, x_n)$ is *skew-symmetric.*

The derivatives $\delta f/\delta C(x)$ and $f\,\delta/\delta C(x)$ of general elements $f \in G$ do not necessarily belong to G. One can consider them naturally as generalized functions, i.e., instead of $\delta f/\delta C(x)$ and $f\,\delta/\delta C(x)$ one can consider the following expressions:

$$\int dx\varphi(x)\frac{\delta}{\delta C(x)} f, \quad \text{or} \quad \int dx\varphi(x)\, f\frac{\delta}{\delta C(x)},$$

where $\varphi(x)$ belongs to some set of test functions. In fact, it may happen that there exists some function $\varphi \in E^1(M)$, $x \in M$ such that

$$\int \mathrm{d}x\varphi(x)[\delta f/\delta C(x)] \in G.$$

In this case, the element f will be called left-differentiable over the functions φ. Similarly one can speak of right-differentiable elements over φ. If, however $\int \mathrm{d}x\varphi(x)\, \delta f/\delta C(x)$ does not belong to G for some function $\varphi(x) \in E^1(M)$ except $\varphi(x) \equiv 0$, the element f is called nondifferentiable from the left. Similarly one understands an element which is nondifferentiable with respect to the right. If the element f is differentiable over all functions φ belonging to some set $D \in E^1(M)$, it is called D-differentiable function.

Further we notice that from the definition of variational derivatives, the following property can be deduced. Let us consider an algebra G with generators $C(x)$ and $B(x)$. Let $F(C) \in G$. Expanding $F(C + B)$ in powers of B, the coefficients of B in the first power are obtained as the left and right derivatives of F;

$$F(C + B) = F(C) + \int B(x) \frac{\delta}{\delta C(x)} F(C)\, \mathrm{d}x + \cdots$$

$$= F(C) + \int \left(F \frac{\delta}{\delta C(x)} \right) B(x)\, \mathrm{d}x + \cdots \tag{5.37}$$

This property may be used for the definition of variational derivatives as in the case of the commuting variables [see formula (5.6)].

Let us consider a rule for change of variables under differentiation. Let now $f \in G$. The following formulas for change of variables under differentiation hold

$$\frac{\delta}{\delta C(x)} f = \int \mathrm{d}y K(x, y) \frac{\delta}{\delta B(y)} f, \qquad f \frac{\delta}{\delta C(x)} = \int \mathrm{d}y K(x, y) f \frac{\delta}{\delta B(y)},$$

when the generators $C(x)$ and $B(x)$ are connected by the relation:

$$\int \mathrm{d}x C(x) K(x, y) = B(y).$$

Here the kernel $K(x, y)$ is the continuous analog of a transition matrix from system $C(x)$ to $B(y)$, and can be interpreted as a generalized function on some space of test functions of x, y.

5.4.4. *Continual (or Functional) Integrals over the Grassmann Algebra (Formal Definition)*

Let G be the Grassmann algebra with involution, and let $E^1 = F + F^*$ be the decomposition of the space E^1 in definition of the algebra with involution. Let us embed a system of nested finite-dimensional subspaces F_n with dimension $2n$ to the space F, such that their closure coincides with F: $\overline{UF_n} = F$. Denote through G_{L_n} a sub-algebra G obtained from the space $L_n = F_n + F_n^*$. We introduce an orthonormalized basis $C_1, \ldots, C_{2n}$ in each F_n in an arbitrary way, and in F_n^* the basis formed by the conjugate elements $C_1^*, \ldots, C_{2n}^*$. Take now an arbitrary element $f(C^*, C)$ from G. Consider its value on G_{L_n}. Denote this value as $f_n(C^*, C)$ and consider the *integral*

$$I_n = (f) = \int f_n(C^*, C)\, \mathrm{d}C_{2n}^*\, \mathrm{d}C_{2n} \cdots \mathrm{d}C_1^*\, \mathrm{d}C_1. \tag{5.38}$$

It is easy to show that the integral (5.38) does not depend on the choice of the orthonormalized basis in F_n. Indeed, let $\{B_k\}$ be another orthonormalized basis in F_n, and $\{B_k^*\}$ the corresponding basis in F_n^*. Then B_k is expressed by C_k by means of a unitary matrix: $B_k = \Sigma_s u_{ks} C_s$. It is clear that B_k^* is related to C_k^* through the complex conjugated matrix: $B_k^* = \Sigma_s \bar{u}_{ks} C_s^*$. Using formula (5.27) we find that

$$\int f_n(B^*, B)\, \mathrm{d}B_{2n}^*\, \mathrm{d}B_{2n} \cdots \mathrm{d}B_1^*\, \mathrm{d}B_1 = \det(u^{-1}\bar{u}^{-1}) \int f_n(C^*, C) \times$$
$$\times \mathrm{d}C_{2n}^* \mathrm{d}C_{2n} \cdots \mathrm{d}C_1^* \mathrm{d}C_1.$$

Since $u = ||u_{ks}||$ is unitary, $\det(u^{-1}\,\bar{u}^{-1}) = 1$, the equality of the integrals is proved.

DEFINITION 5.6. If there exists the limit: $I(f) = \lim_{k\to\infty} I_k(f_k)$ independent of the chosen system of subspaces F_n, then this limit is called the *continual* (*or functional*) *integral* of f on the algebra G. The continual integral over Grassmann algebra in a physical context is called the Fermion continual integral. We shall denote it as in the case of the usual functional integral

$$I(f) = \int f(C^*, C) \prod \mathrm{d}C^*\, \mathrm{d}C. \tag{5.39}$$

Now we explain the rule of partial integration for this case. Let $f_1, f_2 \in G$ be two D-differentiable elements. Then the integral

$$\int f_1\left(\frac{\delta}{\delta C(x)} f_2\right) \prod \mathrm{d}C^*\, \mathrm{d}C \tag{5.40}$$

has an obvious sense of a generalized function on D:

$$\int \mathrm{d}x\varphi(x)\left\{\int f_1\left(\frac{\delta}{\delta C(x)} f_2\right) \prod \mathrm{d}C^*\, \mathrm{d}C\right\} = \int f_1\left\{\int \frac{\delta}{\delta C(x)} f_2 \varphi(x)\, \mathrm{d}x\right\} \prod \mathrm{d}C^*\, \mathrm{d}C.$$

Then the following formulas for integrating by parts hold:

$$\int f_1\left(\frac{\delta}{\delta C(x)} f_2\right) \prod \mathrm{d}C^*\, \mathrm{d}C = \int \left(f_1 \frac{\delta}{\delta C(x)}\right) f_2 \prod \mathrm{d}C^*\, \mathrm{d}C,$$

$$\int f_1\left(f_2 \frac{\delta}{\delta C(x)}\right)\prod \mathrm{d}C^* \,\mathrm{d}C = -\int\left(\frac{\delta}{\delta C(x)} f_1\right) f_2 \prod \mathrm{d}C^* \,\mathrm{d}C,$$

$$\int f_1\left(\frac{\delta}{\delta C^*(x)} f_2\right)\prod \mathrm{d}C^* \,\mathrm{d}C = \int\left(f_1 \frac{\delta}{\delta C^*(x)}\right) f_2 \prod \mathrm{d}C^* \,\mathrm{d}C,$$

$$\int f_1\left(f_2 \frac{\delta}{\delta C^*(x)}\right)\prod \mathrm{d}C^* \,\mathrm{d}C = -\int\left(\frac{\delta}{\delta C^*(x)} f_1\right) f_2 \prod \mathrm{d}C^* \,\mathrm{d}C.$$

5.4.5. *Examples*

(1) Let G be the algebra with involution, $E^1 = F + F^*$, and $C(x)$ be the generalized orthonormalized basis in F. We consider the operator A in E^1, which given naturally by the matrix

$$A = \begin{pmatrix} A_{11}(x, x') & A_{12}(x, x') \\ A_{21}(x, x') & A_{22}(x, x') \end{pmatrix}, \qquad A_{ik}(x, x') = -A_{ki}(x', x).$$

We construct the element $f(C^*, C)$ of space E^2 by the operator A:

$$f(C^*, C) = \int \{C(x)A_{11}(x, x')C(x') + 2C(x)A_{12}(x, x')C^*(x') +$$

$$+ C^*(x)A_{22}(x, x')C^*(x')\} \,\mathrm{d}x \,\mathrm{d}x'$$

or shortly

$$f = CA_{11}C + 2CA_{12}C^* + C^*A_{22}C^*$$

or in the form

$$f = (C, C^*)\begin{pmatrix} A_{11} & A_{12} \\ A_{21} & A_{22} \end{pmatrix}\begin{pmatrix} C \\ C^* \end{pmatrix}, \qquad A_{ij} = -A'_{ji}.$$

In particular, if $A_{12} = E$, $A_{11} = A_{22} = 0$, then $f = 2CC^*$. Let us calculate the *Gaussian integral*

$$\int \exp(\tfrac{1}{2} f(C^*, C)) \prod \mathrm{d}C^* \,\mathrm{d}C.$$

Making use of formula (5.26) and of the definition of continual integral, we find

$$\int \exp\left\{\tfrac{1}{2}(C, C^*)\begin{pmatrix} A_{11} & A_{12} \\ A_{21} & A_{22} \end{pmatrix}\begin{pmatrix} C \\ C^* \end{pmatrix}\right\}\prod \mathrm{d}C^* \,\mathrm{d}C$$

$$= \left[\det\begin{pmatrix} A_{11} & A_{12} \\ A_{21} & A_{22} \end{pmatrix}\right]^{1/2}. \tag{5.41}$$

The determinant on the right-hand side of (5.41) is a *generalized Fredholm determinant*. Power $\frac{1}{2}$ means branching of square root which is positive for real positive values of the argument.

(2) From (5.41), it follows, particularly, that

$$\int \exp(-C^*C) \prod dC^* \, dC = 1.$$

Therefore one can define the *Grassmann measure*

$$d\mu_G = \exp(-C^*C) \prod dC^* \, dC.$$

To conclude this section, we notice that all the three types of functionals (in which we are interested) namely, the *functionals*

$$\tilde{A}(C^*, C) = \sum_{p,q} \frac{1}{\sqrt{p!\, q!}} \int A_{pq}(x_1, \ldots, x_p | y_1, \ldots, y_q) \times$$
$$\times C^*(x_1) \cdots C^*(x_p) C(y_q) \cdots C(y_1) \, d^p x \, d^q y \tag{5.42}$$

and

$$A(C^*, C) = \tilde{A}(C^*, C) \exp\left\{ - \int C^*(x) C(x) \, dx \right\}, \tag{5.43}$$

corresponding to representations of operators $\hat{A}$ in the matrix form and the normal form, i.e., in each term of the type (5.43), creation operators stand left from the annihilation operators, and *functionals*

$$\Phi(C^*) = \sum_n \frac{1}{\sqrt{n!}} \int \Phi_n(x_1, \ldots, x_n) C^*(x_1) \cdots C^*(x_n) \, d^n x,$$

$$\sum_n \int |\Phi_n|^2 \, d^4 x < \infty, \tag{5.44}$$

corresponding to vectors Φ, turn out to be elements of the same Grassmann algebra G. The *scalar product* of operators $\hat{\Phi}$ can be given by the formula

$$(\hat{\Phi}_1, \hat{\Phi}_2) = \int \Phi_1(C^*) \Phi_2^*(C) \exp(-CC^*) \prod dC \, dC^*,$$

or

$$(\hat{\Phi}_1, \hat{\Phi}_2) = \int \Phi_2^*(C) \Phi_1(C^*) \exp(-C^*C) \prod dC^* \, dC,$$

where $\Phi_2^*(C)$ is the conjugate element (functional) to $\Phi_2(C^*)$. Now we have concluded the explanation of operations on functionals in the case of Grassmann algebra, and turn to the application of methods of functional analysis to quantum field theory.

5.5. Functional Integral and the S-Matrix Theory

5.5.1. *Introduction*

There exist several approaches to the quantization of physical systems (in particular, field theory). The operator method (or the *usual canonical* approach) of *quantization* is the most often used. In this method, the operators obeying the canonical commutation rules correspond to physical field configurations (see Chapter 2). However, as we will emphasize below, there is another approach, the so-called *stochastic quantization* method, the basis of which will be expounded in Part II. In Chapter 10 we shall discuss some problems concerning the stochastic quantization of physical fields. There is furthermore a third approach, in which quantum dynamics is described by the *continual* (or *path*, or *functional*) *integral*, that is of the sum over all field configurations. By means of this approach, Feynman was the first to formulate a consistent and explicitly relativistic invariant perturbation theory for quantum electrodynamics. This formalism turns out to be most suitable for the quantization of gauge fields. In this case, integration is carried out not over all field configurations, but over gauge-equivalent classes alone.

Recently, a very interesting link has been discovered between the *stochastic* and *canonical quantization* in perturbation theory for any field theories including gauge ones (Floratos and Iliopoulos, 1983). In the latter case, equivalence is achieved provided one considers gauge-invariant quantities (see Chapter 10).

Guerra (1981) has pointed out that the path integral theory can be clearly reformulated by means of stochastic mechanics considered in Part II of the book. In this section we wish to make use of functional integral techniques in the S-matrix theory. We know that the path integral of Feynman was introduced to evaluate the sum of contribution from all possible paths that a particle can trace moving from a to b. We know that in classical mechanics just one path is important, for which the action is extremal. In quantum mechanics this is not so. *Every path is now important*; it is only that that path which makes the action attain its extremal value, and neighboring paths which give an action within one unit $\hbar$ around the extremal value are the most important. In fact, if $A(x(t))$ is the action for a path $x(t)$ then its amplitude is know to be given by $\exp\{i\hbar^{-1} A(x(t))\}$. Then the *integral over all paths* is given by

$$I = \int \exp\{i\hbar^{-1} A(x(t))\} \prod \mathrm{d}x(t),$$

where $\Pi\, \mathrm{d}x(t)$ as before. The action A is the time integral of the Lagrangian $\mathscr{L}$ where

$$\mathscr{L} = \tfrac{1}{2}m\left(\frac{\mathrm{d}x}{\mathrm{d}t}\right)^2 - U(x).$$

From the Hamiltonian dynamics point of view, quantum field theory is a system with an infinite number of degrees of freedom. For example, in the case of the neutral scalar field $\varphi(x)$, described by the following Lagrangian (see Chapter 2)

$$\mathscr{L} = \tfrac{1}{2}\varphi(x)(\Box - m^2)\varphi(x) - U(\varphi), \tag{5.45}$$

the points of the phase space are pairs of functions $\varphi(x)$, $\pi(x)$ forming an infinite set of canonical variables. In this case state vectors of type (5.44) are functionals $\Phi(\varphi(x))$ of $\varphi(x)$. In quantum field theory the main object is the *S-matrix* and the *generating functional Z* for Green's functions. They are functionals of field $\varphi(x)$, an external source $\eta(x)$ and current $J(x)$, or of a function of 'switching on' the interaction $g(x)$, respectively (see Chapter 2).

Representation of types (5.8) and (5.42) for these quantities is given by the formulas:

$$S[\varphi] = \sum_n \frac{1}{n!} \int d^4x_1 \cdots \int d^4x_n S_n(x_1, \ldots, x_n)\varphi(x_1) \cdots \varphi(x_n), \tag{5.46}$$

and

$$Z(\eta) = \sum_n \frac{1}{n!} \int d^4x_1 \cdots \int d^4x_n G_n(x_1, \ldots, x_n)\eta(x_1) \cdots \eta(x_n). \tag{5.47}$$

In the operator formulation, the coefficient functions $S_n(x_1, \ldots, x_n)$ in (5.46) arise in the expansion of the operator-valued S-matrix in terms of normal products of free fields. The coefficient functions $G_n(x_1, \ldots, x_n)$ in the expansion of the functional (5.47) in powers of $\eta(x)$ (external source) define the Green functions which are the corresponding vacuum expectations of chronological products of the Heisenberg field operators.

Now we turn to derivation of the functional integral representation for the S-matrix and Z. In quantum field theory the action A acquires the form

$$A = \int \mathscr{L}(x)\, d^4x, \tag{5.48}$$

and instead of the volume element $\Pi\, dx(t)$ of the space of functions (paths) the volume element $\Pi_x\, d\varphi(x)$ defined on the space of classical fields appears. The integral for S becomes then, with $\hbar = 1$

$$S = \int \exp\left\{ i \int \mathscr{L}(x)\, d^4x \right\} \delta\varphi. \tag{5.49}$$

where a normalization constant is omitted. The volume element $\Pi_x\, d\varphi(x)$ is usually written $\delta\varphi$ or $\mathscr{D}[\varphi]$.

The content of quantum field theory can be characterized by the set of all its Green's functions $\langle 0|T\{\varphi(x_1)\cdots\varphi(x_n)\}|0\rangle$. These Green's functions can be written compactly as functional integrals. This is accomplished by using a very useful technique due to Schwinger. The technique consists of modifying the action A (5.48) by adding a term $\int d^4x\eta(x)\varphi(x)$ where $\eta(x)$ is not an operator but an ordinary scalar function. The resulting functional integral is defined as $Z(\eta)$, a functional of the source $\eta(x)$,

$$Z(\eta) = \int \delta\varphi \exp\left\{i \int d^4x[\mathcal{L}(x) + \eta(x)\varphi(x)]\right\}. \tag{5.50}$$

Usually, by substituting $U(\varphi) \to U(\varphi + \varphi_{\text{in}})$, another representation of the S-matrix as a functional integral may be obtained. For example, in the case of a neutral scalar field described by Lagrangian (5.45) the S-matrix is given by

$$S[\varphi_{\text{in}}] = \int \delta\varphi \exp\left\{i \int d^4x[\tfrac{1}{2}\varphi(x)(\Box - m^2)\varphi(x) - U(\varphi + \varphi_{\text{in}})]\right\}. \tag{5.51}$$

The obtained formulas (5.49) or (5.51), and (5.50) are attractive in view of their compact form and clarity. So the representation (5.50) for functional $Z(\eta)$ in the form of an integral allows simple formulas of analysis to be used: integration by parts, change of the order of integration; change of variables and calculation by the method of stationary phase.

Before we turn to the definition of functional integral, we shall compute $Z(\eta)$ for the trivial case $U(\varphi(x)) = 0$ and show that $Z(\eta)$ and its functional derivatives $(-1)^n\, \delta^n Z/\delta\eta(x_1)\cdots\delta\eta(x_n)$ give the free particle Green's function. The task of computing $Z(\eta)$ consists entirely of manipulation designed to reduce the argument of the exponential to a quadratic form. The kinetic term is of the form

$$\frac{1}{2}\int d^4x\varphi L(\Box)\varphi, \qquad L(\Box) = \Box - m^2.$$

Then, if we define $\phi' = L^{1/2}(\Box)\varphi$, the resultant term is of the form $\phi'\phi'$. This task is difficult due to the fact that $L(\Box)$ is a differential operator and we do not know how to take its square root. We therefore pass to momentum space and define the function $\tilde{\Delta}^{1/2}(p)$ by the equation

$$(m^2 - p^2)^{1/2}\tilde{\Delta}^{1/2}(p) = 1, \quad \text{or} \quad \tilde{\Delta}^{1/2}(p) = (m^2 - p^2)^{-1/2}.$$

Then we introduce the *Fourier transform* $\Delta^{1/2}(x)$ of $\tilde{\Delta}^{1/2}(p)$. Evidently,

$$\Delta^{1/2}(x) = (2\pi)^{-4}\int d^4p\, e^{ipx}(m^2 - p^2)^{-1/2}.$$

The change of field variable that we require is now clear, it is

$$\varphi(x) = L^{-1/2}(\Box)\phi'(x) = \int d^4y\, \Delta^{1/2}(x - y)\phi'(y).$$

Then after some elementary calculations we get

$$\iiint d^4x\, d^4x_1\, d^4x_2 \phi'(x_1)\, \Delta^{1/2}(x - x_1)(\Box - m^2)\, \Delta^{1/2}(x - x_2)\phi'(x_2)$$

$$= -\int d^4p \tilde{\phi}'(p)\tilde{\phi}'(p) = -\int d^4x \phi'(x)\phi'(x).$$

We have now achieved the result that we required, and we can now change the variable in the functional integral from φ to ϕ. This means that

$$Z(\eta) = K\int \delta\phi' \exp\left\{i\int d^4x\left[-\tfrac{1}{2}\phi'^2(x) + \int d^4x_1 \eta(x)\, \Delta^{1/2}(x - x_1)\phi'(x_1)\right]\right\},$$

where K is the Fredholm determinant for $\Delta^{1/2}$. The last term to be dealt with is the source term. It is straightforward to find a change of variable to eliminate the source term. The choice needed is

$$\phi'(x) = \phi''(x) + \int d^4x_1 \eta(x)\, \Delta^{1/2}(x - x_1).$$

With this change of variable, the argument of the exponential becomes

$$\int d^4x\left[-\frac{1}{2}\phi''^2 + \frac{1}{2}\int d^4y H(y)H(y)\right],$$

where

$$H(y) = \int d^4x_1 \eta(y)\, \Delta^{1/2}(y - x_1).$$

We deal with this term in the same way as above:

$$\frac{1}{2}\int d^4y H(y)H(y) = \frac{1}{2}\int d^4p \tilde{\eta}(p)\, \tilde{\Delta}(p)\tilde{\eta}(p),$$

with

$$\tilde{\Delta}(p) = \tilde{\Delta}^{1/2}(p)\tilde{\Delta}^{1/2}(p) = (m^2 - p^2)^{-1}.$$

Therefore we have

$$\frac{1}{2}\int d^4x H(x)H(x) = \frac{1}{2}\iint d^4x\, d^4y \eta(x)\, \Delta(x - y)\eta(y),$$

where

$$\Delta(x - x') = \int d^4y\, \Delta^{1/2}(x - y)\, \Delta^{1/2}(y - x'),$$

is the *usual scalar particle Green's function* satisfying

$$(\Box - m^2)\,\Delta(x-y) = -\delta^{(4)}(x-y).$$

The functional $Z(\eta)$ is now seen to be

$$Z(\eta) = \exp\left[\frac{i}{2}\iint \mathrm{d}^4x\,\mathrm{d}^4y\eta(x)\,\Delta(x-y)\eta(y)\right]\cdot K\cdot\int\delta\phi' \times$$

$$\times \exp\left\{-\frac{i}{2}\int \mathrm{d}^4x\phi'^2(x)\right\}. \tag{5.52}$$

The factor $K\cdot\int\delta\phi'\exp\{-i/2\int \mathrm{d}^4x\phi'^2(x)\}$ does not depend on η and is simply a number. We eliminate it by redefining the integral that we are interested in as

$$Z(\eta) = \frac{\displaystyle\int\delta\varphi\exp\left\{\frac{i}{2}\int \mathrm{d}^4x[\varphi(x)L(\Box)\varphi(x) + 2\eta\varphi]\right\}}{\displaystyle\int\delta\varphi\exp\left\{\frac{i}{2}\int \mathrm{d}^4y\varphi(y)L(\Box)\varphi(y)\right\}}$$

$$= \exp\left\{\frac{i}{2}\iint \mathrm{d}^4x\,\mathrm{d}^4y\eta(x)\,\Delta\,(x-y)\eta(y)\right\}.$$

Note that $Z(0) = 1$. The functional $Z(\eta)$ has a functional Taylor series obtained by expanding the exponential. The terms in this Taylor series turn out to be the free particle Green's functions. For example, we have

$$\delta^2 Z(\eta)/\delta\eta(x_1)\,\delta\eta(x_2)|_{\eta=0} = i\,\Delta(x_1 - x_2).$$

To summarize the situation for a free field, we write

$$i^n\,\delta^n Z(\eta)/\delta\eta(x_1)\cdots\delta\eta(x_n)|_{\eta=0} = \langle 0|T\{\varphi(x_1)\cdots\varphi(x_n)\}|0\rangle$$

$$= \int\delta\varphi\varphi(x_1)\cdots\times$$

$$\times\ \varphi(x_n)\exp\left\{\frac{i}{2}\int \mathrm{d}^4x\varphi(x)L(\Box)\varphi(x)\right\}\times$$

$$\times\left\{\int\delta\varphi\exp\left[\frac{i}{2}\int \mathrm{d}^4y\varphi(y)L(\Box)\varphi(y)\right]\right\}^{-1}$$

The great importance of this property for free fields is the fact that it remains true for interacting fields. The only change is that the free action is replaced by

$$A_F \to A = \int \mathrm{d}^4x[\tfrac{1}{2}\varphi(x)L(\Box)\varphi(x) - gU(\varphi(x))]$$

and the generating functional for Green's functions is given by

$$Z(\eta) = \frac{\int \delta\varphi \exp\left\{i \int d^4x[\frac{1}{2}\varphi(x)L(\Box)\varphi(x) - gU(\varphi(x)) + \eta\varphi(x)]\right\}}{\int \delta\varphi \exp\left\{i \int d^4x[\frac{1}{2}\varphi(x)L(\Box)\varphi(x) - gU(\varphi(x))]\right\}}.$$

Expansion of this integral in powers of g coincides with the usual perturbation expansion in terms of Feynman graphs.

5.5.2. *Functional Integral over a Bose Field in the Case of Nonlocal-Stochastic Theory (Definition)*

5.5.2a. *Definition of Functional Integral.* We shall investigate here the S-matrix written in the form of functional integral in the Euclidean metric. Therefore all the formulas and integration are written in Euclidean space. The so-called *Wick rotation* between Minkowski space and Euclidean space for Bose (or Fermion) fields was the subject of many studies (in particular, see Schwinger, 1959; Osterwalter and Schrader, 1973, 1975). We will examine the theory of a one-component scalar field $\varphi(x)$ described by the Lagrangian density (2.3) (Chapter 2):

$$\mathscr{L}(x) = \tfrac{1}{2}\varphi(x)L(\Box)\varphi(x) - g[K(l^2\Box)\varphi(x)]^4, \tag{5.53}$$

where $\phi(x) = K(l^2\Box)\varphi(x)$ is a *nonlocal field* resulting from averaging over stochastic space-time $\mathbb{R}^4(\hat{x})$ (see Chapter 1). The operator $K(l^2\Box)$ is nonlocal and satisfies the conditions (see Chapters 1 and 2):

(i) $K(z)$ is an entire analytic function of the order $\rho \geqslant \frac{1}{2}$ in the complex z-plane;
(ii) $K(z)$ decreases rapidly enough when $z = l^2k^2 = l^2(k_0^2 - \mathbf{k}^2) \to -\infty$;
(iii) $K(-m^2l^2) = 1$;
(iv) l is the fundamental length characterizing the region of nonlocal interaction in the given case.

In the paper by Efimov (1977b), the representation of the S-matrix as a functional integral was extended to the case of non-polynomial interactions. In this section, we use the representation obtained above for investigation of the theory described by the Lagrangian (5.53).

For this purpose, the S-matrix as a functional of the scalar field $\varphi_{in}(x)$ in Euclidean space of dimension d is written in the form of the functional integral

$$S[\varphi_{in}] = N^{-1}\int \delta\varphi \exp\left\{-\frac{1}{2}\int d^dx\varphi(x)(m^2 - \Box)\varphi(x) - g \times \right.$$
$$\left. \times \int d^dx[\phi(x) + \varphi_{in}]^4\right\} \tag{5.54}$$

or

$$S[\varphi_{\text{in}}] = C^{-1}\int \delta\varphi \exp\left\{-\frac{1}{2}\iint d^d x_1\, d^d x_2\, \varphi(x_1)\,\Delta^{-1}(x_1 - x_2)\varphi(x_2) - \right.$$

$$\left. - g\int d^d x[\phi(x) + \varphi_{\text{in}}]^4\right\}. \tag{5.55}$$

Here the function $\Delta^{-1}(x)$ is defined by the relation

$$\int d^d y\, \Delta(x_1 - y)\,\Delta^{-1}(y - x_2) = \delta^{(d)}(x_1 - x_2),$$

and normalization constants N and C in (5.54) and (5.55) are chosen so that $S[\varphi_{\text{in}}]|_{g=0} = 1$, where φ_{in} is an arbitrary scalar function (field). In expression (5.55) one can carry out the change of variables

$$\varphi(x) = \int d^d x'\, \Delta_E^{1/2}(x - x')\varphi'(x')$$

where the function $\Delta_E^{1/2}(x)$ is defined by the equality

$$\Delta_E^{1/2}(x) \equiv \Delta^{1/2}(x) = (2\pi)^{-d}\int d^d k\, e^{-ikx}\,\tilde{\Delta}^{1/2}(k^2),$$

$$\tilde{\Delta}^{1/2}(k^2) = (\tilde{\Delta}(k^2))^{1/2} = (m^2 + k^2)^{-1/2}$$

so that the *Euclidean Green's function* $\Delta_E(x)$ is given by

$$\Delta_E(x - x') = \Delta(x - x') = \int d^d y\, \Delta^{1/2}(x - y)\,\Delta^{1/2}(y - x').$$

After this change the S-matrix (5.55) can be represented in the form

$$S[\varphi_{\text{in}}] = C^{-1}\int \delta\varphi \exp\left\{-\frac{1}{2}\int d^d x \varphi^2(x) - g\int d^d x(\phi_d(x) + \varphi_{\text{in}})^4\right\},$$

$$\phi_d(x) = K(l^2\Box)\left\{\int d^d y\, \Delta^{1/2}(x - y)\varphi(y)\right\}. \tag{5.56}$$

The S-matrix in this representation contains a *volume divergence* connected with the *translation invariance* of our theory. Therefore we have to introduce the integration over a finite volume $V \in \mathbb{R}^d$ in (5.56). It is supposed further that in the volume V there is an orthonormal system of functions $\{g_s(x)\}$, $s = 1, 2, \ldots$, such that

$$\int_V d^d x g_s(x) g_{s'}(x) = \delta_{ss'}, \qquad \sum_{s=1}^{\infty} g_s(x) g_s(x') = \delta^{(d)}(x - x').$$

The volume V and the system of functions $\{g_s\}$ will be chosen in the following way

$$V = \{x: -L \leqslant x_j \leqslant L, j = 1, 2, \ldots, d\},$$

$$\{g_s(x)\} = \{g_{s_1 \cdots s_d}(x_1, \ldots, x_d)\} = \left\{\prod_{j=1}^{d} f_{s_j}(x_j)\right\}, \tag{5.57}$$

where

$$f_s(x) = \begin{cases} (2L)^{-1/2} \cos \dfrac{\pi}{L}\left(\dfrac{s-1}{2}\right)x & s \text{ is odd,} \\ (2L)^{-1/2} \sin\left(\dfrac{\pi}{L}\dfrac{s}{2}x\right) & s \text{ is even.} \end{cases}$$

It is assumed that any function $f(x) \in L_2(V)$ can be expanded over this basis

$$f(x) = \sum_{n=0}^{\infty} f_n g_n(x), \qquad f_n = \int_V \mathrm{d}^d x g_n(x) f(x);$$

$$\sum_{n=0}^{\infty} |f_n|^2 = \int_V \mathrm{d}^d x |f(x)|^2 < \infty.$$

After these preliminary remarks, we turn to the definition of functional integral (5.56). In the functional integral (5.56) put

$$\varphi(x) = \sum_{n=0}^{\infty} g_n(x) u_n.$$

Then

$$\phi_d(x) = \phi(u, x) = \sum_{n=0}^{\infty} D_n^{1/2}(x) u_n,$$

where

$$D_n^{1/2}(x) = \int_V \mathrm{d}^d y D_V^{1/2}(x, y) g_n(y), \qquad D_V^{1/2}(x, y) = h_V(x) D^{1/2}(x - y) h_V(y),$$

$$D^{1/2}(x) = (2\pi)^{-d} \int \mathrm{d}^d k \tilde{K}(l^2 k^2)(m^2 + k^2)^{-1/2} \, \mathrm{e}^{ikx}.$$

Here $h_V(x) \in D(V)$ is infinitely differentiable and a positive function and $\lim_{V \to \infty} h_V(x) = 1$. The function $D^{1/2}(x)$ is connected with the *nonlocal causal Green's function* $D_c(x)$ in the following way

$$D_V(x, x') = \int_V \mathrm{d}^d y D_V^{1/2}(x, y) D_V^{1/2}(y, x'),$$

$$D_c(x - x') = \lim_{V \to \infty} D_V(x, x') = \int d^d y D^{1/2}(x - y) D^{1/2}(y - x')$$

$$= (2\pi)^{-d} \int d^d p \, e^{ip(x - x')} [\tilde{K}(l^2 p^2)]^2 (m^2 + p^2)^{-1}.$$

The system $\{g_s\}$ is chosen in such a way that

$$\sum_{n=1}^{\infty} [D_n^{1/2}(x)]^2 = D_V(x, x) \underset{V \to \infty}{\to} D_c(0) = (2\pi)^{-d} \int d^d k \frac{[\tilde{K}(l^2 k^2)]^2}{m^2 + k^2} < \infty,$$

$$\sum_{n=1}^{\infty} n^M [D_n^{1/2}(x)]^2 < \infty, \quad \forall M > 0, V < \infty. \tag{5.58}$$

It is easy to verify that

$$\lim_{V \to \infty} \sum_{n=0}^{\infty} D_n^{1/2}(x) D_n^{1/2}(x') = \int d^d y D^{1/2}(x - y) D^{1/2}(y - x') = D_c(x - x').$$

Let us introduce the following functions:

$$\varphi_N(x) = \sum_{n=0}^{N} g_n(x) u_n, \qquad \phi^{(N)}(u, x) = \sum_{n=0}^{N} D_n^{1/2}(x) u_n$$

and define the measure

$$d^N \mu_u = \prod_{n=0}^{N} \frac{du_n}{\sqrt{2\pi}} \exp(-\tfrac{1}{2} u_n^2), \qquad d\mu_u \equiv d^\infty \mu_u, \tag{5.59}$$

so that

$$\int d\mu_u = \prod_{n=0}^{\infty} \int \frac{du_n}{\sqrt{2\pi}} \exp(-\tfrac{1}{2} u_n^2) = 1.$$

Then the functional integral (5.56) acquires the form

$$S_V[g, \varphi_{in}] = \int d\mu_u \exp\left\{ -g \int_V d^d y [\phi(u, x) + \varphi_{in}(x)]^4 \right\}, \tag{5.60}$$

which is defined as ($\varphi_{in} = \varphi$)

$$S_V[g, \varphi] = \lim_{N \to \infty} S_V^{(N)}[g, \varphi],$$

$$S_V^{(N)}[g, \varphi] = \int d^N \mu_u \exp\left\{ -g \int_V d^d y [\phi^{(N)}(u, x) + \varphi(x)]^4 \right\}, \tag{5.61}$$

where $d^N \mu_u$ is given by expression (5.59).

It is shown (Efimov, 1979) that the *functional integral* (5.60) given by the limit (5.61) *does exist* for any $V < \infty$ and defines an analytic function in the complex g-plane singular at the point $g = 0$. For example, we shall show here that the functional integral (5.60) does exist in the case of $\operatorname{Re} g > 0$, i.e., the limit in (5.61) exists for $\operatorname{Re} g > 0$. In this region, we have

$$|S_V^{(N)}[g, \varphi]| \leqslant 1 \quad (\forall N > 1).$$

Consider further the difference

$$\Delta_N^{N+M} = S_V^{(N+M)}[g, \varphi] - S_V^{(N)}[g, \varphi]$$

$$= \int d^{N+M}\mu_u \left\{ \exp\left[-g \int_V d^d x [\phi^{(N+M)}(u, x) + \varphi(x)]^4 \right] - \right.$$

$$\left. - \exp\left[-g \int_V d^d x (\phi^{(N)}(u, x) + \varphi(x))^4 \right] \right\}. \tag{5.62}$$

We transform this difference in the following way

$$\Delta_N^{N+M} = \int_0^1 d\xi (1 - \xi) \int d^{N+M}\mu_u \frac{\partial^2}{\partial \xi^2} \times$$

$$\times \exp\left\{ -g \int_V d^d x [\phi^{(N)}(u, x) + \xi \psi^{(N, M)}(u, x) + \varphi(x)]^4 \right\}$$

$$= \int_0^1 d\xi (1 - \xi) \int d^{N+M}\mu_u \exp\{-gC(u, \xi)\} \times$$

$$\left\{ -12g \int_V d^d x [\psi^{(N, M)}(u, x)]^2 \times \right.$$

$$\times [\phi^{(N)}(u, x) + \xi \psi^{(N, M)}(u, x) + \varphi(x)]^2 + 16g^2 \left[\int_V d^d x \psi^{(N, M)}(u, x) \times \right.$$

$$\left. \left. \times (\phi^{(N)}(u, x) + \xi \psi^{(N, M)}(u, x) + \varphi(x))^3 \right]^2 \right\},$$

where

$$\psi^{(N, M)}(u, x) = \phi^{(N+M)}(u, x) - \phi^{(N)}(u, x),$$

$$C(u, x) = \int_V d^d x [\phi^{(N)}(u, x) + \xi \psi^{(N, M)}(u, x) + \varphi(x)]^4.$$

Making use of the *Hölder inequalities* for integrals

$$\left| \int d\mu \, A_1 A_2 \right| \leqslant \left[\int d\mu |A_1|^{1/p_1} \right]^{p_1} \left[\int d\mu |A_2|^{1/p_2} \right]^{p_2},$$

where $p_1 + p_2 = 1$, one can obtain

$$|\Delta_N^{N+M}| \leqslant \int_0^1 d\xi(1-\xi) \int d^{N+M}\mu_u \exp\{-\mathrm{Re}\, g \cdot C(u,\xi)\} \times$$
$$\times \{12|g|C^{1/2}(u,\xi) + 16|g|^2 C^{3/2}(u,\xi)\}\left[\int_V d^d x(\psi^{(N,M)}(u,x))^4\right]^{1/2}.$$

Using further the obvious inequalities

$$C^r \exp(-\mathrm{Re}\, g \cdot C) \leqslant [\mathrm{Re}\, g]^{-1} \times r^r \times e^{-r},$$
$$\int d\mu(Q)^s \leqslant \left[\int d\mu Q\right]^s, \quad \left(s < 1, \int d\mu = 1\right),$$

one can obtain finally

$$|\Delta_N^{N+M}| \leqslant \mathrm{const}\left\{\int d^{N+M}\mu_u \int_V d^d x[\psi^{(N,M)}(u,x)]^4\right\}^{1/2}$$
$$= \mathrm{const}\left\{\int_V d^d x\left[\sum_{s=N+1}^{N+M} D_s(x)\right]^2\right\}^{1/2}.$$

Because the series $\Sigma\, D_s(x)$ converges, there exists N_0 for any $\varepsilon > 0$, so that for any $M > 0$ and $N > N_0$, the inequality $|\Delta_N^{N+M}| < \varepsilon$ is valid. This means that the sequences $S_V^{(N)}[g, \varphi]$ for $V < \infty$ and $\mathrm{Re}\, g > 0$ is fundamental.

Thus the limit in (5.61) exists. This limit defines the functional integral (5.60) for any $V < \infty$ and $\mathrm{Re}\, g > 0$. Proof in the case of $\mathrm{Re}\, g < 0$ is given by analytical continuation of the function $S_V^{(N)}[g, \varphi]$ to the region $\mathrm{Re}\, g < 0$. It was done by Efimov (1979).

5.5.2b. *Upper and Lower Bounds of Vacuum Energy E(g) in Nonlocal Theory and in the Anharmonic Oscillator Case.* The representation of the type (5.60) and (5.61) obtained above for the S-matrix as functional integral is very convenient for investigation of many problems of quantum theory. In particular, for the problems of:

- summation of the perturbation series, i.e., of going beyond the scope of the perturbation theory;
- strong coupling ($g \to \infty$), and of the phase transition in different quantum field theory models;
- the behavior of the vacuum energy density in the strong coupling limit, etc.

We shall demonstrate, for example, how to find the upper and lower bounds of *vacuum energy density.*

$$E(g) = -\lim_{V \to \infty} \frac{1}{V} \ln \langle 0 | S_V[g, \varphi] | 0 \rangle,$$

when $g \to \infty$ in the nonlocal $g\varphi_d^4$ – theory as a function of dimension d and of an ultraviolet behavior of the causal Green's function $\tilde{D}_c(k^2)$ in the Euclidean region. We will follow G. V. Efimov (1979). First, the lower bound of $E(g)$ will be obtained. Let us consider the function

$$S_V(g) = \int d\mu_u \exp\left\{-g \int_V d^d x \phi^4(u, x)\right\}. \tag{5.63}$$

The inequality

$$\int_V d^d x \phi^4(u, x) \geqslant \frac{1}{V}\left[\int_V d^d x \phi^2(u, x)\right]^2$$

gives

$$S_V(g) \leqslant \int d\mu_u \exp\left\{-\frac{g}{V}\left[\int_V d^d x \phi^2(u, x)\right]^2\right\}. \tag{5.64}$$

This integral can be calculated in the following way. It can be represented in the form

$$S_V(g) \leqslant \frac{1}{\sqrt{\pi}} \int_{-\infty}^{\infty} dt \exp(-t^2) F_V(g, t), \tag{5.65}$$

where

$$F_V(g, t) = \int d\mu_u \exp\left\{-2it\sqrt{g/V} \int_V d^d x \phi^2(u, x)\right\}.$$

An integral of this type was calculated by Efimov (1979) and his result is given (for $V \to \infty$) by

$$\begin{aligned} Q_V &= \int \delta\varphi \exp\left\{-\frac{1}{2}\int_V d^d x \varphi^2(x) - ia \int_V d^d x \times \right. \\ &\quad \left. \times \left[\int_V d^d y D_V^{1/2}(x, y)\varphi(y)\right]^2 + i\int_V d^d x \alpha(x)\varphi(x)\right\} \\ &= \int \delta u \exp\left\{-\frac{1}{2}\sum_n u_n^2 - ia \int_V d^d x\left[\sum_n D_n^{1/2}(x) u_n\right]^2 + i\sum_n \alpha_n u_n\right\}\Bigg|_{V \to \infty} \\ &= \exp\left\{-\frac{V}{2}\int \frac{d^d k}{(2\pi)^k} \ln[1 + 2ia\tilde{D}_c(k^2)] - \frac{1}{2}\int d^d x \int d^d x' \alpha(x) G(x - x')\alpha(x')\right\} \end{aligned}$$

where

$$G(x - x') = (2\pi)^{-d} \int d^d k \, e^{ik(x - x')}[1 + 2ia\tilde{D}_c(k^2)]^{-1}.$$

Thus, when V is large enough, we get

$$F_V(g,t) = \exp\left\{-\frac{V}{2}\int\left(\frac{\mathrm{d}k}{2\pi}\right)^d \ln\left[1 + 4it\left(\frac{g}{V}\right)^{1/2}\tilde{D}_c(k^2)\right]\right\}. \tag{5.66}$$

Substituting this function $F_V(g, t)$ into (5.65) and introducing the new variable $t = uV^{1/2}$, one obtains

$$S_V(g) = \left(\frac{V}{\pi}\right)^{1/2}\int_{-\infty}^{\infty} \mathrm{d}u \exp\left\{-V\left[u^2 + \frac{1}{2}\int\left(\frac{\mathrm{d}k}{2\pi}\right)^d \ln\left(1 + 4iu\sqrt{g}\,\tilde{D}_c(k^2)\right)\right]\right\}.$$

When $V \to \infty$ this integral can be calculated by the methods of *steepest descent*. The *saddle point* is on the negative imaginary axis in the $(u + iv)$-plane. Putting $u_0 = -iv_0$ we have the equation

$$\frac{\mathrm{d}}{\mathrm{d}v_0}\left\{v_0^2 - \frac{1}{2}\int\left(\frac{\mathrm{d}k}{2\pi}\right)^d \ln[1 + 4v_0\sqrt{g}\,\tilde{D}_c(k^2)]\right\} = 0$$

or

$$v_0 = \sqrt{g}\int\left(\frac{\mathrm{d}k}{2\pi}\right)^d \tilde{D}_c(k^2)[1 + 4v_0\sqrt{g}\,\tilde{D}_c(k^2)]^{-1}. \tag{5.67}$$

Thus the following estimate for $E(g)$ is valid

$$E(g) = -\lim_{V\to\infty}\frac{1}{V}\ln S_V(g) \geqslant E_-(g),$$

$$E_-(g) = -v_0^2 + \frac{1}{2}\int\left(\frac{\mathrm{d}k}{2\pi}\right)^d \ln[1 + 4v_0\sqrt{g}\,\tilde{D}_c(k^2)], \tag{5.68}$$

where v_0 is determined by (5.67). Formulas (5.68) can be written on the basis of equation (5.67) in an alternative form

$$E_-(g) = v_0\sqrt{g}\int\left(\frac{\mathrm{d}k}{2\pi}\right)^d\left[-\frac{4}{d}k^2\tilde{D}_c'(k^2) - \tilde{D}_c(k^2)\right][1 + 4v_0\sqrt{g}\,\tilde{D}_c(k^2)]^{-1} \tag{5.69}$$

Let us now obtain the behavior of $E_-(g)$ for $g \to 0$ and $g \to \infty$. When $g \to 0$

$$v_0 = \sqrt{g}\int\left(\frac{\mathrm{d}k}{2\pi}\right)^d \tilde{D}_c(k^2) = \sqrt{g}\,D_c(0)$$

and $E_-(g) = gD_c^2(0)$. When $g \to \infty$, the behavior of v_0 as a function of g is determined by the ultraviolet behavior of the *causal nonlocal function* $\tilde{D}_c(k^2)$. Further let us consider some particular cases. First, we examine the case when $\tilde{D}_c(k^2)$ decreases as

$$\tilde{D}_c(k^2) = \mathrm{O}((k^2)^{-1-\alpha}) \tag{5.70}$$

for $k^2 \to \infty$, where $2(1 + a) > d$ as a consequence of condition $D(0) < \infty$ in

(5.58). In this case, from (5.67) for $g \to \infty$

$$v_0 = \text{const}\, g^{(d/2)/[4(1+a)-d]}$$

follows, whence

$$E_-(g) = C_- g^{d/[4(1+a)-d]} \tag{5.71}$$

where C_- is some constant.

In the case of *anharmonic oscillator* [$d = 1$ and $a = 1$ in (5.71)] the causal Green's function is $\tilde{D}_c(k^2) = (1 + k^2)^{-1}$ (in units $m = 1$). In this case, the function (5.68) can be calculated in the explicit form:

$$E_-(g) = \max_u \left\{ -\frac{u^2}{g} + \tfrac{1}{2}[(1 + 4u)^{1/2} - 1] \right\}.$$

This expression can be written after simple transformations

$$E_-(g) = \frac{v(16 + 18v + 9v^2)}{8(1 + v)(3v + 4)},$$

where $v(1 + v)(2 + v) = 2g$. Hence it appears that in the limit of large and small g

$$E_-(g) = \begin{cases} \tfrac{1}{2}g & g \to 0, \\ \tfrac{3}{8}(2g)^{1/3} = 0.474\, g^{1/3}, & g \to \infty. \end{cases} \tag{5.72}$$

Second, we consider the case exponential decreasing

$$\tilde{D}_c(k^2) = O(\exp(-k^{2\gamma})) \tag{5.73}$$

for $k^2 \to \infty$. The solution v_0 of equation (5.67) is

$$v_0 = \text{const}(\ln g)^{d/4\gamma}$$

in the limit $g \to \infty$ so that

$$E_-(g) = C_-(\ln g)^{1 + d/2\gamma}[1 + O(\ln \ln g/\ln g)]. \tag{5.74}$$

Here C_- is a certain constant.

Now let us obtain the upper bound to the function $E(g)$ for real positive g. The following obvious inequality is valid for the function $S_V(g)$ in (5.63):

$$S_V(g) \geqslant \int \delta u \exp\left\{ -\frac{1}{2} \sum_{n=1}^{\infty} (1 + q_n) u_n^2 - g \int_V \mathrm{d}^d x \left[\sum_{n=1}^{\infty} D_n^{1/2}(x) u_n \right]^4 \right\}$$

$$= \prod_{n=1}^{\infty} (1 + q_n)^{-1/2} \int \delta u \exp\left\{ -\frac{1}{2} \sum_{s=1}^{\infty} u_s^2 - g \times \right.$$

$$\left. \times \int_V \mathrm{d}^d x \left[\sum_{s=1}^{\infty} \frac{D_s^{1/2}(x)}{\sqrt{1 + q_s}} u_s \right]^4 \right\}. \tag{5.75}$$

Here q_n are positive numbers. Then it is easy to get the following upper bound for $E(g)$:

$$E(g) = -\lim_{V\to\infty}\frac{1}{V}\ln S_V(g) \leqslant \frac{1}{2}\lim_{V\to\infty}\frac{1}{V}\sum_s \ln(1+q_s) - \lim_{V\to\infty}\frac{1}{V}\times$$

$$\times \ln\left\{\int d\mu_u \exp\left[-g\int_V d^dx\left(\sum_s D_s^{1/2}(x)(1+q_s)^{-1/2}u_s\right)^4\right]\right\}$$

$$\leqslant \frac{1}{2}\lim_{V\to\infty}\frac{1}{V}\sum_s \ln(1+q_s) + 3g\lim_{V\to\infty}\frac{1}{V}\int_V d^dx\left[\sum_s D_s(x)(1+q_s)^{-1}\right]^2.$$

The numbers q_s can be chosen in such a way that

$$\lim_{V\to\infty}\frac{1}{V}\sum_s \ln(1+q_s) = \int\left(\frac{dk}{2\pi}\right)^d \ln[1+q(k^2)],$$

$$\lim_{V\to\infty}\sum_s \frac{D_s(x)}{1+q_s} = \int\left(\frac{dk}{2\pi}\right)^d \frac{\tilde{D}_c(k^2)}{1+q(k^2)}. \tag{5.76}$$

For example, in the case of the orthonormal system (5.57) these numbers should be taken as

$$q_s = q_{s_1,\ldots,s_d} = q(L^{-2}\pi^2(s_1^2+\cdots+s_d^2))$$

and $V = (2L)^d$.

Because the numbers q_s and consequently the function $q(k^2)$ are arbitrary, we have $E(g) \leqslant E_+(g)$,

$$E_+(g) = \min_{q>0}\left\{\frac{1}{2}\int\left(\frac{dk}{2\pi}\right)^d \ln[1+q(k^2)] + 3g\left[\int\left(\frac{dk}{2\pi}\right)^d \frac{\tilde{D}_c(k^2)}{1+q(k^2)}\right]^2\right\}. \tag{5.77}$$

When $g\to 0$, the minimum is reached for $q(k^2)=0$. In this limit, we obtain

$$E_+(g) = 3gD_c^2(0), \tag{5.78}$$

i.e., the lowest perturbation order when the interaction Lagrangian is not taken in the normal form.

The behavior of $E_+(g)$ for $g\to\infty$ is determined by the *ultraviolet asymptotic of the Green's function* $\tilde{D}_c(k^2)$. In case (5.70), the function

$$q(k^2) = (A^2/k^2)^\beta \tag{5.79}$$

can be chosen to determine the function $E_+(g)$ in (5.77), where A and $\beta > d/2$ are parameters.

Substituting (5.79) into (5.77) and putting $g\to\infty$, one can see that the minimum is realized for large A. We have

$$E_+(g) \leqslant \min_{\beta>d/2}\min_A\{C_1(\beta)A^d + C_2(\beta)gA^{2d-4(1+a)}\},$$

where

$$C_1(\beta) = \frac{1}{2}\int\left(\frac{dk}{2\pi}\right)^d \ln(1 + k^{-2\beta}),$$

$$C_2(\beta) = 3\left[\int\left(\frac{dk}{2\pi}\right)^d k^{2\beta}\cdot k^{-2(1+a)}(1 + k^{2\beta})^{-1}\right]^2.$$

Then it follows

$$E_+(g) \leqslant C_+ g^{d/[4(1+a)-d]}, \tag{5.80}$$

where the constant C_+ can be calculated as follows

$$C_+ = \min_\beta C_1(\beta)\left[1 + \frac{d\cdot C_2(\beta)}{2(2(1+a)-d)}\right]\left[\frac{2(2(1+a)-d)C_2(\beta)}{d\cdot C_1(\beta)}\right]^{4(1+a)-d}.$$

Thus the two estimates (5.74) and (5.80) show that the nonlocal-stochastic theory when $\tilde{D}_c(k^2) = O((k^2)^{-1-\alpha})$ for $k^2 \to \infty$ in the limit $g \to \infty$, the vacuum energy $E(g)$ is

$$E(g) = Cg^{d/[4(1+a)-d]}, \tag{5.81}$$

where C is a constant satisfying $C_- < C < C_+$. Formula (5.81) can be rewritten in the form

$$E(g) = Cg^{f(d,a)}, \tag{5.82}$$

$$f(d,a) = d[4(1+a)-d]^{-1} = \left\{1 - \left[1 - \frac{d}{2(1+a)}\right]\right\}\left\{1 + \left[1 - \frac{d}{2(1+a)}\right]\right\}^{-1} < 1$$

because of $2(1 + a) > d$ according to the condition $D_c(0) < \infty$.

Note once more that the behavior $E(g)$ for $g \to \infty$ depends on the dimension of space-time d and the ultraviolet asymptotic of the causal nonlocal Green's function $\tilde{D}_c(k^2)$ for $k^2 \to \infty$.

Let us consider $E_+(g)$ for $g \to \infty$ in case (5.73). This estimation can be obtained by choosing $q(k^2) = A\exp(-k^{2\gamma})$. After simple calculations one can get in the limit $g \to \infty$

$$E_+(g) = C_+(\ln g)^{1+d/2\gamma}[1 + O(\ln\ln g/\ln g)].$$

Thus in the φ_d^4-theory with the propagator (5.73), we have

$$E(g) = C(\ln g)^{1+d/2\gamma}[1 + O(\ln\ln g/\ln g)], \tag{5.83}$$

in the limit of strong coupling. In the paper by Bervillier *et al.* (1978), the φ^4-theory with the propagator $\tilde{D}_c(k^2) = \exp(-k^2)$ [i.e., $\gamma = 1$ in (5.73)] was considered. These authors have calculated eight orders of perturbation theory for different characteristics of this model (energy levels, Green's functions). The knowledge of exact asymptotics (5.83) should be useful for different methods of asymptotic summation.

Finally, we turn to the function $E_+(g)$ in (5.77) for the anharmonic oscillator, which reads:

$$E_+(g) = \min_\beta \min_A \left\{ \frac{1}{4\pi} \int_{-\infty}^{\infty} dk \ln\left[1 + \left(\frac{A^2}{k^2}\right)^\beta\right] + \right.$$

$$\left. + 3g\left[\frac{1}{2\pi}\int_{-\infty}^{\infty} dk(1+k^2)^{-1}\left(1+\left(\frac{A^2}{k^2}\right)^\beta\right)^{-1}\right]^2\right\}.$$

In the limit $g \to \infty$, it is easy to obtain

$$E_+(g) = \min_\beta \min_A \left[\frac{A}{2\sin(\pi/2\beta)} + 3gA^{-2}\left(2\beta\sin\frac{\pi}{2\beta}\right)^{-2}\right]$$

$$= g^{1/3}\left\{\frac{3}{2\pi}\left[\max_u \frac{\sin^2 u}{u}\right]^{-1}\right\}^{2/3} = 0.757 \times g^{1/3}. \tag{5.84}$$

Collecting the estimations (5.72) and (5.84), we obtain

$$0.474 \times g^{1/3} < E(g) < 0.757 \times g^{1/3}. \tag{5.85}$$

As a second example, we shall present an elementary asymptotic method which allows a more precise asymptotic value to be obtained instead of (5.85) in the case of the one-dimension *anharmonic oscillator*. For this, let us consider first a numerical function of the form

$$F(g) = \int_{-\infty}^{\infty} \frac{du}{\sqrt{2\pi}} \exp\{-\tfrac{1}{2}u^2 - gu^4\}F(u)\left[\int_{-\infty}^{\infty} \frac{du}{\sqrt{2\pi}} \exp(-\tfrac{1}{2}u^2 - gu^4)\right]^{-1}$$

$$= \frac{\int d\mu_u \exp(-gu^4)F(u)}{\int d\mu_u \exp(-gu^4)},$$

where

$$d\mu_u = \frac{du}{\sqrt{2\pi}}\exp(-\tfrac{1}{2}u^2), \qquad \int d\mu_u u^{2m} = (2m-1)!!, \quad m = 0, 1, \ldots.$$

We now carry out some formal transformations

$$I = \int \frac{du}{\sqrt{2\pi}} \exp\{-\tfrac{1}{2}u^2 - gu^4 - \tfrac{1}{2}(Q-1)u^2 + \tfrac{1}{2}(Q-1)u^2\}F(u)$$

$$= \int \frac{du}{\sqrt{2\pi}} \exp\{-\tfrac{1}{2}Qu^2 + \tfrac{1}{2}(Q-1)u^2 - gu^4\}F(u),$$

and, passing to the variable $u \to u/\sqrt{Q}$, we get

$$I = Q^{-1/2}\int \frac{du}{\sqrt{2\pi}} \exp\{-\tfrac{1}{2}u^2 + \tfrac{1}{2}(1 - 1/Q)u^2 - gQ^{-2}u^4\}F(Q^{-1/2}u).$$

Hence

$$F(g) = \frac{\displaystyle\int d\mu_u \exp(W_Q(u))F(Q^{-1/2}u)}{\displaystyle\int d\mu_u \exp(W_Q(u))},$$

where

$$W_Q = \tfrac{1}{2}(1 - 1/Q)u^2 - (gQ^{-2})u^4.$$

Now the question arises of finding the asymptotics of this function in the limit $g \to \infty$, when $F(u)$ is a non-simple function. The main idea of the solution consists in expanding the exponential $\exp(W_Q(u))$ in a series, and studying the extremal problem for this series. Then we have to find such a procedure of minimalization which allows an expression that would converge to the true asymptotic value of the integral in the limit $g \to \infty$ to be obtained. We shall demonstrate this idea on the simple example when $F = F_1(u) = u^2$. Then the exact asymptotic of the function $F_1(g)$ equals

$$F_1(g)|_{g\to\infty} = \frac{g^{-1/2}\displaystyle\int_0^\infty du\, u^2 \exp(-u^4)}{\displaystyle\int_0^\infty du \exp(-u^4)} = \frac{g^{-1/2}\displaystyle\int_0^\infty dt\, t^{1/2+1/4-1}\, e^{-t}}{\displaystyle\int_0^\infty dt\, t^{1/4-1} e^{-t}}$$

$$= \frac{g^{-1/2}\Gamma(\frac{3}{4})}{\Gamma(\frac{1}{4})} = 0.338 \times g^{-1/2}.$$

Our goal is now to find such extremal procedure for the expression of $F_1(g)$, defined by the series, which approaches the exact value 0.338. We put

$$W_Q(u) = \tfrac{1}{2}(1 - 1/Q)u^2 - (g/Q^2)u^4 \underset{\substack{Q \gg 1 \\ g\to\infty}}{\Rightarrow} W(u) = \tfrac{1}{2}u^2 - \lambda u^4,$$

$$\lambda = g/Q^2, \qquad F_1(Q^{-1/2}u) = (\lambda/g)^{1/2}u^2.$$

Then

$$F(g) = \frac{\displaystyle\int d\mu_u F(u)(1 + W + \tfrac{1}{2}W^2 + \tfrac{1}{6}W^3 + \cdots)}{\displaystyle\int d\mu_u (1 + W + \tfrac{1}{2}W^2 + \cdots)} = \frac{\overline{F} + \overline{FW} + \frac{1}{2}\overline{FW^2} + \cdots}{1 + \overline{W} + \frac{1}{2}\overline{W^2} + \cdots}$$

$$= (\overline{F} + \overline{FW} + \tfrac{1}{2}\overline{FW^2} + \cdots)(1 - \overline{W} - \tfrac{1}{2}\overline{W^2} - \cdots)$$

Collecting terms of the same order, we get

$$F(g) = \overline{F} + [\overline{FW} - \overline{F}\,\overline{W}] + \{\tfrac{1}{2}[\overline{FW^2} - \overline{F}\,\overline{W^2}] - (\overline{FW} - \overline{F}\,\overline{W})\overline{W}\} +$$
$$+ \{\tfrac{1}{6}(\overline{FW^3} - \overline{F}\,\overline{W^3}) - \tfrac{1}{2}(\overline{FW^2} - \overline{F}\,\overline{W^2})\overline{W} + (\overline{FW} - \overline{F}\,\overline{W}) \times$$
$$\times (-\tfrac{1}{2}\overline{W^2} + \overline{W}^2)\} + \{\tfrac{1}{24}(\overline{FW^4} - \overline{FW^4}) - \tfrac{1}{6}(\overline{FW^3}) - \overline{FW^3})\overline{W} +$$
$$+ \tfrac{1}{2}(\overline{FW^2} - \overline{F}\,\overline{W^2})(-\tfrac{1}{2}\overline{W^2} + \overline{W}^2) + (\overline{FW} - \overline{F}\overline{W})(-\tfrac{1}{6}\overline{W^3} +$$
$$+ \overline{W}\,\overline{W^2} - \overline{W}^3)\} + \cdots$$

For example, for $\bar{F}(u) = F_1(u) = u^2$ one can write some terms

$$\overline{W} = \int \mathrm{d}\mu_u(\tfrac{1}{2}u^2 - \lambda u^4) = \tfrac{1}{2} - 3\lambda,$$

$$\overline{W^2} = \int \mathrm{d}\mu_u(\tfrac{1}{2}u^2 - \lambda u^4)^2 = \tfrac{3}{4} - 3 \times 5 \times \lambda + 3 \times 5 \times 7 \cdot \lambda^2,$$

$$\bar{u}^2 = \int \mathrm{d}\mu_u u^2 = 1,$$

$$\overline{u^2 W} = \int \mathrm{d}\mu_u u^2(\tfrac{1}{2}u^2 - \lambda u^4) = \tfrac{3}{2} - 5 \times 3 \times \lambda,$$

$$\overline{u^2 W^2} = \int \mathrm{d}\mu_u u^2(\tfrac{1}{4}u^4 - \lambda u^6 + \lambda^2 u^8) = \tfrac{1}{4}\cdot 3 \times 5 - 3 \times 5 \times 7 \cdot \lambda + 3 \times 5 \times$$
$$\times 7 \times 9 \cdot \lambda^2,$$

$$\overline{W^3} = \int \mathrm{d}\mu_u(\tfrac{1}{2}u^2 - \lambda u^4)^3 = \tfrac{1}{8}5!! - \tfrac{3}{4}\lambda \cdot 7!! + \tfrac{3}{2}\lambda^2 \cdot 9!! - \lambda^3 \cdot 11!!,$$

. . .

Thus

$$F_1(\lambda) = \sqrt{\lambda}\{1 + [1 - 12\lambda] + [1 - 36\lambda + 12 \times 32\lambda^2] + [1 - 72\lambda +$$
$$+ 1920\lambda^2 - 19008\lambda^3] + [1 - 120\lambda + 3 \times 1920\lambda^2 - 9 \times$$
$$\times 14784\lambda^3 + 36 \times 34816\lambda^4]\}g^{-1/2}. \tag{5.86}$$

In the first approximation we have minimalized the term $1 - 12\lambda$ and find $\lambda^{(1)}_{\min} = \frac{1}{12}$. Substituting this value into the first term of expression (5.86), we get

$$F^{(1)}_1(\lambda^{(1)}_{\min}) = 0.289 \times g^{-1/2}.$$

In the second approximation we find a minimum $\min_\lambda[1 - 36\lambda + 12 \times 32\lambda^2]$, $\lambda_{\min}^{(2)} = \frac{3}{64}$. Substituting this quantity in the first two terms of (5.86), we have

$$F_1^{(2)}(\lambda_{\min}^{(2)}) = 0.343 \times g^{-1/2}.$$

Further, as a third approximation, we take the minimum

$$\min_\lambda(1 - 72\lambda + 1920\lambda^2 - 19008\lambda^3), \qquad \lambda_{\min}^{(3)} \sim \tfrac{1}{30}$$

and find

$$F_1^{(3)}(\lambda_{\min}^{(3)}) = 0.329 \times g^{-1/2}.$$

In the fourth approximation,

$$\min_\lambda(1 - 120\lambda + 3 \times 1920\lambda^2 - 9 \times 14784\lambda^3 + 36 \times 34816\lambda^4),$$

we obtain $\lambda_{\min}^{(4)} \sim \frac{1}{30}$ which gives $F_1^{(4)}(\lambda_{\min}^{(4)}) = 0.334 \times g^{-1/2}$. As a result of such manipulations, we see that the formal series obtained after a sensible minimalization procedure may indeed converge to the exact (true) value $0.338 \times g^{-1/2}$.

Now we use this formal method for defining the asymptotic value of the *vacuum energy density for the anharmonic oscillator* with the Lagrangian density:

$$\mathscr{L} = \tfrac{1}{2}\dot{\varphi}^2 + \tfrac{1}{2}\omega^2\varphi^2 - g\varphi^4.$$

In this case, we have

$$\tilde{\Delta}(k^2) = (k^2 + \omega^2)^{-1}, \qquad \tilde{\Delta}^{1/2}(k^2) = (k^2 + \omega^2)^{-1/2}.$$

For simplicity we put $\omega^2 = m^2 = 1$ and find the asymptotics for

$$E(g) = -\lim_{V \to \infty} \frac{1}{V} \ln(\langle 0 | S_V(g) | 0 \rangle), \tag{5.87}$$

where

$$S_V(g) = \prod_{n=1}^{\infty} \int \frac{du_n}{\sqrt{2\pi}} \exp\left\{-\tfrac{1}{2}u_n^2 - g \int_V dx \left[\sum_s \Delta_s^{1/2}(x) u_s\right]^4\right\} \tag{5.88}$$

in the limit $g \to \infty$ (here dimension of space is one, i.e., $d = 1$). In order to accomplish this, we multiply (5.88) by $\exp(-\frac{1}{2}q_n u_n^2 + \frac{1}{2}q_n u_n^2)$ and make the change $u_n \to u_n'(1 + q_n)^{-1/2}$. So we get

$$S_V(g) = \prod_{n=1}^{\infty} (1 + q_n)^{-1/2} \int \frac{du_n}{\sqrt{2\pi}} \times$$

$$\times \exp\left\{-\tfrac{1}{2}u_n^2 + \tfrac{1}{2}\frac{q_n}{1 + q_n} u_n^2 - g \int_V dx \left[\sum_s \frac{\Delta_s^{1/2}(x)}{\sqrt{1 + q_s}} u_s\right]^4\right\}.$$

Let us put now

$$d\mu_u = \prod_{n=1}^{\infty} \frac{du_n}{\sqrt{2\pi}} \exp(-\tfrac{1}{2}u_n^2)$$

and take into account that

$$\int d\mu_u u_n^2 = 1, \qquad \int d\mu_u u_n u_{n'} = \delta_{nn'},$$

$$\int d\mu_u u_{n_1} u_{n_2} u_{n_3} u_{n_4} = C_4^2\, \delta_{n_1 n_2} \delta_{n_3 n_4} + \cdots$$

Here q_n are positive numbers. The numbers q_n can be chosen in such a way that [see, also (5.76)]

$$\lim_{V\to\infty} \frac{1}{V} \sum_n \ln(1 + q_n) = \int \frac{dk}{2\pi} \ln[1 + q(k^2)],$$

$$\lim_{V\to\infty} \frac{1}{V} \sum_n [\Delta_n^{1/2}(x)]^2 (1 + q_n)^{-1} = \int \frac{dk}{2\pi} \frac{\tilde{\Delta}(k^2)}{1 + q(k^2)},$$

$$\lim_{V\to\infty} \frac{1}{V} \sum_n \frac{q_n}{1 + q_n} = \int \frac{dk}{2\pi} \frac{q(k^2)}{1 + q(k^2)}.$$

Then the vacuum energy density (5.87) can be written in the form:

$$E(g) = \lim_{V\to\infty} \frac{1}{2V} \sum_n \ln(1 + q_n) - \lim_{V\to\infty} \frac{1}{V} \ln S_V'(g).$$

We write an expression for $S_V'(g)$ in a convenient form as in the case of the field theory, namely

$$S_V'(g) = \int d\mu \exp(W_q) = 1 + \bar{W}_q + \tfrac{1}{2}\overline{w_q^2} + \tfrac{1}{3}\overline{w_q^3} + \cdots$$

where

$$W_q = \frac{1}{2} \int_V dx \varphi_q^2(x) - g \int_V dx \phi_q^4(x).$$

Here

$$\phi_q(x) = \sum_n \Delta_n^{1/2}(x)(1 + q_n)^{-1/2} u_n, \qquad \varphi_q(x) = \sum_n (q_n)^{1/2} g_n (1 + q_n)^{-1/2} u_n$$

where $g_n = g_n(x)$ is an orthonormalized basis. Then by definition

$$A_q(x - x') = \langle 0|T[\phi_q(x)\phi_q(x')]|0\rangle = \int \frac{dk}{2\pi} e^{-ik(x-x')}\tilde{\Delta}(k^2)[1 + q(k^2)]^{-1},$$

$$B_q(x - x') = \langle 0|T[\varphi_q(x)\varphi_q(x')]|0\rangle = \int \frac{dk}{2\pi} e^{-ik(x-x')} q(k^2)[1 + q(k^2)]^{-1},$$

$$C_q(x - x') = \langle 0|T\{\varphi_q(x)\phi_q(x')\}|0\rangle = \int \frac{dk}{2\pi} e^{-ik(x-x')} \sqrt{q(k^2)}\, \frac{\tilde{\Delta}^{1/2}(k^2)}{1 + q(k^2)}.$$

Expanding $\ln(1 + x)$, we get

$$\frac{1}{V} \ln S'_V = \frac{1}{V}\{\bar{W}_q + \tfrac{1}{2}(\overline{W_q^2} - \bar{W}_q^2) + \tfrac{1}{6}\overline{W_q^3} - \tfrac{1}{2}\bar{W}_q\overline{W_q^2} + \tfrac{1}{3}\bar{W}_q^3 + \cdots\},$$

Then one obtains the following expression for the vacuum energy density

$$E(g) = E_0^q + E_1^q + E_2^q + E_3^q + \cdots \tag{5.89}$$

where

$$E_0^q = \lim_{V\to\infty} \frac{1}{V} \sum_{n=1}^{\infty} \ln(1 + q_n) = \int \frac{dk}{2\pi} \ln[1 + q(k^2)],$$

$$E_1^q = -\lim_{V\to\infty} \frac{1}{V}\bar{W}_q = -\frac{1}{2}\int \frac{dk}{2\pi}\frac{q(k^2)}{1 + q(k^2)} +$$

$$+ 3g\left[\int \frac{dk}{2\pi}(1 + k^2)^{-1}(1 + q(k^2))^{-1}\right]^2,$$

$$E_2^q = -\lim_{V\to\infty} \frac{1}{V}\{\tfrac{1}{2}(\overline{W_q^2} - \bar{W}_q^2)\} = -\frac{1}{4}\int dx B_q^2 - 12g^2 \int dx A_q^4(x) +$$

$$+ 6g \int dx C_q^2(x) A_q(0),$$

$$E_3^q = -\lim_{V\to\infty} \frac{1}{V}\{\tfrac{1}{6}(\overline{W_q^3} - \bar{W}_q^3) - \tfrac{1}{2}\bar{W}_q(\overline{W_q^2} - \bar{W}_q^2)\}.$$

As in the previous example, assuming $E_1^q = 0$, we get

$$-\frac{1}{2}\int \frac{dk}{2\pi}\left[\frac{q}{1 + q} - \frac{6gI}{(1 + k^2)(1 + q)}\right] = 0,$$

where

$$I = \int \frac{\mathrm{d}k}{2\pi}(1 + k^2)^{-1}[1 + q(k^2)]^{-1},$$

whence

$$q(k^2) = \frac{6gI}{1 + k^2} \quad \text{and} \quad I = \int \frac{\mathrm{d}k}{2\pi}(1 + k^2 + 6gI)^{-1} = \tfrac{1}{2}(1 + 6gI)^{-1/2}. \quad (5.90)$$

Substituting this quantity into the first term in the expansion (5.89), we obtain

$$E^{(1)} = \frac{1}{4\pi}\int_{-\infty}^{\infty} \mathrm{d}k \ln\left(1 + \frac{6gI}{1 + k^2}\right) = \tfrac{1}{2}[(1 + 6gI)^{1/2} - 1]\bigg|_{g \to \infty} \Rightarrow \tfrac{1}{2}\, 3^{1/3} g^{1/3}$$

$$= 0.72 \times g^{1/3}.$$

The following terms E_i^q, $i = 2, 3, \ldots$ are calculated explicitly for a concrete form for $q(k^2)$, namely

$$q(k^2) = q_c(k^2) = A^2/k^2, \quad (A > 0).$$

Then, at this value of q_c, we have

$$E_0^{q_c} = \tfrac{1}{2}A, \qquad E_1^{q_c} = -\tfrac{1}{2}B_q(0) + 3gA_q^2(0) = -\tfrac{1}{4}A + \tfrac{3}{4}(g/A^2).$$

According to the above manipulation we must set $E_2^{q_c} = 0$:

$$E_2^{q_c} = -\frac{A}{16} + \frac{23}{40}\frac{g^2}{A^5} + \frac{3}{4}\frac{g}{A^2} = 0,$$

from which it follows

$$A_1 = \{6[1 + (1 - \tfrac{23}{90})^{1/2}]\}^{1/3} g^{1/3}.$$

Substituting this value of A_1 into the two previous terms $E_0^{q_c} + E_1^{q_c}$, we have

$$E^{(2)} = \tfrac{1}{2}A_1 - \tfrac{1}{4}A_1 + \tfrac{3}{4}(g/A_1^2) = 0.71 \times g^{1/3},$$

in the limit $g \to \infty$. The explicit form of the term $E_3^{q_c}$ for the quantity q_c equals to

$$E_3^{q_c} = -\frac{3}{2}\left\{\frac{A}{16} + \frac{33}{8}\frac{g^2}{A^5} - \frac{1}{2}\frac{g}{A^2} - 21.43\frac{g^3}{A^8}\right\}.$$

As a third approximation we put $E_2^{q_c} + E_3^{q_c} = 0$ which gives

$$A_2 = (7.5)^{1/3} \times g^{1/3} = 1.96 \times g^{1/3}.$$

Corresponding to this value, the vacuum energy density in the limit $g \to \infty$, is $E^{(3)} = 0.685 \times g^{1/3}$.

The exact calculations for the anharmonic oscillator using the Schrödinger equation give (Parisi, 1977) $E(g) = 0.668 \times g^{1/3}$. One can see that our estimates are accurate.

5.5.3. *Functional Integrals for Fermions in Quantum Field Theory*

The action for fermions is given by

$$A(\psi, \bar{\psi}) = \int d^4x \bar{\psi}(x)(i\hat{\partial} - m)\psi(x), \quad \hat{\partial} = \gamma_\mu \frac{\partial}{\partial x_\mu}.$$

As might be expected the functional integral over the variables ψ, $\bar{\psi}$ is then the integral of the exponential of $i\hbar^{-1}A(\psi, \bar{\psi})$. In the functional integral, ψ and $\bar{\psi}$ are independent integration variables. This means that the analogue of $Z(\eta)$ (see Section 5.5.1) is the functional $F(\zeta, \bar{\zeta})$, where ($\hbar = 1$)

$$F(\zeta, \bar{\zeta}) = \iint \delta\psi \, \delta\bar{\psi} \exp\left\{i \int d^4x[\bar{\psi}(x)(i\hat{\partial} - m)\psi(x) + \bar{\zeta}\psi + \bar{\psi}\zeta]\right\}, \tag{5.91}$$

here $\zeta(x)$ and $\bar{\zeta}(x)$ are the sources of $\bar{\psi}(x)$ and $\psi(x)$, respectively, and they are elements of the Grassmann algebra. Using exactly the same techniques as those used to evaluate $Z(\eta)$, it may be shown that

$$F(\zeta, \bar{\zeta}) = \exp\left\{i \iint d^4x \, d^4y \bar{\zeta}(x) S_F(x - y)\zeta(x)\right\} K^{-1} \times$$

$$\times \iint \delta\psi \, \delta\bar{\psi} \exp\left\{-i \int d^4x \bar{\psi}\psi\right\}, \tag{5.92}$$

where

$$S_F(x) = (i\hat{\partial} + m)\cdot\Delta(x) \tag{5.93}$$

is the *free fermion propagator*. Note that in all respects this is a very similar formula to that for $Z(\eta)$ in (5.52). An important difference is that the Fredholm determinant K occurs as K^{-1} in (5.92), rather than K, as it does in (5.52). This is because of the change of the variable formula given in (5.27). So that the normalized $F(\zeta, \bar{\zeta})$ is given by

$$F(\zeta, \bar{\zeta}) = \frac{\iint \delta\psi \, \delta\bar{\psi} \exp\left\{i \int d^4x[\bar{\psi}(x)(i\hat{\partial} - m)\psi(x) + \bar{\zeta}\psi + \bar{\psi}\zeta]\right\}}{\iint \delta\psi \, \delta\bar{\psi} \exp\left\{i \int d^4x[\bar{\psi}(x)(i\hat{\partial} - m)\psi(x)]\right\}}. \tag{5.94}$$

Now $F(0, 0) = 1$. We know that the functional integral technique gives rise to the perturbation series of Feynman diagrams (for detail, see Nash, 1978). For example, for quantum electrodynamics the Green function is given by the formula

$$\langle 0 | T\{A_{\mu_1}(x_1) \cdots A_{\mu_n}(x_n)\bar{\psi}(y_1) \cdots \bar{\psi}(y_m)\psi(z_1) \cdots \psi(z_m)\} | 0 \rangle$$

$$= (-i)^{n+2m} \frac{\delta^{n+2m} Z(\zeta, \bar{\zeta}, \eta)}{\delta\eta_{\mu_1}(x_1) \cdots \delta\eta_{\mu_n}(x_n)\delta\zeta(y_1) \cdots \delta\zeta(y_m)\,\delta\bar{\zeta}(z_1) \cdots \delta\bar{\zeta}(z_m)}, \tag{5.95}$$

with $\eta_\mu = \zeta = \bar{\zeta} = 0$, where

$$Z(\zeta, \bar{\zeta}, \eta) = \frac{\displaystyle \iint \delta\psi\, \delta\bar{\psi} \int \delta A \exp\left\{ i \int d^4x [-\tfrac{1}{4} F_{\mu\nu} F^{\mu\nu} + \bar{\psi}(i\,\hat{\partial} - m)\psi - \tfrac{1}{2}(\partial_\lambda A^\lambda)^2 - e{:}\,\bar{\psi}\,\gamma_\mu A^\mu \psi{:} + \eta_\mu A^\mu + \bar{\zeta}\psi + \bar{\psi}\zeta] \right\}}{\displaystyle \left\{ \iint \delta\psi\, \delta\bar{\psi} \int \delta A \exp\left[i \int d^4x (-\tfrac{1}{4} F_{\mu\nu} F^{\mu\nu} + \bar{\psi}(i\,\hat{\partial} - m)\psi - \tfrac{1}{2}(\partial_\lambda A^\lambda)^2 - e{:}\,\bar{\psi}\,\gamma_\mu A^\mu \psi{:}) \right] \right\}}. \tag{5.96}$$

In (5.96), $\eta_\mu(x)$ is a *c*-number source for the photon field $A_\mu(x)$. When $e = 0$, the functional integrals over ψ, $\bar{\psi}$, and A_μ can all be reduced to the integrals over quadratic forms with the result

$$Z(\zeta, \bar{\zeta}, \eta) = \exp\left\{ i \iint d^4x\, d^4y \bar{\zeta}(x) S_F(x - y) \zeta(y) \right\} \times \times \exp\left\{ \frac{i}{2} \iint d^4x\, d^4y \eta_\mu(x) D_c^{\mu\nu}(x - y) \eta_\nu(y) \right\}. \tag{5.97}$$

The function $D_c^{\mu\nu}(x)$ is the photon Green function defined by

$$D_c^{\mu\nu}(x) = -g^{\mu\nu} \Delta_c(x), \tag{5.98}$$

where $\Delta_c(x)$ is the Green function for the zero mass Klein–Gordon equation

$$\Box\, \Delta_c(x - y) = -\delta^{(4)}(x - y). \tag{5.99}$$

In order to give a definition of the functional integral over fermion fields we consider, for example, the *Yukawa model* which describes a pseudoscalar interaction of the fermion field $\psi(x)$ with the boson field $\varphi(x)$ in the Euclidean d-dimensional space $\mathbb{R}_E^d$. It is assumed that d is even. The interaction Lagrangian density is given by

$$\mathcal{L}(x) = ig\bar{\psi}(x)\gamma_5 \psi(x)\varphi(x). \tag{5.100}$$

The Euclidean γ-matrices and spinors ψ, $\bar{\psi}$ are defined so as to ensure, first, the covariance of the theory with respect to the Euclidean rotation, and secondly, complete the coincidence of the *S*-matrix both in the suggested and standard approaches to the series of the perturbation theory.

In the Euclidean $\mathbb{R}_E^d$-space, there exist d anti-commuting hermitian matrices

γ_μ ($\mu = 1, \ldots, d$) with the dimension $d \otimes d$(d is even)

$$\gamma_\mu \gamma_\nu + \gamma_\nu \gamma_\mu = 2\,\delta_{\mu\nu}, \qquad \mu, \nu = 1, \ldots, d, \qquad \gamma_\mu^+ = \gamma_\mu.$$

Introduce also the matrix $\gamma_5 = i^{d(d-1)/2} \gamma_1 \cdots \gamma_d$, which satisfies the following properties:

$$\gamma_5^2 = 1, \qquad \gamma_5^+ = \gamma_5, \qquad \gamma_5 \gamma_\mu \gamma_5 = -\gamma_\mu.$$

For example, in the space $\mathbb{R}^2$ as the γ_μ- and γ_5-matrices one can be chosen

$$\gamma_1 = \sigma_1, \qquad \gamma_2 = \sigma_3, \qquad \gamma_5 = i\sigma_1\sigma_3 = \sigma_2,$$

where σ_j are Pauli's matrices. In the space $\mathbb{R}^4$:

$$\gamma_j = \begin{pmatrix} 0 & -i\sigma_j \\ i\sigma_j & 0 \end{pmatrix}, \quad (j = 1, 2, 3) \qquad \gamma_4 = \begin{pmatrix} I & 0 \\ 0 & -I \end{pmatrix},$$

$$\gamma_5 = -\gamma_1\gamma_2\gamma_3\gamma_4 = \begin{pmatrix} 0 & I \\ I & 0 \end{pmatrix}$$

ψ and $\bar{\psi}$-fields in the Euclidean space are independent and anticommutative.

The causal Green function for the Fermi fields can be written as

$$S_F(x_1 - x_2) = \int \left(\frac{dk}{2\pi}\right)^d e^{ik(x_1 - x_2)}\, \tilde{S}_F(i\hat{k}), \tag{5.101}$$

where $\hat{k} = \gamma_\mu k_\mu$. The integration in (5.101) is undertaken over the Euclidean space $\mathbb{R}^d_E$. The propagator (5.101) satisfies the following conditions

$$S_F^+(x) = S_F(x), \qquad \gamma_5 S_F(x) \gamma_5 = S_F(-x). \tag{5.102}$$

We introduce the following symbol

$$\tilde{G}(k^2) = |S_F(i\hat{k})| = [\tilde{S}_F(i\hat{k})\tilde{S}_F(-i\hat{k})]^{1/2}$$

and assume that

$$\tilde{S}_F(i\hat{k}) = [\tilde{\sigma}(i\hat{k})]^2.$$

Then the following relation holds

$$S_F(x_1 - x_2) = \int d^d y \sigma(x_1 - y) \sigma(y - x_2),$$

where

$$\sigma(x_1 - x_2) = \int \left(\frac{dk}{2\pi}\right)^d \tilde{\sigma}(i\hat{k})\, e^{-ik(x_1 - x_2)}. \tag{5.103}$$

The function $\sigma(x)$ also satisfies relations (5.102). The propagator of the Bose field is

$$\Delta(x_1 - x_2) = \int d^d y\, \Delta^{1/2}(x_1 - y)\, \Delta^{1/2}(y - x_2), \tag{5.104}$$

where

$$\Delta^{1/2}(x) = \int \left(\frac{dk}{2\pi}\right)^d (m^2 + k^2)^{-1/2} e^{-ikx}.$$

The S-matrix depending on the Bose field $\varphi_{in}(x)$ and the Fermi fields $\psi_{in}(x)$, $\bar{\psi}_{in}(x)$ can be written in the form of a functional integral by means of the following formula:

$$S[\varphi_{in}, \psi_{in}, \bar{\psi}_{in}] = C^{-1} \int \delta\varphi \iint \delta\psi\, \delta\bar{\psi} \times$$

$$\times \exp\left\{ -\frac{1}{2} \iint d^d x_1\, d^d x_2\, \varphi(x_1)\, \Delta^{-1}(x_1 - x_2) \varphi(x_2) - \right.$$

$$- \iint d^d x_1\, d^d x_2\, \bar{\psi}(x_1) S_F^{-1}(x_1 - x_2) \psi(x_2) + ig \int d^d x \times$$

$$\left. \times\, (\bar{\psi}(x) + \bar{\psi}_{in}(x)) \gamma_5 (\psi(x) + \psi_{in}(x))(\varphi(x) + \varphi_{in}(x)) \right\}. \quad (5.105)$$

Here the functions $\Delta^{-1}(x)$ and $S_F^{-1}(x)$ are defined by the relations

$$\int d^d y\, \Delta(x_1 - y)\, \Delta^{-1}(y - x_2) = \delta^{(d)}(x_1 - x_2),$$

$$\int d^d y\, S_F(x_1 - y) S_F^{-1}(y - x_2) = \delta^{(d)}(x_1 - x_2).$$

The normalization constant in (5.105) is chosen in such a way that

$$S[\varphi_{in}, \psi_{in}, \bar{\psi}_{in}]|_{g=0} = 1.$$

As stated above, in the case of the Bose field we perform some formal transformations. In (5.105), we carry out the change of variables

$$\varphi(x) = \int d^d x'\, \Delta^{1/2}(x - x') \Phi(x'),$$

$$\psi(x) = \int d^d x' \sigma(x - x') \Psi(x'),$$

$$\bar{\psi}(x) = \int d^d x' \bar{\Psi}(x') \sigma(x - x'), \quad (5.106)$$

where the functions $\Delta^{1/2}(x)$ and $\sigma(x)$ are defined by equations (5.104) and (5.103). As a result of this change, the S-matrix (5.105) can be represented in the form:

$$S[\varphi_{\text{in}}, \psi_{\text{in}}, \bar{\psi}_{\text{in}}] = C^{-1} \int \delta\Phi \iint \delta\Psi\, \delta\bar{\Psi} \exp\left\{ -\frac{1}{2} \int \mathrm{d}^d x \Phi^2(x) - \right.$$

$$- \left. \int \mathrm{d}^d x \bar{\Psi}(x)\Psi(x) \right\} \times \exp\left\{ ig \int \mathrm{d}^d x [\bar{\Psi}(x) + \bar{\psi}_{\text{in}}(x)] \times \right.$$

$$\left. \times\, \gamma_5 [\Psi(x) + \psi_{\text{in}}(x)][\Phi(x) + \varphi_{\text{in}}(x)] \right\}, \tag{5.107}$$

where the fields $\Psi(x)$, $\bar{\Psi}(x)$ and $\Phi(x)$ are connected by relations (5.106) with the fields $\psi(x)$, $\bar{\psi}(x)$ and $\phi(x)$, over which the integration is carried out.

The functional integral over the Bose-field has been defined above. Now we turn to its definition over the fermion fields. We consider the functional integration in (5.107) over the fermion fields in the language of integrals on the Grassmann algebra (see above). Now we define the variables $\{C_\nu, \bar{C}_\nu\}$ as the generators of the Grassmann algebra. Here the index $\nu = (n, \alpha)$, $n = 1, 2, \ldots,$ and $\alpha = 1, \ldots, d$ is a spinor one. The generators $\{C_\nu, \bar{C}_\nu\}$ satisfy the following relations:

$$\{C_\nu, C_{\nu'}\} = \{\bar{C}_\nu, \bar{C}_{\nu'}\} = \{C_\nu, \bar{C}_{\nu'}\} = 0. \tag{5.108}$$

As above, we introduce the symbols $(\mathrm{d}C_\nu, \mathrm{d}\bar{C}_\nu)$ and assume that they are also the elements of the Grassmann algebra, i.e., they satisfy relations (5.108). Besides, they anti-commute with all the elements of the algebra $\{C_\nu, C_\nu\}$. Integrals over the Grassmann algebra are

$$\int \mathrm{d}C_\nu = \int \mathrm{d}\bar{C}_\nu = 0, \qquad \int \mathrm{d}C_\nu C_{\nu'} = -\int \mathrm{d}\bar{C}_\nu \bar{C}_{\nu'} = \delta_{\nu\nu'} = \delta_{nn'}\delta_{\alpha\alpha'}. \tag{5.109}$$

Omitting for simplicity the spinor indexes, we put into (5.107)

$$\Psi(x) = \sum_{n=0}^{\infty} g_n(x) C_\nu, \qquad \bar{\Psi}(x) = \sum_{n=0}^{\infty} \bar{C}_\nu g_n(x), \quad (\nu = n, \alpha).$$

Then, according to (5.106), we obtain

$$\Psi(x) = \sum_{n=0}^{\infty} \sigma_n(x) C_\nu, \qquad \bar{\Psi}(x) = \sum_{n=0}^{\infty} \bar{C}_\nu \bar{\sigma}_n(x), \tag{5.110}$$

where

$$\sigma_n(x) = \int_V \mathrm{d}^d x' \sigma(x - x') g_n(x'), \qquad \bar{\sigma}_n(x) = \int_V \mathrm{d}^d x' g_n(x') \bar{\sigma}(x' - x).$$

The following relations hold

$$\lim_{V \to \infty} \sum_{n=0}^{\infty} \sigma_n(x) \bar{\sigma}_n(x') = S_{\mathrm{F}}(x - x'), \qquad \lim_{V \to \infty} \sum_{n=0}^{\infty} \sigma_n(x) \sigma_n(x') = G(x - x').$$

We introduce the quantities

$$\Psi_N(x) = \sum_{n=0}^{N} \sigma_n(x) C_\nu, \qquad \bar{\Psi}_N(x) = \sum_{n=0}^{N} \bar{C}_\nu \bar{\sigma}_n(x), \qquad \Phi_N(x) = \sum_{n=0}^{N} \Delta_n^{1/2}(x) u_n$$

and define the *Grassmann measure*

$$\mathrm{d}^N \mu_G = \prod_{n=0}^{N} \prod_{\alpha=1}^{d} \mathrm{d}C_{n\alpha} \mathrm{d}\bar{C}_{n\alpha} \exp(-\bar{C}_{n\alpha} C_{n\alpha})$$

and

$$\mathrm{d}^N \mu_\Phi = \prod_{n=0}^{N} \frac{\mathrm{d}u_n}{\sqrt{2\pi}} \exp(-\tfrac{1}{2} u_n^2)$$

so that $\int \mathrm{d}^N \mu_G = 1$. Now we are able to define the functional integral in (5.107). By the definition:

$$S_V[\varphi_{\text{in}}, \psi_{\text{in}}, \bar{\psi}_{\text{in}}] = \lim_{N \to \infty} \int \mathrm{d}^N \mu_\Phi \int \mathrm{d}^N \mu_G \exp\left\{ ig \int_V \mathrm{d}^d x (\bar{\Psi}_N(x) + \bar{\psi}_{\text{in}}(x)) \times \right.$$

$$\left. \times \gamma_5 (\Psi_N(x) + \psi_{\text{in}}(x))(\Phi_N(x) + \varphi_{\text{in}}(x)) \right\}. \tag{5.111}$$

Thus, the functional integral in the finite volume is understood as the limit of the multiple integrals. From (5.111) it is evident that the obtained expression coincides completely with (5.107) at the formal limit $N \to \infty$ and $V \to \infty$.

The series of the perturbation theory for the S-matrix elements obtained from (5.111) coincide completely with the usual series of the perturbation theory, since

$$\lim_{V \to \infty} \int \mathrm{d}\mu_\Phi \Phi(x) \Phi(x') = \lim_{V \to \infty} \sum_n \Delta_n^{1/2}(x) \Delta_n^{1/2}(x') = \Delta(x - x'),$$

$$\lim_{V \to \infty} \int d\mu_G \Psi(x) \bar{\Psi}(x') = \lim_{V \to \infty} \sum_n \sigma_n(x) \bar{\sigma}_n(x') = S_F(x - x').$$

As was mentioned above, the representation obtained for the S-matrix in the form of the functional integral is convenient for investigation of the *vacuum energy density*. In the given model, it is expressed by the formula

$$E(g) = -\lim_{V \to \infty} \frac{1}{V} \ln(\langle 0 | S_V(g) | 0 \rangle). \tag{5.112}$$

According to the above definition, the value of $\langle 0 | S_V(g) | 0 \rangle$ can be written in the form of the functional integral

$$\langle 0 | S_V(g) | 0 \rangle = \int \mathrm{d}\mu_\Phi \int \mathrm{d}\mu_G \exp\left\{ ig \int_V \mathrm{d}^d x \bar{\Psi}(x) \gamma_5 \Psi(x) \Phi(x) \right\}. \tag{5.113}$$

Usually the problem is reduced to the study of the function $E(g)$ in (5.112) for the pseudoscalar interaction (5.100) beyond the scope of the perturbation theory. Here we carry out an integration of expression (5.113) over the Grassmann variables only. For this expression (5.113) can be rewritten in the convenient form

$$\langle 0|S_V(g)|0\rangle = \int d\mu_\Phi \int dC \int d\bar{C} \exp\{-(\bar{C}C) + ig(\bar{C}HC)\},$$

$$dC\, d\bar{C} = \prod_{n=0}^{\infty} \prod_{\alpha_n=1}^{d} d\, C_{n\alpha_n} d\, \bar{C}_{n\alpha_n}, \tag{5.114}$$

$$(\bar{C}C) = \sum_{n,\alpha} (\bar{C}_{n\alpha} C_{n\alpha}), \qquad (\bar{C}HC) = \sum_{n,\alpha} \sum_{n',\alpha'} (\bar{C}_{n\alpha} H_{n\alpha,n'\alpha'} C_{n'\alpha'}).$$

A matrix H equals to

$$H = \int_V d^d x \bar{\sigma}(x) \gamma_5 \sigma(x) \Phi(x) \tag{5.115}$$

and its matrix elements have the following form:

$$H_{n\alpha,n'\alpha'} = \int_V d^d x (\bar{\sigma}_n(x) \gamma_5 \sigma_{n'}(x) \Phi(x))_{\alpha\alpha'}.$$

Now we carry out the transformation identical with the one in the case of the Bose-fields (see Section 5.5.2) in integral (5.114):

$$\langle 0|S_V(g)|0\rangle = \int d\mu_\Phi \int dC \int d\bar{C} \exp\{-[\bar{C}(1+q)C] + (\bar{C}qC) + ig(\bar{C}HC)\} \tag{5.116}$$

where q is a diagonal matrix with matrix elements

$$(q)_{n\alpha,n'\alpha'} = q_n \delta_{nn'} \delta_{\alpha\alpha'}.$$

The numbers q_n are positive and satisfy the conditions $\Sigma_{n=0} q_n < \infty$. By changing the integration variables in (5.116) $C_{n\alpha} \to C_{n\alpha}(1+q_n)^{-1/2}$, we get

$$\langle 0|S_V(g)|0\rangle = \prod_{n=0}^{\infty} (1+q_n)^d \int d\mu_\Phi \int d\mu_G \exp\left\{\left(\bar{C}\frac{q}{1+q}C\right) + ig(\bar{C}H_q C)\right\},$$

where

$$H_q = (1+q)^{-1/2} H (1+q)^{-1/2} = \int_V d^d x \bar{\sigma}_q(x) \gamma_5 \Phi(x) \sigma_q(x),$$

$$\sigma_q(x) = \sigma(x)(1+q)^{-1/2},$$

i.e.,

$$(H_q)_{n\alpha, n'\alpha'} = \int_V \mathrm{d}^d x \left\{ \frac{\bar{\sigma}_n(x)}{\sqrt{1 + q_n}} \gamma_5 \Phi(x) \frac{\sigma_{n'}(x)}{\sqrt{1 + q_{n'}}} \right\}_{\alpha\alpha'}.$$

The functional integral over the Grassmann variables is easily calculated and we obtain

$$\langle 0 | S_V(g) | 0 \rangle = \prod_{n=0}^{\infty} (1 + q_n)^d \int \mathrm{d}\mu_\Phi \det(I - J_q - igH_q), \tag{5.117}$$

where $J_q = q/(1 + q)$ is a diagonal matrix. The next problem is to estimate the determinant in (5.117) and to obtain an estimation for the vacuum energy density in different cases. This problem has been solved by Efimov and Ivanov (1981).

Part II

Stochastic Quantum Mechanics and Fields

Part II presents an elementary introduction to stochastic mechanics and the stochastic quantization of physical systems. In Chapter 6 we give the basic concepts of the probability theory, and random processes such as the Wiener, Markov and Gaussian processes and martingales. Elementary stochastic calculus which may play an important role in stochastic quantization procedure is also given. In Chapter 7 we give the main ideas of stochastic quantization proposed by I. Fényes, E. Nelson, F. Guerra, etc. Here we have attempted to give a physical origin of stochasticity in physics based on the hypothesis of the stochastic or fluctuational nature of space-time and the physical vacuum in the microworld. Chapters 8 and 9 deal with the construction of both the nonrelativistic and the relativistic dynamics of a stochastic particle including two bodies, and with a study of the equations of motion and the Cauchy problem for stochastic mechanics. They deal also with some specific problems, such as self-turbulent phenomena, Feynman-type processes and position-momentum uncertainty relations in stochastic mechanics.

New developments in stochastic quantization of different fields including gauge fields are discussed in Chapter 10. Further, we briefly review other methods of stochastic quantization of gauge fields, vacuum tunneling phenomena, and stochastic solutions of the classical Yang–Mills equations obtained recently by Matinyan *et al.* (1981).

In Chapter 11 we consider some consequences of the hypothesis of stochastic space-time and of fundamental length. It is shown that the introduction of the value of fundamental length leads to the restriction on the density of matter and on the life-time of the proton. We close this book with some comments on gravitational considerations connected with a hypothesis of nonlocality of space-time metric and its possible consequences, and the origin of cosmic rays and the value of fundamental length.

Chapter 6

The Basic Concepts of Random Processes and Stochastic Calculus

In this chapter we briefly give some basic probabilistic concepts and definitions of random processes such as the Markov, Gaussian and Wiener processes (see Gihman and Skorokhod, 1975; Dynkin, 1965), and therefore there is a large number of theoretically-probabilistic terms. There is a reason for introducing the terms: a 'random variable' and 'random process' for the equivalent 'measurable function' and a 'random field', namely an implied change of view-point. We use the word 'random' although in theoretically-probabilistic literature the words 'random' and 'stochastic' are used equally.

The basic concepts of probability theory are the notions of an experiment, event and the probability of an event. At the formal descriptive level of these notions it is accepted to start with the theoretically-set model of probability theory, proposed by Kolmogorov (1933, 1956).

6.1. Events

In probability theory (stochastic) experiments are carried out by the observation of certain conditions. However, this set of conditions does not define the outcome or the final result of the experiment. This means that in repeated experiments with a precise observation of complex conditions, generally speaking, the results will differ. The first fundamental assumption on the formalization of theoretically-probabilistic concepts is that the results of the given set of experiments can be described by means of some set Ω. A definite subset $A \in \Omega$ corresponds to each *event*.

At this, points $\omega \in \Omega$ play the role of atoms; *any event* is a sum of points but each point ω is not presented in the form of the sum of other events. Therefore, points (or an arbitrary subset) of Ω are called elementary events. However, from a practical point of view there is no logic in regarding an arbitrary subset of Ω as events. Therefore, it should be taken in Ω class of events of interest.

DEFINITION 6.1. A *class of events* is called an algebra of events, if it contains the reliable Ω and impossible $\emptyset$ events, together with an arbitrary pair of the events A and B from this class (it contains also their sum and events opposite to A). Two events Ω and $\varnothing$ define the trivial algebra. Minimal algebra containing the event A consists of four events: Ω, $\varnothing$, A and $\bar{A}$.

DEFINITION 6.2. The *algebra of events* is called a *σ-algebra*, if together with an arbitrary sequence of events it contains their sum. Usually, a σ-algebra of all considerable events in a given situation is denoted by the letter Σ.

6.2. Probability

A *probability* (measure) *space* is a triplet $\{\Omega, \Sigma, \mu\}$ of a set Ω, a σ-algebra Σ of subsets of Ω, and a positive measure μ on Ω with $\mu(\Omega) = 1$. Such a measure is called a *probability measure.* A probability (a base or sample) space Ω is the initial object in probability theory. Some simple well-known properties of the *probability measure* are:

(1) $\mu(\varnothing) = 0$.

(2) If $S_k \cap S_r = \varnothing$, $k \neq r$ then $\mu\left(\bigcup_1^\infty S_k\right) = \sum_1^\infty \mu(S_k)$.

(3) If $S_1 \subset S_2$, then $\mu(S_2 \backslash S_1) = \mu(S_2) - \mu(S_1)$.

(4) $\mu(\bar{S}) = 1 - \mu(S)$.

(5) If $S_n \subset S_{n+1}$, $n = 1, 2, \ldots.$, then $\mu\left(\bigcup_1^\infty S_n\right) = \lim \mu(S_n)$.

(6) If $S_n \supset S_{n+1}$, $n = 1, 2, \ldots$, then $\mu\left(\bigcap_1^\infty S_n\right) = \lim \mu(S_n)$.

6.3. Random Variable

Representation of a stochastic experiment consisting of numerical value ξ measuring, corresponds to the concept of a random variable. It is assumed that for an arbitrary pair of numbers a, $b (a < b)$ the event $A(a, b)$ consisting of $\xi \subset (a, b)$, is observed. Let A_x, $-\infty < x < \infty$ denote the event $\xi = x$. It is measurable. Indeed,

$$A_x = \prod_{n=1}^{\infty} A\left(x - \frac{1}{n}, x + \frac{1}{n}\right).$$

At this, if $x_1 \neq x_2$, the events A_{x_1} and A_{x_2} are incompatible (it follows from the fact that the result of measuring is synonymous) and union of all the A_x, $-\infty < x < \infty$ gives Ω. We now define the synonymous real function $f(\omega)$, $\omega \in \Omega$ by setting $f(\omega) = x$ if $x \in A_x$. From the definition it follows that in each experiment $\xi = f(\omega)$ and at this a set

$$\{\omega: a < f(\omega) < b\} = A(a, b),$$

is measurable.

An important example of random variable is the *indicator* of events. The indicator of event A is called a random variable (quantity) $\chi_A = \chi_A(\omega)$:

$$\chi_A(\omega) = \begin{cases} 1 & \text{if } \omega \in A, \\ 0 & \text{if } \omega \notin A. \end{cases}$$

If $A \in \Sigma$, the $\chi(\omega)$ is Σ-measurable.

We recall that the real-valued function $f(\omega)$ determined on the measurable space $\{\Omega, \Sigma\}$ is called *measurable* (or Σ-measurable), if for arbitrary real numbers a and b we have $\{\omega: a < f(\omega) < b\} \in \Sigma$. Thus, a real-valued measurable function on $\{\Omega, \Sigma\}$ is called a *random variable*. More precisely, we have the following definition in terms of a measurable map language. Let $\{\Omega, \Sigma\}$ and $\{S, \mathscr{B}\}$ be two measurable spaces. A *map* $g: \omega \mapsto x$ $(x \in S)$ is called a *measurable map* $\{\Omega, \Sigma\}$ onto $\{S, \mathscr{B}\}$ if

$$g^{-1}(B) = \{\omega: g(\omega) \in B\} \in \Sigma,$$

for arbitrary set $B \in \mathscr{B}$.

DEFINITION 6.3. According to the concept of the random variable $\xi = f(\omega)$ we define a measurable map from a measurable space $\{\Omega, \Sigma\}$ to another one $\{S, \mathscr{B}\}$, i.e., $\xi: \Omega \mapsto S$ such that $\xi^{-1}(\mathscr{B}) \in \Sigma$.

In other words, a *random variable is a measurable function* f mapping Ω to a set of real numbers $\mathbb{R}$ (where one regards Borel sets as measurable ones). For the given random variable f, the measure μ_f on $(-\infty, \infty)$ defined by

$$\mu_f(M) = \mu(f^{-1}[M]), \quad M \subset \mathbb{R}$$

is called the *probability distribution* of f. The (nonnormalized) *Fourier transform* of this measure

$$c_f(t) = \int_{\mathbb{R}} \mathrm{d}\mu_f(x)\, \mathrm{e}^{itx} = \int_{\Omega} \mathrm{d}\mu \exp(itf(\omega))$$

is called the *characteristic function* of f. The characteristic function of the random variable f is a wonderful object.

THEOREM 6.1. (*The Bochner theorem*). *A function c (from* $\mathbb{R}$ *or* $\mathbb{C}$*) is a characteristic function of a random variable, if and only if it satisfies the following conditions:*

1. $c(0) = 1$.
2. *A function* $t \mapsto c(t)$ *is continuous.*

3. *For any* $\{t_i\}_{i=1}^n$ *from* $\mathbb{R}$ *and* $\{z_i\}_{i=1}^n$ *from* $\mathbb{C}$

$$\sum_{i,j} z_i \bar{z}_j (t_i - t_j) \geqslant 0.$$

(The proof of the Bochner theorem can be found in many text books, for example, Reed and Simon, 1975.)

Moments of the random function f are given by their characteristic functions; nth moments of f are called

$$\int_\Omega \mathrm{d}\mu f^n = \int \mathrm{d}\mu_f(x) x^n$$

if this integral converges absolutely. If f has moments of all the orders if and only if $c_f(t)$ is a function of the class $\mathbb{C}^\infty$, and in this case

$$\int_\Omega \mathrm{d}\mu f^n = (-1)^n \left(\frac{\mathrm{d}}{\mathrm{d}t}\right)^n c_f(t) \bigg|_{t=0}. \tag{6.1}$$

DEFINITION 6.4. *A Gaussian random variable* with mean value 0 is called the random variable f, of which the characteristic function has the form

$$c_f(t) = \exp(-at^2/2), \quad a \geqslant 0 \tag{6.2}$$

Constant $\frac{1}{2}$ is chosen so that

$$a = \int_\Omega \mathrm{d}\mu f^2, \tag{6.3}$$

[in accordance with (6.1)]. The quantity a is called the *dispersion* of f. Note that the probability distribution density of the Gaussian random variable is also the Gaussian function:

$$\mathrm{d}\mu_f(x) = \begin{cases} \delta(x)\,\mathrm{d}x & a = 0, \\ (2\pi a)^{-1/2} \exp(-x^2/2a)\mathrm{d}x, & a > 0. \end{cases} \tag{6.4}$$

Expanding (6.2) into power series and using (6.1) it is easy to obtain the value of all the moments of the Gaussian random variable with the dispersion a:

$$\int \mathrm{d}\mu f^{2n+1} = 0, \qquad \int \mathrm{d}\mu f^{2n} = 2^{-n} \frac{(2n)!}{n!} a^n, \quad n = 0, 1, 2, \ldots.$$

DEFINITION 6.5. Let $f_1, f_2, \ldots, f_n$ be n random variables. Their *joint probability distribution* $\mathrm{d}\mu_{f_1,\ldots,f_n}$ and the *joint characteristic function* $c_{f_1,\ldots,f_n}(t_1, \ldots, t_n)$ on $\mathbb{R}^n$ are given by the relations

$$\mathrm{d}\mu_{f_1,\ldots,f_n}(M) = \mathrm{d}\mu\{[f_1 \otimes \cdots \otimes f_n]^{-1}(M)\},$$

and

$$c_{f_1,\ldots,f_n}(t_1, \ldots, t_n) = \int_\Omega \mathrm{d}\mu \exp\left[i \sum_{j=1}^n t_j f_j(\omega)\right],$$

respectively, where the map $f_1 \otimes \cdots \otimes f_n : \Omega \mapsto \mathbb{R}^n$ is defined by the formula

$$(f_1 \otimes \cdots \otimes f_n)(\omega) = [f_1(\omega), \ldots, f_n(\omega)].$$

DEFINITION 6.6. A set of the random variables $f_1, \ldots, f_n$ is called the *joint Gaussian random variables*, if the joint characteristic function of this set has the form

$$c_{f_1,\ldots,f_n}(t_1, \ldots, t_n) = \exp\left\{ -\frac{1}{2} \sum_{i,j} a_{ij} t_i t_j \right\},$$

where $\{a_{ij}\}_{i,j=1}^n$ is (symmetric) real *positive definite matrix.*

From this, we obtain the analogs of the formulas (6.3) and (6.4):

$$a_{ij} = \int \mathrm{d}\mu f_i f_j,$$

$$\mathrm{d}\mu_{f_1,\ldots,f_n} = (2\pi)^{-n/2} (\det a)^{-1/2} \exp\left\{ -\frac{1}{2} \sum_{i,j} b_{ij} x_i x_j \right\} \mathrm{d}^n x,$$

where $a = \{a_{ij}\}$ is called a *covariance matrix* of the set $\{f_i\}$, and the matrix a is assumed to have the inverse one which is denoted through $b \equiv a^{-1}$.

6.4. Expectation and Concept of Convergence over the Probability

DEFINITION 6.7. *Expectation or mean* value of the random variable $\xi = f(\omega)$ is defined by

$$\mathbb{E}\{\xi\} = \int_\Omega \mu(\mathrm{d}\omega) f(\omega).$$

DEFINITION 6.8. If a sequence of random variables $\xi_n = f_n(\omega)$, $n = 1, 2, \ldots,$ converges to $\xi = f(\omega)$ for each ω, with the exception of the set N such that $\mu(N) = 0$, then it is said that ξ_n converges to ξ with the probability one; or that

$$\xi = \lim \xi_n (\mathrm{mod}\ \mu).$$

It is said that the sequence $\{\xi_n;\ n = 1, 2, \ldots\}$ *converges over the probability* to the random variable ξ: $\xi = \mu - \lim \xi_n$ if there exists the random variable ξ such that

$$\mu\{|\xi_n - \xi| > \varepsilon\} \to 0, \quad \text{at } n \to \infty \text{ for any } \varepsilon > 0.$$

In the theory of measure, convergence over measure corresponds to the convergence over probability. A necessary and sufficient condition for the *convergence over probability* of sequence of random variables is that: for any $\varepsilon > 0$ and $\delta > 0$ it may be found such $n_0 = n_0(\varepsilon, \delta)$ that at n and $n' > n_0$; $\mu\{|\xi_{n'} - \xi_n| > \varepsilon\} < \delta$. This condition is called a fundamental condition over the probability of sequence $\{\xi_n,\ n = 1, 2, \ldots\}$.

DEFINITION 6.9. Let $L_p = L_p(\Omega, \Sigma, \mu)$, $p \geqslant 1$ be a linear normalized space of the random variables ξ on $\{\Omega, \Sigma, \mu\}$, for which $\mathbb{E}\{|\xi|^p\} < \infty$. In L_p a *norm* is introduced by the relation

$$\|\xi\|_p = [\mathbb{E}\{|\xi|^p\}]^{1/p}.$$

The convergence of the sequence ξ_n to the limit ξ in L_p (L_p-convergence) means that

$$\mathbb{E}\{|\xi - \xi_n|^p\} \to 0, \quad \text{at } n \to \infty.$$

The convergence over the probability results from the L_p-convergence. It follows immediately from the *Chebyshev inequality*:

$$\mu\{|\xi_n - \xi| > \varepsilon\} \leqslant \varepsilon^{-p}\mathbb{E}\{|\xi - \xi_n|^p\}.$$

The space L_p is dense. Of the spaces L_p, the spaces L_1 and L_2 are most important. The space $L_2 = L_2(\Omega, \Sigma, \mu)$ of complex random variables becomes a Hilbert space if a scalar product of the random variables ξ, η pair in L_2 is defined by the formula $\mathbb{E}\{\xi\bar{\eta}\}$.

Two random variables ξ and η are *orthodiagonal*, if $\mathbb{E}\{\xi\bar{\eta}\} = 0$. In this case, when ξ and η are real and $\mathbb{E}\{\xi\} = E\{\eta\} = 0$, the orthodiagonality means noncorrelativity. The convergence of the sequence $\{\xi_n, n = 1, 2, \ldots\}$ in L_p to the random variable ξ, means that

$$\|\xi - \xi_n\|^2 = \mathbb{E}\{|\xi - \xi_n|^2\} \to 0, \quad \text{at } n \to \infty.$$

Such a convergence form is called the mean square convergence and is written $\xi = \lim \xi_n$. In some cases it is convenient to express the convergence condition in L_p by means of the covariance of a set of random variables.

DEFINITION 6.10. The *covariance* $B(t_1, t_2)$, $t_i \in T$, of a set of random variables $\{\xi_t : t \in T\}$, $\xi_t \in L_2$ is called the function $B(t_1, t_2) = \mathbb{E}\{\xi_{t_1} \bar{\xi}_{t_2}\}$.

THEOREM 6.2. *Let $\mathscr{H}$ be the (separable) real Hilbert space. Then there exists the probability measure space $\{\Omega, \Sigma, \mu\}$ and for each $\xi \in \mathscr{H}$ a random variable, $\phi(\xi)$, so that $\xi \mapsto \phi(\xi)$ is linear and so that for any $\xi_1, \ldots, \xi_n \in \mathscr{H}$, $[\phi(\xi_1), \ldots, \phi(\xi_n)]$ are jointly Gaussian with the covariance $\langle \xi_i, \xi_j \rangle = \mathbf{B}(\xi_i, \xi_j)[\langle \cdot, \cdot \rangle$ is an inner product on $\mathscr{H}$]. Note that $\{\phi(\xi)\}$ is called the Gaussian process with the covariance $\langle \cdot, \cdot \rangle$.*

Proof (for details, see Simon, 1979). Let $\xi_1, \ldots, \xi_n$ be an orthonormal basis for $\mathscr{H}$. Let $c(t)$ be defined for any sequence $t_1, \ldots, t_n, \ldots$ eventually zero, by $c(t) = \exp\{-\frac{1}{2}\sum_i t_i^2\}$. Owing to the above, when $\exp(-\frac{1}{2}\sum_{i,j} a_{ij} t_i t_j)$ is *positive definite* and $\phi_1, \ldots, \phi_n, \ldots$, jointly Gaussian with the covariance δ_{ij} (this is somewhat circumlocutory, one can just take $\Omega = \mathbb{R}^\infty$, $\phi_i = \omega_i$, and $\mu = \Pi_{n=1}^\infty \times$ $\times\ (2\pi)^{-1/2} \exp(-\frac{1}{2}\omega_n^2)\, d\omega_n$, directly). Now there is given a finite sum

$$\xi = \sum_{i=1}^{N} \alpha_i \xi_i, \quad \text{and set } \phi(\xi) = \sum_{i=1}^{N} \alpha_i \phi_i.$$

Then

$$\int d\mu |\phi(\xi)|^2 = \|\xi\|^2,$$

so, by continuity, ϕ extends from finite sum to a map from $\mathscr{H}$ to $L_2(\Omega, d\mu)$. It is easy to use continuity to see that

$$\int d\mu \exp[i\phi(\xi)] = \exp\{-\tfrac{1}{2}\|\xi\|^2\},$$

so that ϕ's are jointly Gaussian with the proper *covariance*. The uniqueness of the process follows from the uniqueness aspect of Kolmogorov's theorem and the L_2-continuity. ■

6.5. Independence

DEFINITION 6.11. n random variables $f_1, \ldots, f_n$ are called *independent* if and only if their joint distribution $d\mu_{f_1,\ldots,f_n}$ is the product measure $d\mu_{f_1} \otimes \cdots \otimes d\mu_{f_n}$. For two events A and B it is easy to see that an independence means $\mu(A \cap B) = \mu(A)\mu(B)$. From the definition it immediately follows that:

1. Ω and A are independent, where A is an arbitrary event.
2. If $\mu(N) = 0$ and A is any event, then N and A are independent.
3. If A and B_i, $i = 1, 2$, are independent, and $B_1 \supset B_2$, then events A and $B_1 \setminus B_2$ are independent.
4. Let A and B_i, $i = 1, 2, \ldots, n$, be the independent events, while $B_1, \ldots, B_n$ are *incompatible* in pairs, then A and $\cup_1^n B_i$ also are independent.
5. $\bar{A}$ is not dependent on A, if and only if $\mu(A) = 0$ or $\mu(A) = 1$. Let I be some set and $\{M_i, i \in I\}$ be a set of event classes indexed by $i \in I$.

DEFINITION 6.12. The classes of events $\{M_i, i \in I\}$ are called independent (or independent in totality), if for any inequalities in pairs $i_1, i_2, \ldots, i_n$ $(i_k \in I)$ and any A_{i_k}, $A_{i_k} \in M_{i_k}$, $k = 1, 2, \ldots, n$;

$$\mu(A_{i_1} \cap A_{i_2} \cap \cdots \cap A_{i_n}) = \mu(A_{i_1})\mu(A_{i_2}) \cdots \mu(A_{i_n}).$$

Let $\xi_i = f_i(\omega)$, $i \in I$ be some set of the random variables in $\{S_i, \mathscr{B}_i\}$. The quantities $\{\xi_i, i \in I\}$ are called independent (or independent in totality), if for any n, $n = 1, 2, \ldots$ and any $B_k \in \mathscr{B}_{i_k}$, $i_k \in I$:

$$\mu\left(\bigcap_{k=1}^{n} \{\xi_{i_k} \in B_{i_k}\}\right) = \prod_{k=1}^{n} \mu(\xi_{i_k} \in B_{i_k}). \tag{6.5}$$

PROPOSITION 6.1 *Two Gaussian random variables* f *and* g *are independent if and only if their* covariance $(f, g) = \int d\mu f \cdot g$ *is zero.*

Proof. The *Fourier transform* of their joint distribution

$$M(s, t) = \mathbb{E}\{\exp(itf + isg)\} = \exp(-\tfrac{1}{2}\|tf + sg\|_2^2)$$

$$= M(t, 0)M(0, s)\exp[-ts(f, g)]$$

is a product if and only if $(f, g) = 0$. ■

6.6. Conditional Probability and Conditional (Mathematical) Expectation

First of all, we recall the definition of conditional probability and conditional expectation in the elementary case. The *conditional probability* $\mathbb{C}\mathrm{ond}(A \mid B)$ of any event A under the hypothesis B, is given by

$$\mathbb{C}\mathrm{ond}(A|B) = \mu(A \cap B)/\mu(B)$$

if $\mu(B) \neq 0$. The conditional probability $\mathbb{C}\mathrm{ond}(A \mid B)$ at fixed B is the normalized measure defined on the same σ-algebra of a set, as for the "unconditional" probability $\mu(A)$. In accordance with this, *conditional expectation* of some random variable $\xi = f(\omega)$ under the hypothesis B, is given by the formula

$$\mathbb{E}\{\xi \mid B\} = \int_\Omega f(\omega)\, \mathbb{C}\mathrm{ond}(\mathrm{d}\omega|B). \tag{6.6}$$

Using the conditional probability definition, this relation can be rewritten in the following way:

$$\mu(B)\mathbb{E}\{\xi \mid B\} = \int_B \mathrm{d}\mu \cdot \xi.$$

We note that if $\xi = \chi_A(\omega)$ is the (the characteristic function) indicator of the event A then $\mathbb{E}\{\xi \mid B\} = \mathbb{C}\mathrm{ond}(A \mid B)$. Thus, conditional probabilities are the particular case of the conditional (mathematical) expectations.

Let $\mathscr{B}$ be a sub-σ-algebra of Σ and ξ a random variable with a finite mean, i.e., $\xi \in L_1(\Omega, \mu)$.

DEFINITION 6.13. A conditional expectation of $\xi = f(\omega)$ with respect to $\mathscr{B}$ is defined to be a $\mathscr{B}$-measurable function $\mathbb{E}\{\xi \mid \mathscr{B}\}$ on Ω such that

$$\int_B \mathrm{d}\mu\, \mathbb{E}\{\xi \mid \mathscr{B}\} = \int_B \mathrm{d}\mu(\mathrm{d}\omega) f(\omega), \tag{6.7}$$

holds for $\forall B \in \mathscr{B}$. The conditional expectation $\mathbb{E}\{\xi \mid \mathscr{B}\}$ is nothing but the *Radon–Nikodym derivative* of the σ-additive function on $\mathscr{B}$

$$\mu(B) = \int_B \mu(\mathrm{d}\omega) f(\omega), \quad \forall B \in \mathscr{B}$$

with respect to the probability measure μ. Therefore, the conditional expectation, if it exists, is unique with the probability one. We prefer to write $\mathbb{E} \times \times \{\xi|\zeta\}$ in spite of $\mathbb{E}\{\xi|\mathscr{B}\}$ if the sub-σ-algebra $\mathscr{B}$ is generated by a random variable ζ. If $\zeta_1, \ldots, \zeta_n$ are random variables, $\mathbb{E}\{\xi|\zeta_1, \ldots, \zeta_n\}$ denotes the conditional expectation of ξ with respect to the σ-algebra generated by $\zeta_1, \ldots, \zeta_n$.

Now we list the basic properties of the *conditional expectation*:

(1) If $\xi \geqslant 0$, then $\mathbb{E}\{\xi \mid \mathscr{B}\} \geqslant 0$.
(2) If ξ is a $\mathscr{B}$-measurable random variable, then $\mathbb{E}\{\xi \mid \mathscr{B}\} = \xi$.
(3) $\mathbb{E}\{\mathbb{E}\{\xi \mid \mathscr{B}\}\} = \mathbb{E}\{\xi\}$.
(4) $\mathbb{E}\{\xi_i\} \geqslant 0$, $i = 1, 2$, then

$$\mathbb{E}\{a\xi_1 + b\xi_2|\mathscr{B}\} = a\mathbb{E}\{\xi_1|\mathscr{B}\} + b\mathbb{E}\{\xi_2|\mathscr{B}\}.$$

(5) If the random variable ξ is independent of the σ-algebra $\mathscr{B}$, then

$$\mathbb{E}\{\xi \mid \mathscr{B}\} = \mathbb{E}\{\xi\}.$$

From the definition, the independence of the random variable ξ from the σ-algebra $\mathscr{B}$ means that the σ-algebra $\sigma(\xi)$ and $\mathscr{B}$ are independent. Therefore, for any $B \in \mathscr{B}$

$$\int_B \mathrm{d}\mu \cdot \xi = \mathbb{E}\{\xi\chi_B\} = \mathbb{E}\{\xi \cdot \mu(B)\}.$$

So that equality (6.7) holds, if to put $\mathbb{E}\{\xi \mid \mathscr{B}\} = \mathbb{E}\{\xi\}$.

(6) If η is $\mathscr{B}$-measurable, then

$$\mathbb{E}\{\xi\eta \mid \mathscr{B}\} = \eta\mathbb{E}\{\xi \mid \mathscr{B}\}, \tag{6.8}$$

or

$$\int \mathrm{d}\mu \cdot \xi \cdot \eta = \int \mathrm{d}\mu\eta\mathbb{E}\{\xi \mid \mathscr{B}\}.$$

For proof of this property, it is sufficient to assume that $\eta \geqslant 0$. If $\eta = \chi_{B_1}$, $B_1 \in \mathscr{B}$, then

$$\int_B \mathrm{d}\mu \cdot \eta \cdot \mathbb{E}\{\xi \mid \mathscr{B}\} = \int_{B \cap B_1} \mathrm{d}\mu \cdot \mathbb{E}\{\xi \mid \mathscr{B}\} = \int_{B \cap B_1} \xi \cdot \mathrm{d}\mu = \int_B \mathrm{d}\mu \cdot \eta \cdot \xi$$

so that equality (6.8) is satisfied.

Since the conditional expectation $\mathbb{E}\{\xi \mid \mathscr{B}\}$ is a random variable, one may consider the conditional expectation of this variable with respect to another σ-algebra $\mathscr{B}_1$, which leads to the iterated conditional expectation $\mathbb{E}\{\mathbb{E}\{\xi|\mathscr{B}\} \mid \mathscr{B}\}$.

(7) Let $\mathscr{B} \subset \mathscr{B}_1$, then

$$\mathbb{E}\{\mathbb{E}\{\xi \mid \mathscr{B}_1\} \mid \mathscr{B}\} = \mathbb{E}\{\xi \mid \mathscr{B}\}. \tag{6.9}$$

Indeed, if $B \in \mathscr{B}$, then $B \in \mathscr{B}_1$, therefore

$$\int_B d\mu\, \mathbb{E}\{\mathbb{E}\{\xi \mid \mathscr{B}_1\} \mid \mathscr{B}\} = \int_B d\mu\, \mathbb{E}\{\xi \mid \mathscr{B}_1\} = \int_B d\mu \cdot \xi = \int_B d\mu\, \mathbb{E}\{\xi \mid \mathscr{B}\}.$$

From this we have equality (6.9).

We now give some definition of the conditional expectation with respect to random variables. Let ξ be a random variable taking the values $z_1, z_2, \ldots, z_n, \ldots$, $\mu(\xi = z_n) > 0$, B_n denotes the event $\{\xi = z_n\}$ and $\mu_n(A) = \mu(A \cap B_n)/\mu(B_n)$ is the conditional probability of A under the hypothesis $\xi = z_n$. Then, the *conditional expectation* of the variable η under the hypothesis $\xi = z_n$ is given by the formula

$$\mathbb{E}\{\eta \mid \xi = z_n\} = \int_\Omega d\mu_n \cdot \eta = \frac{1}{\mu(B_n)} \int_{B_n} \eta \cdot d\mu.$$

Let $\zeta = g(\omega)$ be a measurable map: $\{\Omega, \Sigma\} \mapsto \{S, \mathscr{B}\}$. Thus, ζ is a random variable belonging to S. Let $\mathscr{F}_\zeta$ be a σ-algebra given by the map ζ:

$$\mathscr{F}_\zeta = \{M : M = g^{-1}(B), B \in \mathscr{B}\}.$$

DEFINITION 6.14. The conditional expectation $\mathbb{E}\{\xi \mid \zeta\}$ of the random variable ξ with respect to another random variable ζ is called the random variable $\mathbb{E}\{\xi \mid \mathscr{F}_\zeta\}$. This definition is equivalent to the following: for any $B \subset \mathscr{B}$

$$\int_{M = g^{-1}(B)} d\mu\, \mathbb{E}\{\xi \mid \zeta\} = \int_{M = g^{-1}(B)} \xi \cdot d\mu.$$

THEOREM 6.3 *The conditional expectation with respect to the random variable ζ is a $\mathscr{B}$-measurable function of ζ; $\mathbb{E}\{\xi \mid \zeta\} = s(\zeta)$, where $s(\zeta)$ is a $\mathscr{B}$-measurable function.*

Proof. Let $\xi \geqslant 0$. Then from the definition we have

$$\int_{g^{-1}(B)} d\mu\, \mathbb{E}\{\xi \mid \zeta\} = \int_{g^{-1}(B)} d\mu \cdot \xi = q(B). \tag{6.10}$$

It is evident that $q(B)$ is a σ-finite measure on $\mathscr{B}$. Besides, $q(B) = 0$ if $\mu(g^{-1}(B)) = 0$, i.e., q is absolutely continuous with respect to measure μ_g, where $\mu_g(A) = \mu(g^{-1}(A))$. On the basis of the Radon–Nikodym theorem there exists such $\mathscr{B}$-measurable nonnegative function $s(\omega)$ that

$$q(B) = \int_B \mu_g(d\omega) s(\omega).$$

Using the rule for the change of a variable, we get

$$q(B) = \int_{g^{-1}(B)} \mu(\mathrm{d}\omega)s(g(\omega)).$$

Comparing with (6.10) one obtains the equality $\mathbb{E}\{\xi\,|\,\zeta\} = s(g(\omega)) = s(\zeta)$. ■

DEFINITION 6.15. (*Conditional density*). Let $\{\mathscr{X}, \mathscr{U}, m\}$ be some space with measure m and $\xi = g(\omega)$ be a measurable map: $\{\Omega, \Sigma\}$, to $\{\mathscr{X}, \mathscr{U}\}$. Then it is said that a random variable has the probability distribution $\rho(x)$ (over a measure m), if for any $A \in \mathscr{U}$

$$\mu(\xi \in A) = \int_A m(\mathrm{d}x)\rho(x).$$

According to the *Radon–Nikodym theorem*, the random variable $\xi = g(\omega)$ possesses a probability distribution if and only if the measure μ_g is absolutely continuous with respect to m.

PROPOSITION 6.2. *ξ is independent of ζ if and only if $\mathbb{E}\{e^{i\alpha\xi}\,|\,\zeta\}$ is a constant for each real α.*

Proof. If ξ is independent of ζ, then $\mathbb{E}\{F(\xi)\,|\,\zeta\}$ is constant for any F. Conversely if $\mathbb{E}\{e^{i\alpha\xi}\,|\,\zeta\}$ is constant for all α, then the constant is necessarily ζ $\{e^{i\alpha\xi}\}$ and

$$\mathbb{E}\{\exp(i\alpha\xi + i\beta\zeta)\} = \mathbb{E}\{\mathbb{E}\{e^{i\alpha\xi}\,|\,\zeta\}\,|\,e^{i\beta\zeta}\} = \mathbb{E}\{e^{i\alpha\xi}\}\mathbb{E}\{e^{i\alpha\zeta}\}.$$

for any α and β so, taking the Fourier transforms $\mu_{\xi\zeta} = \mu_\xi \otimes \mu_\zeta$.

We now define the conditional probability $\mathbb{C}\mathrm{ond}(A\,|\,B)$ with respect to the σ-algebra $\mathscr{B}$ as a particular case of the conditional expectations, assuming $\xi = \chi_A(\omega)$.

DEFINITION 6.16. The *conditional probability* $\mathbb{C}\mathrm{ond}(A\,|\,B)$ at fixed A is a $\mathscr{B}$-measurable random variable satisfying the following equation:

$$\int_B \mathrm{d}\mu\,\mathbb{C}\mathrm{ond}(A\,|\,\mathscr{B}) = \mu(A \cap B)$$

at any $B \in \mathscr{B}$. Some properties of the conditional probabilities are:

(a) If event B is $\mathscr{B}$-measurable, then $\mathbb{C}\mathrm{ond}\ (B\,|\,\mathscr{B}) = \chi_B$, which follows from property (2) for the conditional expectations.
(b) Assuming $\xi_i = \chi_{B_i}$, $B_1 \cap B_2 = \emptyset$, then from property (4) we get the additivity of the conditional probability:

$$\mathbb{C}\mathrm{ond}(B_1 \cup B_2\,|\,\mathscr{B}) = \mathbb{C}\mathrm{ond}(B_1\,|\,\mathscr{B}) + \mathbb{C}\mathrm{ond}(B_2\,|\,\mathscr{B}).$$

(c) From property (5) it follows that if the event A does not depend on the σ-algebra $\mathscr{B}$, then $\mathbb{C}\mathrm{ond}(A\,|\,\mathscr{B}) = \mu(A)$.
(d) In the simplest case, $\mathscr{B} = \{\Omega, \emptyset, F, \bar{F}\}$ for the subset $F \in \Sigma$ such that

$0 < \mu(F) < 1$, one obtains

$$\mathbb{C}\mathrm{ond}(B \mid \mathscr{B}) = \begin{cases} \mu(B \cap F)/\mu(F) & \text{on } F, \\ \mu(B \cap \bar{F})/\mu(\bar{F}) & \text{on } \bar{F}. \end{cases} \tag{6.11}$$

THEOREM 6.4. (*Bayes*).

$$\mathbb{C}\mathrm{ond}(B \mid F) = \mu(B)\, \mathbb{C}\mathrm{ond}(F \mid B)/\mu(F).$$

Proof. By equation (6.11), we have

$$\mu(F)\, \mathbb{C}\mathrm{ond}(B \mid F) = \mu(B \cap F) = \mathbb{C}\mathrm{ond}(F \mid B)\mu(B). \qquad \blacksquare$$

6.7. Martingales

DEFINITION 6.17. Let $\xi_1, \ldots. \xi_n$ be a sequence of random variables. We call them a *martingale* (respectively, submartingale and supermartingale) if and only if for each m, $\mathbb{E}\{|\xi_m|\} < \infty$ and

$$\mathbb{E}\{\xi_m \mid \xi_1, \ldots, \xi_{m-1}\} = \xi_{m-1}$$

[respectively, $\mathbb{E}\{\xi_m | \xi_1, \ldots, \xi_{m-1}\} \geqslant \xi_{m-1}$ and $\mathbb{E}\{\xi_m | \xi_1, \ldots, \xi_{m-1}\} \leqslant \xi_{m-1}$]. It follows by induction that for $j < m$

$$\mathbb{E}\{\xi_m \mid \xi_1, \ldots, \xi_j\} = \xi_j, \tag{6.12}$$

(respectively, $\geqslant \xi_j$ and $\leqslant \xi_j$). If $\zeta_1, \ldots, \zeta_n$ are random variables which have a mean zero and are independent, then

$$\xi_j = \sum_{i=1}^{j} \zeta_i,$$

is a martingale.

THEOREM 6.5. (*Doob's*, 1953 *inequality*) *Let* $\{\xi_n\}$ *be a submartingale. Then for each* $\lambda > 0$ *and* n*:*

$$\mathbb{E}\left\{\max_{0 \leqslant j \leqslant n} \xi_j \geqslant \lambda\right\} \leqslant \lambda^{-1}\, \mathbb{E}\{\xi_n^+\},$$

where $\xi_n^+ = \max(\xi_n, 0)$.

Proof. Let χ_j be the characteristic function of the sets A_j of points, where $\xi_1, \ldots, \xi_{j-1} < \lambda$ and $\xi_j \geqslant \lambda$. Let χ be the characteristic function of the set A with $\max_{0 \leqslant j \leqslant n} \xi_j \geqslant \lambda$. Since the A_j's are disjoint sets with union A, we have

$$\lambda \mathbb{E}\{A\} = \sum_{j=1}^{n} \lambda \mathbb{E}\{A_j\} \leqslant \sum_{j=1}^{n} \mathbb{E}\{\xi_j \chi_j\} \leqslant \sum_{j=1}^{n} \mathbb{E}\{\mathbb{E}\{\xi_n \mid \xi_1, \ldots, \xi_j\} | \chi_j\}$$

$$= \sum_{j=1}^{n} \mathbb{E}\{\xi_n \chi_j\} = \mathbb{E}\{\xi_n \chi\} \leqslant \mathbb{E}\{\xi_n^+\}.$$

Above, we have used $\xi_j \geqslant \lambda$ in the inequality, the submartingale relation (6.12) in the second inequality, the fact that χ_j is $\xi_1, \ldots, \xi_j$ measurable and (6.8) in the next equality and the calculation

$$\mathbb{E}\{\xi_n \chi\} \leqslant \mathbb{E}\{\xi_n^+ \chi\} \leqslant \mathbb{E}\{\xi_n^+\} \qquad \blacksquare$$

in the final step.

DEFINITION 6.18. Suppose that $\{q_t\}$, $t \in [a, b]$ is a family of random variables. We say it is a *martingale* (respectively, submartingale and supermartingale) if and only if

$$\mathbb{E}\{q_t \mid [q_s \mid s \leqslant u]\} = q_u,$$

(respectively, $\geqslant q_u$ and $\leqslant q_u$) for all $a \leqslant u \leqslant t \leqslant b$. Note that, a fortiori, one has

$$\mathbb{E}\{q_t \mid q_{s_1}, \ldots, q_{s_n}\} = q_{s_n},$$

(respectively, $\geqslant q_{s_n}$ and $\leqslant q_{s_n}$) if $s_1 \leqslant s_2 \leqslant \cdots \leqslant s_n \leqslant t$ so that for $s_1 < \cdots < s_n < \cdots$, the sequence $\xi_n = q_{s_n}$ is a (sub) and (super) martingale. As a result,

$$\mathbb{E}\left\{\max_{0 \leqslant j \leqslant n} q_{s_n} \geqslant \lambda\right\} \leqslant \lambda^{-1}\, \mathbb{E}\{q_{s_n}^+\}.$$

Taking the mesh of the points s_i to zero we have the following:

COROLLARY 6.1. *Let* $\{q_t\}$, $t \in [a, b]$ *be a submartingale with continuous sample paths (i.e.,* $t \mapsto q_t$ *is continuous almost everywhere; this is a 'version dependent'). Then*

$$\mathbb{E}\left\{\max_{0 \leqslant s \leqslant t} q_s \geqslant \lambda\right\} \leqslant \lambda^{-1}\, \mathbb{E}\{q_t^+\}.$$

The following basic properties of martingales are:

(i) *If* q_t, $-\infty < t < \infty$, *is a submartingale (supermartingale), then* $m(t) = \mathbb{E}\{q_t\}$ *is an increasing (decreasing) function of* t.
(ii) $m(t) =$ *const, if the function* q_t *is a martingale.*

6.8. Definition of Random Processes and Gaussian Processes

Let $\{\Omega, \Sigma, \mu\}$ be a given probability space. If realization of some experiment is described by a function $f(y)$ of a definite argument y, $y \in Y$, then we say that a *random function* is given on $\{\Omega, \Sigma, \mu\}$. Thus, a random function is a map: $\omega \mapsto f(y) = f(y, \omega)$, $\omega \in \Omega$. In addition it is required that the function $f(y, \omega)$ should be a *random variable* at fixed y.

DEFINITION 6.19. (*General Definition*). Let Y be some set and $\{S, \mathscr{B}\}$ be a measurable space. The random map $f(y): Y \mapsto \{S, \mathscr{B}\}$ is called a map of $Y \otimes \Omega$

to S (this is a measurable map: $\{\Omega, \Sigma\} \mapsto \{S, \mathscr{B}\}$ at any fixed y), such that

$$\{\omega: f(y) \in B\} \in \Sigma$$

for any $B \in \mathscr{B}$. Instead of a 'random map' the term a 'random function' is usually used with a given value in S. In this case we call 'y' the argument of a random function.

In the case when $Y = T$ and $y = t$, t is time and a random function is called a *random process*. If $Y = T_+ = \{0, 1, 2, \ldots, n, \ldots\}$ or $Y = T = \{\ldots, -n, -n+1, \ldots, -1, 0, +1, \ldots, n, \ldots\}$, then we deal with a random process with a discrete time. If Y is a finite-dimensional Euclidean space $\mathbb{R}^d$ or a domain in $\mathbb{R}^d$, then $f(y)$ is sometimes called a 'random field'.

Further, in this section we are interested in the case, when $Y = T$, i.e., we consider only random processes. Notice that random processes are often called stochastic processes.

In most cases it is convenient to identify the *random process* $f(t, \omega) = \omega(t)$ with the probability space $\{\Omega, \Sigma, \mu\}$. Indeed, we assume that Ω is a functional space, $\omega = \omega(t)$, $t \in T$; and the σ-algebra Σ contains the whole set Ω of the type $\{\omega: \omega(t_0) \in B\}$ for any $t_0 \in T$ and $B \in \mathscr{B}$, and μ is any probability measure on Ω. Thus, we see that one can naturally compare the random process $f(t, \omega) = \omega(t)$ with the probability space.

DEFINITION 6.20. Let $q(t) = q_t$ be a *random process*. Then by the notion of random process q_t, $-\infty < t < \infty$ in $\mathbb{R}^d$ we denote the *triplet* $\{\Omega, \Sigma, \mu\}$:

$$\Omega = \prod_{-\infty < t < \infty} \mathbb{R}^d$$

is the totality of paths γ in $\mathbb{R}^d$, i.e., $\gamma: \mathbb{R} \mapsto \mathbb{R}^d$. For example,

$$q_s(\omega) = \omega_s = \omega(s) \in \mathbb{R}^d,$$

for each $s \in T \equiv \mathbb{R}$, defines an $\mathbb{R}^d$-valued random variable q_s, where ω_s denotes a cross section of the *sample path* ω at the time s. The random process q_t, $-\infty < t < \infty$, is equivalent to the family of random variables $\{q_s\}$, $s \in T \equiv \mathbb{R}$.

Any random process would be specified by giving the probability measure on Ω; different probability measures define random processes of different natures.

First we consider Gaussian processes. A *Gaussian measure* on $Y \equiv \mathbb{R}^d$, $d = n = 1, 2, \ldots$ is a measure which is the transform of the measure with density $(2\pi l^2)^{-d/2} \exp(-|x|^2/2l^2)$, under an affine transformation.

DEFINITION 6.21. The *real random process* q_t, $t \in T$ is called the *Gaussian* if for any even $m \geqslant 1$ and any t_k, $k = 1, 2, \ldots, m$, $t_k \in T$; the sequence $\{q_{t_1}, \ldots, q_{t_m}\}$ has a joint normal distribution.

From the definition it follows that the *characteristic function* of this distribution has the form:

$$c_q(t_1,\ldots,t_m;u^1,\ldots,u^m)=\exp\left\{i\sum_{k=1}^{m}u^k a_k-\frac{1}{2}\sum_{k,r=1}^{m}b_{kr}u^k u^r\right\},$$

and

$$a_k=\mathbb{E}\{q_{t_k}\},\qquad b_{kr}=\mathbb{E}\{(q_{t_k}-a_k)(q_{t_r}-a_r)\}.$$

Thus, all the particular distributions of the Gaussian process are defined by two real functions: the mean value $a(t)$ and the *correlation function* (*covariance*) $b(t_1,t_2)$:

$$a(t)=\mathbb{E}\{q_t\},\qquad b(t_1,t_2)=\mathbb{E}\{(q_{t_1}-a(t_1))(q_{t_2}-a(t_2))\}.$$

The *correlation function* $b(t_1,t_2)$ possesses the following properties:

(1) $b(t_1,t_2)=b(t_2,t_1)$
(2) For any m, and real numbers u_k and points $t_k\in T$

$$\sum_{k,r=1}^{n}b(t_k,t_r)u_k u_r\geqslant 0.$$

Real functions possessing these properties are called the *positive definite kernels* on T^2. This definition is equivalent to the requirement that for any t_r and $t_k\in T$ the matrix $\|b(t_k,t_r)\|$, $k,r=1,\ldots,m$ should be real symmetric and non-negatively definite.

A similar definition of the vector Gaussian process with real components can be easily formulated, the characteristic function of which has the form

$$c(t_1,\ldots,t_m;u^{(1)},\ldots,u^{(m)})$$

$$=\exp\left\{i\sum_{k=1}^{m}(u^{(k)},a(t_k))-\frac{1}{2}\sum_{k,l=1}^{m}(b(t_k,t_l)u^{(l)},u^{(k)})\right\},$$

where

$$u^{(l)}=\{u_1^{(l)},\ldots,u_n^{(l)}\};\qquad a(t)=\{a_1(t),\ldots,a_n(t)\}$$

are vectors belonging to $\mathbb{R}^n$ and $b(t_1,t_2)$ is a matrix with the elements $b^{rs}(t_1,t_2)$, $r,s=1,2,\ldots,n$.

For some problems it is interesting to consider the *moments of the Gaussian process*. These moments may be obtained by the expansion of the characteristic function to series. We restrict ourselves to the consideration of the *central moments* of the scalar Gaussian process. Put

$$a(t)=0;\qquad \mathbf{u}=(u^{(1)},\ldots,u^{(m)});\qquad \mathbf{B}=\|b(t_k,t_r)\|\quad(k,r=1,\ldots,m).$$

Then

$$c_q(t_1, \dots, t_m; \rho\mathbf{u}) = \exp\left\{-\frac{\rho^2}{2}(\mathbf{Bu}, \mathbf{u})\right\}$$

$$= 1 - \frac{\rho^2}{2}(\mathbf{Bu}, \mathbf{u}) + \frac{\rho^4}{2!\,2^2}(\mathbf{Bu}, \mathbf{u})^2 +$$

$$+ \cdots + (-1)^m \frac{\rho^{2m}}{2^m m!}(\mathbf{Bu}, \mathbf{u})^m + \cdots$$

from this it follows that:

$$\mathbb{E}\left\{\left[\sum_{k=1}^{n} u^{(k)} q_{t_k}\right]^{2m}\right\} = (2m-1)!!(\mathbf{Bu}, \mathbf{u})^m \quad \text{and}$$

$$\mathbb{E}\left\{\left[\sum_{k=1}^{m} u^{(k)} q_{t_k}\right]^{2m-1}\right\} = 0. \tag{6.13}$$

Now we introduce the *m*-point momentum functions of *t*

$$M_{j_1, \dots, j_m}(t_1, \dots, t_m) = \mathbb{E}\{[q_{t_1}]^{j_1} \cdots [q_{t_m}]^{j_m}\}.$$

The value of $j_1 + j_2 + \cdots + j_m$ is called the *order* of momentum function. The momentum functions of the odd order equal zero

$$M_{j_1, \dots, j_m}(t_1, \dots, t_m) = 0; \qquad at \sum_{k=1}^{m} j_k = 2s - 1, \quad s = 1, 2, \dots$$

Formula (6.13) can be rewritten in the form:

$$M_{j_1, \dots, j_m}(t_1, \dots, t_m) = (\partial^{2m}/\partial u_1^{j_1} \dots \partial u_m^{j_m}) \frac{1}{2^m m!}(\mathbf{Bu}, \mathbf{u})^m. \tag{6.14}$$

The momentum function of the second order coincide with the correlation function

$$M_{11}(t_1, t_2) = b(t_1, t_2); \qquad M_2(t) = M_{11}(t, t) = b(t, t).$$

For the momentum functions of the fourth order, we have the following expressions:

$$M_4(t) = 3b^2(t, t); \qquad M_2(t_1, t_2) = 2b^2(t_1, t_2);$$
$$M_{31}(t_1, t_2) = 3b(t_1, t_1)b(t_1, t_2);$$
$$M_{211}(t_1, t_2, t_3) = b(t_1, t_1)b(t_2, t_3) + 2b(t_1, t_2)b(t_1, t_3);$$
$$M_{1111}(t_1, t_2, t_3, t_4) = b(t_1, t_2)b(t_3, t_4) + b(t_1, t_3)b(t_2, t_4) +$$
$$+ b(t_1, t_4)b(t_2, t_3)$$

etc.

DEFINITION 6.22. The *oscillator process* is the family $\{q(t)\}$, $-\infty < t < \infty$, of the *Gaussian random variables* with the covariance

$$\mathbb{E}\{q(t)q(s)\} = \tfrac{1}{2}\exp(-|t-s|).$$

LEMMA 6.1. *Let* $s_1, \ldots, s_n$ *be real and* $z_1, \ldots, z_n \in \mathbb{C}$. *Then*

$$\sum_{i,j=1}^{n} \bar{z}_i z_j [\tfrac{1}{2}\exp(-|s_i - s_j|)] \geqslant 0.$$

Proof. By a direct calculation of the Fourier transform of $\exp(-|t|)$ and the Fourier inversion formula,

$$\tfrac{1}{2}\exp(-|t-s|) = (2\pi)^{-1}\int dk\, e^{ik(t-s)}(1+k^2)^{-1}$$

so that

$$\sum_{j,i=1}^{n} \bar{z}_i z_j [\tfrac{1}{2}\exp(-|s_i - s_j|)] = (2\pi)^{-1}\int dk \left|\sum_{j=1}^{n} z_j e^{iks_j}\right|^2 (1+k^2)^{-1}$$

which is clearly *positive*. ■

Often, the process we have called the *oscillator process* is called the '*Ornstein–Uhlenbeck velocity process*' since Uhlenbeck and Ornstein (1930) introduced a process $f(t)$ with differentiable paths so that $df(t)/dt = q(t)$.

6.9. Stochastic Processes with Independent Increments

DEFINITION 6.23. A random process $\{q(t), t \in T\} \in \mathbb{R}^n$ is called the *process* with *independent increments*, if for any m, $t_k \in T$; $t_1 < t_2 < \cdots < t_m$, the random vectors $q(a)$, $q(t_1) - q(a), \ldots, q(t_m) - q(t_{m-1})$ are mutually independent. The vector $q(a)$ is called the initial state (value) of the process $q(t)$, and its distribution is called the initial distribution of $q(t)$. In order to define the process with independent increments it is sufficient to give an initial distribution $P_0(B)$ and a set of distributions $P(t, h, B)$ ($t \geqslant 0$, $h > 0$, $B \in \mathscr{B}^n$), where $\mathscr{B}^n$ is the σ-algebra of the Borel sets $\mathbb{R}^n$ and $P(t, h, B)$ is the distribution of the vector $q(t+h) - q(t)$. Indeed, if these distributions are given, then any joint distribution of vectors $q(t_1), q(t_2), \ldots, q(t_m)$ is synonymously defined by the formula

$$P\left(\bigcap_{k=0}^{m} \{q(t_k) \in B_k\}\right) = \int_{B_0} P_0(dy_0) \int_{B_1 - y_0} P(0, t_1, dy_1) \times$$

$$\times \int_{B_2 - (y_0 + y_1)} P(t_1, t_2 - t_1, dy_2) \cdots \times$$

$$\times \int_{B_m - (y_0 + \cdots + y_m)} P(t_{m-1}, t_m - t_{m-1}, \mathrm{d}y_m). \tag{6.15}$$

Here $B - z$ denotes the set: $\{x: x = y - z, y \in B\}$. As concerns the initial distribution $P_0(B)$, it can be arbitrary. On the other hand, it cannot be guaranteed that to any given set of the distribution $P(t, h, B)$ there corresponds some process with independent increments. In order to have this, it is necessary and sufficient that $P(t, h, B)$ should possess the following property: at any m and any $a = t_0 < t_1 < \cdots < t_m = h + t$; $P(t, h, B)$ is a distribution of the sum of the independent vectors $q(t_1), q(t_2), \ldots, q(t_m)$, where $q(t_k)$ has the distribution $P(t_{k-1}, t_k - t_{k-1}, B)$. Indeed, if this condition is fulfilled, then the set of distribution (6.15) is satisfied by the conditions of co-ordination. Therefore, the *Kolmogorov theorem* is used and there exists a random process with finite-dimensional distribution (6.15). From the form of these distributions it follows that this process has independent increments.

It is convenient to study the processes with independent increments by means of the characteristic functions. Let us define

$$c(t, h, u) = \int_{\mathbb{R}^n} P(t, h, \mathrm{d}x) \exp[i(u, x)].$$

We have called the function $c(\cdots)$ the *characteristic function* of the process with independent increments. It defines fully a joint distribution of the differences:

$$q(t_1) - q(a), q(t_2) - q(t_1), \ldots, q(t_m) - q(t_{m-1}). \tag{6.16}$$

Indeed, joint distribution of the sequence of vectors (6.16) has the characteristic function $c(t_1, \ldots, t_m; u^1, \ldots, u^m)$ determined by the expression:

$$c(t_1, \ldots, t_m; u^1, \ldots, u^m) = \prod_{k=1}^{m} c(t_{k-1}, \Delta t_k, u^k),$$

$$(\Delta t_k = t_k - t_{k-1}, t_0 = a).$$

Thus, in order to give the process with independent increments, in the wide sense, in addition to $P_0(B)$, it is sufficient to define $c(t, h, B)$. The condition

$$c(t, h_1 + h_2, u) = c(t, h_1, u)\, c(t + h_1, h_2, u)$$

is necessary and sufficient for $c(t, h, u)$ to be the characteristic function of the process with independent increments.

DEFINITION 6.24. The process with independent increments is called *homogeneous*, if the differences $q(t + h) - q(t)$ have the distributions which do not depend on time, i.e., $P(t, h, B) = P(h, B)$. The homogeneous process is called *stochastic continuous*, if

$$\lim_{h \to 0} P(h, \bar{\cup}_{\varepsilon}) = 0,$$

for any sphere $\cup_{\varepsilon} = \{x\colon |x| < \varepsilon\}$, $\varepsilon > 0$.

Now we give some of the properties of the characteristic function of the homogeneous stochastic continuous process with independent increments:

(a) The *characteristic function* of the homogeneous process with independent increments satisfies the equation:

$$c(h_1 + h_2, u) = c(h_1, u)\, c(h_2, u). \tag{6.17}$$

In particular, at any even n: $c(n \cdot h, u) = [c(h, u)]^n$.
(b) The characteristic function of the homogeneous stochastic process is non-zero everywhere and its value can be represented by

$$c(t, u) = \exp[tg(u)], \tag{6.18}$$

where $g(u)$ is a synonymous and continuous function. From form (6.17) it follows that

$$g(u) = \lim_{t \to 0} [c(t, u) - 1]/t \tag{6.19}$$

and convergence is uniform in each restricted sphere $|u| \leqslant N$, $0 < N < \infty$.

THEOREM 6.6 *Let $c(t, u)$, $t > 0$, $u \in \mathbb{R}^n$, be a family of the characteristic functions such that limit* (6.19) *will exist uniformly in an arbitrary sphere $|u| \leqslant N$, $N > 0$. Then in $\{\mathbb{R}^n, \mathcal{B}^n\}$ there exists a finite measure $M(B)$, $B \in \mathcal{B}$'and nonnegatively defined in $\mathbb{R}^n$ operator b and vector a such that*

$$g(u) = i(a, u) - \tfrac{1}{2}(bu, u) + \int_{\mathbb{R}^n} M(\mathrm{d}z)\left[\mathrm{e}^{i(u, z)} - 1 - \frac{i(u, z)}{1 + |z|^2}\right] \times$$

$$\times (1 + |z|^2)|z|^{-2}. \tag{6.20}$$

Proof was given by Gihman and Skorokhod (1975).

THEOREM 6.7. *If $q(t) \in \mathbb{R}^n$, $t \geqslant 0$ is a homogeneous and stochastic continuous process, then the* characteristic function $c(t, u)$ *of difference $q(t + s) - q(s)$ has the form*

$$c(t, u) = \exp[tg(u)], \tag{6.21}$$

where $g(u)$ is given by formula (6.20).

Now we consider some particular cases of formula (6.20):

(a) $b = 0$ and $M(B) \equiv 0$. In this case $c(t, u) = \exp[it(a, u)]$ which corresponds to the characteristic function of the degenerated distribution located at the point $ta \in \mathbb{R}^n$. Thus, with probability one, $q(t) = q_0 + at$ and the point $q(t)$ would be found in the *uniform motion* with the velocity a.

(b) $a = 0$, $b = 0$, and M represents the measure located at the point z_0 mass of the value m. Characteristic function (6.21) in this case has the form

$$c(t, u) = \exp\left\{\frac{mt(1 + |z_0|^2)}{|z_0|^2}\left[e^{i(u, z_0)} - 1 - \frac{i(u, z_0)}{1 + |z_0|^2}\right]\right\}.$$

It is easy to see that the increment $q(t) - q(0)$ can be represented in the form

$$q(t) - q(0) = z_0[\nu(t) - mt|z_0|^{-2}],$$

where $\nu(t)$ is the *Poisson process* with the mean value

$$\mathbb{E}\{\nu(t)\} = mt(1 + |z_0|^2)|z_0|^{-2}.$$

(c) $M(B) \equiv 0$. In this case the difference $q(t + s) - q(s)$ has the normal distribution with the mean value a and the (covariance) correlation matrix bt, so that, if, for example, $q(0) = 0$, then the process $q(t)$ is the *Gaussian.* The considerable process is also called the process of the *Brownian motion.* Thus the Brownian motion can be considered as a continuous process with independent increments. It is also a Gaussian process. If $q(t)$ is a one-dimensional process, and $b = 1$, $a = 0$, then the process of the Brownian motion is called the *Wiener process.*

6.10. Markov Processes

Before giving a formal definition of the Markov processes we recall a simple model starting from the theory of the Brownian motion. In 1828, an English botanist R. Brow observed under a microscope that small particles, pollen of some flower, suspended in a liquid move chaotically, changing position and direction incessantly. To describe such a phenomenon, we shall consider the *transition probability* $P(t, x; t_0, \Gamma)$ that a particle starting from the position x at time t_0 belongs to the set Γ at a later time t.

The introduction of the *transition probability* (sometimes it is called the transition function) $P(t, x; t_0, \Gamma)$ is based on the fundamental hypothesis that the (stochastic) chaotic motion of the particle after the time moment t is entirely independent of its past history before the time moment t. The hypothesis that the particle has no memory of the past implies that the transition probability P satisfies the equation

$$P(t, x; s, \Gamma) = \int_S P(t, x; u, dy)\, P(u, y; s, \Gamma), \tag{6.22}$$

for $t < u < s$, where the integration is performed over the entire space S of the chaotic movement of the particle.

The process of evolution in time governed by a transition probability satisfying (6.22) is called the *Markov process*, and equation (6.22) is called in mathematics (in physics) the *Chapman–Kolmogorov equation* (*the Smoluchowski equation*). The Markov process is a natural generalization of the deterministic process for which $P(t, x; s, \Gamma) = 1$ or $= 0$ according to $y \in \Gamma$ or $y \notin \Gamma$; that is, the process in which the particle at the position x at the time moment t moves to a definite position $y = y(x, t, s)$ with the probability of one at every fixed later time moment s. The Markov process P is said to be *temporally homogeneous* if $P(t, x; s, \Gamma)$ is the function of difference $(s - t)$ independent of t. In such a case, we are dealing with the transition probability $P(t, x; \Gamma)$ that a particle at the position x is transferred into the set Γ after the lapse of t units of time. Equation (6.22) then becomes

$$P(t + s, x; \Gamma) = \int_S P(t, x, \mathrm{d}y)\, P(s, y; \Gamma) \tag{6.23}$$

for $t, s > 0$.

A more developed theory deals with the object of the motion trajectory $x(t) = x_t$. The random character of motion is expressed by the mathematical assumption that $x_t = x_t(\omega)$, where ω belongs to some set Ω on which there is given a family of probability measures μ_x (as the set Ω above is called the space of elementary events). A set A for which the value of $\mu_x(A)$ is determined is called 'events' and the quantity of $\mu_x(A)$ is interpreted as the probability event A by the assumption that the motion is started from the point x. Then a *transition probability* is given by the formula: $P(t, x; \Gamma) = \mu_x\ (x_t \in \Gamma)$. In the case of the Brownian motion the phase space, on which the value of the function $x(t)$ is given, is some domain of the three-dimensional Euclidean space. Generally speaking, this space is the set S in which some system of 'measurable sub-sets' $\mathscr{B}$ is picked.

The basic condition connecting the function $x_t(\omega)$ and the measure μ_x is the *Markov principle*: the future depends on the past only through the present. The mathematical (equation) expression of this principle is the Chapman–Kolmogorov equation (6.23). More precisely it means that at the given value of x_t prognosis for further motion of a particle does not depend on the character of motion before the moment t. We recall that the Markov processes, for which the condition of independence of the future on the past known at the present is fulfilled not only for the constant moment but for the definite class of random moments $\zeta(\omega)$, are called *strict Markov processes*.

A wider and more adaptable formulation of the *Markov principle* is obtained if under 'the past' for the moment t one understands the family of all the events Σ_t which are observed before the moment t. Up to now we have assumed that x_t is determined for all $t \geqslant 0$. However, many problems leads to the processes, for which $x_t(\omega)$ is only given on some interval $[0, \zeta(\omega)]$. The random variable $\zeta(\omega)$ is called the *truncation moment*. Thus, we lead to a definition of the Markov process as a family $X_t = \{x_t, \zeta; \Sigma_t, \mu_x\}$.

DEFINITION 6.25. Let us give:

(a) The function $\zeta(\omega) \in \Omega$ taking the value on $[0, \infty)$;
(b) The function $x(t, \omega) = x_t(\omega)$ defined for $\omega \in \Omega$ and $t \in [0, \xi(\omega)]$ taking the value from the phase space $\{S, \mathscr{B}\}$;
(c) For each $t \geqslant 0$, the σ-algebra $\Sigma_t \subset \Omega_t = \{\omega: \zeta(\omega) > t\}$;
(d) For each $x \in S$, the function $\mu_x(A)$ defined on some σ-algebra $\Sigma^0 \subset \Omega$, and $\Sigma_t \subset \Sigma^0$ at all $t \geqslant 0$.

Then we say that these elements define the Markov process $X_t = \{x_t, \zeta, \Sigma_t, \mu_x\}$, if the following conditions are fulfilled:

(1) If $t \leqslant s$ and $A \in \Sigma_t$, then $\{A: \zeta > s\} \in \Sigma_s$.
(2) $\{x_t \in \Gamma\} \in \Sigma_t$ $(t \geqslant 0, \Gamma \in \mathscr{B})$.
(3) μ_x is the probability measure on the σ-algebra Σ^0.
(4) At any $t \geqslant 0$, $\Gamma \in \mathscr{B}$ the function

$$P(t, x, \Gamma) = \mu_x\{x_t \in \Gamma\} \tag{6.24}$$

is the $\mathscr{B}$-measurable function of x.[!]
(5) $P(0, x, S \backslash x) = 0$.
(6) For any $t, h \geqslant 0$, $\Gamma \in \mathscr{B}$, $\mu_x\{x_{t+h} \in \Gamma \mid \Sigma_t\} = P(h, x_t, \Gamma)$ with the probability Ω_t, μ_x.
(7) For every $\omega \in \Omega_t$ there is found such $\omega' \in \Omega$ that

$$x_s(\omega') = x_{t+s}(\omega), \qquad (0 \leqslant s < \zeta(\omega') = \zeta(\omega) - t).$$

According to (4) condition (6) is equivalent to the following requirement:
(4′) For any $t, h \geqslant 0$, $\Gamma \in \mathscr{B}$, $A \in \Sigma_t$

$$\mu_x\{A, x_{t+h} \in \Gamma\} = \int_\Omega \mu_x(d\omega)\chi_A \cdot P(h, x_t(\omega), \Gamma).$$

The *Markov process* is called *without truncation*, if $\zeta(\omega) = +\infty$ for every $\omega \in \Omega$. At the fixed ω the function $x_t(\omega)$, $\{t \in [0, \zeta(\omega))\}$ is defined on the trajectory space S of the process corresponding to the elementary event ω. The σ-algebra Σ_t can be clearly represented as the family of events which are observed during the time interval $[0, t]$. From the above definition, we see that the meaning of conditions (1)–(3), and (5) is clear. Condition (4) has a technical character. Condition (6) means that the prognosis of the particle position through the time moment h with the knowledge of its position at the moment t depends neither on the value of t, nor on the phenomena which have been observed during the time interval $[0, t]$. This is just the Markov principle: the independence of 'the future' on 'the past' by the knowledge of 'the present'. Condition (7) means some homogeneity of the family of trajectories of the process. More precisely, let $\varphi(u) \in S$ be a function defined on some interval $[0, \lambda)$, and let $0 \leqslant t < \lambda$. Then, we define the function $T_t\varphi(u) = \varphi(t + u)$ $(0 \leqslant u < \lambda - t)$ by a displacement of the function φ on t. Condition (7) requires that a set of trajectories of the process should be invariant with respect to all the displacements. In this connection we notice the following. In a suitable function space F, $P(t, x, \Gamma)$ gives rise to the linear transformation T_t:

$$(T_t f)(x) = \int_S P(t, x, \mathrm{d}y) f(y), \quad f \in F,$$

so that, by (6.23), the *semi-group property* holds: $T_{t+s} = T_t T_s$ $(t, s > 0)$.

A fundamental mathematical problem in the statistical mechanics concerns the existence of the *time average* $\lim_{t\to\infty} t^{-1} \int_0^t \mathrm{d}s\, T_s f$. In fact, let S be phase space of a mechanical system governed by the classical Hamiltonian equations whose Hamiltonian does not contain the time variable explicitly. Then the point x of S is moved to the point $y_t(x)$ of S after the lapse of t units of time in such a way that, by the classical theorem due to Liouville, the mapping $x \to y_t(x)$ of S onto S, for each fixed t, is an equi-measure transformation, that is, the mapping $x \to y_t(x)$ leaves the 'phase volume' of S invariant. In such a deterministic case, we have $(T_t f)(x) = f(y_t(x))$ and hence the *ergodic hypothesis of Boltzmann* that the time average of any physical quantity coincides with the space average of this physical quantity, is expressed, assuming $\int \mathrm{d}x < \infty$, by

$$\lim_{t\to\infty} t^{-1} \int_0^t \mathrm{d}s\, f(y_s(x)) = \frac{\int_S \mathrm{d}x\, f(x)}{\int_S \mathrm{d}x}$$

for all $f \in F$, $\mathrm{d}x$ denoting the phase volume element of S. A natural generalization of the equi-measure transformation $x \to y_t(x)$ to the case of the Markov process $P(t, x, \Gamma)$ is the condition of the existence of an invariant measure $\mu(\mathrm{d}x)$:

$$\int_S \mu(\mathrm{d}x) P(t, x, \Gamma) = \mu(\Gamma)$$

for all $t > 0$ and all Γ.

We notice that the transition probability (or transition function) $P(t, x, \Gamma)$ is a considerably simpler object than the Markov process. Therefore, for the Markov process it is very important to know its transition function $P(t, x, \Gamma)$.

DEFINITION 6.26. Let us consider an arbitrary phase space $\{S, \mathscr{B}\}$. The function $P(t, x, \Gamma)$ $(t \geqslant 0,\ x \in S,\ \Gamma \in \mathscr{B})$ is called a *transition function*, if the following conditions are fulfilled:

(i) For the fixed t and x, $P(t, x, \Gamma)$ is a measure on the σ-algebra $\mathscr{B}$;
(ii) For the fixed t and Γ, $P(t, x, \Gamma)$ is $\mathscr{B}$-measurable in x;
(iii) $P(t, x, \Gamma) \geqslant 0$, $\quad P(t, x, S) \leqslant 1$;
(iv) $P(0, x, S \backslash x) = 0$;
(v) $P(t + s, x, \Gamma) = \int_S P(s, x, \mathrm{d}y) P(t, y, \Gamma)$ $(s, t \geqslant 0)$ (the Chapman–Kolmogorov equation).

From (v) and (iii) we have

$$P(u+t, x, S) \leqslant \int_S P(u, x, \mathrm{d}y) = P(u, x, S) \quad (u, t \geqslant 0).$$

Hence, $P(t, x, S)$ is an unincreased function of t and there exists the limit $P(+0, x, S)$. We can say that the transition function $P(t, x, \Gamma)$ is *normal*, if $P(+0, x, S) = 1$ for any $x \in S$. We call it *conservative*, if $P(t, x, S) = 1$ for any $t, x \in S$.

The function $P(t, x, \Gamma)$ defined by formula (6.24) satisfies conditions (i)–(v). It is called the transition function of the Markov process. We introduce the following conditions:

Condition $L(\Gamma)$: For any $u \geqslant 0$

$$\lim_{y \to \infty} \operatorname{Sup}_{t \leqslant u} P(t, y, \Gamma) = 0.$$

Condition $N(\Gamma)$: For any $\varepsilon > 0$

$$\lim_{t \to 0} t^{-1} \operatorname{Sup}_{x \in \Gamma} P(t, x, \bar{\cup}_{\varepsilon}(x)) = 0$$

(through $\bar{\cup}_{\varepsilon}(x)$ we denote a complement of ε-neighborhood of the point x, i.e., the set $\{y: \rho(x, y) \geqslant \varepsilon\}$, where $\rho(x, y)$ is called the metric or the distance function).

THEOREM 6.8. *Let $P(t, x, \Gamma)$ be the normal transition function on the half-compact $\{S, \mathbb{C}\}$ and for every compact subset of Γ the conditions $L(\Gamma)$ and $N(\Gamma)$ are fulfilled. Then, there exists the continuous Markov process with the transition function $P(t, x, \Gamma)$.*

Proof. This can be found in any textbook on the theory of the Markov process, in particular, see Dynkin (1965).

DEFINITION 6.27. Let μ be some measure on $\{S, \mathscr{B}\}$. The function $p(t, x, y)$ $(t > 0, x, y \in S)$ is called the *transition density*, if the following conditions are fulfilled:

(1) $p(t, x, y) \geqslant 0 \quad (t > 0; x, y \in S)$.
(2) For fixed t, $p(t, x, y)$ is $\mathscr{B} \otimes \mathscr{B}$-measurable function in x and y

(3) $$\int_S \mu(\mathrm{d}y)\, p(t, x, y) = 1, \quad (t \geqslant 0, x \in S),$$

(4) $$p(u+t, x, z) = \int_S \mathrm{d}y\, p(u, x, y)\, p(t, y, z), \quad (u, t \geqslant 0; x, z \in S).$$

It is easily verified that if $p(t, x, y)$ is the transition density, then the formula

$$P(t, x, \Gamma) = \begin{cases} \int_\Gamma \mathrm{d}y p(t, x, y) & \text{at } t > 0 \\ \chi_\Gamma(x) & \text{at } t = 0 \end{cases} \quad (x \in S, \Gamma \in \mathscr{B})$$

defines the transition function. This function is normal if

$$\lim_{t \to 0} \int_S \mathrm{d}y\, p(t, x, y) = 1, \quad (x \in S),$$

and conservative if

$$\int_S \mathrm{d}y\, p(t, x, y) = 1, \quad (t > 0, x \in S).$$

EXAMPLE 6.1. Let $\{S, \mathscr{B}\}$ be any measurable space and let $\Pi(x, \Gamma)$ $(x \in S, \Gamma \in \mathscr{B})$ be the function satisfying the following conditions:

1. For any $x \in S$, $\Pi(x, \Gamma)$ is a probability measure on $\mathscr{B}$.
2. For any $\Gamma \in \mathscr{B}$, $\Pi(x, \Gamma)$ is a $\mathscr{B}$-measurable function on S.

Define the function $\Pi_n(x, \Gamma)$ by formulas

$$\Pi_0(x, \Gamma) = \chi_\Gamma(x),$$

$$\Pi_n(x, \Gamma) = \int_S \Pi(x, \mathrm{d}y)\Pi_{n-1}(y, \Gamma), \quad (n \geqslant 1)$$

and put

$$P(t, x, \Gamma) = \sum_{n=0}^{\infty} \frac{a^n t^n}{n!} \mathrm{e}^{-at}\, \Pi_n(x, \Gamma), \tag{6.25}$$

where a is some positive constant. It is verified that the series on the right-hand side of (6.25) converges and determines the transition function. We call it the *Poisson transition function.*

EXAMPLE 6.2. Let S be a numerical (axis) line, and let $\mathscr{B}$ be the σ-algebra, obtained by a family of all the intervals, and v is some constant. The formula $P(t, x, \Gamma) = \chi_\Gamma(x + vt)$ gives the conservative transition function. It is said that this function corresponds to a uniform motion with the velocity v.

EXAMPLE 6.3. Let S be the n-dimensional Euclidean space $\mathbb{R}^n$ and let $\mathscr{B}$ be the σ-algebra of all the Borel sets of this space. Then put

$$p(t, x, y) = (2\pi t)^{-n/2} \exp\{-(y - x)^2/2t\}. \tag{6.26}$$

It is easy to verify that $p(t, x, y)$ is a transition density with respect to the Lebesque measure. We call it the *Wiener transition density* and the corresponding transition function – the *Wiener transition* (function) *probability*.

EXAMPLE 6.4. The formula

$$p(t, x, y) = (2\pi t)^{-1/2}\{\exp[-(y-x)^2/2t] + \exp[-(y+x)^2/2t]\} \tag{6.27}$$

defines the transition density on half-axis $[0, \infty)$ (with respect to the Lebesque measure). The corresponding transition function is conservative.

EXAMPLE 6.5. The formula

$$p(t, x, y) = (2\pi t)^{-1/2}\{\exp[-(y-x)^2/2t] - \exp[-(y+x)^2/2t]\} \tag{6.28}$$

gives the transition density on the half-axis $(0, \infty)$. The corresponding transition function is not conservative but is normal.

Finally, we want to give some formal definition of the *Markov processes* and the *time-reversed* one. The Markov process in $\mathbb{R}^n$, $X(t)$, $-\infty < t < \infty$, is a random process such that

$$\mathbb{C}\mathrm{ond}\{X_t \in B \mid X(t_m) = x_m, \ldots, X(t_1) = x_1\} = \mathbb{C}\mathrm{ond}\{X_t \in B \mid X(t_m) = x_m\},$$
$$X_t \equiv X(t)$$

holds with the probability one for any time series $-\infty < t_1 < \cdots < t_m < t$, where B is a Borel set of $\mathbb{R}^n$. Since one can regard the condition probability with respect to a random variable as a function of this variable, then we denote

$$P(t, x; u, B) = \mathbb{C}\mathrm{ond}\{X_t \in B \mid x_u = X(u) \equiv X_u = x\}, \quad t > u.$$

This is nothing but the transition probability or the transition function of the Markov process defined above, which, therefore, satisfies the following Chapman–Kolmogorov equation:

$$P(t, x; u, B) = \int P(t, y; s, B)\, P(s, x; u, \mathrm{d}y), \quad t > s > u.$$

Further, we shall prove that the time-reversed process of the Markov process is also the Markov one. The *time-reversed process* of the Markov process is defined to be a random process X_t^*, $-\infty < t < \infty$ such that $X_t^* = X(-t) = X_{-t}$ holds with the probability one.

For technical simplicity we replace $\mathbb{R}^n$ by $\mathbb{Z}^n$ ($\mathbb{Z}$ denotes the totality of integers): this corresponds to a lattice approximation.

THEOREM 6.9. *Let X_t, $-\infty < t < \infty$, be the Markov process in $\mathbb{Z}^n$, i.e.,*

$$\mathbb{C}\mathrm{ond}\{X(t_m) = x_m \mid X(t_{m-1}) = x_{m-1}, \ldots, X(t_1) = x_1\}$$
$$= \mathbb{C}\mathrm{ond}\{X(t_m) = x_m \mid X(t_{m-1}) = x_{m-1}\}$$

for any $-\infty < t_1 < \cdots < t_m < \infty$ and $x_1, \ldots, x_m \in \mathbb{Z}^n$. Then we have

$$\mathbb{C}\mathrm{ond}\{X(t_1) = x_1 \mid X(t_2) = x_2, \ldots, X(t_m) = x_m\}$$
$$= \mathbb{C}\mathrm{ond}\{X(t_1) = x_1 \mid X(t_2) = x_2\}, \tag{6.29}$$

i.e., the time-reversed process X_t^*, $-\infty < t < \infty$, is also the Markov one.

Proof. The left-hand side of equation (6.29) can be manipulated as

$$\frac{\mu\{X(t_1) = x_1, \ldots, X(t_m) = x_m\}}{\mu\{X(t_2) = x_2, \ldots, X(t_m) = x_m\}}$$

$$= \mathbb{C}\text{ond}\{X(t_m) = x_m \mid X(t_{m-1}) = x_{m-1}\} \cdots \times$$

$$\times\, \mathbb{C}\text{ond}\{X(t_2) = x_2 \mid X(t_1) = x_1\} \times$$

$$\times \frac{\mu\{X(t_1) = x_1\}}{\dfrac{\mathbb{C}\text{ond}\{X(t_m) = x_m \mid X(t_{m-1}) = x_{m-1}\} \ldots \times}{\times\, \mathbb{C}\text{ond}\{X(t_3) = x_3 \mid X(t_2) = x_2\} \cdot \mu\{X(t_2) = x_2\}}}$$

$$= \mathbb{C}\text{ond}\{X(t_2) = x_2 \mid X(t_1) = x_1\} \cdot \frac{\mu\{X(t_1) = x_1\}}{\mu\{X(t_2) = x_2\}}$$

$$= \mathbb{C}\text{ond}\{X(t_1) = x_1 \mid X(t_2) = x_2\}$$

which is identical to the right-hand side of equation (6.29). ■

THEOREM 6.10. *Let $P(t, x; s, y)$, $t > s$, be a transition probability of the Markov process X_t, $-\infty < t < \infty$; $\rho(t, x)$ be a probability distribution of X_t. Then a* transition probability of the time-reversed Markov process X_t^*, $-\infty < t < \infty$; *$P^*(t, y; s, x)$, $t > s$, is given by*

$$P^*(t, y; s, x) = \rho(-t, y)P(-s, x; -t, y)\rho(-s, x)^{-1}. \tag{6.30}$$

Proof. A straightforward manipulation yields

$$P^*(t, y; s, x) = \mathbb{C}\text{ond}\{X_t^* = y \mid X_s^* = x\} = \mathbb{C}\text{ond}\{X_{-t} = y \mid X_{-s} = x\}$$

$$= \frac{\mu\{X_{-t} = y, X_{-s} = x\}}{\mu\{X_{-s} = x\}} = \frac{\mu\{X_{-s} = x, X_{-t} = y\}}{\mu\{X_{-t} = y\}} \frac{\mu\{X_{-t} = y\}}{\mu\{X_{-s} = x\}}$$

$$= \mu\{X_{-t} = y\} \frac{\mathbb{C}\text{ond}\{X_{-s} = x \mid X_{-t} = y\}}{\mu\{X_{-s} = x\}}$$

$$= \rho(-t, y)P(-s, x; -t, y)\rho(-s, x)^{-1}.$$ ■

6.11. Wiener Processes

We begin by giving some simple definition of this process, based on its transition probability (or function) $P(\cdot)$. We consider the *Wiener transition function* $P(\cdot)$ (6.26) defined in Example 6.3 of Theorem 6.8. We show that for every compact Γ this function is satisfied by the conditions $L(\Gamma)$ and $N(\Gamma)$ (see

Section 6.10). Indeed, as is easily seen

$$P(t, x, \Gamma) \leqslant 2(\pi t)^{-n/2} \exp\{-(|x| - a)^2/2t\} \int_\Gamma dy,$$

for $|x| > a$, where $a = \mathrm{Sup}_{y \in \Gamma} |y|$. From this we see that the condition $L(\Gamma)$ is fulfilled.

Further

$$P(t, x, \bar{\cup}_\varepsilon(x)) = (2\pi t)^{-n/2} \int_{y > \varepsilon} dy \exp(-y^2/2t).$$

After simple transformations we get

$$P(t, x, \bar{\cup}_\varepsilon(x)) = c_n F(\varepsilon/\sqrt{t}),$$

where c_n is some constant and

$$F(r) = \int_r^\infty dx\, x^{n-1} \exp(-x^2/2).$$

By the *Lopital rule*

$$F(r) \sim r^{n-2} \exp(-\tfrac{1}{2}r^2), \quad \text{for } r \to \infty$$

and therefore at $t \to 0$

$$\mathop{\mathrm{Sup}}_{x \in S} P(t, x, \bar{\cup}_\varepsilon(x)) \sim c_n \varepsilon^{n-2} t^{-(n-2)/2} \exp(-\varepsilon^2/2t).$$

From this we see that $P(t, x, \Gamma)$ satisfies condition $N(\Gamma)$. According to Theorem 6.8 some continuous Markov process without truncation corresponds to this Wiener's function. We call it the *Wiener process.*

Transition function (6.27) from Example 6.4 of Theorem 6.8 is also conservative and satisfies the conditions of Theorem 6.8. The corresponding continous Markov process without truncation is called the Wiener process with reflection at point 0. We now consider transition function (6.28) determined on the half-axis $(0, \infty)$ (Example 6.5). A direct calculation shows that for every compact Γ the conditions $L(\Gamma)$ and $N(\Gamma)$ are fulfilled in this case. From Theorem 6.8 we conclude that some continuous Markov process $X_t = \{x_t, \zeta, \Sigma_t, \mu_x\}$ corresponds to function (6.28). We shall call this process the Wiener process with a truncation at point 0. Thus, we arrive at a definition of the Wiener process.

DEFINITION 6.28. We denote through $S \equiv \mathbb{R}^n$ the n-dimensional Euclidean space and through $\mathscr{B}$ a σ-algebra of all the Borel subsets of this space. Then under the *Wiener process* one can understand the standard continuous process X_t without-truncation in space $\{S, \mathscr{B}\}$ with the transition function

$$P(t, x, \Gamma) = \int_\Gamma dy\, p_t(y - x), \quad (t \geqslant 0, x \in S, \Gamma \in \mathscr{B}),$$

where

$$p_t(x) = (2\pi t)^{-n/2} \exp(-x^2/2t).$$

According to Example (b) of Theorem 6.7 we can regard the Wiener process as a Gaussian process with a mean value 0 and the correlation t. By definition, it is also the Brownian motion. To the Wiener process an object-definition based on the notion of *random walks* can be given. It is known that the Wiener process $X_t = q(t)$ is the limit of elementary random walks. Let y_n be the nth coordinate function and $X_n = \Sigma_{m=1}^n y_m$. The family of random variables $\{X_n\}_{n=1}^\infty$ is called a *random walk*. Since $\mathbb{E}\{y_n y_m\} = \delta_{nm}$ we have $\mathbb{E}\{X_n^2\} = n$ so that the variable $\tilde{X}_n = n^{-1/2} X_n$ at least have a chance of having a limit. Due to the well-known De Moirre–Laplace limit theorem $\tilde{X}_n$ approaches a Gaussian.

THEOREM 6.11 (*Central limit thoerem*). *Let* $\{y_n\}_{n=1}^\infty$ *be a family of independent, identically distributed random variables with* $\mathbb{E}\{y_n\} = 0$, $\mathbb{E}\{y_n^2\} = 1$. *Let* $\tilde{X}_n = n^{-1/2} \Sigma_{m=1}^n y_m$. *Then* $\tilde{X}_n$ *approaches a Gaussian random variable* $\tilde{X}_\infty$ *with the covariance (correlation) one in the sense, that for any continuous, bounded function on* $\mathbb{R}$,

$$\mathbb{E}\{f(\tilde{X}_n)\} \to (2\pi)^{-1/2} \int dy\, f(y) \exp(-y^2/2).$$

Proof. By a simple limiting procedure (going through $f \in L$ as an intermediate step), it is sufficient to prove that

$$\mathbb{E}\{\exp(i\alpha \tilde{X}_n)\} \to \exp(-\alpha^2/2)$$

for all real α. But

$$\mathbb{E}\{\exp(i\alpha \tilde{X}_n)\} = [F(\alpha/\sqrt{n})]^n,$$

where

$$F(\beta) = \mathbb{E}\{e^{i\beta y}\}.$$

Since $\mathbb{E}\{y_n\} = 0$, $\mathbb{E}\{y_n^2\} = 1$, we have that

$$[F(\beta) - 1]\beta^{-2} = -\tfrac{1}{2}$$

as $\beta \to 0$ (by using the dominated convergence theorem) so that

$$\mathbb{E}\{\exp(i\alpha \tilde{X}_n)\} = [1 - \tfrac{1}{2}(\alpha^2/n) + o(1/n)]^n = \exp(-\alpha^2/2) + o(1)$$

by the compound formula of interest. ■

Intuitively $q(t)$ is $\lim_{n\to\infty} n^{-1/2} X_{[nt]}$, where $[a]$ is an integral part of a. Thus, each $q(t)$ should be Gaussian with covariance t. To find the covariance, we note that $X_n - X_m$ and X_m are independent if $m < n$. Thus, we expect that $q(t) - q(s)$ should be independent of $q(s)$ for $s < t$, i.e.,

$$\mathbb{E}\{q(s)[q(t) - q(s)]\} = 0 \quad \text{so} \quad \mathbb{E}\{q(t)q(s)\} = s \quad \text{for } s < t.$$

Therefore, we define the following

DEFINITION 6.29 The *Wiener process* (or Brownian motion) is the family $\{q(t)\}_{0<t}$ of Gaussian random variables with the *covariance*

$$\mathbb{E}\{q(t)q(s)\} = \min(s, t).$$

LEMMA 6.2. *Let* $0 \leqslant s_1 \leqslant s_2 \leqslant \cdots \leqslant s_n$ *and* $z_1, \ldots, z_n \in \mathbb{C}$. *Then*

$$\sum_{i,j=1}^{n} \bar{z}_i z_j \min(s_i, s_j) \geqslant 0.$$

Proof. Note that (with $s_0 = 0$)

$$\sum_{i,j=1}^{n} \bar{z}_i z_j \min(s_i, s_j) = \sum_{i=1}^{n} (s_i - s_{i-1}) \left| \sum_{j=1}^{n} z_j \right|^2$$

is obviously positive. ■

This lemma can also be proved by noting that $q(t)$ is really the limit of $n^{-1/2} X_{[nt]}$ as far as the joint probability distributions of the finitely great number of q's are concerned. Notice that

$$\mathbb{E}\{[q(t) - q(s)]^2\} = t - s, \quad (t > s).$$

Moreover, by writting $q(t) = q(s) + [q(t) - q(s)]$ and using the independence of $q(t) - q(s)$ from $\{q(u)\}_{0 \leqslant u \leqslant s}$, we see that

$$\mathbb{E}\{q(t) \mid \{q(u)\}, 0 \leqslant u \leqslant s\} = q(s), \quad (t > s),$$

so that $q(t)$ is a *martingale*.

THEOREM 6.12. *If* $x(t)$ *is a process for* $0 \leqslant t$ *with a continuous version so that for* $0 \leqslant s \leqslant t$

$$\mathbb{E}\{x(t) - x(s) \mid \{x(u)\}, 0 \leqslant u \leqslant s\} = 0,$$

$$\mathbb{E}\{[x(t) - x(s)]^2 \mid \{x(u)\}, 0 \leqslant u \leqslant s\} = t - s.$$

Then $x(t)$ *is the* Brownian motion.

An interesting realization of the Brownian motion comes from the fact that if χ_s is the characteristic function$[0, s]$, then

$$\langle \chi_s, \chi_t \rangle = \min(s, t).$$

Thus, if ϕ is the Gaussian process associated to $L_2(0, \infty)$, then $q(s) = \phi(\chi_s)$ is the Brownian motion. ϕ is often called a *white noise* and the formal relation $dq/ds = \phi(s)$ is often expressed by saying that '*the derivative of Brownian motion is a white noise*'.

Finally, we shall give some construction of the Wiener process in $\mathbb{R}^n$ due to Nelson (1964). The *base* (*sample*) *space of the Wiener process* is taken $\Omega = \Pi_t \dot{\mathbb{R}}^n$, $-\infty < t < \infty$, where $\dot{\mathbb{R}}^n$ denotes a one-point compactification of

$\dot{\mathbb{R}}^n$. With the usual product topology, Ω becomes compact Hausdorff. Let

$$p_t(x, d^n y) = (4\pi \nu t)^{-n/2} \exp\{-|x-y|^2/4\nu t\}\, d^n y,$$

$t > 0$, be the n-dimensional *Gaussian measure* centered around $x \in \mathbb{R}^n$, where ν stands for the diffusion constant.

By the concept of a *cylinder function* on Ω we define a function $f: \Omega \mapsto R$ of the form

$$f(\omega) = F(\omega(t_{-m}), \ldots, \omega(t_{-1}), \omega(t_1), \ldots, \omega(t_m)),$$

for $t_{-m} < \cdots < t_{-1} < 0 < t_1 < \cdots < t_m$. First, we define the *Wiener integral* of a cylinder function f by

$$I_W(f) = \int p_{t_{-m+1}-t_{-m}}(x_{-m+1}, dx_{-m}) \cdots p_{-t_{-1}}(x, dx_{-1}) \times$$

$$\times\, p_{t_1}(x, dx_1) \cdots p_{t_m - t_{m-1}}(x_{m-1}, dx_m) F(x_{-m}, \ldots, x_m). \tag{6.31}$$

For continous F; $\dot{\mathbb{R}}^{2m} \mapsto \mathbb{R}$, there exists integral (6.31) because we have $I_W(1) = 1$).

Let $C(\Omega)$ be the totality of continuous functions on Ω and $C_{cyl}(\Omega)$ that of continuous cylinder functions on Ω (a cylinder function f is continuous if the function F is continuous). $C(\Omega)$ with the usual supremum norm is a Banach space. Secondly, we define the Wiener integral of a continuous function $g \in C(\Omega)$ as follows.

By the Stone–Weierstrass theorem, we can approximate any function in $C(\Omega)$ uniformly by those in $C_{cyl}(\Omega)$. The mapping $I_W: f \mapsto I_W(f)$ defines a bounded linear functional on $C_{cyl}(\Omega)$ of a positive type, i.e.,

$$|I_W(f)| \leqslant \|f\| = \operatorname*{Sup}_{\omega \in \Omega} |f(\omega)|, \quad I_W(f) \geqslant 0$$

if function $f \geqslant 0$. Then, the mapping I_W has a unique extension to a bounded positive linear functional on $C(\Omega)$. The extension should also be denoted over I_W.

The *Wiener integral* of $g \in C(\Omega)$ is defined as $I_W(g)$. By the Riesz–Markov–Kakutani theorem, we find that there exists a regular normalized measure $\mu^x{}_W$ on Ω, indexed by $x \in \dot{\mathbb{R}}^n$, such that

$$I_W(g) = \int_\Omega \mu_W^x(d\omega) g(\omega)$$

holds for $\forall g \in C(\Omega)$. This is the *Wiener measure.*

Adopting the Wiener measure as a probability measure we can define the Wiener process W_t, $-\infty < t < \infty$ in $\dot{\mathbb{R}}^n$ by the triplet $\{\Omega, \Sigma(\mu_W^x), \mu_W^x\}$.

Concerning the domain of the Wiener measure μ_W^x we have the following theorem.

THEOREM 6.13 [*Wiener* (*Nelson*, 1964)]. *Let* ϕ *be the totally of continuous paths* γ *such that* $\gamma(t) \in \mathbb{R}^n$ *for* $-\infty < t < \infty$; *then* $\mu_W^x(\phi) = 1$ *for* x *in* $\mathbb{R}^n$. *Consequently*, $\Sigma(\mu_W^x)$ *is taken as a* σ*-algebra of Borel sets of* ϕ, *i.e.*, $\mathscr{B}(\phi)$.

We conclude this section with the following two theorems.

THEOREM 6.14. *The function of the form*

$$u(t, x) = \int \mu_W^x(d\omega)\, u_0(\omega_t)$$

is a solution to the Cauchy problem *of the following equation*

$$\frac{\partial u}{\partial t} = \nu \operatorname{div} \operatorname{grad} u$$

with the initial condition $u(0, x) = u_0(x)$.

THEOREM 6.15. (*Feynman-Kac*). *A solution to the* Cauchy problem *of the equation*

$$\frac{\partial u}{\partial t} = \nu \operatorname{div} \operatorname{grad} u + U(t, u)u \tag{6.32}$$

$$u(0, x) = u_0(x), \tag{6.33}$$

is given by the Wiener integral

$$u(t, x) = \int \mu_W^x(d\omega)\, u_0(\omega_t) \exp\left\{\int_0^t U(\omega_s, s) \frac{ds}{2\nu}\right\}. \tag{6.34}$$

6.12. Functionals of Stochastic Processes and Stochastic Calculus

Interest in the study of the functionals of stochastic processes has grown in recent years. One of the powerful methods in this field is stochastic integrals and based on them stochastic differential equations. They allow wide classes of random processes to be constructed starting from the simple process – the Brownian motion – and give a wide possibility of applications of the probability methods to stochastic partial differential equations. The methods of stochastic integrals are intensively used in the optimum control theory.

The theory of stochastic differential equations was constructed by K. Itô and I. I. Gihman. N. Wiener and S. N. Bernstein were their predecessors. The main results of the stochastic integrals and stochastic differential equations were expounded in a series of monographs (Doob, 1953; Dynkin, 1965; Gihman and Skorokhod, 1972, 1975; McKean, 1969). The functional integration and its application in quantum physics was considered by Simon (1979); Gel'fand and Yaglom (1956); Glimm and Jaffe (1981). Therefore our consideration of these problems will be very brief.

Here we present an elementary treatment of some calculation aspects of stochastic integrals which may play an important role in the stochastic quantization procedure.

Notice that a stochastic integral may be understood as a functional of the Wiener process. Thus, our interest in the Wiener process $b(t)$ derives from the fact that it is intimately connected with the definition of *stochastic integrals via the diffusion process.* The diffusion process corresponding to the operator $H_0 = \frac{1}{2}\Delta$ (Δ is the Laplacian) was constructed by Wiener (1923) and it is often called the Wiener process. According to Wiener this construction may extent to any elliptical operator L which succeeds in constructing a fundamental solution of the parabolic differential equation

$$\partial u_t/\partial t = L u_t.$$

The simplest probability method for construction of the *diffusion process without truncation* was proposed by Itô. Let X_t be a one-dimensional Wiener process and $\tilde{X}_t$ be the diffusion process with a generating differential operator

$$L = \tfrac{1}{2}\sigma^2(x)\frac{\mathrm{d}^2}{\mathrm{d}x^2} + a(x)\frac{\mathrm{d}}{\mathrm{d}x}.$$

Itô showed that both processes can be realized on the *same space* Ω of elementary events so that $\tilde{\mu}_X = \mu_X$ and the trajectories $\tilde{x}(t)$ and $x(t)$ are connected by the relation

$$\tilde{x}(t) = \tilde{x}_0 + \int_0^t \mathrm{d}s\, a(\tilde{x}(s)) + \int_0^t \mathrm{d}x(s)\sigma(\tilde{x}(s)). \tag{6.35}$$

On the right-hand side of this equation stands the so-called *stochastic integral*, the general definition of which was formulated by Itô.

Equation (6.35) can be written in the differential form:

$$\mathrm{d}\tilde{x}(t) = a(\tilde{x}(t))\,\mathrm{d}t + \sigma(\tilde{x}(s))\,\mathrm{d}x(s). \tag{6.36}$$

The last relation represents a *stochastic analog* of the usual differential equation. The diffusion process theory based on Wiener's and Itô's ideas, was given in many works, particularly in McKean's (1969) monograph. Roughly speaking, the diffusion process theory comprises differential and integral calculi based on the Brownian motion, with the only difference being that for the calculation of a *differential* of a smooth function f depending on the Brownian trajectory $t \to x(t) = b(t)$ two terms of the power expansion should be taken and $\mathrm{d}b^2$ replaced by $\mathrm{d}t$:

$$\mathrm{d}f(b) = f'(b)\,\mathrm{d}b + \tfrac{1}{2}f''(b)\,\mathrm{d}b^2 = f'(b)\,\mathrm{d}b + \tfrac{1}{2}f''(b)\,\mathrm{d}t$$

or

$$\int_0^t \mathrm{d}b f'(b) = f(b)\Big|_0^t - \frac{1}{2}\int_0^t \mathrm{d}t f''(b).$$

This calculus possesses some *new outlines*, for example, the role of exponential is played by $\exp(b - \frac{1}{2}t)$ instead of the usual e^b and the ordinary formal rules of calculus do not hold for the stochastic integral, for example,

$$\int_0^1 db(s) \cdot b(s) = \tfrac{1}{2}[b^2(1) - 1],$$

not $b^2(1)/2$. Note that $b(0) = 0$. The main advantage of this method is that any smooth diffusion $t \to \tilde{x}(t)$ can be considered as a nonanticipatory functional of the Brownian trajectory $x(t)$, so that $\tilde{x}(t)$ is a solution to stochastic differential equation (6.36) with smooth coefficients a and σ.

The main goal of the *stochastic calculus* is to define $\int dbf(s)$ or more generally $\int dbg(b(s))$. This integral cannot be understood as a Stieltjes integral, since the curve $b(t)$ is not rectifiable. Itô's idea plays an important role in the definition of integrals of this type. The basic idea of Itô's definition is

$$\int_0^1 dbf(b(s)) = \lim_{n \to \infty} \sum_{m=1}^{2n} f\left(b\left(\frac{m-1}{2^n}\right)\right)\left[b\left(\frac{m}{2^n}\right) - b\left(\frac{m-1}{2^n}\right)\right]. \qquad (6.37)$$

This integral illustrates the *difference between differentials pointing into the past and the future.* This fact is very important in Nelson's stochastic quantization method. We define

$$I_+(n) = \sum_{m=1}^{2n} b\left(\frac{m}{2^n}\right)\left[b\left(\frac{m}{2^n}\right) - b\left(\frac{m-1}{2^n}\right)\right],$$

$$I_-(n) = \sum_{m=1}^{2n} b\left(\frac{m-1}{2^n}\right)\left[b\left(\frac{m}{2^n}\right) - b\left(\frac{m-1}{2^n}\right)\right].$$

We have that for any n, $I_+(n) + I_-(n) = [b(1)]^2$, since

$$\left[b\left(\frac{m}{2^n}\right) + b\left(\frac{m-1}{2^n}\right)\right]\left[b\left(\frac{m}{2^n}\right) - b\left(\frac{m-1}{2^n}\right)\right] = b^2\left(\frac{m}{2^n}\right) - b^2\left(\frac{m-1}{2^n}\right)$$

and the sum telescopes. On the other hand, by

$$\lim_{N \to \infty} f(b, n, 2) = \lim_{n \to \infty} \sum_{k=1}^{2n} \left|b\left(\frac{k}{2^n}\right) - b\left(\frac{k-1}{2^n}\right)\right|^2 \to 1$$

we have $I_+(n) - I_-(n) \to 1$ pointwise as $n \to \infty$. It follows that

$$\lim_{n \to \infty} I_\pm(n) = \tfrac{1}{2}[b^2(1) \pm 1].$$

This equality shows the difference between I_+ and I_-. From this we see that the ordinary formal rules of calculus do not hold for the stochastic integral:

$$\int_0^1 dbb(s) = \tfrac{1}{2}[b^2(1) - 1], \quad \text{at } b(0) = 0.$$

For suitable f's functions of s alone, Paley *et al.* (1933) succeeded in defining integral:

$$G(f) \equiv \int_0^1 db(s)\, f(s),$$

in the following way. If $f \in \mathbb{C}^1$ and $f(1) = f(0) = 0$, then we define

$$G(f) = -\int_0^1 ds\, f'(s) \cdot b(s),$$

by a formal integration by parts. With this definition

$$\mathbb{E}\{G^2(f)\} = \int_0^1 ds \int_0^1 dt\, f'(s) f'(t) \min(t, s) = \int_0^1 ds |f(s)|^2,$$

since $\partial^2/\partial s\, \partial t[\min(s, t)] = \delta(s - t)$.

Let f be the $\mathbb{C}^1$ function on $\mathbb{R}$ with $f' = df/dt$ and f bounded. Then the limit on the right-hand side of (6.37) exists in L_2-sense and ($\|\cdot\| = L_2$-norm)

$$\left\| \int_0^1 db\, f(b(s)) \right\|_2^2 = \mathbb{E}\left\{ \int_0^1 ds\, f^2(b(s)) \right\}. \tag{6.38}$$

This was proved by Simon (1979). Now we extend the definition of the integral $\int ds f(b(s))$ from $\mathbb{C}^1$-function to an arbitrary function f with

$$\mathbb{E}\left\{ \int_0^1 ds\, f^2(b(s)) \right\} = \int_0^1 ds \left\{ (2\pi s)^{-1/2} \int dx\, f^2(x) \exp\left(\frac{-x^2}{2s} \right) \right\}$$

which is finite; for example f is bounded near zero and L_2 will certainly suffice. This integral is in L_2 for any such f. We can actually go one step further: Let χ_a be a characteristic function of $(-a, a)$ and suppose that

$$\mathbb{E}\left\{ \int_0^1 ds\, \chi_a(b(s))\, f^2(b(s)) \right\} < \infty.$$

Then for each a, we can define $\int_0^1 ds\, f \cdot \chi_a(b(s))$. We claim that for virtually every $b(s)$ there exists an integral

$$\lim_{a \to \infty} \int_0^1 ds\, \chi_a(b(s)) f(b(s)), \tag{6.39}$$

pointwise, and which is finite. For $b(s)$ integral (6.39) is continuous on $[0, 1]$ and so bounded. Thus for any fixed $b(s)$ expression (6.39) is independent of a for $a \to \infty$.

Now we can clearly use the d-*dimensional Brownian motion* and define $\int_0^1 d\mathbf{b}\, f(\mathbf{b}(s))$ and we can easily define $\int_s^t$ in place of $\int_0^1$. It is more important for the applications we have in mind, that we can define $\int_0^1 d\omega\, f(\omega(s))$, for the $d\mu_W$-Wiener (measure) paths. The construction is identical with the one stated above, but now

$$\mathbb{E}\left\{\int_0^1 \mathrm{d}s|f(\omega(s))|^2\right\} = \int \mathrm{d}^d x|f(\mathbf{x})|^2,$$

where $\omega(s)$ is the Wiener (process) path. This equality follows from the fact that $\omega(s)$ has the distribution of the Lebesque measure. This (random variable) integral is called the *Itô integral.*

ITÔ'S LEMMA. *Let* $\mathrm{f}(\mathbf{x}, \mathrm{t})$ *be a function on* $\mathbb{R}^d \otimes [0, \mathrm{t}]$ *and use* $\nabla \mathrm{f}$, $\nabla^2 \mathrm{f}, \ldots$ *for* $\mathbf{x}$ *derivative and* $\dot{\mathrm{f}}$ *for* t *derivatives. Then*

$$f(\omega(t), t) - f(\omega(0), 0) = \int_0^t \mathrm{d}\omega \cdot (\nabla f)(\omega(s), s) +$$

$$+ \frac{1}{2}\int_0^t \mathrm{d}s(\nabla^2 f)(\omega(s), s) + \int_0^t \mathrm{d}s \dot{f}(\omega(s), s) \tag{6.40}$$

for almost every ω.

Proof was shown by Simon (1979). This expression is often written in an *infinitesimal form* as the formal one

$$\mathrm{d}f = \nabla f \cdot \mathrm{d}\omega + (\tfrac{1}{2}\Delta f + \dot{f})\mathrm{d}t, \quad (\nabla^2 \equiv \Delta). \tag{6.41}$$

It is surprising that the second derivative of the function f appear in the first-order differential. Equation (6.41) can be expressed even more succinctly by $\mathrm{d}\omega_i \, \mathrm{d}\omega_j = \delta_{ij} \, \mathrm{d}t$. Expression (6.40) can be used for calculational purposes:

EXAMPLE 6.6 ($d = 1$). f's with $\frac{1}{2}f'' + \dot{f} = 0$ will have particularly simple stochastic integrals. For example, if

$$f(x, t) = \exp(\alpha x - \tfrac{1}{2}\alpha^2 t),$$

then $\mathrm{d}f = \nabla f \cdot \mathrm{d}\omega$. Recognizing the *Wick-ordered exponential*

$$f(\omega(s), s) = :\exp(\alpha\omega(s)):$$

and taking ω to b, we see that

$$\int_0^t \mathrm{d}b : \exp(\alpha b(s)): = \alpha^{-1} : \exp \alpha b(t): - \alpha^{-1}. \tag{6.42}$$

If (6.42) is expanded in α one finds that

$$\int_0^t \mathrm{d}b : b^n(s): = (n + 1)^{-1} : b^{n+1}(t): .$$

If we take $n = 1$ we find that

$$\int_0^t \mathrm{d}b \, b(s) = \tfrac{1}{2} : b^2(t): = \tfrac{1}{2}[b^2(t) - t]$$

a result already obtained. Itô's lemma directly provides the connection of the *Brownian motion* and $-\frac{1}{2}\Delta = H_0$ and the *Feynman–Kac formula.* A key observation is that because the differentials point into the future, we have

$$\mathbb{E}\{f(\mathbf{b}(s))\,\mathrm{d}\mathbf{b}\} = \mathbb{E}\{f(\mathbf{b}(s))\}\mathbb{E}\{d\mathbf{b}\} = 0$$

since $\mathbb{E}\{\mathrm{d}\mathbf{b}\} = 0$; i.e.,

$$\mathbb{E}\left\{\int_a^b \mathrm{d}\mathbf{b}\, f(\mathbf{b}(s))\right\} = 0,$$

more strongly

$$\mathbb{E}\left\{g\int_a^b \mathrm{d}\mathbf{b}\, f(\mathbf{b}(s))\right\} = 0 \tag{6.43}$$

if g is a measurable function with respect to $\{\mathbf{b}(u) \mid u \leqslant a\}$. Choose

$$f(\mathbf{x}, s) = \{\exp[-(t-s)H_0], g\}(\mathbf{x}),$$

so $\dot{f} = H_0 f$. Thus $\dot{f} + \frac{1}{2}\Delta f = 0$, so that Itô's lemma implies that

$$f(\mathbf{b}(t), t) = f(\mathbf{b}(0), 0) + \int_0^t \mathrm{d}\mathbf{b}\cdot(\nabla f)(\mathbf{b}(s), s).$$

Using (6.43), we see that

$$\mathbb{E}\{f(\mathbf{b}(t), t)\} = \mathbb{E}\{f(\mathbf{b}(0), 0)\}.$$

Since $f(\mathbf{b}(t), t) = g(\mathbf{b}(t))$ and $f(\mathbf{b}(0), 0) = (\exp(-tH_0), g)(\mathbf{0})$, we have proved once again that

$$\mathbb{E}\{g(\mathbf{b}(t))\} = (exp(-tH_0), g)(\mathbf{0}).$$

The proof of the *Feynman–Kac formula* is similar but more complicated (see Simon, 1979). Here, we take $H = H_0 + U$ and

$$f(\mathbf{x}, s) = (\exp[-(t-s)H], g)(\mathbf{x}).$$

Then, at least for the sufficiently well-defined U's (say $U \in \mathbb{C}_0^\infty$), we have

$$f(\mathbf{b}(s), s) = f(\mathbf{b}(t), t) - \int_s^t \mathrm{d}u\, U(\mathbf{b}(u)) f(\mathbf{b}(u), u) - \int_s^t \mathrm{d}\mathbf{b}\cdot(\nabla f)(\mathbf{b}(u), u).$$

Notice that according to (6.43) the last term has a zero expectation. Reiterating this equation, we obtain

$$f(\mathbf{b}(0), 0) = f(\mathbf{b}(t), t) - \int_0^t \mathrm{d}u\, U(\mathbf{b}(u)) f(\mathbf{b}(t), t) +$$

$$+ \int_0^t \mathrm{d}u\, U(\mathbf{b}(u)) \int_u^t \mathrm{d}s\, U(\mathbf{b}(s)) f(\mathbf{b}(s), s) + G,$$

where G also has a zero expectation due to (6.43). Continuing in this way, one finds that

$$\mathbb{E}\{f(\mathbf{b}(0), 0)\} = \mathbb{E}\left\{\left[\sum_{n=0}^{\infty} (-1)^n \int_{0 \leqslant s_1 \leqslant \cdots < s_n < t} \prod \mathrm{d}s_i\, U(\mathbf{b}(s_i))\right] f(\mathbf{b}(t), t)\right\}.$$

If one recognizes the object in $[\cdots]$ as $\exp\{-\int_0^t \mathrm{d}s\, U(\mathbf{b}(s))\}$, this will prove that

$$(\exp(-tH), g)(\mathbf{0}) = \mathbb{E}\left\{\exp\left[-\int_0^t \mathrm{d}s\, U(\mathbf{b}(s))\right] g(\mathbf{b}(s))\right\}. \tag{6.44}$$

Now we turn to stochastic integrals. A *stochastic integral* is a random function $c(\mathbf{b}, t)$ obeying

$$c(\mathbf{b}, t) = c(\mathbf{b}, 0) + \int_0^t \mathrm{d}\mathbf{b} \cdot f(\mathbf{b}, s) + \int_0^t \mathrm{d}s\, g(\mathbf{b}, s), \tag{6.45}$$

where f and g are nonanticipatory functionals (i.e., functions $f(\mathbf{b}, s)$ with the property that $f(\cdot, s)$ is only a function of $\{\mathbf{b}(t) \mid t \leqslant s\}$, i.e., is $\mathscr{B}_s$-measurable) with suitable L_p properties and $c(\mathbf{b}, 0)$ is independent of $\mathbf{b}$. One writes (6.45) in the short form

$$\mathrm{d}c = \mathbf{f} \cdot \mathrm{d}\mathbf{b} + g\, \mathrm{d}s. \tag{6.46}$$

The following result is crucial:

THEOREM 6.16 (*Itô's lemma*). *If* $c_1, \ldots, c_m$ *are stochastic integrals and* u *is a* $\mathbb{C}^2$*-function on* $\mathbb{R}^m \otimes [0, \infty)$ *with some mild restrictions imposed on growth at infinity, then* $x = u(c_1, \ldots, c_m, t)$ *is a stochastic integral, and*

$$\mathrm{d}x = \sum_{i=1}^{m} \frac{\partial u}{\partial y_i} \mathrm{d}c_i + \frac{\partial u}{\partial t} \mathrm{d}s + \frac{1}{2} \sum_{i,j=1}^{m} \frac{\partial^2 u}{\partial y_i\, \partial y_j} \mathrm{d}c_i\, \mathrm{d}c_j \tag{6.47}$$

where

$$\mathrm{d}c_i\, \mathrm{d}c_j = \mathbf{f}_i \mathbf{f}_j\, \mathrm{d}s; \qquad \mathrm{d}c_i = \mathbf{f}_i \cdot \mathrm{d}\mathbf{b} + g_i\, \mathrm{d}s, \quad \text{i.e., } \mathrm{d}b_k\, \mathrm{d}b_i = \delta_{ki}\, \mathrm{d}s.$$

For the proof of this theorem, see McKean (1969).

Now we consider some consequences of this theorem.

EXAMPLE 6.7 (*Product rule*). If $\mathrm{d}c_i = \mathbf{f}_i \cdot \mathrm{d}\mathbf{b} + g_i \mathrm{d}s$, then

$$d(c_1 c_2) = c_1\, \mathrm{d}c_2 + c_2\, \mathrm{d}c_1 + \mathbf{f}_1 \mathbf{f}_2\, \mathrm{d}s. \tag{6.48}$$

EXAMPLE 6.8 (*Feynman–Kac formula*). Let us compute the differential of

$$c = \exp\left\{-\int_0^t \mathrm{d}s\, U(\mathbf{b}(s))\right\} f(\mathbf{b}(t)),$$

where $U, f \in \mathbb{C}_0^\infty$. Since the function $z = \int_0^t \mathrm{d}s\, U(\mathbf{b}(s))$ is a stochastic integral with $\mathrm{d}z = U\, \mathrm{d}s$, we have according to Itô's lemma and $(ds)^2 = 0$ that

$\mathrm{d}(\mathrm{e}^{-z}) = -\mathrm{e}^{-z}\, U\, \mathrm{d}s$. Again by Ito's lemma,

$$\mathrm{d}f = \nabla f \cdot \mathrm{d}\mathbf{b} + \tfrac{1}{2}(\Delta f)\mathrm{d}s.$$

Thus, using (6.48), we obtain

$$\mathrm{d}c = \exp\left\{-\int_0^t \mathrm{d}s\, U(\mathbf{b}(s))\right\}[\nabla f \cdot \mathrm{d}\mathbf{b} - (Hf)\,\mathrm{d}s],$$

where $H = -\frac{1}{2}\Delta + U$. Thus, if we define

$$(Q_t f)(\mathbf{x}) = \mathbb{E}\left\{\exp\left[-\int_0^t \mathrm{d}s\, U(\mathbf{x} + \mathbf{b}(s))\right] f(\mathbf{x} + \mathbf{b}(s))\right\},$$

and we use (6.43) that $\mathbb{E}\{\mathrm{d}\mathbf{b}\} = 0$, we see that

$$(Q_t f)(\mathbf{x}) = f(\mathbf{x}) - \int_0^t \mathrm{d}s[Q_s(Hf)](\mathbf{x}),$$

which yields $Q_t = \exp(-tH)$ and so we have a new proof of the *Feynman–Kac formula.*

EXAMPLE 6.9 (*Feynman–Kac–Itô formula*). Let us compute the differential of

$$c = \exp\left\{-i\int_0^t \mathrm{d}\mathbf{b} \cdot \mathbf{a}(\mathbf{b}(s)) - \tfrac{1}{2}i\int_0^t \mathrm{d}s(\operatorname{div}\mathbf{a})(\mathbf{b}(s))\right\} f(\mathbf{b}(t)) = \mathrm{e}^{-iz} f(\mathbf{b}(t))$$

with $\mathbf{a}, f \in C_0^\infty$. Then, according to Itô's lemma, we have

$$\mathrm{d}(\mathrm{e}^{-iz}) = -i\,\mathrm{e}^{-iz}\,\mathrm{d}z - \tfrac{1}{2}\mathrm{e}^{-iz}(\mathrm{d}z)^2$$

$$= -i\,\mathrm{e}^{-iz}\{\mathbf{a}\cdot\mathrm{d}\mathbf{b} + \tfrac{1}{2}(\nabla\cdot\mathbf{a})\mathrm{d}s\} - \tfrac{1}{2}\mathrm{e}^{-iz}\,\mathbf{a}^2\,\mathrm{d}s.$$

Further, we use (6.48)

$$\mathrm{d}c = \mathrm{e}^{-iz}\{(\cdots)\mathrm{d}\mathbf{b} + [\tfrac{1}{2}\Delta f - i\mathbf{a}\cdot\nabla f - \tfrac{1}{2}(\nabla\cdot\mathbf{a})f - \tfrac{1}{2}\mathbf{a}^2 f]\mathrm{d}s\}.$$

So taking into account

$$H_0(\mathbf{a})f = \tfrac{1}{2}(-i\nabla - \mathbf{a})^2 f$$

and writing

$$(Q_t f)(\mathbf{x}) = \mathbb{E}\left\{\exp\left[-i\int_0^t \mathrm{d}\mathbf{b}\cdot\mathbf{a} + \tfrac{1}{2}i\int_0^t \mathrm{d}s\,\operatorname{div}\mathbf{a}\right] f(\mathbf{x} + \mathbf{b}(s))\right\},$$

we find that

$$(Q_t f)(\mathbf{x}) = f(\mathbf{x}) - \int_0^t \mathrm{d}s[Q_s(H_0(\mathbf{a})f)](\mathbf{x}) \tag{6.49}$$

which yields the Feynman–Kac–Itô formula.

EXAMPLE 6.10 (*Drift*). Let $a \in C_0^\infty$ and

$$\mathbf{x}(t) = \mathbf{b}(t) + \int_0^t \mathrm{d}s\, \mathbf{a}(\mathbf{b}(s))$$

which for obvious reasons is often called the *drift* generated by a. Then $\mathrm{d}\mathbf{x} = \mathrm{d}\mathbf{b} + \mathbf{a}\,\mathrm{d}s$ so that

$$\mathrm{d}[f(\mathbf{x}(t))] = \nabla f \cdot \mathrm{d}\mathbf{b} + (\mathbf{a} \cdot \nabla f + \tfrac{1}{2}\Delta f)\,\mathrm{d}s.$$

Following our usual procedure

$$\mathbb{E}\{f(\mathbf{x}(t))\} = \exp(-tX), f)(0),$$

where X is the differential operator $Xf = -\frac{1}{2}\Delta f - \mathbf{a} \cdot \nabla f$ which is thus called the *generator* of the drift.

EXAMPLE 6.11 (*Cameron–Martin's* (1949) *formula*). Let $\mathrm{d}z = \mathbf{a}\cdot\mathrm{d}\mathbf{b} - -\frac{1}{2}\mathbf{a}^2\,\mathrm{d}s$, and $c = \mathrm{e}^z f(\mathbf{b}(t))$. Then

$$\mathrm{d}c = \mathrm{e}^z\{[\nabla f + \mathbf{a}f]\cdot\mathrm{d}\mathbf{b} + [\mathbf{a}\cdot\nabla f + \tfrac{1}{2}\Delta f]\,\mathrm{d}s\},$$

so in terms of the generator X in Example 6.10, we obtain

$$\mathbb{E}\{\mathrm{e}^z f(\mathbf{b}(t))\} = (\exp(-tX), f)(0).$$

This is usually summarized in the Cameron–Martin formula

$$\frac{\mathrm{D}x}{\mathrm{D}b}\left(\begin{matrix}\textit{Radon–Nikodym}\\ \textit{derivative}\end{matrix}\right) = \exp\left\{\int_0^t \mathrm{d}\mathbf{b}\cdot\mathbf{a}(\mathbf{b}(s)) - \tfrac{1}{2}\int_0^t \mathrm{d}s\mathbf{a}^2(\mathbf{b}(s))\right\} \tag{6.50}$$

If one writes

$$H_0(\mathbf{a}) = -\tfrac{1}{2}\Delta + i\mathbf{a}\cdot\nabla + \tfrac{1}{2}\mathbf{a}^2 + \tfrac{1}{2}i\,\mathrm{div}\,\mathbf{a},$$

then one sees that $H_0(\mathbf{a})$ is the generator of a (complex) drift $-i\mathbf{a}$ and a potential $\frac{1}{2}\mathbf{a}^2 + \frac{1}{2}i\cdot\mathrm{div}\,\mathbf{a}$. Then the combined Cameron–Martin formula for $\mathbf{a}\cdot\nabla$ and the Feynman–Kac one for $\frac{1}{2}\mathbf{a}^2 + i\cdot\mathrm{div}\,\mathbf{a}$ yields the Feynman–Kac–Itô formula. There is a connection between the Cameron–Martin formula and the Feynman–Kac formula (see Simon, 1979).

REMARK 6.1. The F. Riesz–A. Markov–S. Kakutani, and M. Stone–C. Weierstrass theorems listed in Section 6.11 are the fundamental theorems in topological spaces and measures. For these theorems, the reader is referred to standard text books on given fields, for example, Dunford and Schwartz (1958), and Alexandrov and Hopf (1935).

Chapter 7

Basic Ideas of Stochastic Quantization

7.1. Introduction

In recent years interest has significantly increased in the investigation of stochastic processes and fields. This is due in the first place to the fact that it has been possible to establish an intimate connection between the theory of stochastic processes, quantum mechanics (Fényes, 1952; Kershaw, 1964; Nelson, 1966, 1967; de la Peña-Auerbach, 1969; de la Peña-Auerbach and Cetto, 1974, 1975, 1982; Davidson, 1979; Moore, 1979; Yasue, 1979; Lee, 1980; Santamato and Lavenda, 1981; Claverie and Diner, 1973, 1975, and others), and Euclidean quantum field theory (Nelson, 1973; Dohrn and Guerra, 1978; Guerra and Ruggiero, 1973), known under the general name of the stochastic quantization of systems (stochastic mechanics) stochastic quantum mechanics. Studies have been made on the generalization of the ideas of the stochastic quantization of Nelson and Fényes for continuous systems and fields (i.e., for systems with an infinite number of degrees of freedom) (Guerra and Ruggiero, 1973; Dohrn and Guerra, 1978; Dankel, 1970, 1977; Moore, 1980 and Davidson, 1980–82), and also for particles with spin (Dankel, 1970, 1977; de la Peña-Auerbach, 1971) and relativistic mechanics (Caubet, 1976; Aron, 1966; Lehr and Park, 1977; Guerra and Ruggiero, 1978; Vigier, 1979 and Namsrai, 1980a).

From the mathematical point of view the most interesting of the results obtained is that the dynamics equations of stochastic mechanics are *nonlinear partial differential equations* which admit linearization and that the obtained linear equations are formally identical with the *Schrödinger equation* if the *diffusion coefficient* ν is taken as equal to $\hbar/2m$.

This common character of the mathematical formalism of the two theories suggests the existence of a deep connection between the theory of stochastic processes and quantum mechanics. This problem is not solved completely and requires further detailed investigation (in this connection see Ghirardi *et al.*, 1978; Onofri, 1979; Grabert *et al.*, 1979; Lavenda, 1980; Aron, 1981, and Mielnik and Tengstrand, 1980). With regard to the problem of the connection between the stochastic (Markov) processes and the Euclidean quantum field theory, Nelson (1971, 1973) has finally formulated the Euclidean quantum field theory in terms of random processes, and we now have the solution to the problem of the unique correspondence between the *Euclidean and pseudo-Euclidean Green's functions* (Osterwalder and Schrader, 1973, 1975 and Glaser, 1974), the study of which was already begun by Schwinger (1959); Nakano (1959), Fradkin (1954), Symanzik (1966), Taylor (1966), and Efimov (1963).

Besides this approach, other directions are also being developed in the study of stochastic processes and fields (see Chapter 10 and reviews of Skagerstam, 1975; Moore, 1979; Boyer, 1975; Surdin, 1971, 1978; Blokhintsev, 1975; de la Peña-Auerbach and Cetto, 1982). Some of them take as their starting point the *hypothesis of the stochastic properties of the electromagnetic vacuum* (Braffort and Tzara, 1954; Bourrett, 1964; Braffort *et al.*, 1965; Marshall, 1963, 1965; Santos, 1974; Rueda, 1981, and Cavalleri, 1981). This approach is called *stochastic electrodynamics.* The reviews (Boyer, 1975; Surdin, 1971, 1978; Sandos, 1974; Claverie and Diner, 1976, 1978; Rueda, 1978, and de la Peña-Auerbach and Cetto, 1978, 1982) with most of the relevant references in the field are included and present the basic elements and ideas of stochastic electrodynamics and its connection with quantum theory. Stochastic electrodynamics is constructed as classical electrodynamics with radiation damping of charged particles and their interaction with a background electromagnetic (vacuum) field with *spectral density*

$$\rho(\omega) = (\hbar/2\pi^2 c^3)\omega^3.$$

Other directions are based (to some extent or after a fashion) on the concept of *stochastic space-time* (see Chapter 1 and references therein). In this case, it is assumed that the origin of the *random behavior* of particles (like *Brownian motion*) is the *stochastic nature of physical space-time.* In other words, the stochastic nature of the processes is regarded as the result of the effect (or influence) of space-time itself on the physical system. Our approach presented in this part of the book is concerned with this direction in the theory of stochastic processes and the main attention is devoted to a generalization of Nelson's stochastic mechanics to the relativistic case.

This brief list shows that the theory provides the most attractive formulation of quantum mechanics and relativistic quantum field theory by means of a stochastic context, which still offers interacting possibilities for development and application to the inquiring nonorthodox mind.

Roughly speaking, these directions arose from introducing the de Broglie–Bohm–Vigier *hidden variables* in quantum mechanics, related to an attempt to give the *physical interpretation of quantum phenomena at the subquantum level,* or, at a deeper level, to the *very nature of the quantum system,* which sometimes seems to force the renunciation of such basic concepts as *objectivity* or *causality.* We shall not dwell on these questions which are widely discussed in the literature (see Bunge, 1956; Jammer, 1974; Ballentine, 1970; de la Peña-Auerbach and Cetto, 1977; and Vigier, 1982).

7.2. The Hypothesis of Space-Time Stochasticity as the Origin of Stochasticity in Physics

Before turning to a description of the methods of stochastic mechanics and its relation to quantum mechanics, we discuss briefly the idea of stochasticity in quantum mechanics and its *possible origin* (or physical cause).

The idea that the electron moves *stochastically* is almost as old as quantum theory itself (for historical discussion of this problem see Jammer, 1974). Schrödinger (1931) and Fürth (1933) first considered the formal analogy between *Schrödinger's equation* and the *diffusion equation*. The proposition remained a purely formal one, until Fényes (1952), Kershaw (1964), and Nelson (1966, 1967) took it seriously and proposed to interpret quantum mechanics as a *Markov process in configuration space*, i.e., that the quantum mechanical system is assumed to execute a sort of *Brownian movement* which is fundamentally a quantum theory of the same structure conceptually as the Schrödinger theory of the electron. This is the basic view of the first school mentioned above. As a necessary *ingredient* of quantum mechanics, stochasticity has also been considered by de Broglie (1967).

Accepting that quantum mechanics describes the behavior of matter only statistically, the essence of the so-called *statistical interpretation* of quantum mechanics (see, for example, Jammer, 1974; de la Peña-Auerbach and Cetto, 1977; Ballentine, 1970; Rosen, 1974, and Ross-Bonney, 1975), strongly defended by Einstein, Podolsky and Rosen (1935), consists in the assumption that the description refers to an *ensemble* of similarly prepared systems, not to an *individual* member of it, and that each one of these systems behaves stochastically, thus contributing to unavoidable dispersions in all the dynamical variables that are not uniquely fixed by the preparation process.

It deals with the meaning of statistics as well as the question of possible limits on scientific knowledge. As is known, Einstein, de Broglie, and their present followers never accepted Bohr's assumption that quantum probabilities represented an *ultimate limit to human knowledge*. In this sense, they have always held that probability distributions should result from the *real action of randomly correlated subquantum causal motions* – a conception which implies the introduction of subquantal hidden variables to interpret the probabilities correctly described by field (and field equations) of quantum mechanics. This view is the basis of Vigier's statement: 'New stochastic variables which imply a superimposed random set of motions (reminiscent of Feynman's paths) resulting from the particle's interactions with a real 'vacuum'; a vacuum now considered as a deeper (of course, covariant) subquantum distribution of violent motions and thus alive with fluctuations and randomness...' (Vigier, 1982b, p. 924).

By leaving aside the philosophical aspect of the problem, we assume that if the stochastic motion of particles is indeed responsible for the statistical nature of quantum mechanics, then it should be possible to employ this fact as a *cornerstone* for the construction of a more explicit – and perhaps more fundamental – theory of quantum systems. The above-mentioned directions along these lines have been developed with different degrees of success. Among them, stochastic mechanics preserves the idea of random trajectories for the configurational degrees of freedom of the mechanical system, but interprets these *random trajectories* as effective trajectories of *well-defined stochastic processes*, satisfying stochastic differential equations well-known in statistical mechanics and control theory. The evolution of the random processes is ruled by a stochastic version of the Newton second principle of dynamics proposed by Nelson (1966, 1967) (see Section 7.4 and Chapter 8). The link with quantum mechanics is provided by the proof that the distribution of the relevant physical observables at each time is identical in stochastic mechanics and quantum mechanics, as a consequence of the respective time evolution equations.

Obviously, the next step in this direction is to search for the *physical cause* (or the *origin*) of *stochasticity*. Thus the question: what are 'carriers' (or what is the physical cause) of stochasticity, is of great importance in the stochastic interpretation of quantum mechanics. Without answering this question the theory would remain for ever a phenomenological and unsatisfactory one, at least philosophically, as is the theory with the hidden variables.

We would like to emphasize from the beginning that the stochastic theory is not a *hidden variables theory* and should not be considered as such. Moreover, one would like to have to hand a more fundamental stochastic theory of matter, from which quantum mechanics would emerge under certain circumstances and approximations. If such a theory could be constructed, it would not only provide us with a deeper understanding of matter, but also broaden the perspectives of the quantum theory itself.

As listed above, various independent efforts in this direction have contributed to the groundwork of a theory of this kind: stochastic mechanics, stochastic electrodynamics, the stochastic theory based on the hypothesis of space-time stochasticity.

In stochastic electrodynamics the basic question concerning the *origin of stochasticity is answered by assigning a real character* to the *random zero-point radiation field or vacuum field*, which in usual quantum electrodynamics is treated as a virtual – i.e., formal – field. This idea is already discussed in several early papers, notably those by Welton (1948) and Weisskopf (1949) (for more details, see de la Peña-Auerbach and Cetto, 1982 and references therein).

With regard to the first school, i.e., stochastic mechanics, unfortunately it cannot explain what causes the stochastic nature of the microscopic world. In this approach either the problem was not discussed or no clear advantage of it was obtained. There are differing points of view. For example, Bohm (1952) interpreted the deviation from the Newtonian equations of motion as due to a quantum-mechanical potential associated with the wave function. Fényes (1952) showed that instead of assuming a quantum-mechanical potential the *motion* could be understood in terms of a *Markov process.* Weizel (1953) also proposed a model for the random aspects of the motion in terms of the interaction with hypothetical particles, which he calls zerons. Nelson's (1966, 1967) aim has been to show how close it is to the classical theories of Brownian motion and Newtonian mechanics, and how the Schrödinger equation might have been discovered from this point of view. He examined the hypothesis that particles in empty space, say aether, are subject to Brownian motion. De la Peña-Auerbach and Cetto (1975) dealt with particles owing their stochastic behavior to the action of a stochastic surrounding medium acting in addition to the external (applied) forces.

An interpretation of the Schrödinger equation in terms of particle trajectories was first proposed by de Broglie. Bohm and Vigier (1954) introduced the concept of random fluctuations arising from the interaction with a sub-quantum medium. Recently, this idea has developed at a deeper level and obtained a new formulation by Vigier and his team (see Vigier, 1982a, b; Cufaro Petroni and Vigier, 1983b).

Another possibility concerning the third of the above-listed approaches remains. The idea to use the *stochastic space-time* or *random fluctuations* of the *metric* $g_{\mu\nu}$ as the *origin of stochasticity* (i.e., the real quantum forces) is a most attractive and interesting conjecture considered first by Einstein (1924) (see Chapter 1 and references therein). We propose that the origin of stochasticity in the microworld is caused by the effects due to space-time, i.e., gravitation (exactly, '*gravitational vacuum*') on material systems (particles). There exists a standard technique for considering these effects when spinor fields may be presented (Weinberg, 1972). In particular, these effects give rise to the *cosmological term* $\Lambda g_{\mu\nu}$ in the Einstein equation:

$$R_{\mu\nu} - \tfrac{1}{2}g_{\mu\nu}R + \Lambda g_{\mu\nu} = \kappa T_{\mu\nu}, \tag{7.1}$$

as a representation of the energy momentum tensor distribution of the random background gravitational aether or noise, i.e., the 'gravitational vacuum'. Here $\kappa = 8\pi G/c^4$, G is the *Newtonian constant.*

We discuss the appearance of the cosmological term in (7.1). It is proved that gravitation of homogenuous substances situated outside some sphere and extending to infinity does not contribute to the force inside this sphere, and thereby its action tended to zero. Thus attraction to the center of the sphere of Galaxies situated on it is only caused by substances of the sphere and acceleration of these Galaxies with respect to the sphere center is $a_g = -GM/R^2$. Let us give the mass of the sphere through the mean density of matter ρ: $M = (4/3)\pi R^3\rho$. Then

$$a_g = -(4/3)\pi G R\rho. \tag{7.2}$$

Thus, even if at some moment Galaxies are stationary, at the next moment they will begin to move with acceleration towards each other. As a result, the Universe will compress and become unstable.

In order to balance the gravitational forces and to make the Universe stationary it is necessary to introduce a *repulsive force* independent of substance. Gravitational and repulsive accelerations must be equal and opposite to each other with respect to modulus and sign: $a_g = -a_r$.

On the basis of such consideration Einstein introduced the cosmological repulsive force which makes the Universe stationary. This force is universal: it is not dependent on the masses of the bodies involved, but depends only on the distance between them. Taking into account the fact that the mean value of the matter density in the Universe is $\rho \approx 2 \times 10^{-29}\,\mathrm{g\,cm^{-3}}$ one can calculate the numerical quantity of the repulsive acceleration which is $a_r \approx 5 \times 10^{-37} R\,\mathrm{cm\,sec^{-2}}$, or

$$\Lambda \approx 5 \times 10^{-37}\,\mathrm{sec}^{-2} c^{-2} \approx 5 \times 10^{-58}\,\mathrm{cm}^{-2},$$

where $c = 3 \times 10^{10}\,\mathrm{cm\,sec^{-1}}$.

Now we show that according to our idea of the space-time stochasticity presented in the concrete form (1.1) in Chapter 1, a test particle acquires an additional energy caused by the gravitational vacuum. As shown above (Chapter 1), our proposal about the stochasticity of space-time has led to the change of the space-time metric form (1.4) and (1.12) in the cases of flat and curve space-time, respectively. On the basis of these formulas in Chapter 11 we will obtain the following expression

$$E_l^0 = m(1 + \tfrac{1}{2}am^2l^2) \tag{7.3}$$

for the rest energy of the test particle [see formula (11.54)], in the units of $c = \hbar = 1$. Thus we see that the Einstein formula $E = mc^2$ is changed here. Assuming $l \to l_{\mathrm{gr}} = G^{1/2}$ [the case $l \gg l_{\mathrm{gr}}$ formally corresponds to some strong gravitation near particles which are regarded as 'small black holes' (see Salam and Strathdee, 1976, and Li, 1982)], in the second term of (7.3) we get the gravitational interaction energy of the particles:

$$E_{\mathrm{gr}} = \tfrac{1}{2}am^3G = \tfrac{1}{2}am^2G/\lambda,$$

where $\lambda = \hbar/mc$ is the *Compton wavelength* of the test particle with mass m and a as some constant. In order to obtain the energy density we divide E_{gr} by the one-particle volume $\sim\lambda^3$:

$$\varepsilon_{\mathrm{vac}} = \tfrac{1}{2}a(Gm^2/\lambda) : \lambda^3 = \tfrac{1}{2}aGm^6c^4\hbar^{-4}. \tag{7.4}$$

This formula coincides with the *vacuum energy density* proposed by the Soviet physicist Zel'dovich. His argument for the occurence of non-zero vacuum energy density may be presented in the following way: Let virtual particles with mass m be created and annihilated in a vacuum. According to the quantum theory, the mass of the particle defines the characteristic distance $\lambda = \hbar/mc$. Virtual particles are generated from each other at this mean characteristic distance. Then, by a usual formula the gravitational interaction energy of surrounding particles is given by $E_{\mathrm{gr}} = Gm^2/\lambda$. This energy raises the non-zero vacuum energy density or mass density: $\rho_{vac} = \varepsilon_{vac}/c^2$. Hence

$$\varepsilon_{vac} = (Gm^2/\lambda)/\lambda^3 = Gm^6c^4\hbar^{-4}, \quad (\lambda = \hbar/mc).$$

We see that this formula differs from (7.4) by the constant $a/2$.

Moreover, it transpires that the theory requires that a 'vacuum fluid' should possess an unusual negative pressure, according to our consideration above. Therefore, it is better to speak about tension not pressure. Thus we assume $P = -\varepsilon$ or, using Einstein's relation $P = -\rho c^2$, we get $P/c^2 = -\rho$. This last equality is very important. Namely, it leads, as shown below, to the *gravitation repulsion* of the vacuum.

We recall that the mass of a uniform ball with the radius R is equal to $M = (4/3)\pi R^3 \rho$. But, in the Einstein theory it holds only in the case when the pressure of matter is very small with respect to the value ρc^2. If it is amiss, then $M = (4/3)\pi R^3(\rho + 3P/c^2)$. Usually ρ is much larger than P/c^2 and the term $3P/c^2$ may be omitted. However, in the case of the gravitational interaction of virtual particles, the pressure P and energy density are related: $\rho = P/c^2$. The 'vacuum fluid' fills the whole space naturally. Therefore, for calculation of the gravitation acceleration caused by the vacuum, one can use formula (7.2) obtained for the usual matter, but ρ should be replaced by the sum $(\rho + 3P/c^2)$. Taking into account that $\rho_{vac} = -P/c^2$, we get

$$a_{vac} = -\tfrac{4}{3}\pi G(\rho_{vac} + 3P/c^2)R = -\tfrac{4}{3}\pi G(\rho_{vac} - 3\rho_{vac})R = \tfrac{4}{3}\pi G \cdot 2\rho_{vac} R. \tag{7.5}$$

Thus, for the gravitational vacuum we do not obtain attraction as in the case of usual matter [minus sign in (7.2)], but repulsion [plus sign in (7.5)]. This repulsion is caused by negative vacuum pressure which should be taken into account in calculations of the gravitational force, according to the Einstein theory.

With regard to the numerical quantity of the gravitational repulsion of the vacuum, the theory is too crude to yield any precise value.

It should be noted that properties of space-time and physical vacuum including the gravitational vacuum, and fluctuations and instabilities in them have recently acquired the most important significance in contemporary physics. Thus, for example, a vacuum is properly determined by *evolutionary*, not merely energetic, considerations. This point is relevant for all proposed models (especially in the *Yang–Mills gauge theory*) of elementary particle interactions. The concept of the *metastable* and *false vacuum*, and gravitational vacuum (Turner and Wilczek, 1982; Guth, 1981, 1982; Linde, 1982; Kirzhnits and Linde, 1972), properties of vacuum as a physical medium (Lee, 1981), and fluctuations in the Yang–Mills vacuum (Migdal, 1981) assist in understanding very important physical problems:

- phase transitions of *grand unified theories*;
- very early evolution of our Universe described by the *inflationary model* which can solve the monopole, horizon, and flatness problems of the standard hot *Big-Bang cosmology*;
- *quark confinement*, and so on.

It is important to remark that interrelation between particles and space-time or exactly a profound connection between the particle-wave dualism and the structure of space-time has developed further (see Fujiwara, 1980 and Efinger, 1981). According to Efinger, the idea is that the particles in question are *geometrical objects* on a *Riemannian space-time*, i.e., one can take the view that freely moving particles are represented by *solitary waves* which, by definition, preserve their shape. Then one could hypothesize that associated with these waves is a *Riemannian metric* $g_{\mu\nu}$ which is nearly singular (or has some difference from the Minkowski metric) in a region corresponding to the width of the solitary wave-amplitude (or a region of confinement of these particles – an idea that recalls an old concept by Einstein, 1967; for detail see Chapter 11). Outside this region one expects $g_{\mu\nu} \approx \eta_{\mu\nu} = \text{diag}(1, -1, -1, -1)$, where $\eta_{\mu\nu}$ is the *Minkowski metric.* Then by our concept of space-time stochasticity, we associate a nonvanishing random curvature $R = g^{\mu\nu} R_{\mu\nu} = R_0 + R_s$ (or $g_{\mu\nu} = \eta_{\mu\nu} + \delta g_{\mu\nu}$) with every point in space-time, such that $R_s \to 0$ (for $\langle \delta g_{\mu\nu} \rangle_R \to 0$) outside the region of confinement of these particles. For example, in Chapter 11 we will use the metric form $g_s^{\mu\nu}(x) = \phi^2(x, b_E)\eta_{\mu\nu}$, ($b_E = i\tau, \mathbf{b}$) in the *conformally flat space-time.* After averaging the space-time $\mathbb{R}^4(\hat{x})$ (Chapter 1) this metric form will be

$$\langle g_s^{\mu\nu}(\hat{x}) \rangle_{\mathbb{R}^4(\hat{x})} = \phi^2(l^2/x^2)\eta_{\mu\nu},$$

in particular, a special form

$$\phi(l^2/x^2) = 1 + l^2/(\mathbf{x}^2 + \lambda^2)$$

gives an interesting result which will be presented in Chapter 11.

We close this section with some hope that the hypothesis of space-time stochasticity will help us eventually to obtain a deeper and more intuitive physical picture of the quantum system and stochastic processes.

7.3. Stochastic Space and Random Walk

Let us consider random walk from the point of view of stochastic space and give some primitive reply to the question: How to understand that the particle subjects *Brownian-type motion* is caused by the stochasticity of space? We assume that the stochastic particle of interest is regarded as a classical punctiform mass, occupying at every instant a single point in space and traveling, therefore, along a trajectory. The probabilistic element, which is essential to provide a link with quantum mechanics, is introduced by assuming that this trajectory is continually influenced by a 'vacuum fluid', i.e., the gravitational vacuum force induced around the particle. As a result, the particle subjects random motion. While the exact properties of the gravitational vacuum interaction are unknown, the fluctuations of the particle position resulting from this interaction are presumed to be describable as a *Markov process.*

Thus we consider the motion of the particle in the space $\mathbb{R}^3(\hat{\mathbf{x}})$ (Chapter 1). We assume that the particle executes N displacements; then its position $\mathbf{R}_0$ after N displacements is given by the expression

$$\mathbf{R}_0 = \sum_{j=1}^{N} \mathbf{r}_j^0.$$

If the motion is uniform and rectilinear, then $\mathbf{r}_j^0 = \mathbf{v}^0 t_j$. We shall now assume that in each displacement $\mathbf{r}_j^0$ the particle makes a random displacement because of the stochastic nature of space: $\mathbf{b}_j = \alpha_j \mathbf{b}$, where α_j is an arbitrary sequence of real numbers, and $\mathbf{b}$ is a universal stochastic variable, subject to the probability distribution (1.6) (Chapter 1). Then after N displacements in the stochastic space $\mathbb{R}^3(\hat{\mathbf{x}})$, the position of the particle is determined by

$$\boldsymbol{\Phi} = \mathbf{R}_0 + \mathbf{B} = \sum_{j=1}^{N} (\mathbf{r}_j^0 + \alpha_j \mathbf{b}). \tag{7.6}$$

What is the probability $W_N(\boldsymbol{\Phi})\,\mathrm{d}^3\boldsymbol{\Phi}$ that the particle's coordinates after N displacements lie in the interval between $\boldsymbol{\Phi}$ and $\boldsymbol{\Phi} + \mathrm{d}\boldsymbol{\Phi}$ where $\boldsymbol{\Phi}$ is determined by equation (7.6)? In such a formulation, the problem can be solved by the method proposed by A. A. Markov. The application of Markov's method to the random-walk problem can be found in Chandrasekhar's (1943) paper.

The probability $W_N(\mathbf{B})\,\mathrm{d}\mathbf{B}$ that after N displacements the particle is in the interval $(\mathbf{B}, \mathbf{B} + \mathrm{d}\mathbf{B})$

$$\mathbf{B} = \boldsymbol{\Phi} - \mathbf{R}_0, \qquad \mathbf{B} = \sum_{j=1}^{N} \alpha_j \mathbf{b} \tag{7.7}$$

is given by

$$W_N(\mathbf{B}) = \frac{1}{8\pi^3} \int \mathrm{d}^3\rho \, \exp(-i\boldsymbol{\rho}\mathbf{B})\, A_N(\boldsymbol{\rho}), \tag{7.8}$$

where

$$A_N(\boldsymbol{\rho}) = \prod_{j=1}^{N} \int \mathrm{d}^3 b w(\mathbf{b}) \exp(i\boldsymbol{\rho}\alpha_j \mathbf{b}) = \prod_{j=1}^{N} (2\pi l^2)^{-3/2} \int \mathrm{d}^3 b \exp\left(i\boldsymbol{\rho}\alpha_j \mathbf{b} - \frac{|\mathbf{b}|^2}{2l^2} \right).$$

To estimate $A_N(\boldsymbol{\rho})$, we calculate the value of the typical integral

$$I = (2\pi l^2)^{-3/2} \int \mathrm{d}^3 b \exp(i\boldsymbol{\rho}\alpha_j \mathbf{b} - |\mathbf{b}|^2/2l^2) = \exp(-l^2 \alpha_j^2 \rho^2/2)$$

and we then have

$$A_N(\boldsymbol{\rho}) = \exp\left(-l^2 \frac{1}{2} \sum_{j=1}^{N} \alpha_j^2 \rho^2 \right) = \exp(-\tfrac{1}{2} P_N |\boldsymbol{\rho}|^2), \tag{7.9}$$

where

$$P_N = l^2 \times \sum_{j=1}^{N} \alpha_j^2.$$

Substituting (7.9) in (7.8), we obtain

$$W_N(\mathbf{B}) = (2\pi P_N)^{-3/2} \exp(-\tfrac{1}{2}|\mathbf{B}|^2/P_N). \tag{7.10}$$

We now determine ν, which is called the *diffusion coefficient*. This vanishes, $\nu = 0$, when $l = 0$, i.e., when the space is not stochastic. Setting

$$\begin{aligned}\overline{B_x^2} = \overline{B_y^2} = \overline{B_z^2} &= \lim_{N\to\infty} P_N = \lim_{N\to\infty} l^2 \sum_{j=1}^{N} \alpha_j^2 = \lim_{N\to\infty} Nl^2 \frac{1}{N} \sum_{j=1}^{N} \alpha_j^2 \\ &= \lim_{N\to\infty} ntl^2 \frac{1}{N} \sum_{j=1}^{N} \alpha_j^2 = t \lim_{N\to\infty} \frac{1}{N} \sum_{j=1}^{N} (\Delta B_x^j)^2,\end{aligned}$$

and denoting by ν the quantity

$$\lim_{N\to\infty} \frac{1}{N} \frac{1}{2} \sum_{j=1}^{N} (\Delta B_x^j)^2,$$

where $\Delta B_x^j = l\sqrt{n}\,\alpha_j$, in which n is the number of displacements per unit time, we obtain $2\nu t = \overline{B_x^2}$. Thus,

$$\nu = \frac{1}{2} \lim_{N\to\infty} \frac{1}{N} \sum_{j=1}^{N} (\Delta B_x^j)^2. \tag{7.11}$$

After the determination of ν, expression (7.10) can be written in the form

$$\rho(\mathbf{B}) = \lim_{N\to\infty} W_N(\mathbf{B}) = (4\pi\nu t)^{-3/2} \exp(-|\mathbf{B}|^2/4\nu t). \tag{7.12}$$

Since $\mathbf{B} = \mathbf{\Phi} - \mathbf{R}_0$,

$$\rho(\mathbf{\Phi}) = (4\pi\nu t)^{-3/2} \exp(-|\mathbf{\Phi} - \mathbf{R}_0|^2/4\nu t). \tag{7.13}$$

As one would expect, our problem has been reduced to a random-walk problem. In accordance with (7.12), a particle which makes n displacements per unit time with each of these displacements $\mathbf{b}_j = \alpha_j \mathbf{b}$ subject to the probability distribution $w(|\mathbf{b}|^2)$ is situated after time t in the element of volume between $\mathbf{B}$ and $\mathbf{B} + \mathrm{d}\mathbf{B}$ with probability $\rho(\mathbf{B})\,\mathrm{d}^3B$ or $\rho(\mathbf{\Phi})\,\mathrm{d}^3\Phi$ (when $\mathbf{R}_0 \neq 0$). It is known from the theory of Brownian motion that the distribution $\rho(\mathbf{r}, t; \mathbf{r}_0, \mathbf{u}_0)$, which determines the probability of finding the position $\mathbf{r}$ of the Brownian particle at the time t if $\mathbf{r} = \mathbf{r}_0$ and $\mathbf{u} = \mathbf{u}_0$ at $t = 0$, is determined by an expression just like (7.13)

$$\rho(\mathbf{r}, t; \mathbf{r}_0, \mathbf{u}_0) = (4\pi\nu_\delta t)^{-3/2} \exp(-|\mathbf{r} - \mathbf{r}_0|^2/4\nu_\delta t),$$

for $t \gg \beta^{-1}$, where $\mathbf{u}_0$ is the initial velocity of the Brownian particle, $\beta = 6\pi a\eta/m$, $\nu_\delta = kT/m\beta = kT/6\pi a\eta$, a is the radius of the Brownian particle, η

is the coefficient of viscosity of the surrounding medium (the fluid), k is Boltzmann's constant, and T is the absolute temperature. Thus, the introduction of the stochastic nature of space leads to a Brownian type of motion of a particle which in the absence of the stochastic properties of space would move along a definite preassigned trajectory.

We now turn to the derivation of a differential equation for the *probability density* $\rho(r, t)$. The derivation of the equation is the same as in the random-walk theory. Therefore, we shall here give only the results. As usual, we make important assumptions:

(1) The time interval Δt is chosen sufficiently large for the particle to make a large number of displacements in this interval, but small enough that during this interval the mean-square increment of $\mathbf{r}$, i.e., $\langle|\Delta\mathbf{r}|^2\rangle$ remains small; under these conditions the probability that the particle is displaced by $\Delta\mathbf{r}$ in the time Δt is given by

$$P(\Delta\mathbf{r}, \Delta t) = (4\pi\nu\,\Delta t)^{-3/2}\exp(-|\Delta\mathbf{r}|^2/4\nu\,\Delta t), \tag{7.14}$$

and does not depend on $\mathbf{r}$. The function $P(\mathbf{x}, t; \mathbf{x}', t')$, where $\Delta r = \mathbf{x}' - \mathbf{x}$, $\Delta t = t' - t$, $t' \geqslant t$, is called the *transition probability* (see Chapter 6). The physical meaning of this quantity is that $P\,\mathrm{d}^3x'$ represents the probability that the particle performing Brownian motion will reach the volume d^3x' around $\mathbf{x}'$ at time t' if it started from the point $\mathbf{x}$ at the initial time t.

(2) It is assumed that the displacement of the particle in space at a given time does not depend on the previous displacements; then

$$\rho(r, t + \Delta t) = \int \mathrm{d}^3(\Delta r)\rho(\mathbf{r} - \Delta\mathbf{r}, t)P(\Delta\mathbf{r}, \Delta t). \tag{7.15}$$

An equation of type (7.15) is called a *Smoluchowski equation*. Since $\langle|\Delta r|^2\rangle$ is small by hypothesis, we can expand $\rho(\mathbf{r} - \Delta\mathbf{r}, t)$ in the integrand in a Taylor series and integrate the obtained expression term by term. Going to the limit $\Delta t \to 0$, we obtain

$$\partial\rho/\partial t = \nu\,\Delta\rho, \quad (\Delta = \partial^2/\partial x^2 + \partial^2/\partial y^2 + \partial^2/\partial z^2). \tag{7.16}$$

On the basis of these results obtained, we would like to construct stochastic mechanics in the following chapter by means of Smoluchowski's equations of type (7.15) for nonrelativistic and relativistic cases.

7.4. The Main Prescriptions of Stochastic Quantization

In this section we briefly present the main ideas of stochastic quantization of dynamical systems and give a simple introduction of the probabilistic terms used in our consideration (for more detail, see Chapter 6). We also refer the reader to works of Nelson (1979) and Guerra (1981) for a deep and extensive review of the interplay between quantum mechanics and the theory of stochastic processes. At the same time, an important step in this trend was made by those who gave a precise mathematical definition of stochastic mechanics and field quantization.

Thus, let us consider a system with n independent degrees of freedom described by the classical Lagrangian

$$L(Q, \dot{Q}) = \frac{1}{2}\sum_{i=1}^{n} \dot{Q}_i^2 + \sum_{i=1}^{n} A_i(Q, t)\dot{Q}_i - U(Q, t), \tag{7.17}$$

where Q_i are the canonical co-ordinates of the system. The *Euler–Lagrange equations* of motion are given by

$$\ddot{Q}_i = \sum_{j=1}^{n} (\partial A_j/\partial Q_i - \partial A_i/\partial Q_j)\dot{Q}_j - \partial A_i/\partial t - \partial U/\partial Q_i, \quad (i = 1, 2, \ldots, n)$$

and are invariant under the *gauge transformation*:

$$A_i(Q, t) \to A_i(Q, t) - \partial f(Q, t)/\partial Q_i, \qquad U(Q, t) \to U(Q, t) - \partial f(Q, t)/\partial t.$$

Here A_i and U are electromagnetic and potential fields, and $f(\cdot)$ is an arbitrary scalar function. The canonical momenta of our system are

$$P_i = \partial L/\partial \dot{Q}_i = \dot{Q}_i + A_i, \quad i = 1, 2, \ldots, n.$$

The procedure of *stochastic quantization* is based on the (first) assumption that each of the *classical configuration* variables $\mathbf{Q}(t) = \{Q_i\}$, $i = 1, 2, \ldots, n$ is interpreted as a *Markov random process* and satisfies a *stochastic differential equation* of the form

$$\mathrm{d}Q_i(t) = b_i(\mathbf{Q}(t), t)\,\mathrm{d}t + \mathrm{d}W_i(t), \tag{7.18}$$

where $b_i(\mathbf{x}, t)$, $\mathbf{x} \in \mathbb{R}^n$, $t \in \mathbb{R}$, $i = 1, 2, \ldots, n$ is a velocity field to be determined and the W's are *Wiener process* which satisfy

$$\mathbb{E}\{\mathrm{d}W_i(t)\} = 0, \qquad \mathbb{E}\{\mathrm{d}W_i(t)\,\mathrm{d}W_j(t)\} = 2\nu\delta_{ij}\,\mathrm{d}t, \tag{7.19}$$

where ν is the diffusion coefficient (as usual it is assumed that $\nu = \hbar/2m$, another possibility will be discussed below). The solutions to equation (7.18) are called *Itô's processes* or the *generalized Brownian motion*. Here we assume that $\mathrm{d}W_i(t)$ do not depend on $\mathbf{Q}(s)$ for $s \leqslant t$.

If the b's satisfy a global *Lipschitz condition* (Nelson, 1967, p. 43) then a unique solution to (7.18) is ensured in the finite-dimensional case. The restriction to finite-dimensionality is not of major concern since any field theory can be made finite-dimensional by imposing a momentum cutoff and working in a finite volume (see Chapter 10). However, the global Lipschitz condition presents more serious difficulties since it is not satisfied when the quantum wave function has *zero* as it always does in excited stationary states. This problem has been discussed by Albeverio and Høegh-Krohn (1974).

These complications can be avoided by working only with wave functions which do not have exact zeros.

Further, it is convenient to introduce according to Nelson, the *mean forward derivative* D_+ and the *mean backward derivative* D_- through the definitions

$$D_+ f(\mathbf{Q}, t) = \lim_{h \to 0_+} h^{-1}\, \mathbb{E}\{f(\mathbf{Q}(t+h), t+h) - f(\mathbf{Q}(t), t) \mid \mathbf{Q}(t) = \mathbf{Q}\},$$

$$D_- f(\mathbf{Q}, t) = \lim_{h \to 0_+} h^{-1}\, \mathbb{E}\{f(\mathbf{Q}(t), t) - f(\mathbf{Q}(t-h), t-h) \mid \mathbf{Q}(t) = \mathbf{Q}\}, \quad (7.20)$$

where $\mathbb{E}$ denotes the *conditional expectation* with respect to the random variables $Q_i(t)$ (see Chapter 6).

It is easily seen from the equality

$$\frac{\mathrm{d}}{\mathrm{d}t}\mathbb{E}\{f(\mathbf{Q}(t), t)\, g(\mathbf{Q}(t), t)\}$$

$$= \mathbb{E}\{[D_+ f(\mathbf{Q}(t), t)] g(\mathbf{Q}(t), t) + f(\mathbf{Q}(t), t)[D_- g(\mathbf{Q}(t), t)]\},$$

and from definitions that the following equality holds for the expectations:

$$\int_{-\infty}^{\infty} \mathbb{E}\{f(\mathbf{Q}(t), t)[D_+ g(\mathbf{Q}(t), t)]\}\mathrm{d}t = -\int_{-\infty}^{\infty} \mathbb{E}\{[D_- f(\mathbf{Q}(t), t)] g(\mathbf{Q}(t), t)\}\mathrm{d}t$$

if f and g are sufficiently regular and at least one of them has *compact support* with respect to the time variable t.

If we introduce the probability density $\rho(\mathbf{x}, t)$, $\mathbf{x} \in \mathbb{R}^n$, $t \in \mathbb{R} = T$, of the random variables $Q_i(t)$, so that

$$\mathbb{E}\{f(\mathbf{Q}(t), t)\} = \int \mathrm{d}^n x\, f(\mathbf{x}, t)\rho(\mathbf{x}, t),$$

then we can immediately prove the following relations

$$D_+ Q_i(t) = b_i^+(\mathbf{Q}(t), t) \qquad D_- Q_i(t) = b_i^-(\mathbf{Q}(t), t)$$

with

$$b_i^-(\mathbf{x}, t) - b_i^+(\mathbf{x}, t) = -2\nu \frac{\partial}{\partial x_i} \ln \rho(\mathbf{x}, t)$$

and, more generally,

$$D_+ f(\mathbf{Q}(t), t) = \left(\frac{\partial}{\partial t} + \sum_{i=1}^{n} b_i^+ \frac{\partial}{\partial Q_i} + \nu\Delta\right) f(\mathbf{Q}(t), t),$$

$$D_- f(\mathbf{Q}(t), t) = \left(\frac{\partial}{\partial t} + \sum_{i=1}^{n} b_i^- \frac{\partial}{\partial Q_i} - \nu\Delta\right) f(\mathbf{Q}(t), t),$$

where

$$\Delta = \sum_{i=1}^{n} \left(\frac{\partial^2}{\partial Q_i^2} \right).$$

The probability density $\rho(\mathbf{x}, t)$ satisfies the *Fokker–Planck equations*

$$\frac{\partial \rho}{\partial t} = -\sum_{i=1}^{n} \frac{\partial}{\partial x_i}(\rho b_i^+) + \nu \Delta \rho, \tag{7.21}$$

or

$$\frac{\partial \rho}{\partial t} = -\sum_{i=1}^{n} \frac{\partial}{\partial x_i}(\rho b_i^-) - \nu \Delta \rho. \tag{7.22}$$

If we define

$$v_i(\mathbf{x}, t) = \tfrac{1}{2}[b_i^+(\mathbf{x}, t) + b_i^-(\mathbf{x}, t)],$$

we have the *equation of continuity*

$$\frac{\partial \rho}{\partial t} = -\sum_{i=1}^{n} \frac{\partial}{\partial x_i}(\rho v_i) = -\operatorname{div}(\rho \mathbf{v}) \tag{7.23}$$

obtained by averaging (7.21) and (7.22). We call $\mathbf{v}$ the *current velocity*. If we make the definition of so-called *stochastic velocity* $u_i = \frac{1}{2}(b_i^+ - b_i^-)$, we have

$$u_i = \nu \frac{\partial}{\partial x_i} \ln \rho = \nu(\text{grad})_i \cdot \ln \rho.$$

Therefore

$$\frac{\partial u}{\partial t} = \nu \mathbf{grad} \frac{\partial}{\partial t} \ln \rho = \nu\, \mathbf{grad}\left[\frac{\partial \rho}{\partial t}\frac{1}{\rho}\right] = \nu\, \mathbf{grad}\left[\frac{-\operatorname{div} \rho \mathbf{v}}{\rho}\right]$$

$$= -\nu \mathbf{grad}\left[\operatorname{div} \mathbf{v} + \mathbf{v}\frac{\mathbf{grad}\, \rho}{\rho}\right] = -\nu\, \mathbf{grad}\, \operatorname{div} \mathbf{v} - \mathbf{grad}(\mathbf{vu}).$$

That is

$$\frac{\partial u}{\partial t} = -\nu\, \mathbf{grad}\, \operatorname{div} \mathbf{v} - \mathbf{grad}(\mathbf{vu}).$$

Now we define as *mean acceleration*

$$a_i(\mathbf{Q}(t), t) = \tfrac{1}{2}(D_+ D_- + D_- D_+) Q_i(t). \tag{7.24}$$

Then the second Nelson assumption is

$$m a_i(\mathbf{Q}(t), t) = F_i \tag{7.25}$$

by the *Newton law*; where F_i is some external force, in particular,

$$F_i = -\partial U / \partial Q_i.$$

Equations (7.23) and (7.25) lead to a system of equations which can be transformed into Schrödinger's equation

$$i\hbar\frac{\partial}{\partial t}\psi(\mathbf{x},t) = \left(-\frac{\hbar^2}{2m}\nabla^2 + U\right)\psi(\mathbf{x},t),$$

where

$$\psi = \exp(R + iS), \tag{7.26}$$

$$b_i^+ = 2\nu\frac{\partial}{\partial Q_i}(R+S), \qquad b_i^- = 2\nu\frac{\partial}{\partial Q_i}(-R+S). \tag{7.27}$$

It was established by Davidson (1979) that Nelson's dynamical assumption [(7.24) and (7.25)] is not the only possibility. The following equation [instead of (7.25)]

$$\tfrac{1}{2}(D_+D_- + D_-D_+)Q_i(t) + \tfrac{1}{8}\beta(D_+ - D_-)^2Q_i(t) = -\partial U/\partial Q_i(t)$$

also leads to Schrödinger's equation where

$$\psi = \exp(R + izS),$$

$$z = (1 - \tfrac{1}{2}\beta)^{-1/2}, \qquad \nu = \tfrac{1}{2}\hbar(1 - \tfrac{1}{2}\beta)^{-1/2}. \tag{7.28}$$

Here β is a constant and b_i^+ and b_i^- are still given by equation (7.27) but with the value of ν in equation (7.28). If ν in equation (7.28) is real, which is necessary for a physical interpretation, then β must be real and satisfy condition $\beta < 2$.

By choosing β appropriately, any value of the diffusion coefficient can be used in a stochastic model of quantum mechanics. It is possible to choose ν arbitrarily small by letting β be large in magnitude but negative. In the limit of zero ν the model becomes deterministic and in fact turns to the familiar hidden variable model of Bohm (1952), a point first made Shucker (1980a).

In the case of (7.17) with the formulas previously given, we can easily prove that

$$a_i(\mathbf{x},t) = \frac{\partial v_i}{\partial t} + \sum_{j=1}^{n} v_j\frac{\partial}{\partial x_j}v_i - 2\nu^2\frac{\partial}{\partial x_i}(\rho^{-1/2}\Delta\cdot\rho^{1/2}).$$

Nelson's second basic assumption is the validity of the Euler–Lagrange equations of motion with the substitutions $\ddot{Q}_i \to a_i$, $\dot{Q}_i \to v_i$. Therefore we have

$$\frac{\partial v_i}{\partial t} = -\sum_{j=1}^{n} v_j\frac{\partial}{\partial x_j}v_i + \sum_{j=1}^{n}\left(\frac{\partial A_j}{\partial x_i} - \frac{\partial A_i}{\partial x_j}\right)v_j - \frac{\partial A_i}{\partial t} - \frac{\partial U}{\partial x_i} +$$

$$+\, 2\cdot\nu^2\frac{\partial}{\partial x_i}(\rho^{-1/2}\Delta\cdot\rho^{1/2}), \quad i = 1, 2, \ldots, n.$$

The process is completely described by the probability density $\rho(\mathbf{x}, t)$ and the current velocity $v_i(\mathbf{x}, t)$, which satisfy a system of nonlinear differential equations for the first order with respect to time [continuity equation (7.23) and the Euler–Lagrange equations]. Therefore, if ρ and $\mathbf{v}$ are known at some initial time t_0, we can solve the *Cauchy Problem* and know the Markov process $\mathbf{Q}(t)$ completely.

With the additional kinematical assumption that the generalized momentum $v_i + A_i$ is a gradient, therefore $v_i + A_i = \partial S/\partial x_i$ for some function $S(\mathbf{x}, t)$, we can show the complete equivalence of our scheme of quantization with the usual one. In fact, if we introduce the wave function

$$\psi(\mathbf{x}, t) = [\rho(\mathbf{x}, t)]^{1/2} \exp[i\hbar^{-1} S(\mathbf{x}, t)],$$

then we can check that when ρ and S envolve according to the stochastic differential equations then ψ evolves according to the Schrödinger equation

$$i\hbar \frac{\partial \psi}{\partial t} = \frac{1}{2} \sum_{j=1}^{n} \left(\frac{\hbar}{i} \frac{\partial}{\partial x_j} - A_j \right)^2 \psi + U\psi,$$

associated with the same classical system. On the other hand, if we know the wave function $\psi(\mathbf{x}, t)$, then we can recover the stochastic process.

7.5. Stochastic Field Theory and its Connection with Euclidean Field Theory

First of all consider the simple example of the *harmonic oscillator* with Lagrangian

$$L(Q, \dot{Q}) = \tfrac{1}{2}\dot{Q}^2 - \tfrac{1}{2}\omega^2 Q^2.$$

The normalized ground state wave function is given by

$$\psi(x, t) = (\omega/\hbar\pi)^{1/4} \exp(-\omega x^2/2\hbar) \exp(-i\hbar\omega t/2),$$

$x \in \mathbb{R}$. Now we consider the stochastic process $Q(t)$ associated with the ground state. The density is

$$\rho(x, t) = |\psi(x, t)|^2 = (\omega/\hbar\pi)^{1/2} \exp(-\omega x^2/\hbar).$$

The current velocity vanishes because the phase function $S = i\hbar\omega t/2$ does not depend on x.

On the other hand, for the *mean forward derivative* $D_+ Q(t) = b^+$ and the *mean backward derivative* $D_- Q(t) = b^-$ we have

$$b^+(x, t) = -b^-(x, t) = -\omega x.$$

Therefore the basic *stochastic differential equation* of the process corresponding to the harmonic oscillator is

$$\mathrm{d}Q(t) = -\omega Q(t)\,\mathrm{d}t + \mathrm{d}W.$$

Taking into account the Gaussian form of the density we find that the process

$Q(t)$ is a *Gaussian process* characterized by the expectations

$$\mathbb{E}\{Q(t)\} = 0, \qquad \mathbb{E}\{Q(t)Q(t')\} = (\hbar/2\omega)\exp(-\omega|t - t'|).$$

This Gaussian–Markov process is known as the *Ornstein–Uhlenbeck process* (see Chapter 6).

We can consider now the *Klein–Gordon equation* for a real classical field $\varphi(\mathbf{x}, t)$, $\mathbf{x} \in \mathbb{R}^3$, $t \in \mathbb{R} = T$

$$\left(\frac{\partial^2}{\partial t^2} - \Delta + m^2\right)\varphi(\mathbf{x}, t) = 0,$$

where m is a mass parameter with the dimensions L^{-1}. The Hamiltonian of the field is

$$H = \frac{1}{2}\int\left[\left(\frac{\partial\varphi}{\partial t}\right)^2 + (\nabla\varphi)^2 + m^2\varphi^2\right]\mathrm{d}^3x.$$

In order to use the stochastic quantization it is convenient to enclose, as usual, the field in a cubic box of the length L, with periodic conditions at the boundary. In this way we reduce the system to a family of the *independent harmonic oscillators*, with canonical coordinates $\{\varphi_{\mathbf{k}}(t)\}$ and Hamiltonian

$$H = \tfrac{1}{2}\sum_{\mathbf{k}}[\dot{\varphi}_{\mathbf{k}}^2(t) + (\mathbf{k}^2 + m^2)\varphi_{\mathbf{k}}^2(t)],$$

where $\mathbf{k} = \{k_1, k_2, k_3\}$, $k_i = 2\pi n_i/L$, n_i are relative integers. For each oscillator of the frequency $\omega^2 = \mathbf{k}^2 + m^2$, we consider the ground state and associated stochastic process explained before. Then going to the limit $L \to \infty$, we find immediately that the ground state stochastic process associated with the Klein–Gordon equation is a *Gaussian generalized process* $\varphi(\mathbf{x}, t)$, with *mean zero* and a *covariance* (for detail, see Chapters 6 and 10)

$$\mathbb{E}\{\varphi(\mathbf{x}, t)\varphi(\mathbf{x}', t')\}$$

$$= \hbar\int\left(\frac{\mathrm{d}p}{2\pi}\right)^4(\mathbf{p}^2 + m^2)^{-1}\exp[ip_4(t - t') + i\mathbf{p}(\mathbf{x} - \mathbf{x}')], \tag{7.29}$$

where d^4p is the Lebesque measure in $\mathbb{R}^4$ and $p^2 = p_4^2 + \mathbf{p}^2$ is the squared Euclidean norm in $\mathbb{R}^4$.

Therefore we can recognize that the covariance is nothing but the *two-point Schwinger function* of a free field with mass m (in a system of units in which $\hbar = 1$), i.e., the *two-point Wightman function* continued to a purely imaginary time (see the following section).

The important fact, discovered by Guerra and Ruggiero (1973), is that for the process we have constructed, the parameter t is the real physical time. In conclusion, we would like to say that if the Euclidean field identified with the ground state stochastic process is associated with the classical field theory, then the underlying four-dimensional manifolds $(\mathbf{x}, t)$ on which the field is defined can be considered as the *Minkowski physical space-time*, and the powerful method of the Euclidean field theory can also be applied to the study of the stochastic field theory.

7.6. Euclidean Quantum Field Theory

Because of the fact that there is a profound connection between the Euclidean quantum field theory and the theory of stochastic processes, we give here some elementary definitions of the Euclidean QFT. The Euclidean QFT is based on the idea of replacing the usual description of quantum field in the Minkowski space-time with a description in some auxiliary *Euclidean space.* References in this direction are presented in Sections 1.1 and 7.1.

First of all, we consider a free scalar quantum field $\varphi(x)$ of mass m, completely defined through its vacuum expectation values:

$$W_{2n} = W(x_1, \ldots, x_{2n}) = \langle 0|\varphi(x_1)\cdots\varphi(x_{2n})|0\rangle$$

$$= \sum_{j=2}^{2n} W_1(x_1, x_j)W(x_2, \ldots, x_{j-1}, x_{j+1}, \ldots, x_{2n}), \quad (x = x_0, \mathbf{x}) \quad (7.30)$$

with the expectation values of an odd number of fields vanishing. The two-point function is given by (see Section 1.4.4)

$$W(x, y) = W(x - y) = (2\pi)^{-3} \int \frac{d^3k}{2\omega(\mathbf{k})} \exp[-i\omega(\mathbf{k})(x_0 - y_0) + $$

$$+ i\mathbf{k}(\mathbf{x} - \mathbf{y})], \quad (7.31)$$

where $\omega(\mathbf{k}) = (\mathbf{k}^2 + m^2)^{1/2} \geqslant m$. If we perform an analytic continuation on x_0, y_0 to the Schwinger points $x_0 = -ix_4$, $y_0 = -iy_4$, $x_4 \geqslant y_4$, then we obtain the *Schwinger function*

$$S(x_E, y_E) = S(x_E - y_E) = (2\pi)^{-3} \int \frac{d^3k}{2\omega(\mathbf{k})} \exp[-\omega(\mathbf{k})(x_4 - y_4) + $$

$$+ i\mathbf{k}(\mathbf{x} - \mathbf{y})], \quad (7.32)$$

$(x_E = x_4, \mathbf{x})$. The elementary formula

$$\frac{1}{2}\frac{1}{\pi}\int dk_4 \exp[ik_4(x_4 - y_4)][k_4^2 + \omega^2(\mathbf{k})]^{-1}$$

$$= \frac{1}{2}\frac{1}{\omega(\mathbf{k})} \exp\{-\omega(\mathbf{k})|x_4 - y_4|\} \quad (7.33)$$

allows one to write (7.32) in the form

$$S(x, y) = \int \left(\frac{dk}{2\pi}\right)^4 \exp[ik(x - y)](k^2 + m^2)^{-1}. \tag{7.34}$$

where x, k are points with coordinates $(x_4, \mathbf{x})$, $(k_4, \mathbf{k})$, $x, y \in \mathbb{R}^4$, $kx = k_4 x_4 + \mathbf{kx}$, $k^2 = k_4^2 + \mathbf{k}^2$, $d^4k = dk_4\, d^3k$, $(dk)^4 = d^4k$. For convenience the index E is omitted here.

According to the symmetry and the positive definiteness of (7.34) *Euclidean fields* are introduced as *Gaussian random fields* with mean zero and *covariance*

$$\mathbb{E}\{\varphi(x)\varphi(y)\} = S(x, y), \quad x, y \in \mathbb{R}^4. \tag{7.35}$$

Therefore, we see that the Euclidean theory can be described through stochastic fields as opposed to the non-commutative Heisenberg fields of the Minkowski theory.

As an example, we now consider the Schwinger function (7.34) in two-dimensional space-time:

$$S_2(y, m_0^2) = (2\pi)^{-2} \int d^2k\, e^{iky}(k^2 + m_0^2)^{-1}. \tag{7.36}$$

An explicit form of this function is

$$S_2(x) = (2\pi)^{-1} K_0(m_0|x|) = i(4)^{-1} H_0^{(1)}(im_0|x|),$$

where $K_0(x)$ is the Mac'Donald function connected with the Hankel one by the formula

$$K_0(x) = \tfrac{1}{2}\pi i H_0^{(1)}(ix).$$

From formulas (7.36) and (7.33) it follows, in particular, that

$$S_2(x, s = 0; m_0^2) = (4\pi)^{-1} \int dk\, e^{ikx}(k^2 + m_0^2)^{-1/2} = W_2(t = 0, x; m_0^2).$$

Thus, the *Schwinger function* S_2 and the Wightman function W_2 coincide at time $t = s = 0$. Since the function S_2 is real-analytic, then according to equality (7.36) it co-ordinates with the analytical continuation of the function W_2 in the extended future (cone) tube. For the function S_2 all the Euclidean regions are exhausted by this. Having an analytic continuation for the function W_2, it is easy to obtain it and the functions W_n given by formula (7.30).

THEOREM 7.1. *Let* $[xy] = S_2(x - y, m_0^2)$, *where* S_2 *is given by formula* (7.36). *Then the functions* (*for notations see Section* 1.4.4 *in Chapter* 1):

$$S_n(y_1, \ldots, y_n) = [y_1 \cdots y_n] \tag{7.37}$$

are the Schwinger functions for the free field with mass m_0. *The proof is given by Simon* (1974).

Notice that the Euclidean field must commute at different points of space. Indeed, from the *locality* it follows that $[\varphi(x,0), \varphi(y,0)]_- = 0$, because the Euclidean field at an imaginary time must coincide with the Minkowski field at the zero time. However, any two points of the Euclidean region can be reduced to a zero (imaginary) time by means of the Euclidean (rotational) motion.

Nelson (1971, 1973) pointed out the importance of the *Markov property* (see Chapter 6) of the free Euclidean field and also of the interpretation of the Euclidean field as the functional integrals (see Simon, 1974, 1979).

We notice that there exists a deeper connection between the field formulated by Gårding and Wightman, and the Euclidean QFT. Thus, one consequence of the *Wightman axioms* is that W_n is the boundary value of the analytic function $W_n(z_1, \ldots, z_n)$ in a certain region G. G includes all the points of the form $z_j = (x_j, is_j)$ with $(x_j, s_j) \in \mathbb{R}^d$ and with $z_k \neq z_l$ (all k, l). The functions

$$S_n(y_1, \ldots, y_n) = W_n(y_1, is_1; \ldots; y_n, is_n)$$

are called the *Schwinger functions*.

Rather than the construction of the Wightman field $\Phi(x)$, one constructs a measure $\mathrm{d}\mu$ on a space $\mathcal{U}'(\mathbb{R}^d)$ ($\mathcal{U}$ is the Schwartz space of tempered functions) whose moments are the candidates for the Schwinger functions; i.e., one assumes $\{\int \mathrm{d}^d y\, \Phi(y) f(y)\}(T) = T(f)$ as random variables and

$$S_n(y_1, \ldots, y_n) = \int \mathrm{d}\mu\, \Phi(y_1) \cdots \Phi(y_n), \tag{7.38}$$

that is

$$\int \mathrm{d}^d y_1 f_1(y_1) \cdots \int \mathrm{d}^d y_n f_n(y_n) S_n(y_1, \ldots, y_n) = \int T(f_1) \cdots T(f_n)\, \mathrm{d}\mu(T).$$

Thus one needs some conditions on the measure $\mathrm{d}\mu$ or on S_n which allow us to reconstruct Φ. The earliest such conditions are due to Nelson (1973), but they are difficult to verify in practice. Osterwalder-Schrader (1973, 1975) gave a set of conditions on the S_n's whose modulo growth conditions with n are equivalent to the Gårding–Wightman axioms and, in particular, isolated a condition now known as OS (Osterwalder-Schrader) positivity.

THEOREM 7.2 (*Fröhlich's* (1974) *reconstruction theorem*). *Let* $\mathrm{d}\mu$ *be a cylinder measure on* $\mathcal{U}'(\mathbb{R}^d)$ *which satisfies the following conditions.*

(i) *Proper Euclidean motions* [*i.e.*, $T(x) \mapsto T(Ax + b)$, $b \in \mathbb{R}^d$, $A \in SO(d)$] *leave* $\mathrm{d}\mu$ *invariant*;

(ii) *OS positivity, i.e., given a real-valued* $f \in \mathcal{U}(\mathbb{R}^d)$ *with* Supp $f \in \{\mathbf{x}, s), s > 0\}$, *let* $(\theta f)(\mathbf{x}, s) = f(\mathbf{x}, -s)$. *Then for real-valued* $f_1, \ldots, f_n$ *with the above support and* $z_1 \ldots, z_n \in \mathbb{C}$:

$$\sum_{j,k=1}^{n} \bar{z}_k z_j \int \mathrm{d}\mu \exp\{i[\Phi(f_k) - \Phi(\theta f_j)]\} \geqslant 0;$$

(iii) *For any* $f \in \mathcal{U}(\mathbb{R}^d)$, $\int d\mu \exp[\Phi(f)] < \infty$.
(iv) *The action of the translations* $(\mathbf{x}, s) \to (\mathbf{x}, s + t)$ is ergodic. *Then, there is a unique (scalar) field theory obeying the Garding-Wightman axioms whose Schwinger functions are given by* (7.38).

The most natural way to construct measures is to try Gaussian $d\mu$'s. It transpires that these measures describe 'trivial' field theories in that they describe particles without interactions – they are the analog of the harmonic oscillator.

THEOREM 7.3. *A* Gaussian measure $d\mu$ *on* $\mathcal{U}'(\mathbb{R}^d)$ *obeys conditions* (i) (*iv*) *of Theorem* 7.2 *if and only if the* covariance $S_2(x, y) = \int d\mu \Phi(x)\Phi(y)$ *is of the form*

$$S_2(x, y) = (2\pi)^{-d} \int d^d k \, e^{ik(x-y)} \tilde{S}_2(k)$$

with

$$\tilde{S}_2(k) = \int d\rho(m^2)(k^2 + m^2)^{-1} + P(k^2), \tag{7.39}$$

where $d\rho$ *is a polynomially bounded positive measure on* $[0, \infty)$, *so that the integral lies in* $\mathcal{U}'$, *and* P *is a polynomial which is positive on* $[0, \infty)$.

Proof. Sketch of proof (Simon, 1979) is the following: the condition (i) is equivalent to $S_2(x, y) = f(x - y)$ with f rotation invariant, (iii) is automatic for the Gaussian process, and (iv) can be seen to be equivalent to $f(x) \to 0$ as $x \to \infty$. That leaves the analysis of (ii). This is clearly equivalent to

$$\sum_{j,k=1}^{n} \bar{z}_k z_j \exp\{-\tfrac{1}{2}S_2(f_k - \theta f_j, f_k - \theta f_j)\} \geqslant 0.$$

It can be shown that this, in turn, is equivalent to $S_2(\theta f, f) \geqslant 0$, for all f with the proper support property. This is an analog to the usual *positive definiteness* condition that leads to Bochner's theorem. The unitary group of Bochner's theorem (see Reed and Simon, 1975) is an unitary group in the $\mathbf{x}$ variables and a self-adjoint semigroup in the $\mathbf{s}$-variable. The temperedness of S_2 leads to the boundedness of the semigroup. Thus for $s > 0$ and $d\mu$ a tempered measure:

$$S_2(\mathbf{x}, s; \mathbf{0}, 0) = 2(2\pi)^{1-d} \int_{E \geqslant 0} d\mu(E, \mathbf{k}) \, e^{i\mathbf{kx}} \, e^{-sE}.$$

Since S_2 is rotationally invariant, $[x_i(\partial/\partial s) - s(\partial/x_i)]S_2 = 0$. This translates into Lorentz invariance of $d\mu$. Since $d\mu$ is supported in the region $E \geqslant 0$, it follows (see, for example, Paley *et al.*, 1933) that

$$d\mu(E, \mathbf{k}) = C_0 \delta(E, \mathbf{k}) + \int d\rho(m^2)[E^{-1} \, \delta(E - (k^2 + m^2)^{1/2}) \, d^{d-1}k] \, dE.$$

Using formula (7.33) we find that for $s > 0$

$$S_2(\mathbf{x}, s; \mathbf{0}, 0) = 2C_0(2\pi)^{1-d} + (2\pi)^{-d} \int \mathrm{d}^d k \, \mathrm{e}^{ikx} \left\{ \int \mathrm{d}\rho(m^2)(k^2 + m^2)^{-1} \right\}.$$

Since $S_2 \to 0$ as $s \to \infty$, $C_0 = 0$. Rotation invariance yields the applicability of the formula for all $(\mathbf{x}, s) \neq 0$. All that remains is an ambiguity of a positive-definite rotation invariant distribution supported at zero. This gives the polynomial $P(k^2)$ for the Fourier transform.

As usual, it is assumed that $P = 0$, $\mathrm{d}\rho(m^2) = \delta(m^2 - m_0^2)\,\mathrm{d}(m^2)$ in expression (7.39). The resulting *Gaussian process* is called the *free Euclidean field* of mass m_0.

With this we finish a very brief consideration of Euclidean field. The detailed study of this problem may be found in the monographs by Simon (1974), Glimm and Jaffe (1981).

Chapter 8

Stochastic Mechanics

8.1. Introduction

As indicated by recent studies (Nelson, 1979; Guerra, 1981) stochastic mechanics (its name is probably due to Guerra) is a very simple and clear theory, with enormous possibilities of physical and mathematical effectiveness in the exploration of properties of quantum mechanical systems, especially when a large number of degrees of freedom are involved, as the experience of the Euclidean methods in constructive quantum field theory has shown. The main advantage of stochastic mechanics, from the point of view of mathematical control, relies on the possibility of exploiting the well-developed mathematical methods of probability theory and stochastic processes (see Chapter 6). However stochastic mechanics is not free from criticism. We would like to stress a peculiar aspect of stochastic mechanics which is the basis of typical misunderstanding and unjustified criticism. It is remarked that a very natural objection arises in connection with any proposal of exploiting stochastic processes for the formulation of a theoretical scheme related to quantum mechanics. As usual, it is suggested that quantum mechanics may be interpreted as a Markov process, but *irreducible* to a Brownian-type stochastic motion. Or, more clearly it can be phrased as follows: 'Stochastic processes are typical expressions of diffusion phenomena and therefore share a substantial time *irreversible* character, while quantum mechanics is time *reversible* and therefore cannot have anything to do with stochastic processes' (see Guerra, 1981). This distinction must be clearly borne in mind, since it is the failure to take it seriously which leads to most of the criticisms (Kracklauer, 1974; Ghirardi *et al.*, 1978; Grabert *et al.*, 1979; Mielnik and Tengstrand, 1980) usually made against the stochastic interpretation. However, as has been explained by Guerra (1981), the scheme of stochastic mechanics is time reversible as it should be.

On the other hand, it should be noted that in stochastic mechanics there are several open conceptual problems. Two of them are related to canonical transformations and to the Fermi statistics (see Guerra, 1981). With regard to the problem of the general covariance properties of stochastic mechanics, we believe that it may be successfully investigated in our approach based on the hypothesis of space-time stochasticity.

Thus the attraction of the approach based on the hypothesis of stochastic space-time to the description of stochastic processes is that we have succeeded in generalizing stochastic mechanics to the relativistic case and defining rigorously, in the mathematical sense, relativistic integrals of the Feynman-type (Namsrai, 1980c). In addition, in this scheme, as one would expect, we encounter the self-turbulent phenomenon (Namsrai, 1980b) which is characteristic of a nonlinear system.

By assumption about the form of the stochastic space-time $\mathbb{R}^4(\hat{x})$ presented in Chapter 1, one can construct an *equation of the Smoluchowski type* for the probability density $\rho(x_\mu, s)$ in the relativistic case; for $\rho(x_\mu, s)$ there is a representation of the form

$$\rho(x_\mu, s + \Delta s) = \int \mathrm{d}^4 b_{\mathrm{E}}\, a(b_{\mathrm{E}}^2, \Delta s)\rho(x_0 + ib_4, \mathbf{x} - \mathbf{b}, s), \tag{8.1}$$

where s is some invariant parameter which is, as usual, interpreted as a proper time. The integration in (8.1) is over a four-dimensional Euclidean space and $a(b_{\mathrm{E}}^2, \Delta s)$ is some integrable function of the variable $b_{\mathrm{E}}^2 = b_4^2 + b_1^2 + b_2^2 + b_3^2$. Relativistic invariance is ensured by the fact that the algebra of the Lorentz group and the algebra of the group of four-dimensional Euclidean rotations are identical in the complex domain (see Section 1.4.6 in Chapter 1).

There is a hope that if we begin the construction of a theory in the stochastic Euclidean space $\mathbb{E}^4(\hat{\mathbf{x}}, \tau = b_4)$, a relativistic invariant description of the motion of particles in stochastic spacetime can be realized.

Before starting the construction of dynamics of stochastic particles, we give here the physical meaning of the stochastic nature of space-time in the nonrelativistic and relativistic cases. In the framework of our approach in the nonrelativistic case, the stochastic nature of space leads to the impossibility of determining the coordinates of a particle with an accuracy exceeding at least the value of the particle's Compton wavelength. In the relativistic case, it is harder to give this property a physical interpretation. Formally, it can be interpreted as the presence of *fluctuations* of the four-dimensional co-ordinates of the particle *in the Euclidean space* $\mathbb{E}^4(\hat{\mathbf{x}}, \tau = b_4)$ (i.e., as the presence of a random walk in the imaginary time). By itself, such an interpretation has no physical meaning, but the method has value as one method of relativistic description of random processes in terms of an equation of the Smoluchowski type. It should be noted that equations (8.20) and (8.21) shown below do not in general bear any relation to the ordinary Smoluchowski equations which describe a stochastic consequence, and therefore the quantities $P_\pm$ which occur in these equations cannot be interpreted as transition probabilities (see Chapter 6).

By a *universal* (*or fundamental*) *length*, we shall understand the following. Physically, the universal length l is some characteristic distance over which the corresponding notions about spacetime and locality (causal connection) begin to break down; in particular, stochastic properties and nonlocality can be manifested if they exist. Recent experimental data show $l \lesssim 10^{-15} - 10^{-16}$ cm (see Part 1). Other possibilities for introducing the concept of a fundamental length in physics will be discussed in Chapter 11 (see also Chapter 1 and references therein).

Our exposition is as follows. In Sections 8.2 and 8.3 we study the nonrelativistic motion of a particle in a force field and the motion of a relativistic particle in a four-dimensional stochastic space-time, and obtain equations of motion for a particle that are formally *equivalent* to the *Klein–Gordon equation*. The two-body problem, Section 8.4, is studied in the nonrelativistic and relativistic cases.

8.2. Equations of Motion of a Nonrelativistic Particle

In constructing the dynamics of stochastic particles, it is customary to use the mathematical concepts of left and right derivatives (or *stochastic*, or Itô's *derivatives*) (see Chapter 6), by means of which the stochastic and current (or systematic) velocities of the particle are constructed. Newton's law is used as the dynamical equation (see the preceding chapter). We recall here that the stochastic derivatives of a function $f(x)$ are given by Itô's (1961) rule:

$$D_+ f(x) = (\partial/\partial t + \mathbf{v}_+ \cdot \nabla + \nu \nabla^2) f(x)$$

and

$$D_- f(x) = (\partial/\partial t + v_- \cdot \nabla - \nu \nabla^2) f(x), \quad (x = t, \mathbf{x})$$

where $\mathbf{v}_+$ and $\mathbf{v}_-$ are called *forward and backward velocities*, respectively. In this case the *Newton law* takes the form: $m\mathbf{a}(t) = \mathbf{F}$. Here $\mathbf{a}(t) = \frac{1}{2}(D_+ D_- + D_- D_+)\mathbf{Q}(t)$ is called the mean second derivative or *mean acceleration*. $\mathbf{Q}(t)$ is a stochastic process representing the stochastic motion of a particle with mass m in an external force $\mathbf{F}$.

However, Kershaw's (1964) approach (see also Lehr and Park, 1977) was based on equations of the Smoluchowski type for the probability density $\rho(\mathbf{x}, t)$ and the mean particle velocity $\mathbf{v}(\mathbf{x}, t)$.

Despite the difference interpretation of the stochastic behavior of the system (the particle), the mathematical method of description of the stochastic processes in our scheme will be the same as in the studies of Kershaw, and Lehr and Park which are based on the theory of Bohm (1952) and de Broglie (1964).

In Section 7.3 we considered the random-walk problem from the point of view of a stochastic space. The greatest interest attaches to the study of the motion of a particle in a space whose properties are assumed to be stochastic. We now turn to this question.

If it is assumed that the displacement of the particle in space at a given time does not depend on the previous displacements, then the following relation holds for the probability density $\rho(\mathbf{x}, t)$:

$$\rho(\mathbf{x}, t + \Delta t) = \int d^3(\delta x_+)\rho(\mathbf{x} - \delta\mathbf{x}_+, t)P_+(\mathbf{x} - \delta\mathbf{x}_+, t; \delta\mathbf{x}_+, \Delta t), \tag{8.2}$$

where $P_+(\mathbf{x} - \delta\mathbf{x}_+, t: \delta\mathbf{x}_+, \Delta t)$ can be interpreted as the probability that a particle at the position $\mathbf{x} - \delta\mathbf{x}_+$ at the time t is displaced through $\delta\mathbf{x}_+$ during the time Δt and, therefore, it reaches the point $\mathbf{x}$ at the time $t + \Delta t$. In the stochastic theory (Nelson, 1966, 1967; Lehr and Park, 1977; Namsrai, 1980a) with twice the number of transition probabilities, one uses $P_-(\mathbf{x} + \delta\mathbf{x}_-, t; \delta\mathbf{x}_-, \Delta t)$, which is the probability that a particle has moved from the position $\mathbf{x} + \delta\mathbf{x}_-$ by $\delta\mathbf{x}_-$ in the time interval Δt prior to the time t and thus occupied the position at the point $\mathbf{x}$ at the earlier time $t - \Delta t$. Thus, the equation analogous to (8.2) takes in this case the form

$$\rho(\mathbf{x}, t - \Delta t) = \int d^3(\delta\mathbf{x}_-)\rho(\mathbf{x} + \delta\mathbf{x}_-, t)P_-(\mathbf{x} + \delta\mathbf{x}_-, t; \delta\mathbf{x}_-, \Delta t). \tag{8.3}$$

For $P_\pm$, we can here choose an expression of the form

$$P_\pm = (4\pi v_\pm \cdot \Delta t)^{-3/2} \exp\left\{\frac{-[\delta\mathbf{x}_\pm - \mathbf{v}_\pm(\mathbf{x}, t)\cdot\Delta t]^2}{4v_\pm \cdot \Delta t}\right\},$$

where $\delta\mathbf{x}_\pm = \mathbf{v}_\pm(\mathbf{x}, t)\cdot\Delta t + \Delta\mathbf{x}_\pm$ is the total displacement of the particle in the time Δt and $v_\pm$ are certain constants of the type of diffusion coefficient. Setting $v_+ = v_- = v$, expanding ρ and $P_\pm$ in Taylor series, integrating over $\delta\mathbf{x}_\pm$, and retaining terms of order Δt, we obtain *Fokker–Planck equations* for $\rho(\mathbf{x}, t)$

$$\partial\rho/\partial t = -\nabla\cdot(\rho\mathbf{v}_+) + v\nabla^2\cdot\rho, \qquad \partial\rho/\partial t = -\nabla\cdot(\rho\mathbf{v}_-) - v\cdot\nabla^2\cdot\rho, \tag{8.4}$$

or

$$\begin{aligned} \partial\rho/\partial t &= -\nabla(\rho\mathbf{v}); \qquad \mathbf{u} = v\nabla\ln\rho; \\ \mathbf{v} &= \tfrac{1}{2}(\mathbf{v}_+ + \mathbf{v}_-); \quad \mathbf{u} = \tfrac{1}{2}(\mathbf{v}_+ - \mathbf{v}_-). \end{aligned} \tag{8.5}$$

In stochastic theory, $\mathbf{v}$ and $\mathbf{u}$ are called the ordinary (*current*), and *stochastic velocity* of the particle, respectively.

We now consider the motion of a particle in an external force field $\mathbf{F} = -\nabla U$. Following Kershaw (1964), we can derive equations of the Smoluchowski type for the mean velocities $\mathbf{v}_+$ and $\mathbf{v}_-$ of the particles in accordance with the formulas

$$\mathbf{v}_\pm(\mathbf{x}, t \pm \Delta t) = \frac{1}{N^\pm}\int\left[\mathbf{v}_\pm(\mathbf{x} \mp \delta\mathbf{x}_\pm, t) \pm \frac{\Delta t}{m}\mathbf{f}_\pm(\mathbf{x} \mp \delta\mathbf{x}_\pm, t)\right] \times$$

$$\times\, \rho(\mathbf{x} \mp \delta\mathbf{x}_\pm, t) P_\pm(\mathbf{x} \mp \delta\mathbf{x}_\pm, t; \delta\mathbf{x}_\pm, \Delta t)\, \mathrm{d}^3(\delta\mathbf{x}_\pm); \tag{8.6}$$

$$\mathbf{v}_\pm(\mathbf{x}, t \mp \Delta t) = \frac{1}{N^\mp}\int\left[\mathbf{v}_\pm(\mathbf{x} \pm \delta\mathbf{x}_\mp, t) \mp \frac{\Delta t}{m}\mathbf{f}'_\pm(\mathbf{x} \pm \delta x_\mp, t)\right]\times$$

$$\times\, \rho(\mathbf{x} \pm \delta\mathbf{x}_\mp, t) P_\mp(\mathbf{x} \pm \delta x_\mp, t; \delta\mathbf{x}_\mp, \Delta t)\, \mathrm{d}^3(\delta x_\mp), \tag{8.7}$$

where $\mathbf{f}_\pm$ and $\mathbf{f}'_\pm$ are certain external forces, and

$$N^\pm = \int \rho(\mathbf{x} \mp \delta\mathbf{x}_\pm, t) P_\pm(\mathbf{x} \mp \delta\mathbf{x}_\pm, t; \delta\mathbf{x}_\pm, \Delta t)\, \mathrm{d}^3(\delta\mathbf{x}_\pm)$$

and normalization factors. The upper and lower signs correspond to $\mathbf{v}_+$ and $\mathbf{v}_-$, respectively.

Expanding $\mathbf{v}_\pm$, ρ, $P_\pm$, $\mathbf{f}_\pm$, and $\mathbf{f}'_\pm$ in the Taylor series, integrating over $\delta\mathbf{x}_\pm$, and going to the limit $\Delta t \to 0$, we obtain from (8.6) and (8.7) four possible equations:

$$m\left[\frac{\partial\mathbf{v}_\pm}{\partial t} + (\mathbf{v}_\pm\cdot\nabla)\mathbf{v}_\pm\right] = \mathbf{f}_\pm \pm m\nu\left[\frac{2}{\rho}(\nabla\rho\cdot\nabla)\mathbf{v}_\pm + \nabla^2\mathbf{v}_\pm\right],$$

$$m\left[\frac{\partial\mathbf{v}_\pm}{\partial t} + (\mathbf{v}_\mp\cdot\nabla)\mathbf{v}_\pm\right] = \mathbf{f}'_\pm \mp m\nu\left[\frac{2}{\rho}(\nabla\rho\cdot\nabla)\mathbf{v}_\pm + \nabla^2\mathbf{v}_\pm\right]. \tag{8.8}$$

Going over to the variables $\mathbf{v}$ and $\mathbf{u}$ and adding and subtracting equations (8.8), we obtain the following equations, which describe entirely different processes:

$$\mathrm{d}_c\mathbf{v} - \lambda\mathrm{d}_s\mathbf{u} = (1/m)\mathbf{F}_\lambda^+; \qquad \mathrm{d}_c\mathbf{u} + \lambda\mathrm{d}_s\mathbf{v} = (1/m)\mathbf{F}_\lambda^- \tag{8.9}$$

and

$$\mathrm{d}_c\mathbf{v} - \lambda\mathrm{d}_s\mathbf{v} = (1/m)\mathbf{F}_\lambda; \qquad \mathrm{d}_c\mathbf{u} + \lambda\mathrm{d}_s\mathbf{u} = (1/m)\mathbf{F}'_\lambda, \tag{8.10}$$

where

$$\mathrm{d}_c = \partial/\partial t + (\mathbf{v}\cdot\nabla); \qquad \mathrm{d}_s = (\mathbf{u}\cdot\nabla) + \nu\nabla^2; \qquad \lambda = \pm 1;$$

$$\mathbf{F}_1^+ = \tfrac{1}{2}(\mathbf{f}_+ + \mathbf{f}_-); \qquad \mathbf{F}_{(-1)}^+ = \tfrac{1}{2}(\mathbf{f}'_+ + \mathbf{f}'_-); \qquad \mathbf{F}_1^- = \tfrac{1}{2}(\mathbf{f}'_+ - \mathbf{f}'_-);$$

$$\mathbf{F}_{(-1)}^- = \tfrac{1}{2}(\mathbf{f}_+ - \mathbf{f}_-); \qquad \mathbf{F}_1 = \tfrac{1}{2}(\mathbf{f}_+ + \mathbf{f}'_-); \qquad \mathbf{F}_{(-1)} = \tfrac{1}{2}(\mathbf{f}'_+ + \mathbf{f}_-);$$

$$\mathbf{F}'_1 = \tfrac{1}{2}(\mathbf{f}'_+ - \mathbf{f}_-); \qquad \mathbf{F}'_{(-1)} = \tfrac{1}{2}(\mathbf{f}_+ - \mathbf{f}'_-).$$

The left-hand sides of the obtained equations have a definite parity under the *time-reversal transformation*. Indeed, under $t \to -t$

$$\mathbf{v} \to -\mathbf{v}; \qquad \mathrm{d}_c \to -\mathrm{d}_c$$

$$\mathbf{u} \to \mathbf{u}; \qquad \mathrm{d}_s \to \mathrm{d}_s$$

and we readily conclude that the expression $d_c\mathbf{v} - \lambda d_s\mathbf{u}$ does not change, while $d_c\mathbf{u} + \lambda\, d_s\mathbf{v}$ $(\lambda = \pm 1)$ changes signs under the transformation $t \to -t$. Therefore, the right-hand side of the corresponding equations, which is the force, must be chosen such that the individual equation as a whole remains invariant under $t \to -t$. This requirement is satisfied if we assume that

$$\mathbf{f}_+ \to \mathbf{f}_-, \qquad \mathbf{f}'_+ \to \mathbf{f}'_- \quad \text{as } t \to -t.$$

Then $\mathbf{F}_\lambda^+$ does not change, while $\mathbf{F}_\lambda^-$ reverses its sign, and $\mathbf{F}_1 \to \mathbf{F}_{(-1)}$ and $\mathbf{F}'_1 \to -\mathbf{F}'_{(-1)}$ as $t \to -t$ and, therefore, the four equations (8.10) actually reduce to two equations. We also write the equation

$$\partial\mathbf{u}/\partial t = -\nabla(\mathbf{uv}) - \nu\nabla\cdot(\nabla\cdot\mathbf{v}), \tag{8.11}$$

which follows from the continuity equation and expression (8.5) for the velocity $\mathbf{u}$.

We see that, on the basis of the hypothesis of a stochastic space and Smoluchowski's equations, we have obtained in the framework of Kershaw's approach the same fundamental equations that were obtained by Nelson (1966–1967) and de la Peña-Auerbach and Cetto (1975) (see also the review of Skagerstam, 1975 and the lecture of Santos, 1972) by different routes. Note that Kershaw could not obtain these equations, since he considered only one transition probability, namely $P_+(\Delta\mathbf{x}, \Delta t)$.

Detailed investigations in equations (8.9)–(8.11) are made in stochastic mechanics. In Section 9.3 we consider a question associated with the requirements imposed on the equations of stochastic mechanics. In particular, on the basis of some mathematical assumptions about stochastic processes, Nelson (1966) obtained the first equation in (8.9) with $\lambda = 1$ and (8.11), and showed that if one considers a charged particle and takes the forces

$$\mathbf{F}_1^+ = e\mathbf{E} + \tfrac{e}{e}\mathbf{v} \times \mathbf{H} \quad \text{and} \quad \mathbf{F}_1^+ = -\nabla U,$$

then these equations are equivalent to the Schrödinger equations

$$\frac{\partial\psi}{\partial t} = -\frac{i}{2m\hbar}(-i\hbar\nabla - \tfrac{e}{c}\mathbf{A})^2\psi - i\frac{e}{\hbar}U\psi$$

and

$$i\hbar\frac{\partial\psi}{\partial t} = \left(-\frac{\hbar^2}{2m}\nabla^2 + U\right)\psi$$

respectively. Here $\nu = \hbar/2m$, $\psi = \exp(R + iS)$,

$$R = \tfrac{1}{2}\ln\rho, \qquad \mathbf{grad}\ S = \frac{m}{\hbar}\left(\mathbf{v} + \frac{e}{mc}\mathbf{A}\right),$$

and $\mathbf{A}$, the vector potential, is related to $\mathbf{E}$ and $\mathbf{H}$ by

$$\mathbf{H} = \operatorname{curl} \mathbf{A}, \qquad \mathbf{E} = -\frac{1}{c}\frac{\partial \mathbf{A}}{\partial t} - \nabla \phi.$$

Further, on the basis of the description of the mean local motion of the elements of continuous media (ensembles), which leads to two different kinds of motion during a short time interval Δt, de la Peña-Auerbach and Cetto (1975) obtained equations (8.9) with $\lambda = 1$. These equations can also be reduced to Schrödinger equations for the wave functions ψ and ψ^* if the Lorentz force is chosen as follows:

$$\mathbf{F}_1^+ = -e\nabla\phi - \frac{e}{c}\frac{\partial \mathbf{A}}{\partial t} + \frac{e}{c}\mathbf{v} \times \mathbf{H} \quad \text{and} \quad \mathbf{F}_1^- = \frac{e}{c}\mathbf{u} \times \mathbf{H} + \frac{e}{c}\nu\nabla \times \mathbf{H}.$$

They also discussed the case $\lambda = -1$.

Skagerstam (1976) investigated equations (8.9) and (8.11) from the point of view of the theory of stochastic processes, defined them in a classical configuration space. Recently, Davidson (1979) showed that in the Fènyes–Nelson stochastic model there is an entire class of different dynamical schemes which lead to a Schrödinger equation as the solution of a Markov diffusion problem. As a result, the *diffusion* coefficient ν in the Markov theory is an *arbitrary positive parameter* (see Sections 7.4 and 10.2).

The remaining equations (8.10) have the same form as the Navier–Stokes equations for 'fluids' with velocities $\mathbf{v}$ and $\mathbf{u}$ if ν is formally identified with the coefficient of viscosity, they are studied in Chapter 9.

Finally, we determine

$$\lim_{\Delta t \to 0} \frac{1}{2} \frac{\overline{(\Delta \mathbf{x})^2}}{\Delta t}.$$

By definition,

$$\overline{(\Delta x_+)^2} = \overline{(\Delta x_-)^2} = \lim_{N \to \infty} l^2 \sum_{j=1}^{N} \alpha_j^2 = \Delta t \lim_{N \to \infty} \frac{1}{N} \sum_{j=1}^{N} (\Delta x_j)^2,$$

where $\Delta x_j = l\sqrt{n}\,\alpha_j$, and n is the number of displacements per unit time (see Section 7.3). We also have

$$\overline{(\Delta x_+)^2} = 2\nu_+ \cdot \Delta t = \overline{(\Delta x_-)^2} = 2\nu_- \cdot \Delta t$$

and hence

$$\lim_{\Delta t \to 0} \frac{1}{2} \frac{\overline{(\Delta \mathbf{x})^2}}{\Delta t} = \frac{3}{2} \lim_{N \to \infty} \frac{1}{N} \sum_{j=1}^{N} (\Delta x_j)^2 = \frac{3}{2}\frac{\hbar}{m}.$$

We have arrived at Nelson's relation.

Thus, from the hypothesis of the stochastic nature of space we have obtained stochastic mechanics, identical with quantum mechanics, at least in the mathematical formulation. The connection between the Schrödinger equation and the stochastic theory based on the hypothesis of a stochastic space can be demonstrated by the following simple example. Suppose that at $t = 0$ the wave function has the form

$$\varphi(\mathbf{x}, t = 0) = N \exp(-\mathbf{x}^2/2a^2), \tag{8.12}$$

where N is normalization constant, and a is a positive number; then for $t > 0$ the wave function is determined by means of the Green function of the Schrödinger equation:

$$\varphi(\mathbf{x}, t) = \int d^3x' \, K(\mathbf{x} - \mathbf{x}', t)\varphi(\mathbf{x}', 0),$$

where

$$K(\mathbf{x}, t) = i^{-1}(4\pi\nu t)^{-3/2} \exp(i\mathbf{x}^2/4\nu t), \quad \nu = \hbar/2m.$$

After simple calculations, we obtain

$$|\varphi(\mathbf{x}, t)|^2 = N^2\left[1 + \frac{\hbar^2 t^2}{m^2 a^4}\right]^{-3/2} \exp\left\{-\frac{\mathbf{x}^2}{a^2}\left(1 + \frac{\hbar^2 t^2}{m^2 a^4}\right)^{-1}\right\}. \tag{8.13}$$

We now consider the probability density corresponding to function (8.12), i.e.,

$$\rho(\mathbf{x}, t = 0) = N^2 \exp(-\mathbf{x}^2/a^2) \tag{8.14}$$

and we attempt to determine $\rho(\mathbf{x}, t)$ by means of the stochastic mechanics considered above. Since $\mathbf{u} = \nu \nabla \ln \rho$ the velocity corresponding to expression (8.14) is $\mathbf{u}^0(\mathbf{x}, t = 0) = -(\hbar/ma^2)\mathbf{x}$ and the solution of an equation of type (8.4) for $\mathbf{u}^0 = \mathbf{v}_+$, ($\mathbf{v}_- = 0$) is determined by the relation

$$\rho(\mathbf{x}, t) = N^2(4\pi\nu t)^{-3/2} \int d^3x' \exp\left\{-\frac{[\mathbf{x} - \mathbf{x}' - \mathbf{u}^0(\mathbf{x}')t]^2}{4\nu t} - \frac{\mathbf{x}'^2}{a^2}\right\}$$

$$= N^2\left(1 + \frac{\hbar^2 t^2}{m^2 a^4}\right)^{-3/2} \exp\left\{-\frac{\mathbf{x}^2}{a^2}\left(1 + \frac{\hbar^2 t^2}{m^2 a^4}\right)^{-1}\right\}, \tag{8.15}$$

which is identical to the quantum-mechanical quantity (8.13). The kernel

$$G(\mathbf{x}, t; \mathbf{x}', t' = 0) = (4\pi\nu t)^{-3/2} \exp\{-[\mathbf{x} - \mathbf{x}' - \mathbf{u}^0(\mathbf{x}')t]^2/4\nu t\}$$

of equation (8.15) satisfies the *Fokker–Planck equation*

$$\frac{\partial G}{\partial t} = \beta\frac{\partial \mathbf{x} G}{\partial \mathbf{x}} + \nu\frac{\partial^2}{\partial \mathbf{x}^2} G,$$

where

$$\mathbf{x}\beta = -\mathbf{u}^0(\mathbf{x}), \quad \beta = \hbar/ma^2.$$

8.3. Relativistic Dynamics of Stochastic Particles

As was point out in the Introduction (Section 8.1), in the relativistic case we shall formally consider the motion of a particle which executes a random walk due to the stochastic nature of the four-dimensional Euclidean space $\mathbb{E}^4(\mathbf{x}, \tau)$. Suppose the particle executes N displacements; then its coordinates are determined by the expression $B_\mu = \Sigma_{j=1}^N \beta_j b_\mu$, where β_j is a sequence of N numbers, and the vector $b_\mu = (b_4, \mathbf{b})$ is distributed with the probability

$$w(b_\mu)\,\mathrm{d}^4 b = w(b_\mu b^\mu)\,\mathrm{d}^4 b.$$

Here $b^2 = b_\mu b^\mu = b_4^2 + b^2$.

The probability $W_N(B_\mu)\,\mathrm{d}^4 B$ that B_μ lies in the interval between B_μ, $B_\mu + \mathrm{d}B_\mu$ is given by

$$W_N(B_\mu)\,\mathrm{d}^4 B = \mathrm{d}^4 B(2\pi P_N)^{-2} \exp\left(\frac{-B^2}{2P_N}\right), \qquad P_N = \sum_{j=1}^{N} l^2 \beta_j^2.$$

Defining

$$2\nu s = \overline{B_x^2} = \overline{B_y^2} = \overline{B_z^2} = \overline{B_4^2}$$

or

$$\nu = \frac{1}{2} \lim_{N \to \infty} \frac{1}{N} l^2 \sum_{j=1}^{N} n\beta_j^2, \quad (n = N/s)$$

we obtain

$$\rho(B_\mu) = \lim_{N \to \infty} W_N(B_\mu) = (4\pi\nu s)^{-2} \exp\left(\frac{-B^2}{4\nu s}\right),$$

where s is some positive invariant parameter which can be understood in the present case as the proper time of the particle. Further, as in the three-dimensional case, we define the 'transition probability'

$$P(\Delta y_\mathrm{E}, \Delta s) = (4\pi\nu \cdot \Delta s)^{-2} \exp(-|\Delta y_\mathrm{E}|^2/4\nu \cdot \Delta s)$$

and then the 'Smoluchowski equation' takes the form

$$\rho_\mathrm{E}(x_\mu^E, s + \Delta s) = \int \mathrm{d}^4 y_\mathrm{E}\, \rho_\mathrm{E}(x_\mu^E - y_\mu^E, s) P(y_\mathrm{E}, \Delta s), \tag{8.16}$$

where for convenience we have set $y_\mathrm{E} = \Delta y_\mathrm{E}$, and the diffusion equation becomes

$$\partial \rho_\mathrm{E}/\partial s = \nu \Box_\mathrm{E} \rho_\mathrm{E}, \qquad \Box_\mathrm{E} = \partial^2/\partial x_4^2 + \partial^2/\partial \mathbf{x}^2. \tag{8.17}$$

If $P(y_\mathrm{E}, \Delta s)$ depends on the point x_μ^E, for example,

$$P(x_\mu^E; y_\mathrm{E}, \Delta s) = (4\pi\nu \cdot \Delta s)^{-2} \exp\{-[y_\mathrm{E} - y_\mathrm{E}^0(x^E)]^2/4\nu \cdot \Delta s\},$$

where $(y_E^0)^\mu = u_E^\mu \cdot \Delta s$ and $u_E^\mu = (u_4, \mathbf{u})$ is the four-dimensional Euclidean velocity of the particle, then instead of (8.17) we obtain an equation of the Fokker–Planck type in Euclidean space:

$$\partial \rho_E/\partial s = -\partial_\mu(\rho u_E^\mu) + \nu \Box_E \cdot \rho_E, \qquad \partial_\mu = (\partial/\partial x_4, \nabla).$$

The question of the transition to pseudo-Euclidean space (Minkowski space) now arises. As shown above (Chapters 1 and 2), the idea of space-time stochasticity and the construction of the nonlocal theory of quantized (extended) fields can be realized by a displacement of the co-ordinates such that x_0 acquires a purely imaginary addition ($x_0 \to x_0 + i\tau$), while the co-ordinates $\mathbf{x}$ remain real. It is found that the procedure for displacing the co-ordinate x_0 is deeply connected to the transition between Euclidean and pseudo-Euclidean quantum field theories, and evidently it also has a direct bearing on the relativistically invariant description of extended objects (see Chapters 1, 2 and Efimov, 1977a).

Using this idea, we can write equation (8.16) in the form

$$\rho(x_\mu, s \pm \Delta s) = \int d^4 y_E P_\pm (y_E, \Delta s; \mathbf{x} \mp \mathbf{y}, x_0 + iy_4)\rho(\mathbf{x} \mp \mathbf{y}, x_0 + iy_4, s),$$

where the variables $x_\mu = (x_0, \mathbf{x})$ are pseudo-Euclidean and $P_\pm$ can be chosen in the form

$$P_\pm = (4\pi\nu_\pm \cdot \Delta s)^{-2} \exp\{-(y_E - y_\pm^p)^2/4\nu_\pm \cdot \Delta s\},$$

$$y_\pm^p = (\pm i u_\pm^0 \cdot \Delta s, \mathbf{u}_\pm \cdot \Delta s), \qquad \nu_- = \nu_+ = \nu,$$

where $u_\pm^\mu$ are four-dimensional velocity vectors. From this we readily obtain the two Fokker–Planck equations

$$\partial\rho/\partial s = -\partial_\mu(\rho u_+^\mu) + \nu\Box\rho, \qquad \partial\rho/\partial s = -\partial_\mu(\rho u^\mu) - \nu\Box\rho, \tag{8.18}$$

or, in terms of v^μ and u^μ,

$$\partial\rho/\partial s = -\partial_\mu(\rho v^\mu); \qquad u^\mu = \tfrac{1}{2}(u_+^\mu - u_-^\mu) = -\nu\, \partial^\mu \ln \rho, \tag{8.19}$$

where

$$v^\mu = \tfrac{1}{2}(u_+^\mu + u^\mu); \qquad \partial_\mu = (\partial/\partial x_0, \nabla); \qquad \Box = -\partial^2/\partial x_0^2 + \partial^2/\partial \mathbf{x}^2.$$

We now attempt to obtain equations of the motion of a particle which in accordance with the correspondence principle must take the form (8.8) depending on the choice of the force field in the nonrelativistic limit. In the framework of our scheme, relativistic equations for $u_\pm^\mu$ are obtained from the integral equations

$$u_\pm^\mu(x_\nu, s + \varepsilon\,\Delta s) = \frac{1}{N^\pm} \iint \Big[u^\mu{}_\pm(\mathbf{x} - \varepsilon\mathbf{y}, x_0 + iy_4, s) +$$

$$+ \varepsilon \frac{\Delta s}{m} F^{\mu}_{\pm}(\mathbf{x} - \varepsilon \mathbf{y}, x_0 + iy_4) \Big] \times$$

$$\times \rho(\mathbf{x} - \varepsilon \mathbf{y}, x_0 + iy_4, s) P_{\pm}(\mathbf{x} - \varepsilon \mathbf{y}, x_0 + iy_4, s; y_{\mathrm{E}}, \Delta s)\, \mathrm{d}^4 y_{\mathrm{E}}, \tag{8.20}$$

$$u^{\mu}_{\pm}(x_{\nu}, s - \varepsilon \Delta s) = \frac{1}{N^{\mp}} \int \Big[u_{\pm}(\mathbf{x} + \varepsilon \mathbf{y}, x_0 + iy_4, s) -$$

$$- \varepsilon \frac{\Delta s}{m} F'^{\mu}_{\pm}(\mathbf{x} + \varepsilon \mathbf{y}, x_0 + iy_4) \Big] \times$$

$$\times \rho(\mathbf{x} + \varepsilon \mathbf{y}, x_0 + iy_4, s) P_{\pm}(\mathbf{x} + \varepsilon \mathbf{y}, x_0 + iy_4, s; y_{\mathrm{E}}, \Delta s)\, \mathrm{d}^4 y_{\mathrm{E}}, \tag{8.21}$$

where

$$N^{\pm} = \int \mathrm{d}^4 y_{\mathrm{E}} P_{\pm}(\mathbf{x} - \varepsilon \mathbf{y}, x_0 + iy_4, s; y_{\mathrm{E}}, \Delta s) \rho(\mathbf{x} - \varepsilon \mathbf{y}, x_0 + iy_4, s),$$

$$\varepsilon = \begin{cases} 1 & \text{for } u^{\mu}_{+} \\ -1 & \text{for } u^{\mu}_{-} \end{cases}$$

and $F^{\mu}_{\pm}$ and F'^{μ}_{+} are certain forces.

Expanding the expressions in (8.20) and (8.21) in the Taylor series, integration over $\mathrm{d}^4 y_{\mathrm{E}}$, going to the limit $\Delta s \to 0$, and making some calculations, we obtain the equations

$$m\left(\frac{\partial u^{\mu}_{\pm}}{\partial s} + u^{\lambda}_{\pm}\, \partial_{\lambda} u^{\mu}_{\pm}\right) = F^{\mu}_{\pm} \pm \nu m\left(\frac{2}{\nu} u^{\lambda}\, \partial_{\lambda}\, u^{\mu}_{\pm} + \Box u^{\mu}_{\pm}\right),$$

$$m\left(\frac{\partial u^{\mu}_{\pm}}{\partial s} + u^{\lambda}_{\mp}\, \partial_{\lambda} u^{\mu}_{\pm}\right) = F'^{\mu}_{\pm} \mp \nu m\left(\frac{2}{\nu} u^{\lambda}\, \partial_{\lambda} u^{\mu}_{\pm} + \Box u^{\mu}_{\pm}\right). \tag{8.22}$$

From this we obtain relativistic equations for v^{μ} and u^{μ}:

$$D_c v^{\mu} - \lambda D_s u^{\mu} = \frac{1}{m} \Phi^{(+)\mu}_{\lambda}; \qquad D_c u^{\mu} + \lambda D_s v^{\mu} = \frac{1}{m} \Phi^{(-)\mu}_{\lambda}; \tag{8.23}$$

and

$$D_c v^{\mu} - \lambda D_s v^{\mu} = \frac{1}{m} \Phi^{\mu}_{\lambda}; \qquad D_c u^{\mu} + \lambda D_s u^{\mu} = \frac{1}{m} \Phi'^{\mu}_{\lambda}; \tag{8.24}$$

where

$$D_c = \partial/\partial s + v^\lambda \partial_\lambda; \qquad D_s = u^\lambda \partial_\lambda + \nu\Box, \quad \lambda = \pm 1.$$

The functions $\Phi_\lambda^{(+)\mu}, \ldots, \Phi_\lambda'^{\mu}$ can be expressed in terms of $F_+^\mu, \ldots, F_-'^{\mu}$ in the same way as in the nonrelativistic case. $\Phi_\lambda^{(+)\mu}$ $[\Phi_\lambda^{(-)\mu}]$ does not (does) change sign, and

$$\Phi_1^\mu \Rightarrow \Phi_{(-1)}^\mu \quad \text{and} \quad \Phi_1'^\mu \Rightarrow -\Phi_{(-1)}'^\mu \quad \text{under } s \to -s.$$

Equations (8.23) and (8.24) in conjunction with equation (8.19) are the covariant analog of (8.9), (8.10), and (8.5) in the relativistic case.

Note that the left-hand side of the first equation in (8.23) for $\lambda = 1$ is identical to the expression for the acceleration obtained on the basis of some assumptions in the framework of the mathematical approach of Nelson (see Lehr and Park, 1977; Guerra and Ruggiero, 1978).

We can consider the special case when only the Lorentz force $\Phi_{\lambda=1}^{(+)\mu}$ is present; this is related to the electromagnetic field $F^{\mu\nu D}$ by

$$\Phi_1^{(+)\mu} = \frac{e}{c} F^{\mu\nu} v_\nu = \frac{e}{c}(\partial^\mu A^\lambda - \partial^\lambda A^\mu) v_\lambda.$$

Here A^μ is the electromagnetic potential, for which the *Lorentz condition* is satisfied:

$$\partial_\mu A^\mu = 0.$$

In this case, it is usually assumed that the generalized momentum is a four-gradient of a world scalar S:

$$\partial^\mu S = m v^\mu + \frac{e}{c} A^\mu$$

and then the first equation in (8.23) with $\lambda = 1$ and (8.19) are equivalent to the *Klein–Gordon equation*

$$\left(\partial^\mu - i\frac{e}{c} A^\mu\right)^2 \varphi - \frac{m^2 c^2}{\hbar^2} \varphi = 0. \tag{8.25}$$

The proof is given in Nelson's (1966, 1967) and Leht and Park's (1977) papers, and therefore we shall not give it here.

The equations of (8.23) with $\lambda = 1$ were also investigated by Vigier (1979) in the framework of the approach of de la Peña-Auerbach and Cetto (1975). The external field in this case was chosen in the form

$$\Phi_{\lambda=1}^{(+)\mu} = \frac{e}{c} F^{\mu\rho} v_\rho; \qquad \Phi_{\lambda=1}^{(-)\mu} = \frac{e}{c} F^{\mu\rho} u_\rho + \frac{e}{c} \nu\, \partial^\lambda F^{\lambda\mu}.$$

Such a choice of the field $\Phi_1^{(-)\mu}$ ensures consistency of the second equation in

(8.23) with (8.19) for $\lambda = 1$, as a result of which there is no overdetermination of the physical quantities ρ and v^μ (see Section 9.3).

The covariant analog of (8.10) is provided by equations (8.24), which will be considered in Chapter 9.

8.4. The Two-Body Problem in Stochastic Theory

8.4.1. *The Nonrelativistic Case*

We consider first two interacting nonrelativistic particles. We shall investigate the problem in the framework of a stochastic theory based on Smoluchowski equations for the probability density $\rho(\mathbf{x}_1, \mathbf{x}_2, t)$ of finding the first particle at the point $\mathbf{x}_1$ and the second at the point $\mathbf{x}_2$ at the time t with relative velocities $\mathbf{v}_1(\mathbf{x}_1, \mathbf{x}_2, t)$ and $\mathbf{v}_2(\mathbf{x}_1, \mathbf{x}_2, t)$. It is assumed that the interaction potential $U(\mathbf{x}_1, \mathbf{x}_2)$ between two particles will depend only on the difference $|\mathbf{x}_1 - \mathbf{x}_2|$ between their co-ordinates, i.e., $U(\mathbf{x}_1, \mathbf{x}_2) = U(|\mathbf{x}_1 - \mathbf{x}_2|)$.

Description of the two-body problem in stochastic mechanics is not only a problem of real physical interest, but may play a very important role in the causal action-at-a-distance interpretation of the Aspect–Rapisarda experiments (Aspect *et al.*, 1982a, b) on singlet photon pairs (see Cufaro Petroni and Vigier, 1979, 1983a).

The problem we consider was first studied by Kershaw (1964) for the transition probability $P_+(\Delta\mathbf{x}, \Delta t)$, and he obtained equations that describe the motion of the center of mass of the two particles and their relative motion. As above, we shall investigate the problem of two bodies in the framework of two transition probabilities P_+ and P_-. Here, for the first particle

$$P^1_\pm(\Delta\mathbf{x}^\pm_1, \Delta t) = (2\pi\tau_1 \Delta t/m_1)^{-3/2} \exp[-m_1(\Delta\mathbf{x}^\pm_1)^2/2\tau_1\Delta t],$$

and for the second

$$P^2_\pm(\Delta\mathbf{x}^\pm_2, \Delta t) = (2\pi\tau_2\, \Delta t/m_2)^{-3/2} \exp[-m_2(\Delta\mathbf{x}^\pm_2)^2/2\tau_2\Delta t],$$

where $\tau_1 = 2v_1 m_1$ and $\tau_2 = 2v_2 m_2$.

Without loss of generality, we take $\tau_1 = \tau_2 = \tau$. In terms of the variables

$$\mathbf{r}^+ = \mathbf{x}^+_1 - \mathbf{x}^+_2, \qquad \mathbf{R}^+ = (m_1\mathbf{x}^+_1 + m_2\mathbf{x}^+_2)/(m_1 + m_2),$$

P^1_+ and P^2_+ take the forms $P_+(\Delta\mathbf{r}^+, \Delta t)$ and $P_+(\Delta\mathbf{R}^+, \Delta t)$, where

$$P_+(\Delta\mathbf{r}^+, \Delta t) = \frac{\partial(\mathbf{x}^+_2)}{\partial(\mathbf{r}^+)} \int d^3(\Delta x^+_1) P^1_+(\Delta\mathbf{x}^+_1, \Delta t) P^2_+(\Delta\mathbf{x}^+_1 - \Delta\mathbf{r}^+, \Delta t)$$

$$= (2\pi\tau\, \Delta t/\mu)^{-3/2} \exp[-\mu(\Delta\mathbf{r}^+)^2/2\tau\, \Delta t].$$

Here $\partial(\mathbf{x}^+_2)/\partial(\mathbf{r}^+)$ is the Jacobian of the transition from $\mathbf{x}^+_2$ to $\mathbf{r}^+$, and

$\mu = m_1 m_2/(m_1 + m_2)$. Similarly,

$$P_+(\Delta\mathbf{R}^+, \Delta t) = \frac{\partial(\mathbf{x}_2^+)}{\partial(\mathbf{R}^+)} \int d^3(\Delta x) P_+^1(\Delta\mathbf{x}, \Delta t) P_+^2\left(\frac{M}{m_2}\Delta\mathbf{R}^+ - \frac{m_1}{m_2}\Delta\mathbf{x}, \Delta t\right)$$

$$= (2\pi\tau\,\Delta t/M)^{-3/2}\exp\{-M(\Delta\mathbf{R}^+)^2/2\tau\,\Delta t\},$$

where $M = m_1 + m_2$ is the total mass of the particles. Suppose

$$\mathbf{V}^+(\mathbf{r}, \mathbf{R}, t) = (m_1\,\mathbf{v}_1^+ + m_2\,\mathbf{v}_2^+)/M; \qquad \mathbf{C}^+(\mathbf{r}, \mathbf{R}, t) = \mathbf{v}_1^+ - \mathbf{v}_2^+,$$

here

$$\mathbf{r} = \mathbf{x}_1^+ + \mathbf{x}_1^- - \mathbf{x}_2^+ - \mathbf{x}_2^- = \mathbf{x}_1 - \mathbf{x}_2;$$

$$\mathbf{R} = [m_1(\mathbf{x}_1^+ + \mathbf{x}_1^-) + m_2(\mathbf{x}_2^+ + \mathbf{x}_2^-)]/M = (m_1\mathbf{x}_1 + m_2\mathbf{x}_2)/M.$$

Then the total variations $\delta\mathbf{R}^+$ and $\delta\mathbf{r}^+$ are determined by the equations

$$\delta\mathbf{R}^+ = \mathbf{V}^+\,\Delta t + \Delta\mathbf{R}^+ \quad \text{and} \quad \delta\mathbf{r}^+ = \mathbf{C}^+\,\Delta t + \Delta\mathbf{r}^+$$

respectively. Analogously with the earlier investigations,

$$P_R^+(\delta\mathbf{R}^+, \Delta t; \mathbf{r}, \mathbf{R}, t) = (2\pi\tau\,\Delta t/M)^{-3/2}\exp\{-M(\delta\mathbf{R}^+ - \mathbf{V}^+\,\Delta t)^2/2\tau\,\Delta t\},$$

and

$$P_r^+(\delta\mathbf{r}^+, \Delta t; \mathbf{r}, \mathbf{R}, t) = (2\pi\tau\,\Delta t/\mu)^{-3/2}\exp\{-\mu(\delta\mathbf{r}^+ - \mathbf{C}^+\,\Delta t)^2/2\tau\,\Delta t\}$$

and at the same time

$$\rho(\mathbf{r}, \mathbf{R}, t) = \rho(\mathbf{x}_1, \mathbf{x}_2, t)\,\partial(\mathbf{x}_1, \mathbf{x}_2)/\partial(\mathbf{r}, \mathbf{R})$$

satisfies the equation

$$\rho(\mathbf{r}, \mathbf{R}, t + \Delta t) = \iint d^3(\delta\mathbf{r}^+)\,d^3(\delta R^+)\rho(\mathbf{r} - \delta\mathbf{r}^+, \mathbf{R} - \delta\mathbf{R}^+, t) \times$$

$$\times P_r^+(\delta\mathbf{r}^+, \Delta t; \mathbf{r} - \delta\mathbf{r}^+, \mathbf{R} - \delta\mathbf{R}^+, t) \times$$

$$\times P_R^+(\delta\mathbf{R}^+, \Delta t; \mathbf{r} - \delta\mathbf{r}^+, \mathbf{R} - \delta\mathbf{R}^+, t)$$

$$= \rho - \Delta t\{\nabla_r(\rho\mathbf{C}^+) + \nabla_R(\rho\mathbf{V}^+)\} + \Delta t\frac{\tau}{2}\left\{\frac{\nabla_r^2\rho}{\mu} + \frac{\nabla_R^2}{M}\right\},$$

whence

$$\partial\rho/\partial t = -\nabla_r(\rho\mathbf{C}^+) - \nabla_R(\rho\mathbf{V}^+) + \nu_\mu\nabla_r^2\rho + \nu_M\nabla_R^2\rho \tag{8.26}$$

where

$$\nu_\mu = \tau/2\mu, \qquad \nu_M = \tau/2M.$$

The corresponding equations for ρ obtained on the basis of the concept of the transition probabilities P^1_- and P^2_- take the form

$$\rho(\mathbf{r},\mathbf{R},t-\Delta t)=\iint d^3(\delta r^-)\,d^3(\delta R^-)\rho(\mathbf{r}+\delta\mathbf{r}^-,\mathbf{R}+\delta\mathbf{R}^-,t)\times$$

$$\times\, P_r^-(\delta\mathbf{r}^-,\Delta t;\mathbf{r}+\delta\mathbf{r}^-,\mathbf{R}+\delta\mathbf{R}^-,t)\times$$

$$\times\, P_R^-(\delta\mathbf{R}^-,\Delta t;\mathbf{r}+\delta\mathbf{r}^-,\mathbf{R}+\delta\mathbf{R}^-,t)$$

or

$$\partial\rho/\partial t=-\nabla_r(\rho\mathbf{C}^-)-\nabla_R(\rho\mathbf{V}^-)-\nu_\mu\nabla_r^2\rho-\nu_M\nabla_R^2\rho, \tag{8.27}$$

where

$$\mathbf{C}^-=\mathbf{v}_1^- -\mathbf{v}_2^-;\qquad \mathbf{V}^-=(m_1\mathbf{v}_1^- + m_2\mathbf{v}_2^-)/M.$$

From equations (8.26) and (8.27), we obtain

$$\partial\rho/\partial t=-\nabla_r(\rho\mathbf{V}_r)-\nabla_R(\rho\mathbf{V}_c);$$

$$\mathbf{u}_c=\nu_M\nabla_R\ln\rho;\qquad \mathbf{u}_r=\nu_\mu\nabla_r\ln\rho. \tag{8.28}$$

Here

$$\mathbf{V}_c=\tfrac{1}{2}(\mathbf{V}^+ +\mathbf{V}^-)=(m_1\mathbf{v}_1+m_2\mathbf{v}_2)/M,\qquad \mathbf{V}_r=\tfrac{1}{2}(\mathbf{C}^+ +\mathbf{C}^-)=\mathbf{v}_1-\mathbf{v}_2$$

,

$$\mathbf{u}_c=\tfrac{1}{2}(\mathbf{V}^+ -\mathbf{V}^-)=(m_1\mathbf{u}_1+m_2\mathbf{u}_2)/M,\qquad \mathbf{u}_r=\tfrac{1}{2}(\mathbf{C}^+ -\mathbf{C}^-)=\mathbf{u}_1-\mathbf{u}_2.$$

We recall that the potential $U(r)$ acts only on the velocity $\mathbf{C}$, then the equations for $\mathbf{C}^\pm$ and $\mathbf{V}^\pm$ have the form

$$\mathbf{V}^\pm(\mathbf{r},\mathbf{R},t\pm\Delta t)=\frac{1}{N^\pm}\iint d^3(\delta r^\pm)\,d^3(\delta R^\pm)\mathbf{V}^\pm(\mathbf{r}\mp\delta\mathbf{r}^\pm,\mathbf{R}\mp\delta\mathbf{R}^\pm,t)\times$$

$$\times\,\rho(\mathbf{r}\mp\delta\mathbf{r}^\pm,\mathbf{R}\mp\delta\mathbf{R}^\pm,t)P_r^\pm P_R^\pm;$$

$$\mathbf{C}^\pm(\mathbf{r},\mathbf{R},t\pm\Delta t)=\frac{1}{N^\pm}\iint d^3(\delta r^\pm)\,d^3(\delta R^\pm)\times$$

$$\times\left[\mathbf{C}^\pm(\mathbf{r}\mp\delta\mathbf{r}^\pm,\mathbf{R}\mp\delta\mathbf{R}^\pm,t)\mp\Delta t\frac{1}{\mu}\nabla_r U(r\mp\delta r^\pm)\right]\times$$

$$\times\,\rho(\mathbf{r}\mp\delta\mathbf{r}^\pm,\mathbf{R}\mp\delta R^\pm,t)P_r^\pm P_R^\pm.$$

where

$$N^\pm=\iint d^3(\delta r^\pm)\,d^3(\delta R^\pm)\rho(r\mp\delta r^\pm,R\mp\delta R^\pm,t)P_r^\pm P_R^\pm.$$

As usual, expanding $\mathbf{V}^\pm,\mathbf{C}^\pm,\mathbf{P}_r^\pm,\mathbf{P}_R^\pm$, ρ and U in Taylor series, integrating and retaining terms only of order in Δt, we obtain

$$M[d\mathbf{V}^{\pm}/dt + (\mathbf{V}^{\pm}\nabla_R)\mathbf{V}^{\pm} + (\mathbf{C}^{\pm}\nabla_r)\mathbf{V}^{\pm}]$$
$$= \pm M\left\{\nu_\mu \rho^{-1}[\nabla_r^2(\rho\mathbf{V}^{\pm}) - \mathbf{V}^{\pm}\nabla_r^2\rho] + \nu_M \rho^{-1}[\nabla_R^2(\rho\mathbf{V}^{\pm}) - \right.$$
$$\left. - \mathbf{V}^{\pm}\nabla_R^2\rho]\right\};$$

$$\mu[\partial\mathbf{C}^{\pm}/\partial t + (\mathbf{V}^{\pm}\nabla_R)\mathbf{C}^{\pm} + (\mathbf{C}^{\pm}\nabla_r)\mathbf{C}^{\pm}]$$
$$= -\nabla_r U(r) \pm \mu\{\nu_\mu \rho^{-1} \times [\nabla_r^2(\rho\mathbf{C}^{\pm}) - \mathbf{C}^{\pm}\nabla_r^2\rho] +$$
$$+ \nu_M \rho^{-1}[\nabla_R^2(\rho\mathbf{C}^{\pm}) - \mathbf{C}^{\pm}\nabla_R^2\rho]\}.$$

Going over to the variables $\mathbf{V}_c, \mathbf{u}_c, \mathbf{V}_r$ and $\mathbf{u}_r$ we obtain the equations

$$d_s'\mathbf{V}_c - d_u'\mathbf{u}_c = 0; \qquad d_s'\mathbf{V}_r - d_u'\mathbf{u}_r = -\nabla_r U/\mu, \tag{8.29}$$

where

$$d_s' = \frac{\partial}{\partial t} + (\mathbf{V}_c\nabla_R) + (\mathbf{V}_r\nabla_r); \qquad d_u' = (\mathbf{u}_c\nabla_R) + (\mathbf{u}_r\nabla_r) + \nu_\mu\nabla_r^2 + \nu_M\nabla_R^2.$$

Thus, the operators d_s' and d_u' decompose into a sum of two independent parts. Accordingly, we can seek $\rho(\mathbf{r}, \mathbf{R}, t)$ in the form of the product

$$\rho(\mathbf{r}, \mathbf{R}, t) = \rho_r(\mathbf{r}, t)\rho_R(\mathbf{R}, t),$$

and the variables $\mathbf{V}_{c,r}$ and $\mathbf{u}_{c,r}$ can be set equal to

$$\mathbf{V}_r(\mathbf{r}, \mathbf{R}, t) = \mathbf{V}_r(\mathbf{r}, t); \qquad \mathbf{u}_r(\mathbf{r}, \mathbf{R}, t) = \mathbf{u}_r(\mathbf{r}, t);$$

$$\mathbf{V}_c(\mathbf{r}, \mathbf{R}, t) = \mathbf{V}_c(\mathbf{R}, t); \qquad \mathbf{u}_c(\mathbf{r}, \mathbf{R}, t) = \mathbf{u}_c(\mathbf{R}, t),$$

where $\rho_R, \mathbf{V}_c$ and $\mathbf{u}_c$ describe the motion of the center of mass (as the free motion of a particle with mass $m_1 + m_2$) and ρ_r, $\mathbf{V}_r$, and $\mathbf{u}_r$ describe the relative motion of the particle [as the motion of a particle of mass μ in the centrally symmetric field $U = U(r)$]; then equations (8.28) and (8.29) take the form

$$\partial\rho_R/\partial t = -\operatorname{div}(\rho_R\mathbf{V}_c); \qquad d_R'\mathbf{V}_c - d_R''\mathbf{u}_c = 0 \tag{8.30}$$

and

$$\partial\rho_r/\partial t = -\operatorname{div}(\rho_r\mathbf{V}_r); \qquad d_r'\mathbf{V}_r - d_r''\mathbf{u}_r = -\nabla_r U(r)/\mu, \tag{8.31}$$

where

$$d_R' = \partial/\partial t + (\mathbf{V}_c\nabla_R), \qquad d_R'' = (\mathbf{u}_c\nabla_R) + \nu_M\nabla_R^2,$$
$$d_r' = \partial/\partial t + (\mathbf{V}_r\nabla_r), \qquad d_r'' = (\mathbf{u}_r\nabla_r) + \nu_\mu\nabla_r^2.$$

Following Nelson, equations (8.30) and (8.31) can be *linearized* by the substitutions

$$q_r = \tfrac{1}{2}\ln\rho_r, \qquad q_R = \tfrac{1}{2}\ln\rho_R, \qquad \mathbf{u}_r = 2\nu_\mu \nabla_r q_r,$$

$$\mathbf{u}_c = 2\nu_M \nabla_R q_R, \qquad \mathbf{V}_r = 2\nu_\mu \nabla_r \delta_r; \qquad \mathbf{V}_c = 2\nu_M \nabla_R \delta_R;$$

$$\varphi_r = \varphi_r(\mathbf{r}, t) = \exp(q_r + i\,\delta_r); \qquad \varphi_R = \varphi_R(\mathbf{R}, t) = \exp(q_R + i\,\delta_R).$$

We then obtain the following two equations for φ_r and φ_R:

$$\partial\varphi_r/\partial t = i\nu_\mu \nabla_r^2 \varphi_r - i(1/2\mu\nu_\mu)U(r)\varphi_r \tag{8.32}$$

and

$$\partial\varphi_R/\partial t = i\nu_M \nabla_R^2 \varphi_R \tag{8.33}$$

respectively. The last two equations are formally identical with Schrödinger equations for φ_r and φ_R if we set $\tau = \hbar$, i.e.,

$$\nu_M = \hbar/2M, \qquad \nu_\mu = \hbar/2\mu.$$

8.4.2. The Relativistic Case

We now turn to the problem of two relativistic particles. In this section, we consider the stochastic behavior of two identical and correlated relativistic scalar particles, since this problem has real physical interest. Our investigation is based on equations of the Smoluchowski type for the probability $\rho(x_1^\nu, x_2^\mu, s_1, s_2)$ for finding the first particle at the point x_1^ν and the second at the point x_2^μ at the 'time' s_1 and s_2 and for their relative velocities v_1^μ $(x_1^\lambda, x_2^\nu, s_1, s_2)$ and $v_2^\mu(x_1^\lambda, x_2^\nu, s_1, s_2)$, respectively.

For the direct generalization of the results obtained above for the single-particle model to the relativistic two-particle case, it was found to be mathematically more convenient to introduce the eight-dimensional configuration space considered by Cufaro Petroni and Vigier (1979).

In this space, the positions of two particles and their relative velocities are determined by the eight-component vectors X^i and $v^i (i = 1, \ldots, 8)$, respectively, where

$$\{X^i\}_{i=1,\ldots,8} \equiv \{x_1^\mu, x_2^\nu\}_{\mu,\nu=0,1,2,3}; \qquad \{v^i(X^j, s_1, s_2)\} \equiv \{v_1^\mu, v_2^\nu\}.$$

Here, x_1^μ and x_2^ν are the co-ordinate vectors of each particle. The metric tensor

g_{ij} in this space is defined as

$$g_{ij} = \begin{pmatrix} 1 & 0 & 0 & 0 & 0 & 0 & 0 & 0 \\ 0 & -1 & 0 & 0 & 0 & 0 & 0 & 0 \\ 0 & 0 & -1 & 0 & 0 & 0 & 0 & 0 \\ 0 & 0 & 0 & -1 & 0 & 0 & 0 & 0 \\ 0 & 0 & 0 & 0 & 1 & 0 & 0 & 0 \\ 0 & 0 & 0 & 0 & 0 & -1 & 0 & 0 \\ 0 & 0 & 0 & 0 & 0 & 0 & -1 & 0 \\ 0 & 0 & 0 & 0 & 0 & 0 & 0 & -1 \end{pmatrix}$$

and

$$X^2 = X_i X^i = g_{ij} X^i X^j = (x_1)^2 + (x_2)^2.$$

If $x_1^\mu(s_1)$ and $x_2^\nu(s_2)$ are the trajectors of each particle, then in the eight-dimensional space their common trajectory will be $X^j(s_1, s_2)$.

As in the three- and four-dimensional cases, we introduce the two-particle analog $P(X^j, s_1, s_2, \Delta s_1, \Delta s_2)$ of the single-particle transition probabilities $P(\mathbf{x}, t, \Delta t)$ and $P(x^\mu, s, \Delta s)$ used above. If the particles are not correlated, $P(X^j, s_1, s_2, \Delta s_1, \Delta s_2)$ can be factorized:

$$P(X^j, s_1, s_2, \Delta s_1, \Delta s_2) = P_1(x_1^\mu, s_1, \Delta s_1) P_2(x_2^\nu, s_2, \Delta s_2).$$

Without loss of generality, we choose the gauge $\Delta s_1 = \Delta s_2 = \Delta s$; then the equation of *Smoluchowski type* for $\rho(X^j, s_1, s_2)$ takes the form

$$\begin{aligned} \rho(X^j, s_1 \pm \Delta s, s_2 \pm \Delta s) = \int & \mathrm{d}^8 y_{\mathrm{E}}\, \rho(X^I \mp Y^I, X^1 + iY_{\mathrm{E}}^1, X^5 + iY_{\mathrm{E}}^5, s_1, s_2) \times \\ & \times P^{\pm}(X^I \mp Y^I, X^1 + iY_{\mathrm{E}}^1, X^5 + \\ & + iY_{\mathrm{E}}^5, s_1, s_2; \Delta s, Y_{\mathrm{E}}^i), \end{aligned} \tag{8.34}$$

where X^I and Y^I ($I = 2, 3, 4, 6, 7, 8$) are the spatial parts of the vectors X^i and Y^i, respectively.

Taking into account the explicit form of $P^{\pm}$,

$$\begin{aligned} P^{\pm} &= (4\pi v_{\pm} \cdot \Delta s)^{-4} \exp\{-(Y_{\mathrm{E}}^i - Y_{\pm}^i)^2 / 4 v_{\pm} \Delta s\}, \\ Y_{\pm}^i &= (\pm i v_{\pm}^1 \Delta s, \pm i v_{\pm}^5 \Delta s, v_{\pm}^I \Delta s), \end{aligned} \tag{8.35}$$

we obtain from (8.34) the differential equations

$$\begin{aligned} &\partial\rho/\partial s_1 + \partial\rho/\partial s_2 + \partial_i(\rho v_+^i) - v_+ \Box \rho = 0, \\ &\partial\rho/\partial s_1 + \partial\rho/\partial s_2 + \partial_i(\rho v_-^i) + v_- \Box \rho = 0, \\ &\partial_i = \partial/\partial X^i, \qquad -\partial_i \partial^i = \Box = \Box_1 + \Box_2. \end{aligned} \tag{8.36}$$

Here, we have set $\nu_- = \nu_+ = \nu$, where ν is the diffusion coefficient, and v^i_+ and v^i_- are the forward and backward velocities. Going over to the variables

$$v^i = \tfrac{1}{2}(v^i_+ + v^i_-); \qquad u^i = \tfrac{1}{2}(v^i_+ - v^i_-)$$

and adding and subtracting equations (8.36), we obtain

$$\partial\rho/\partial s_1 + \partial\rho/\partial s_2 + \partial_i(\rho v^i) = 0; \qquad u^i = -\nu\,\partial^i \ln \rho, \tag{8.37}$$

where v^i is the ordinary (regular or current) and u^i the stochastic velocity of the two-particle system.

In our scheme, conservation of mass (of the probability density multiplied by the volume) means that there is no loss of mass through any hypersurfaces characterized by the vectors v_1^μ and v_2^ν; in this connection, we adopt the physical hypothesis that the total number of particles (i.e., the pair in the real space-time) is conserved. Then we can write

$$\begin{aligned}\partial/\partial s_1 + \partial\rho/\partial s_2 &= (\mathbf{d}\rho\cdot\mathbf{v}_1) + (\mathbf{d}\rho\cdot\mathbf{v}_2)\\ &= (\partial\rho/\partial x_1^0)v_1^0 + (1-\beta_1^2)^{-1/2}(\mathbf{v}_1\cdot\partial\rho/\partial\mathbf{x}_1) +\\ &\quad + (\partial\rho/\partial x_2^0)v_2^0 + (1-\beta_2^2)^{-1/2}(\mathbf{v}_2\cdot\partial\rho/\partial\mathbf{x}_2) = 0,\end{aligned}$$

$(\beta_i^2 = v_i^2/c^2,\ i = 1, 2)$ and our continuity equation in the configuration space becomes

$$\partial_i(\rho v^i) = 0 \tag{8.38}$$

or, in terms of v^i and u^i

$$-u^i v_i + \nu\cdot\partial^i v_i = 0. \tag{8.39}$$

In the case of a two-particle system, following Kershaw, we can derive equations of type (8.34) for the mean velocities $v^i_\pm(X^j, s_1, s_2)$ in some external field $F^i_\pm(X^j, s_1, s_2)$:

$$\begin{aligned}&v^i_\pm(X^j, s_1 + \varepsilon\,\Delta s, s_2 + \varepsilon\,\Delta s)\\ &= \frac{1}{N^\pm}\int \mathrm{d}^8 Y_\mathrm{E}\Big[v^i_\pm(X^I - \varepsilon Y^I, X^1 + iY^1_\mathrm{E}, X^5 + iY^5_\mathrm{E}, s_1, s_2)\\ &\quad + \varepsilon\frac{\Delta s}{M}F^i_\pm(X^I - \varepsilon Y^I, X^1 + iY^1_\mathrm{E}, X^5 + Y^5_\mathrm{E}, s_1, s_2)\Big]\times\\ &\quad\times \rho(X^I - \varepsilon Y^I, X^1 + iY^1_\mathrm{E}, X^5 + iY^5_\mathrm{E}, s_1, s_2)\times\\ &\quad\times P^\pm(X^I - \varepsilon Y^I, X^1 + iY^1_\mathrm{E}, X^5 + iY^5_\mathrm{E}, s_1, s_2; Y^i_\mathrm{E}, \Delta s),\end{aligned} \tag{8.40}$$

where

$$N^\pm = \int \mathrm{d}^8 Y_\mathrm{E}\, P^\pm(X^I - \varepsilon Y^I, X^1 + iY^1_\mathrm{E}, \ldots, Y^i_\mathrm{E}, \Delta s)\,\rho(X^I - \varepsilon Y^I, \ldots, s_2)$$

are normalization factors; M is some effective mass (it can be of a matrix form with respect to m – mass of the scalar particle) of our two-particle system; and

$$\varepsilon = \begin{cases} 1 & \text{for } v^i_+, \\ -1 & \text{for } v^i_-. \end{cases}$$

In our case, equations (8.40) lead to the differential equations

$$\frac{\partial u^i_\pm}{\partial s_1} + \frac{\partial u^i_\pm}{\partial s_2} + u^j_\pm\, \partial_j u^i_\pm = \frac{F^i_\pm}{M} \pm \nu\left(\frac{2}{\nu} u^j\, \partial_j u^i_\pm + \Box u^i_\pm\right). \tag{8.41}$$

Adding equation (8.41), we obtain

$$D_c v^i - D_s u^i = \frac{1}{2M}(F^i_+ + F^i_-) = F^i/M;$$

$$D_c = \partial/\partial s_1 + \partial/\partial s_2 + v^i\, \partial_i; \qquad D_s = u^i\, \partial_i + \nu\Box. \tag{8.42}$$

Equation (8.42) in conjunction with the continuity equation (8.38) is the covariant analog of the single-particle case for a two-particle system.

Note that the left-hand side of equation (8.42) is identical to the expression for the 'acceleration' obtained by Cufaro Petroni and Vigier (1979) on the basis of some assumptions in the framework of the mathematical approach of Nelson (1966, 1967), and Guerra and Ruggiero (1978).

The coupled pair of *nonlinear differential equations* (8.39) and (8.42) can be linearized if we set

$$v^i = \frac{1}{m}\, \partial^i \phi \tag{8.43}$$

as before, where $\phi(X^j, s_1, s_2)$ is the phase function determined by the equation

$$\phi(X^j, s_1, s_2) = \tfrac{1}{2}mc^2(s_1 + s_2) + S(X^j). \tag{8.44}$$

Using expressions (8.37), (8.43), and (8.44), on the basis of equations (8.38) and (8.42) we obtain the *equation of Hamilton–Jacobi* type

$$(\partial_i\, \partial^i - \partial_i S\, \partial^i S/\hbar - 2m^2 c^2/\hbar^2)R = 0 \tag{8.45}$$

for the two-particle system in the case when there is no external force: $F^i \equiv 0$. Here, we have set $R = \rho^{1/2}$, $\nu = \hbar/2m$. From equation (8.45) there follows a continuity equation of the form

$$2\, \partial_i R\, \partial^i S + R\, \partial_i\, \partial^i S = 0.$$

Finally, we have a formal equation for $\psi = R \exp(i\hbar^{-1} S)$:

$$(\Box - 2m^2c^2/\hbar^2)\psi = 0. \tag{8.46}$$

In the nonrelativistic limit equation (8.46) leads to an ordinary two-particle

Schrödinger equation for

$$\psi(\mathbf{x}_1, \mathbf{x}_2, t) = R(\mathbf{x}_1, \mathbf{x}_2, t) \exp(i\hbar^{-1} S)$$

which decomposes into two equations:

$$\partial P/\partial t + \nabla_1(P \nabla_1 S/m) + \nabla_2(P \nabla_2 S/m) = 0,$$

where $P = R^2 = \psi^* \psi$, and

$$\partial S/\partial t + (1/m)(\nabla_1 S)^2 + (1/m)(\nabla_2 S)^2 + Q = 0.$$

Here,

$$Q = -(\hbar^2/2m)(\nabla_1^2 R/R + \nabla_2^2 R/R)$$

is some potential, and it is usually called the *nonlocal quantum potential* of the two bodies (see, for example, Cufaro Petroni and Vigier, 1979).

We note in conclusion that the physical consequences of the results obtained above are discussed by Cufaro Petroni and Vigier (1979, 1983a).

Chapter 9

Selected Topics in Stochastic Mechanics

It should be noted that because of the fact that stochastic mechanics is a nonlinear system, it will contain much physical information, and in particular may admit some interesting solutions describing such phenomena as solitary or soliton-like waves, turbulent and Feynman type processes, and so on. Recently, studies in this direction have been initiated. A role of the Feynman path integral theory in stochastic mechanics has been discussed by Guerra (1981) and Moore (1980). We believe that the full clarification of the deep connection between these two approaches will be a major step toward a better understanding of the physical foundations of quantum mechanics.

There are some studies of solitary waves (Efinger, 1981) and soliton waves (Gueret and Vigier, 1982) associated with the space-time structures near particles and to the stochastic interpretation of quantum mechanics, respectively. A possibility of the origin of processes of such types will be discussed in Section 11.8.

Another interesting problem in stochastic mechanics is how to construct a comprehensive stochastic scheme allowing stochastic processes in momentum space to be considered. Some results are given by Shucker (1980b) and de Falso *et al.* (1982, 1983), where it is found that quantum-mechanical momentum can be read from the asymptotic behavior of Nelson's sample paths and the position-momentum uncertainty product is obtained too.

In this chapter we shall investigate the equation of motion, the Cauchy problem for stochastic mechanics, and some its specific problems such as the *self-turbulent phenomenon* and *Feynman-type processes.* We also attempt to describe quantum-mechanical momentum from the point of view of the stochastic processes under consideration.

9.1. A Stochastic Derivation of the Sivashinsky Equation for the Self-Turbulent Motion of a Free Particle

Sivashinsky (1978, 1980) noted a formal analog between the equation of motion for a flame front and the Hamilton–Jacobi equation for the motion of a free particle. He showed that if in the equation for the flame front one introduces terms with higher derivatives, describing the structural perturbation of the front, the front is unstable with respect to long-wavelength perturbations (Sivashinsky, 1977; Michelson and Sivashinsky, 1977). As a result, the original deterministic equation can generate solutions of a stochastic type. An attempt was made to interpret the equation with higher derivatives as a *Hamilton–Jacobi equation* describing the motion of a 'quantum' particle.

However, in the framework of Sivashinsky's approach the choice of the potential (the self-interaction potential of the particle) which 'generates the turbulence' in the Hamilton–Jacobi equation is not unique and does not have a clear physical justification.

This section is devoted to the derivation of a Sivashinsky equation for the self-turbulent motion of a free particle in the framework of stochastic theory on the basis of the hypothesis of a stochastic space-time (see Chapter 1). We are therefore attempting to justify the occurrence of the potential which generates the turbulence in the equation of motion of the particle.

We consider first the nonrelativistic motion of a single scalar particle in the stochastic space $\mathbb{R}^3(\hat{\mathbf{x}})$ with coordinates $\hat{\mathbf{x}} = \mathbf{x} + \mathbf{b}$. In accordance with the results in Chapter 1, by the physical quantity $f(\mathbf{x}, t)$ we shall understand $\langle f(\hat{\mathbf{x}}, t)\rangle_{\mathbb{R}^3(\hat{\mathbf{x}})}$, averaged with respect to the measure $w(\mathbf{b})$ at a given time t (in particular, see formula (1.6) in Chapter 1). The assumption that the stochastic component of the space $\mathbb{R}^3(\hat{\mathbf{x}})$ is small means that

$$\begin{aligned} F(\mathbf{x}, t) &= \langle f(\mathbf{x} + \mathbf{b}, t)\rangle_{\mathbb{R}^3(\hat{\mathbf{x}})} = \int \mathrm{d}^3 b w(\mathbf{b}) f(\mathbf{x} + \mathbf{b}, t) \\ &= \langle f(\mathbf{x}, t) + b_i(\partial/\partial x_i) f(\mathbf{x}, t) + \tfrac{1}{2} b_i b_j (\partial^2/\partial x_i\, \partial x_j) f(\mathbf{x}, t) + \cdots \rangle \\ &\approx f(\mathbf{x}, t) + l^2\, \Delta f(\mathbf{x}, t). \end{aligned} \tag{9.1}$$

It is assumed that $w(\mathbf{b}) = w(-\mathbf{b})$.

It should be noted that with a suitable choice of the function $w(\mathbf{b})$ the method of averaging (9.1), as shown above in Chapter 1, which ensures the transition from the small to the large scale, leads to an entirely new object:

$$F_n(\mathbf{x}, t) = \sum_{n=0}^{\infty} \frac{c_n}{(2n)!} (l^2 \nabla^2)^n f(\mathbf{x}, t) = \int \mathrm{d}^3 y\, K(\mathbf{x} - \mathbf{y}) f(\mathbf{y}, t),$$

and in the relativistic case

$$F_n(x_\mu, s) = \sum_{n=0}^{\infty} \frac{c_n'}{(2n)!} (l^2 \Box)^n f(x_\mu, s) = \int \mathrm{d}^4 y\, K(x - y) f(y, s),$$

where

$$K(\mathbf{x}) = \sum_{n=0}^{\infty} \frac{c_n}{(2n)!} (l^2 \nabla^2)^n\, \delta^{(3)}(x); \qquad K(x_\mu) = \sum_{n=0}^{\infty} \frac{c_n'}{(2n)!} (l^2 \Box)^n\, \delta^{(4)}(x),$$

and the coefficient c_n and (c'_n) depend on the particular choice of the measure $w(\mathbf{b})$ $[w(b^\mu_E)]$. An object of this kind was constructed for the first time by Blokhintsev (1973a) in quantum field theory for the special case when $b_\mu = a\gamma_\mu$, where a is a stochastic variable and γ_μ are the ordinary Dirac matrices. In quantum field theory, such an object has been investigated in Chapters 1 and 2 from the point of view of generalized functions (or distributions) $K(x)$ whose space-time properties depend essentially on different sequences of the coefficient $\{c'_n\}$. It is shown (in Chapter 1) that an object constructed by means of such generalized functions $K(x)$ is smeared (nonlocal) in space (see also Efimov, 1977a).

Thus, the procedure for averaging (9.1) in the framework of our hypothesis leads to a nonlocal object. However, such averaging cannot change the *physical nature* of the object (for example, if it were stochastic before the averaging, it remains so after it) but merely the spatial structure of the object is changed, which is smeared in a certain region determined by l. In calculations of local and nonlocal objects such as the velocity $\mathbf{v}(\mathbf{x}, t)$, the randomness effect which arises from the fluctuation of the spatial coordinates accumulates with the course of time, is manifested only in the dynamical aspect, and does not depend on the intermediate averaging procedure (9.1). Then, in our case, the local $f(\mathbf{x}, t)$, and the nonlocal, $F_n(\mathbf{x}, t)$, objects can be assumed to be stochastic quantities. At the present time, it is not clear how one should write down the probability equation (an equation of the *Chapman–Kolmogorov type*; see Chapter 6) for a nonlocal stochastic quantity $F_n(\mathbf{x}, t)$. Therefore, in a rough approximation, when the parameter l is infinitesimally small, we shall assume that $F_n(\mathbf{x}, t)$ is determined by a finite sum of the form (9.1). The object

$$F_n(\mathbf{x}, t) = f(\mathbf{x}, t) + f_l(\mathbf{x}, t), \tag{9.2}$$

where $f_l(\mathbf{x}, t)$ is determined by a finite sum of a series in the parameter l, as shown above in Section 1.4.6, possesses like $f(\mathbf{x}, t)$ a local property and, therefore, for $F_n(\mathbf{x}, t)$ in (9.2) we can write an equation of the Chapman–Kolmogorov type.

As a first approximation in the parameter l, we shall ignore the second term in (9.2), i.e.,

$$\langle f(\hat{\mathbf{x}}, t) \rangle = F_n(\mathbf{x}, t) \approx f(\mathbf{x}, t).$$

It is to this approximation that our previous results apply (Chapter 8), and in this section we shall not ignore the second term in expression (9.2).

We now attempt to obtain the general form of the dynamical equations of a scalar particle in the stochastic space when a term of order l^2 is present in the expressions for the velocity and the force in accordance with equation (9.2):

$$\mathbf{v}_\pm \to \mathbf{v}_\pm + \mathbf{v}^l_\pm; \qquad \mathbf{f}_\pm \to \mathbf{f}_\pm + \mathbf{f}^l_\pm.$$

It is assumed that in equations (8.6) and (8.7) [(8.20) and (8.21)] of the Smoluchowski type for $\mathbf{v}_\pm$ $(v^\mu_\pm)$ the small corrections $\mathbf{v}^l_\pm$ and $\mathbf{f}^l_\pm$ $(v^{\mu,l}_\pm$ and $F^{\mu,l}_\pm)$

occur only in a symmetric combination with respect to Δt and $\delta\mathbf{x}$ (Δs and $\delta x_E^\mu = y_E^\mu$), i.e., they are even functions under the transformations $\Delta t \to -\Delta t$ and $\delta\mathbf{x} \to -\delta\mathbf{x}$ ($\Delta s \to -\Delta s$ and $\delta x_E^\mu \to -\delta x_E^\mu$). In this case, $\mathbf{v}_\pm$ and $\mathbf{f}_\pm$ ($v_\pm^\mu$ and $F_\pm^\mu$), which occur in equations (8.6) and (8.7) [(8.20) and (8.21)], are replaced by the expressions

$$\mathbf{v}_\pm + \sum_{\{\Delta t,\, \delta\mathbf{x}\}} \mathbf{v}_\pm^l \quad \text{and} \quad \mathbf{f}_\pm + \sum_{\{\delta\mathbf{x}\}} \mathbf{f}_\pm^l,$$

$$\left(v_\pm^\mu + \sum_{\{\Delta s,\, y_E\}} v_\pm^{\mu,l} \quad \text{and} \quad F_\pm^\mu + \sum_{\{y_E\}} F_\pm^{\mu,l}\right),$$

where the symbol $\Sigma_{\{\cdots\}}$ denotes the operation of symmetrization with respect to the variables $\{\cdots\}$; for example,

$$\sum_{\{\mathbf{y}\}} \mathbf{v}_\pm^l(\mathbf{x}-\mathbf{y}) = \sum_{\{\mathbf{y}\}} \mathbf{v}_\pm^l(\mathbf{x}+\mathbf{y}) \equiv \tfrac{1}{2}[\mathbf{v}_\pm^l(\mathbf{x}+\mathbf{y}) + \mathbf{v}_\pm^l(\mathbf{x}-\mathbf{y})].$$

As a result, we obtain equations analogous to (8.9), (8.10), (8.23) and (8.24):

$$\mathrm{d}_c\mathbf{v} - \lambda\, \mathrm{d}_s'\mathbf{u} = (1/m)\mathbf{F}_\lambda^+, \qquad \mathrm{d}_c\mathbf{u} + \lambda\, \mathrm{d}_s'\mathbf{v} = (1/m)\mathbf{F}_\lambda^-, \tag{9.3}$$

$$\mathrm{d}_c\mathbf{v} - \lambda\, \mathrm{d}_s'\mathbf{v} = (1/m)\mathbf{F}_\lambda, \qquad \mathrm{d}_c\mathbf{u} + \lambda\, \mathrm{d}_s'\mathbf{u} = (1/m)\mathbf{F}_\lambda', \tag{9.4}$$

and

$$\mathrm{D}_c v^\mu - \lambda\, \mathrm{D}_s' u^\mu = (1/m)\Phi_\lambda^{(+)\mu}, \qquad \mathrm{D}_c u^\mu + \lambda\, \mathrm{D}_s' v^\mu = (1/m)\Phi_\lambda^{(-)\mu}, \tag{9.5}$$

$$\mathrm{D}_c v^\mu - \lambda\, \mathrm{D}_s' v^\mu = (1/m)\Phi_\lambda^\mu, \qquad \mathrm{D}_c u^\mu + \lambda\, \mathrm{D}_s' u^\mu = (1/m)\Phi_\lambda'^\mu \tag{9.6}$$

in the nonrelativistic and relativistic cases, respectively. Here

$$\mathrm{d}_s' = (\mathbf{u}\nabla) + \nu(\nabla^2 + l^2\nabla^4), \qquad \mathrm{D}_s' = u^\lambda\,\partial_\lambda + \nu(\Box + l^2\Box^2). \tag{9.7}$$

Naturally, if we ignore the terms of order $\nu l^2\nabla^4(\nu l^2\Box^2)$ in expression (9.7) for $\mathrm{d}_s'(\mathrm{D}_s')$, then, as one would expect, we obtain the old equations (8.9) and (8.10) [(8.23) and (8.24)].

In the limit $\mathbf{v}_+ \equiv \mathbf{v}_-$, i.e., $\mathbf{u} \equiv 0$ we obtain from (9.3) and (9.4) *Newton's equation* $\mathrm{d}_c\mathbf{v} = \mathbf{F}/m$ and also the equation

$$\mathrm{d}_c\mathbf{v} - \lambda\nu(\nabla^2 + l^2\nabla^4)\mathbf{v} = \mathbf{F}_\lambda/m$$

for the particle.

For $\mathbf{F}_\lambda = 0$, these last equations are equations of the Sivashinsky type for a free particle. For example, setting $\lambda = -1$, we obtain

$$\mathrm{d}_c\mathbf{v} = -\nu\nabla^2\mathbf{v} - \nu l^2\nabla^4\mathbf{v} \tag{9.8}$$

which is invariant under a Galileo transformation.

Note that for the system of equations (9.3) [or (8.9)] the condition $\nu \to 0$ in the 'classical' limit is consistent, since the first equation of (9.3) goes over into

Newton's equation, and both sides of the second equation of (9.3) vanish in the limit when $\nu \to 0$. We recall that $\mathbf{F}^+$ is the external force, which does not contain a stochastic component, and the second equation of (9.3) for $\lambda = 1$ agrees with the continuity equation (see also Section 9.3) $\partial\rho/\partial t + \nabla(\rho\mathbf{v}) = 0$. Therefore, if the Lorentz force is introduced in a natural manner in our formalism, we can write

$$\mathbf{F}_1^- = -m[\mathbf{u} \times (\nabla \times \mathbf{v}) + \nu\nabla \times (\nabla \times \mathbf{v})].$$

In the case of the electromagnetic field, it takes the form

$$\mathbf{F}_1^- = \frac{e}{c}\mathbf{u} \times \mathbf{H} + \frac{e}{c}\nu\nabla \times \mathbf{H}.$$

Conversely, in the system of equations (9.4) it is impossible to go over formally to the condition $\nu \to 0$ in the 'classical' limit, since in the expressions for the forces $\mathbf{F}_\lambda$ and $\mathbf{F}'_\lambda$ the ordinary and stochastic components of the force are already present. It is entirely possible that the stochastic part of the force does not necessarily vanish even as $\nu \to 0$, and on the other hand the stochastic velocity $\mathbf{u} = \nu\nabla \ln \rho$ vanishes in the limit $\nu \to 0$. Then the stochastic terms on both sides of equation (9.4) do not vanish in an equal way as $\nu \to 0$.

A special method is needed to investigate the *nonlinear equations* (9.3), (9.4), and (9.8) whose solution goes beyond the scope of the present book. In the papers by Sivashinsky (1977); Michelson and Sivashinsky (1977), equation (9.8) was studied numerically and it was shown that its solution behaves like a turbulent processes.

In this section, we shall make some general comments about these equations. First, we note that equation (9.8) is equivalent to the *Hamilton–Jacobi equation* only in the case when the 'flow' $\mathbf{v} = \nabla S$ is irrotational, i.e., curl $\mathbf{v} = 0$. Otherwise, the solution of equation (9.8) grows unboundedly with the time. For example, let us consider the field $\mathbf{v} = (u, \mathrm{v}, \mathrm{w})$, where

$$u(x, y, z, t) = v(x, y, z, t) = 0; \qquad w(x, y, z, t) = \varphi(x, y, t).$$

Note that div $\mathbf{v} = 0$. For such a field, all the nonlinear terms in equation (9.8) vanish, and we obtain a purely linear equation for $\varphi(x, y, t)$:

$$\partial\varphi/\partial t + \nu\nabla^2\varphi + \nu l^2\nabla^4\varphi = 0.$$

This equation has spatially periodic solutions of the type

$$\varphi = A \exp(\sigma t + ik_1 x + ik_2 y),$$

where

$$\sigma = \nu(k_1^2 + k_2^2) - \nu l^2(k_1^2 + k_2^2)^2.$$

If $k_1^2 + k_2^2 < l^{-2}$, then the amplitudes of these solutions grow exponentially.

At present there are difficulties connected with the physical interpretation of such a noncausal behavior of the amplitudes; perhaps only a hypothesis on the existence of tachyons and a nonordinary version of the causal conditions will permit this noncausal behavior of signals at least in a domain determined by length l.

The second question relates to the choice of the sign of the parameter λ. For example, if the terms of order νl^2 are ignored, the second equation of (9.3) for $\lambda = -1$ does not agree with the continuity equation, and, therefore, the case $\lambda = -1$ for (9.3) is eliminated from consideration altogether.

When $\lambda = 1$, the equation

$$\mathrm{d}_c \mathbf{v} = \nu \nabla^2 \mathbf{v} + \nu l^2 \nabla^4 \mathbf{v} \tag{9.9}$$

of type (9.8) obtained from equation (9.4) is such that for it there is no scattering (dissipation) in the short-wavelength regions. It is well known that for such equations an important problem – the initial value problem – is not sufficiently formulated. For this reason, the case $\lambda = 1$, i.e., equation (9.9), simply has no physical meaning if one does not assume the existence of terms with derivatives of higher than the fourth order.

Since our approximation includes only terms with derivatives of the fourth order, the case $\lambda = 1$ [equation (9.9)] is eliminated. Of course, the questions posed above remain valid in the relativistic case.

In the relativistic case, Sivashinsky's equation (9.8) now takes the form

$$\mathrm{D}_c v^\mu = -\nu(\Box + l^2 \Box^2) v^\mu \tag{9.10}$$

which is obtained from equation (9.6) for $u^\mu \equiv 0$ and $\Phi^\mu = 0$.

Thus, adopting the hypothesis of a stochastic space-time, we have obtained Sivashinsky's equation. We have shown that the resulting instability of the uniform and rectilinear motion of a free particle leads to random fluctuations of its trajectory. Despite the purely classical nature of the original equation, typically quantum effects are imitated: the uncertainty relation, de Broglie waves and their interference, discrete energy levels, and zero-point fluctuations (for more details see Sivashinsky, 1978, 1980).

As we have seen above, the self-interaction potential of the particle [the right-hand side of equations (9.8) and (9.10)] which generates the turbulence in the motion of the free particle has a stochastic origin. In other words, the stochasticity, which disappears in the limit $\mathbf{u} \Rightarrow 0$, renders the motion for $\mathbf{v}$ unstable and preserving thus a memory of itself.

9.2. Relativistic Feynman-Type Integrals

In the scheme of quantum mechanics proposed by Feynman (1948) an analog of the probability measure is used which allows one to describe the behavior of quantum mechanical particles. Kac (1959) noticed that this measure is a complex quantity. This complex measure in the Feynman path integral corresponds to the presence of the factor i in the exponential for the Wiener measure (see Chapter 5).

Roughly speaking, the presence of i in the exponential for the Wiener measure causes uncontrollable oscillations in the path integral. This makes it sometimes difficult to understand the Feynman path integral as a well-defined mathematical object. Despite this drawback it is a matter of history that the path integral is an extremely important contribution to quantum theory.

There are many approaches devoted to the generalization of the Feynman integrals to the relativistic case (see Feynman, 1950, 1951; Miura, 1979 and references therein). Interrelation between the Feynman path integrals and stochastic mechanics, and the path integral method as derived from a stochastic variational principle have been presented by Guerra (1981) and Berrondo (1973), respectively.

In this section we shall construct the Feynman-type integrals in the framework of our approach using Smoluchowski equations. Equations obtained allow the Schrödinger, Klein–Gordon, and Dirac equations to be derived easily. The interaction in the Smoluchowski-type equations for fields $\varphi(\mathbf{x}, t)$ is introduced here by means of Weyl's (1929, 1950) gauge theory.

In our model the *Feynman process* may formally be interpreted as a stochastic process in complex time with a real probability measure of the Gaussian type, which occurs in the Euclidean space.

9.2.1. *Diffusion Process in Real Time*

In the language of motion of a stochastic particle the property of the Markov process means that the character of displacement of a particle at a given time does not depend on the property of previous displacements. Accordingly, the probability density $\rho(\mathbf{x}, t)$ must obey the Smoluchowski equation

$$\rho(\mathbf{x}, t + \Delta t) = \int d^3(\delta x)\rho(\mathbf{x} - \delta\mathbf{x}, t)P_0(\mathbf{x} - \delta\mathbf{x}, t; \delta\mathbf{x}, \Delta t), \tag{9.11}$$

where P_0 is the conditional probability density that a particle at position $\mathbf{x} - \delta\mathbf{x}$ at time t will be displaced by $\delta\mathbf{x}$ during the interval Δt, thus reaching position $\mathbf{x}$ at time $t + \Delta t$. The simple form

$$P_0 = (4\pi\nu\, \Delta t)^{-3/2} \exp\{-(\delta\mathbf{x})^2/4\nu\, \Delta t\} \tag{9.12}$$

reduces to the diffusion equation for $\rho(\mathbf{x}, t)$ (see Chapter 7)

$$\partial\rho/\partial t = \nu\nabla^2\rho, \tag{9.13}$$

where ν is the diffusion coefficient.

Following our model (see Chapter 8) in the relativistic case, we consider formally the motion of a particle suffering the random walk due to stochasticity of the four-dimensional Euclidean space $\mathbb{E}^4(\hat{\mathbf{x}}, \tau)$. Then equation (9.11) acquires the following form

$$\rho(x_\mu, s + \Delta s) = \int d^4 y_{\mathrm{E}}\, \rho(\mathbf{x} - \mathbf{y}, x_0 + iy_4, s)P_1(\mathbf{x} - \mathbf{y}, x_0 +$$

$$+ iy_4, s; y_E, \Delta s), \tag{9.14}$$

where the variables $x_\mu = (x_0, \mathbf{x})$ are pseudo-Euclidean and P_1 can be chosen in the form

$$P_1 = (4\pi\nu\,\Delta s)^{-2} \exp(-y_E^2/4\nu\,\Delta s). \tag{9.15}$$

From (9.14) and (9.15) we have

$$\partial\rho/\partial s = \nu\Box\rho; \qquad \Box = -\partial^2/\partial x_0^2 + \partial^2/\partial \mathbf{x}^2. \tag{9.16}$$

Note that a more complicated form of the functions (measures) P_0 and P_1 make Fokker–Planck equations for ρ in both the nonrelativistic and relativistic cases, respectively (see Chapter 8).

9.2.2. *'Diffusion Process' in Complex Time*

Now a basic postulate is that a field $\varphi(\mathbf{x}, t)$ (probability amplitude) associated with a particle suffers a transformation and is defined as a diffusion process in complex time, whenever the particle displaces from point $\mathbf{x} - \delta\mathbf{x}$ at time t to position $\mathbf{x}$ at $t + \Delta t$. Then the corresponding Smoluchowski-type equation for the field $\varphi(\mathbf{x}, t)$ becomes

$$\varphi(\mathbf{x}, t - i\,\Delta t) = \int d^3(\delta x)\varphi(\mathbf{x} - \delta\mathbf{x}, t)P_0(\delta\mathbf{x}, \Delta t), \tag{9.17}$$

where P_0 is given by (9.12). It is easily seen that from equation (9.17), we obtain the Schrödinger equation assuming $\nu = \hbar/2m$.

Substituting $\Delta t \to -i\,\Delta t$ into equation (9.17) gives the Feynman integral, and therefore in the nonrelativistic case our postulate does not make a new result in the method of the Feynman path integrals. An essential difference appears in the construction of the relativistic Feynman-type integrals by using the diffusion processes. We now pass to this question.

Roughly speaking, in the relativistic case the Feynman-type integrals for the probability amplitude is formally replaced by Smoluchowski-type equations at complex time. So, if the Feynman process $\varphi(x^\mu, s)$ is known at one value of s, its value at a slightly larger value $s - i\,\Delta s$ is given by

$$\varphi(x^\mu, s - i\,\Delta s) = \int d^4y_E\, \varphi(\mathbf{x} - \mathbf{y}, x_0 + iy_4, s)P_1(y_E, \Delta s). \tag{9.18}$$

From expression (9.15) and equation (9.18), we obtain the *Klein–Gordon equation* in the parametric form

$$i\,\partial\varphi/\partial s = -\nu\Box\varphi. \tag{9.19}$$

Sometimes, instead of equation (9.19), the following equation is considered

$$i\,\partial\varphi/\partial s = -\nu(\Box - m^2)\varphi \tag{9.20}$$

which is obtained by using the measure:

$$d\mu(y_E, \Delta s) = d^4 y_E (4\pi v\, \Delta s)^{-2} \exp(-m^2 v\, \Delta s - y_E^2/4v\, \Delta s).$$

Formal formulas (9.17) and (9.18) will be interpreted as well-defined mathematical objects – some integral equations with a real probability measure of the Gaussian type. Here quantity v is real always.

9.2.3. *Introduction of Interactions into the Scheme*

As is clear, in the nonrelativistic case our formalism is equivalent to the Feynman integral if $\Delta t \to -i\,\Delta t$. Then due to Feynman we can write equation (9.17) in a potential field $U(\mathbf{x})$ by the following formula

$$\varphi(\mathbf{x}, t - i\,\Delta t) = \int d^3(\delta x) \exp\left[-\frac{\Delta t}{2mv} U(\mathbf{x})\right] \varphi(\mathbf{x} - \delta\mathbf{x}, t) P_0(\delta\mathbf{x}, t).$$

From this we obtain the Schrödinger equation

$$-\frac{\hbar}{i}\frac{\partial\varphi}{\partial t} = \frac{1}{2m}\left(\frac{\hbar}{i}\nabla\right)^2 \varphi + U(\mathbf{x})\varphi$$

if $v = \hbar/2m$.

In the relativistic case we introduce the interactions into our scheme within the framework of Weyle's gauge theory (see Miura, 1979). Following this theory, the field takes the value $\varphi(x_\mu + dx_\mu, \Delta\mathscr{P})$ after the transport connected with the displacement of a particle from a world point $\mathscr{P}$(co-ordinates x_μ) to a position $\mathscr{P}'$ (co-ordinates $x_\mu + dx_\mu$), and therefore the variation $\delta\varphi$ of φ made by this transport is given by

$$\delta\varphi = \varphi(x_\mu + dx_\mu, \Delta\mathscr{P}) - \varphi(x_\mu) = dN\, \varphi(x_\mu). \tag{9.21}$$

If dN is the total differential of a co-ordinate function $ief(x_\mu)/\hbar c$, i.e.,

$$dN = i\frac{e}{\hbar c}\frac{\partial f}{\partial x_\mu} dx_\mu,$$

then (9.21) affects only an arbitrary phase of φ. Generally, dN is given in the form

$$dN = -i\frac{e}{\hbar c} A_\mu\, dx_\mu, \tag{9.22}$$

where A_μ is an electromagnetic potential. We assume $dx_\mu = x_\mu - x'_\mu$ and rewrite (9.21) in the following form

$$\varphi(x_\mu, \Delta\mathscr{P}) = \exp(dN)\varphi(x'_\mu) = \exp\left(-\frac{ie}{\hbar c} A_\mu\, dx_\mu\right)\varphi(x'_\mu). \tag{9.23}$$

By using this formula the Feynman path integral for a Klein–Gordon particle may be defined in the form (Miura, 1979)

$$\varphi(x_\mu, s + \varepsilon) = \int d^4x' \exp\left(\frac{i}{\hbar} S_0\right) \prod_1^4 (\pm 2\pi i\hbar\varepsilon)^{-1/2} \varphi(x_\mu, \Delta\mathscr{P}),$$

where S_0 is the action of a free particle.

Because the displacement of the variables $\mathbf{x}$ and x_0 in our case is different, i.e., it is of the Euclidean character, the corresponding formula (9.23) must be changed in the following manner;

$$\varphi(x_\mu, \Delta\mathscr{P}) = \exp\left\{\frac{ie}{2mvc} A_\mu^E y_E^\mu\right\} \varphi(\mathbf{x} - \mathbf{y}, x_0 + iy_4), \tag{9.24}$$

where $A_\mu^E = (-iA_0, \mathbf{A})$ and $A_\mu^E y_E^\mu = A_4 y_4 + \mathbf{A}\mathbf{y}$. Then we obtain the value of φ at a space-time point after the transformation using averaging over φ shifted by all possible Euclidean displacements with a real probability measure $P_1(y_E, \Delta s) d^4 y_E$ of the Gaussian type and multiplied further by a weight function $\exp\{(ie/2mvc) A_\mu^E y_E^\mu\}$ for an infinitesimal value of y_E^μ:

$$\varphi(x_\mu, s - i\,\Delta s) = \int d\mu(y_E, \Delta s) \exp\left(\frac{ie}{2mvc} A_\mu^E y_E^\mu\right) \varphi(\mathbf{x} - \mathbf{y}, x_0 + iy_4, s), \tag{9.25}$$

where

$$d\mu(y_E, \Delta s) = (4\pi v\,\Delta s)^{-2} \exp(-y_E^2/4v\,\Delta s)\, d^4 y_E.$$

From this we obtain the parametric Klein–Gordon equation in an external field

$$i\frac{\partial\varphi}{\partial s} = v\left(\frac{\partial}{\partial x_\mu} - \frac{ie}{2mvc} A_\mu\right)^2 \varphi$$

if $v = \hbar/2m$.

The generalization of our formalism to a Dirac particle does not present any difficulty, for example, expressions (9.24) and (9.25) take the form

$$\psi(x_\mu, \Delta\mathscr{P}) = \exp\left\{\frac{1}{2} ie \frac{1}{2mvc} [\hat{A}^E y_\nu \gamma_\nu + y_\mu \gamma_\mu \cdot \hat{A}'^E]\right\} \psi(\mathbf{x} - \mathbf{y}, x_0 + iy_4, s), \tag{9.26}$$

and

$$\psi(x_\mu, s - i\,\Delta s) = \int d\mu(y_E, \Delta s) \exp\left\{\frac{1}{2} ie \frac{1}{2mvc} [\hat{A}^E \gamma_\nu y_\nu + y_\mu \gamma_\mu \cdot \hat{A}'^E]\right\} \times$$
$$\times\, \psi(\mathbf{x} - \mathbf{y}, x_0 + iy_4, s), \tag{9.27}$$

where

$$A_\mu'^E = A_\mu^E(\mathbf{x} - \mathbf{y}, x_0 + iy_4) \quad \text{and} \quad \hat{A}^E = A_\mu^E \gamma_\mu;$$

γ are Dirac matrices.

After some calculations we have the *Dirac equation* in the parametric form

$$i\frac{\partial\psi}{\partial s} = v\left[\frac{\partial}{\partial x_\mu} - \left(\frac{ie}{2mvc}\right)A_\mu\right]^2\psi - \frac{e}{4c}\frac{i}{2m}\sigma_{\mu\nu}F_{\mu\nu}\psi,$$

here

$$\sigma_{\mu\nu} = \tfrac{1}{2}(\gamma_\mu\gamma_\nu - \gamma_\nu\gamma_\mu), \qquad F_{\mu\nu} = \partial A_\nu/\partial x_\mu - \partial A_\mu/\partial x_\nu.$$

Finally, notice that equations (9.18), (9.25) and (9.27) we have obtained, generally speaking, have nothing to do with the Smoluchowski-type equations which describe the probability consequence, and therefore the value $P_1(y_E, \Delta s)$ is not interpreted as the transition probability. On the contrary, these equations may be interpreted as a formal exposition of some mathematical objects obtained by integrating with the real probability measure $d\mu(y_E, \Delta s)$ of the Gaussian type. Feynman path integrals themselves are not obtained in our scheme owing to the fact that variables $\mathbf{x}$ and x_0 are shifted in a different way: $\mathbf{x} \to \mathbf{x} - \mathbf{y}$ and $x_0 \to x_0 + iy_4$.

However, our method is interesting as representing the possibility of relativistic generalization of Feynman-type integrals.

9.3. Discussion of the Equations of Motion in Stochastic Mechanics

The description of a stochastic system is complete if its probability density $\rho(\mathbf{x}, t)$ and velocity $\mathbf{v}(\mathbf{x}, t)$ are known, since the stochastic velocity $\mathbf{u}$ is related to $\rho(\mathbf{x}, t)$ by

$$\mathbf{u}(\mathbf{x}, t) = v\nabla \ln \rho(\mathbf{x}, t).$$

For them, we have obtained the equations (Chapter 8)

$$\partial\rho/\partial t + \operatorname{div}(\rho\mathbf{v}) = 0, \qquad m(\mathrm{d}_c\mathbf{v} - \mathrm{d}_s\mathbf{u}) = \mathbf{f}, \tag{9.28}$$

and in the relativistic case

$$\partial_\mu J^\mu(x) = 0, \qquad m(\mathrm{D}_c v^\mu - \mathrm{D}_s u^\mu) = F^\mu,$$

where $J^\mu(x) = \rho v^\mu$ is the four-dimensional current (see Appendix C in this chapter). In addition, there are equations for $\mathbf{u}$ and u^μ:

$$m(\mathrm{d}_c\mathbf{u} + \mathrm{d}_s\mathbf{v}) = \mathbf{f}_s, \qquad m(\mathrm{D}_c u^\mu + \mathrm{D}_s v^\mu) = \mathrm{F}_s^\mu. \tag{9.29}$$

On physical grounds, it can be asserted that these basic equations must satisfy a number of requirements:

(1) To ensure that the physical quantities ρ and $\mathbf{v}$ are not overdetermined, it is necessary to require that the last two equations (9.29) for $\mathbf{u}$ and u^μ be equivalent to the continuity equations for ρ and J^μ, respectively.

(2) We shall assume that in stochastic mechanics the work done by a field

on a particle is associated only with the velocity $\mathbf{v}$ and does not depend on the velocity $\mathbf{u}$, i.e.,

$$\langle (d/dt)(m\mathbf{v}^2)\rangle = (\mathbf{fv}); \qquad d/dt = \partial/\partial t + (\mathbf{v}\nabla).$$

In the absence of an external force, this condition amounts to conservation of the kinetic energy in time.

(3) In the relativistic case, we must have $v_\mu v^\mu = c^2$. Thus, on the basis of the first requirement we can choose the form of the external forces $\mathbf{f}_s$ and F_s^μ, and the requirements (2) and (3) impose the further conditions

$$(\mathbf{v}\, d_s\mathbf{u}) = 0; \qquad v_\mu\, D_s u^\mu = 0. \tag{9.30}$$

Note that in the nonrelativistic limit the last condition goes over into $\mathbf{v}\, d_s\mathbf{u} = 0$, so that these conditions are interrelated.

As an external field, we now consider the electromagnetic field and show that equations (9.29) with external force

$$\mathbf{f}_s = \frac{e}{c}\mathbf{u}\times\mathbf{H} + \frac{e}{c}\nu\nabla\times\mathbf{H}; \qquad \mathbf{H} = \operatorname{curl}\mathbf{A};$$

$$F_s^\mu = \frac{e}{c}F^{\mu\lambda}u_\lambda + \frac{e}{c}\nu\,\partial^\lambda F^{\lambda\mu} \tag{9.31}$$

are equivalent to equations for ρ and J^μ, respectively. In the non-relativistic case, we set

$$m\mathbf{v} + \frac{e}{c}\mathbf{A} = \nabla S,$$

where S is the action. Then the first equation of (9.29) with allowance for (9.31) gives

$$d_c\mathbf{u} + d_s\mathbf{v} = -\mathbf{u}\times\operatorname{curl}\mathbf{v} - \nu\nabla\times\operatorname{curl}\mathbf{v}. \tag{9.32}$$

Applying the operator ∇ to the continuity equation for ρ and using the formula $\mathbf{u} = \nu\nabla\ln\rho$, we obtain

$$\partial\mathbf{u}/\partial t + \nabla(\mathbf{u}\cdot\mathbf{v}) + \nu\nabla(\nabla\cdot\mathbf{v}) = 0. \tag{9.33}$$

Using the well-known formulas of vector analysis,

$$\nabla(\mathbf{u}\cdot\mathbf{v}) = (\mathbf{u}\nabla)\mathbf{v} + (\mathbf{v}\nabla)\mathbf{u} + \mathbf{u}\times\operatorname{curl}\mathbf{v} + \mathbf{v}\times\operatorname{curl}\mathbf{u},$$

$$\nabla(\nabla\cdot\mathbf{v}) = \Delta\mathbf{v} + \nabla\times(\nabla\times\mathbf{v}), \qquad \operatorname{curl}\operatorname{grad} \equiv 0,$$

we arrive at equation (9.32), which is what we wanted to prove.

As in the derivation of equation (9.32), the time component of the second equation in (9.29) can be reduced to the form

$$D_c u_0 + D_s v_0 = (\mathbf{u}\nabla)v_0 - \mathbf{u}\,\partial_0\mathbf{v} - \nu\nabla\,\partial_0\mathbf{v} + \nu\,\Delta v_0. \tag{9.34}$$

Here, we have used equations (9.31) and the equation (see Landau and Lifschitz, 1971):

$$mv^\mu + \frac{e}{c}A^\mu = -\frac{\partial S}{\partial x^\mu}.$$

Differentiating the continuity equation for J^μ, we find

$$\mathrm{D}_c u_0 + \mathrm{D}_s v_0 = (\mathbf{u}\nabla)v_0 - \mathbf{u}\,\partial_0\mathbf{v} - \nu\nabla\,\partial_0\mathbf{v} + \nu\,\Delta v_0 + (\mathbf{v}\nabla)u_0 - \mathbf{v}\,\partial_0\mathbf{u}.$$

Since $u_0 = \nu\partial_0 \ln\rho$ and $\mathbf{u} = \nu\nabla\ln\rho$, and therefore $(\mathbf{v}\nabla)u_0 - \mathbf{v}\partial_0\mathbf{u} = 0$, from which we obtain equation (9.34).

Similarly, we can easily show that the three spatial components of the second equation of (9.29) are identical to the vector equation

$$\mathrm{D}_c\mathbf{u} + \mathrm{D}_s\mathbf{v} = -v_0\nabla u_0 + v_0\,\partial_0\mathbf{u} - u_0\nabla v_0 + u_0\,\partial_0\mathbf{v} - \\ - \mathbf{u}\times\operatorname{curl}\mathbf{v} + \nu\,\partial_0\nabla v_0 - \nu\,\partial_0^2\mathbf{v} - \nu\nabla\times\operatorname{curl}\mathbf{v},$$

which is obtained by differentiating the continuity equation for J^μ by the operator ∇.

We now turn to the additional conditions (9.30). The equation

$$\mathbf{v}\cdot\mathbf{B} = 0, \tag{9.35}$$

where

$$\mathbf{B} = \mathrm{d}_s\mathbf{u} = (\mathbf{u}\nabla)\mathbf{u} + \nu\,\Delta\mathbf{u}$$

has a nondenumerable set of solutions (we assume that **B** is an unknown vector), since it determines only the component of **B** in the direction of the vector **v**, the value of which is $B_v = 0$; however, the component perpendicular to **v** remains entirely arbitrary. Thus, if **B** is regarded as the radius vector of some point M relative to the origin 0, the locus of the ends of all vectors **B** satisfying equation (9.35) will be the plane perpendicular to **v** through 0. In the relativistic case, this question was discussed by Vigier (1979) from the geometrical point of view.

To conclude this section, we note that our equations (9.28), (9.29), and (9.33) are nonlinear differential equations, so that their direct solution is an almost insoluble problem in concrete applications. Finally, we give some special solutions of equations (9.28) for the simplest case of the free and uniform motion of a particle with the initial conditions

$$\rho(x, t=0) = \pi^{-1/2}\cdot l^{-1}\exp[-(x-x_0)^2/l^2]; \quad v(x, t=0) = v_0.$$

These solutions have the form

$$\rho(x,t) = \pi^{-1/2}l^{-1}(1+bt^2)^{-1/2}\times \\ \times\exp[-(x-x_0-v_0t)^2/l^2(1+bt^2)]; \quad b = 4\nu^2 l^{-2},$$

$$v(x,t) = [v_0 + bt(x-x_0)]/(1+bt^2),$$

$$u(x,t) = -\sqrt{b}(x-x_0-v_0t)/(1+bt^2),$$

and their mean value and dispersion are

$$\langle v(x,t)\rangle = \int dx\, v(x,t)\rho(x,t) = v_0,$$

$$\langle u(x,t)\rangle = \int dx\, u(x,t)\rho(x,t) = 0,$$

$$\langle x(t)\rangle = \int dx\, x\rho(x,t) = x_0 + v_0 t,$$

$$\text{Dis. } x^2(t) = \tfrac{1}{2}l^2(1 + bt^2); \qquad \text{Dis. } v^2(x,t) = \tfrac{1}{2}l^2 \frac{b^2t^2}{1 + bt^2};$$

$$\langle v^2(x,t)\rangle = v_0^2 + \tfrac{1}{2}l^2 \frac{b^2t^2}{1 + bt^2}.$$

In the case of a uniform translational and rotational motion of particles ($\mathbf{f}$ = const) solutions similar to those above can be readily obtained.

9.4. Cauchy Problem for the Diffusion Equation

We consider the following equation in two-dimensional space-time (x_1 and x_0) [first equation of (8.18) with $\mathbf{u}_+ = 0$ in Chapter 8]

$$\partial\rho(x_0, x_1, s)/\partial s = \nu\Box_2\rho(x_0, x_1, s); \qquad \Box_2 = -\partial^2/\partial x_0^2 + \partial^2/\partial x_1^2.$$

Since the variable s and $x_0 = ct$ are related by $dx_0/ds = c(1 - \beta^2)^{1/2}$, and in the rest frame of the particle they are simply equal; then this equation can be written in the form (now it is taken $x_1 = x$, $x_0 = t$)

$$(\Box_2 - a^2\partial/\partial t)\rho(x,t) = 0, \qquad a^2 = 1/\nu.$$

We now investigate the *Green function* of the equation

$$(\Box_2 - a^2\,\partial/\partial t)G(x - x_0, t - t_0) = -\delta(x - x_0)\,\delta(t - t_0), \tag{9.36}$$

where (x, t) is the point of observation and (x_0, t_0) is the source point. To construct the function G, we consider first the two functions

$$D_1(x,t) = \frac{1}{2\pi i}\int d^2p\theta(p_0')\,\delta\!\left(ic\frac{p_0}{\nu} - p^2\right)\exp(ip_0c\tau - ip_1R), \tag{9.37}$$

$$D_2(x,t) = \frac{i}{2\pi}\int d^2p\theta(p_0'')\,\delta\!\left(-ic\frac{p_0}{\nu} - p^2\right)\exp(-ip_0c\tau + ip_1R), \tag{9.38}$$

where

$$p_0' = p_0 - ic/2v, \quad p_0'' = p_0 + ic/2v, \quad \tau = t - t_0,$$
$$R = x - x_0, \quad p = (p_1, p_0).$$

After elementary calculations, we obtain ($p_1 = p$)

$$D_1(x, t) = \frac{1}{4\pi i} \exp\left[-c^2 \frac{t - t_0}{2v}\right] \int dp\omega^{-1} \exp(ic\omega\tau + ipR),$$

$$D_2(x, t) = \frac{i}{4\pi} \exp\left[-c^2 \frac{t - t_0}{2v}\right] \int dp\omega^{-1} \exp(-ic\omega\tau + ipR),$$

where

$$\omega = (p^2 - c^2/4v^2)^{1/2}.$$

We now consider the integral (see Morse and Feshbach, 1953)

$$D_2 = \frac{i}{4\pi} \exp\left[-c^2 \frac{t - t_0}{2v}\right] \int_C dp(p^2 - \mu^2)^{1/2} \exp\{i[Rp - (p^2 - \mu^2)^{1/2} c\tau]\},$$
$$\mu^2 = c^2/4v^2. \tag{9.39}$$

(1) If $R > c\tau$, the contour C can be closed along the semicircle in the upper half-plane. Then the integral (9.39) is equal to zero, since there are no singularities inside the contour (Figure 13).

(2) If $R < c\tau$, the contour C can be displaced so that it is stretched along the negative imaginary half-axis (Figure 14). In the last case, the integral (9.39) can be reduced to a well-known form. Let us introduce the new variable θ so that $c\tau = L \operatorname{sh} \theta$,

$$L = |(R^2 - c^2\tau^2)^{1/2}|,$$

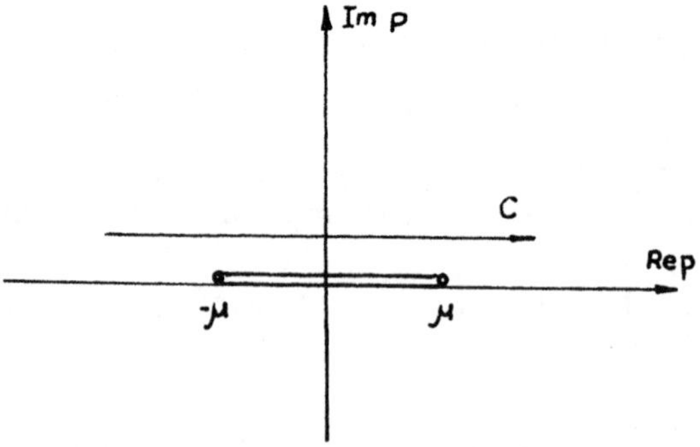

Fig. 13. The contour C for integral (9.39).

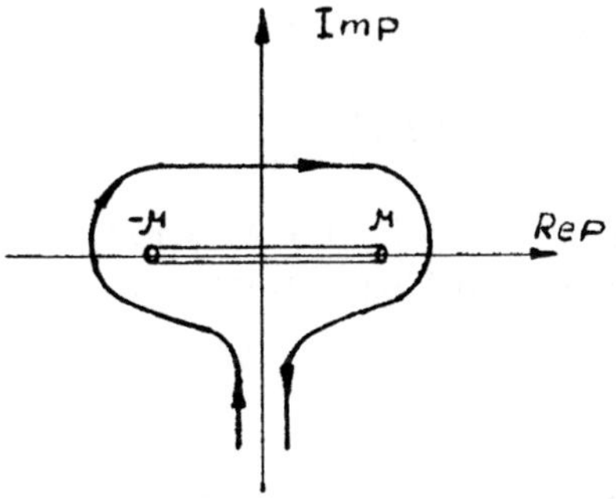

Fig. 14. The contour C for integral (9.39) at $R < c\tau$.

and set $p = \mu \operatorname{ch} x$. Then

$$i[Rp - (p^2 - \mu^2)^{1/2} c\tau] = i\mu(L \operatorname{ch} \theta \operatorname{ch} x - L \operatorname{sh} x \operatorname{sh} \theta) = i\mu L \operatorname{ch}(x - \theta);$$

$$D_2 = \frac{i}{2\pi} \mathrm{e}^{-\mu c\tau} \int_{3\pi i/2 + \infty}^{-i\pi/2 + \infty} \mathrm{d}x \exp[i\mu L \operatorname{ch}(x - \theta)].$$

(3) Setting $x - \theta = iy$, then

$$D_2 = -\frac{1}{4\pi} \mathrm{e}^{-\mu c\tau} \int_{3\pi/2 - i\infty}^{-\pi/2 - i\infty} \mathrm{d}y \exp(i\mu L \cos y).$$

By the equation

$$-\frac{1}{2\pi} \int_{3\pi/2 - i\infty}^{-\pi/2 - i\infty} \mathrm{d}y \exp(i\mu L \cos y) = J_0(\mu(R^2 - c^2\tau^2)^{1/2}),$$

the function D_2 takes the form

$$D_2 = \tfrac{1}{2} \exp(-\mu c\tau) J_0(\mu(R^2 - c^2\tau^2)^{1/2}) \theta(c\tau - R).$$

Let us now consider the integral

$$D_1 = \frac{1}{4\pi i} \mathrm{e}^{-\mu c\tau} \int_C \mathrm{d}p (p^2 - \mu^2)^{-1/2} \exp\{ipR + ic\tau(p^2 - \mu^2)^{1/2}\}.$$

This integral is equal to zero when $R + c\tau > 0$ (we recall that in the one-dimensional case R may be negative), but it differs from zero when $R + c\tau < 0$. In this case we have

$$D_1 = -\tfrac{1}{2} \exp(-\mu c\tau) J_0(\mu(R^2 - c^2\tau^2)^{1/2})(1 - \theta(R + c\tau)).$$

Green's function of equation (9.36) is given by the following expression

$$G(R, \tau) = -c(D_2 - D_1) = \tfrac{1}{2} c \exp(-\mu c\tau) J_0\left(\frac{c}{2\nu}(R^2 - c^2\tau^2)^{1/2}\right) \theta(c\tau - |R|).$$

It is readily seen that as one would expect

$$\lim_{c \to \infty} \frac{1}{v} G(R, \tau) = (4\pi v\tau)^{-1/2} \exp\left(\frac{-R^2}{4v\tau}\right).$$

Further, for comparison we write the Green function of the Schrödinger equation in the proper-time representation:

$$i\,\partial\psi/\partial s = -v\Box\psi, \qquad v = \hbar/2m = 1/a^2,$$

or

$$(\Box_2 + ia^2\,\partial/\partial t)K(x - x_0, t - t_0) = -\delta(x - x_0)\delta(t - t_0),$$

where

$$K(R, \tau) = -i\frac{a^2}{4} c \exp\left[-i\frac{\pi}{4} + \tfrac{1}{2} i\tau a^2 c^2\right][J_0(X) - iN(X)]\theta(c\tau - R),$$

$$X = \tfrac{1}{2} ca^2(c^2\tau^2 - R^2)^{1/2},$$

and in the limit $c \to \infty$ the function $K(R, \tau)$ is equal to the Green function

$$G = i^{-1}(4\pi v\tau)^{-1/2} \exp(iR^2/4v\tau)$$

of the equation

$$i\,\partial\psi/\partial t = -v(\partial^2/\partial x^2)\psi.$$

We now solve the one-dimensional initial-value problem

$$\rho(x, t) = a^2 \int dx_0 [\rho G]_{t_0 = 0} + c^{-2} \int dx_0 \left[G \frac{\partial \rho}{\partial t_0} - \rho \frac{\partial G}{\partial t_0}\right]_{t_0 = 0}, \qquad (a^2 = 1/v),$$

Whence

$$\rho(x, t) = \tfrac{1}{2} e^{-\mu ct}\left\{ f(x + ct) + f(x - ct) + \int_{x - ct}^{x + ct} dx_0 f(x_0) \times \right.$$

$$\left. \times \left[\tfrac{1}{2}\mu I_0(X) + \frac{1}{2}\frac{1}{c}\frac{\partial}{\partial t} I_0(X)\right] + \frac{1}{2}\frac{1}{c}\int_{x - ct}^{x + ct} dx_0 I_0(X) v(x_0)\right\}, \qquad (9.40)$$

where

$$X = \mu[c^2 t^2 - (x - x_0)^2]^{1/2}; \qquad f(x_0) = \rho(x_0, t)|_{t = 0};$$

$$v(x_0) = \partial\rho(x_0, t)/\partial t|_{t = 0}.$$

Here $J_0(x)$ and $I_0(x)$ are the well-known cylindrical functions.

It should be noted that if functions $f(x)$ and $v(x)$ are strictly equal to zero outside some interval $[-a, a]$, i.e., if they are functions with a finite support situated in the interval $[-a, a]$, then the study of the integral (9.40) will not represent any trouble and moreover, the causality condition (local) is strictly

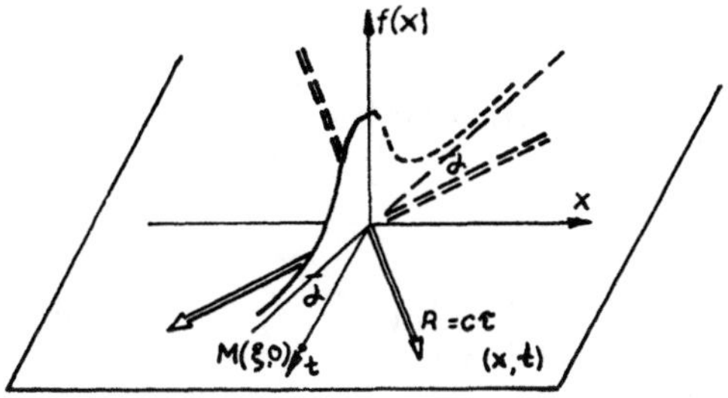

Fig. 15. The illustration of a function with infinite support.

fulfilled (see Chapter 2). The situation changes if we consider the functions $f(x)$ and $v(x)$, support of which is infinite (see, for example, Figure 15). Our aim now consists of studying the behavior of integrals of the type

$$T_1 = \frac{1}{2}\frac{1}{c}\int_{-\infty}^{\infty} dz \exp\left(\frac{-z^2}{a^2}\right) I_0(X),$$

and

$$T_2 = \text{const}\; t \int_{-\infty}^{\infty} dz \exp\left(\frac{-z^2}{a^2}\right) \frac{I_1(X)}{X},$$

$$X = \mu[c^2 t^2 - (z - x)^2]^{1/2},$$

at the limits $t \to \infty$ and $|x| \to \infty$.

In the case of $|x| \to \infty$, we assume the variable t belongs to the future (forward) light cone, i.e., $t \in V_+$. For simplicity, let us take the point M of observation with the co-ordinates $(\zeta, 0)$, i.e., $M = M(\zeta = t, 0)$. Then the integral T_1 takes the form

$$T_1 = \text{const} \int_{-\infty}^{\infty} dz \exp\left(\frac{-z^2}{a^2}\right) I_0(Q),$$

where $Q = \mu(c^2 t^2 - z^2)^{1/2}$.

Making use of the integral representation for the Bessel function $I_n(z)$:

$$I_n(z)(z/2)^{-n} = \Gamma^{-1}(n + \tfrac{1}{2})\Gamma^{-1}(\tfrac{1}{2}) \int_{-1}^{1} dx (1 - x^2)^{n - 1/2} e^{zx},$$

we obtain

$$T_1 = \text{const} \int_{-1}^{1} d\rho (1 - \rho^2)^{-1/2} \int_{-\infty}^{\infty} dz \exp\left(\frac{-z^2}{a^2}\right) \exp(\rho Q).$$

Here

$$\int_{-\infty}^{\infty} dz \exp\left(\frac{-z^2}{a^2}\right) \exp(\rho Q) \sim \exp\left(-\frac{c^2t^2}{a^2} + \tfrac{3}{4}a^2\mu^2\rho^2\right),$$

and, therefore

$$T_1 = \text{const} \exp\left(-\frac{c^2t^2}{a^2}\right) \int_{-1}^{1} d\rho (1-\rho^2)^{-1/2} \exp(\tfrac{3}{4}a^2\mu^2\rho^2)$$

$$= \text{const} \exp\left(-\frac{c^2t^2}{a^2}\right).$$

Similarly,

$$T_2 = \text{const}\left[\frac{t}{\Gamma(\frac{3}{2})\Gamma(\frac{1}{2})}\right] \int_{-1}^{1} d\rho (1-\rho^2)^{-1/2} \int_{-\infty}^{\infty} dz \exp\left(\frac{-z^2}{a^2}\right) \exp(\rho Q)$$

$$= \text{const}\, t \exp\left(\frac{-c^2t^2}{a^2}\right).$$

Thus, in the case of functions $f(x)$, $v(x)$ with an infinite support the signal (as possible, noncausal) attenuates as $\exp(-t^2/a^2)$, when $t \to \infty$.

Let us now consider the behavior of the function $\rho(x, t)$ at $|x| \to \infty$:

(1) Let t be a finite value, then

$$T_1 = \frac{1}{2c} \int_{-ct}^{ct} dz \exp\left[\frac{-(z+x)^2}{a^2}\right] I_0(Q).$$

Going over the substitution $z = cty$, we obtain

$$T_1 = \frac{ct}{2c} \exp\left(\frac{-x^2}{a^2}\right) \int_{-1}^{1} dy \exp\left[-\frac{c^2t^2y^2}{a^2} - \frac{2ct}{a^2}yx\right] I_0(\mu ct(1-y^2)^{1/2}).$$

Setting again $x(y+1) = \rho$ and $y = \rho/x - 1$, $yx = \rho - x$, we obtain

$$T_1 = (t/2x)\exp(-x^2/a^2)S(x),$$

where

$$S(x) = \int_0^{2x} d\rho \exp\left\{-\frac{c^2t^2}{a^2}\left(\frac{\rho}{x} - 1\right)^2 - \frac{2ct}{a^2}(\rho - x)\right\} \times$$

$$\times I_0\left\{\mu ct\left[1 - \left(\frac{\rho}{x} - 1\right)^2\right]^{1/2}\right\},$$

and

$$\lim_{x\to\infty} S(x) \sim \left(\frac{a^2}{2ct}\right) \exp\left(\frac{2ct}{a^2}x - \frac{c^2t^2}{a^2}\right) I_0(0).$$

Therefore,

$$T_1 \sim (a^2/4cx)\exp[-(x-ct)^2/a^2],$$

for $x \to \infty$. Similarly,

$$T_2 = \tfrac{1}{2}\mu ct \int_{x-ct}^{x+ct} dz \exp\left(\frac{-z^2}{a^2}\right)\frac{I_1(X)}{X} \sim \frac{\mu^2 c^2 t^2}{8x} a^2 \exp\left\{\frac{-(x-ct)^2}{a^2}\right\},$$

where

$$X = \mu[c^2t^2 - (z-x)^2]^{1/2}.$$

(2) Let $t \gg 1$; then the region of integration over z can be extended over the whole plane. In this case, the estimates obtained above are not changed. For example,

$$T_1 = \frac{1}{2c}\int_{-ct}^{ct} dz \exp\left[\frac{-(z+x)^2}{a^2}\right] I_0(Q)$$

$$\Rightarrow \frac{1}{2c}\int_{-\infty}^{\infty} dz \exp\left\{\frac{-(z+x)^2}{a^2}\right\} I_0(Q) = A(x) + A(-x),$$

where

$$A(x) = \tfrac{1}{2}t \exp\left(\frac{-x^2}{a^2}\right)\int_0^\infty dy \exp\left\{-\frac{c^2t^2}{a^2}y^2 - \frac{2xcty}{a^2}\right\} I_0(\mu ct\sqrt{1-y^2})$$

$$\Rightarrow (t/2x)\exp\left(\frac{-x^2}{a^2} + \frac{2xct}{a^2}\right) g(x);$$

$$g(x) = \int_x^\infty d\rho \exp\left\{-\frac{c^2t^2}{a^2}\left(\frac{\rho}{x}-1\right)^2 - 2\rho\frac{ct}{a^2}\right\} I_0\left(\mu ct\left[1 - \left(\frac{\rho}{x}-1\right)^2\right]^{1/2}\right),$$

$$(x + xy = \rho).$$

It is easily verified that when $x \to \infty$ the integral $g(x)$ is less then any ε chosen arbitrarily small i.e.,

$$\lim_{x\to\infty} |T_1(x)| \leqslant \varepsilon \exp\{-(x-ct)^2/a^2\}.$$

A similar estimation holds for T_2. As a result, we have

$$|\rho(x,t)| \leqslant \varepsilon \exp\{-(x-ct)^2/a^2\},$$

when $|x| \to \infty$ and $t \gg 1$.

Thus, we see that instead of the condition $\rho = 0$ for the case of functions $f(x)$ and $v(x)$ with a finite support, i.e., located in the interval $[-a, a]$, we have obtained the estimates of the type

$$\rho(x,t) \sim \exp\{-(x-ct)^2/a^2\}$$

when $|x| \to \infty$ and $t \to \infty$ for the case of the functions $f(x)$ and $v(x)$ with an infinite support. In this connection we notice that, in accordance with the macrocausality condition (2.110) or (2.112) discussed in Chapter 2, non-causal signals attenuating as $\exp\{-|x + ct|^2/a^2\}$ are acceptable for the practical applications, in particular, for the description of physical processes taking place in the macroscale.

9.5. Position-Momentum Uncertainly Relations in Stochastic Mechanics

In stochastic mechanics the concept of the configuration space and configuration observables plays a specific role. In contrast with classical mechanics and quantum mechanics where all constituents of phase space have a similar role, this feature is that the relevant random processes have always been defined as taking values on the configuration space of the dynamical system. In this section we make an attempt to construct momentum, and read information about momentum observables corresponding to the stochastic processes investigated in stochastic mechanics. Let us consider the simple case of a point particle of mass m, moving on the real line in a potential U.

As we have seen above, stochastic quantization consists in fact in the suitable reinterpretation of classical *configuration observables* as *random variables* whose time evolution is to be read from the Euler–Lagrange equations of motion, rewritten as dynamic conditions for the corresponding stochastic process $Q(t)$ which is described by the stochastic differential equation (see Chapter 7):

$$dQ(t) = v_+(Q(t), t) + dW(t), \quad (dt > 0). \tag{9.41}$$

This equation describes a random distrubance $dW(t)$ with expectation $\mathbb{E}\{dW(t)\} = 0$ and covariance $\mathbb{E}\{[dW(t)]^2\} = 2\nu\, dt$. In stochastic quantization procedure, the link between stochastic and quantum theories means that the stochastic process $Q(t)$ with diffusion coefficient $\nu = \hbar/2m$ is associated with a quantum state of a particle of mass m described by a normalized wave function $\psi(x, t) = \exp[R(x, t) + iS(x, t)]$. At this, probability density $\rho(x, t)$, current $v(x, t)$ and stochastic $u(x, t)$ velocities of the process $Q(t)$ are given by

$$\rho(x, t) = |\psi(x, t)|^2, \tag{9.42}$$

$$v(x, t) = \tfrac{1}{2}[v_+(x, t) + v_-(x, t)] = \frac{1}{m}\frac{\partial S(x, t)}{\partial x}, \tag{9.43}$$

$$u(x, t) = \tfrac{1}{2}[v_+(x, t) - v_-(x, t)] = \frac{\hbar}{m}\frac{\partial R(x, t)}{\partial x}, \tag{9.44}$$

or

$$v_+(x, t) - v_-(x, t) = 2\nu\frac{\partial}{\partial x}\ln\rho(x, t). \tag{9.45}$$

Here, we recall again definitions:

$$v_+(x,t) = \lim_{\Delta t \to 0+} \mathbb{E}\left\{\frac{Q(t+\Delta t) - Q(t)}{\Delta t}\bigg|_{Q(t)=x}\right\},$$

$$v_-(x,t) = \lim_{\Delta t \to 0+} \mathbb{E}\left\{\frac{Q(t) - Q(t-\Delta t)}{\Delta t}\bigg|_{Q(t)=x}\right\}.$$

The operational meaning of the conditional expectation $\mathbb{E}\{\cdots\}$ appearing in the definition is clear: $v_+(x,t)[v_-(x,t)]$ is the mean slope with which those sample paths at time t in x leave [enter] x.

On the other hand, this interrelation may be interpreted in the following way. With each soluton $\psi(x,t)$ of the Schrödinger equation for a particle, the stochastic quantization procedure associates a diffusion process $Q(t)$ whose expectation, for every configuration observable $f(x)$ satisfies

$$\mathbb{E}\{f(Q(t))\} = \langle \psi(x,t), f(x)\psi(x,t)\rangle. \tag{9.46}$$

We now discuss the following problem: how can one read information about momentum observables from 'measurements' on the sample paths of the process $Q(t)$? The study of the problem has been recently broached by Shucker (1978, 1980b), and de Falso, Martino and Siena (1982, 1983) (see also de la Peña-Auerbach and Cetto, 1972). The given section is based on their studies. They have presented an analysis of how information concerning momentum is encoded in the stochastic processes $Q(t)$, and they have given the quantum-mechanical procedure for reading the momentum distribution from the *Fourier transform* of the wave function. The basic ideas are:

(i) If τ is the time at which the momentum distribution is requested, construct the *free process* $Q_f(t)$ 'tanget' to the process $Q(t)$ at time τ, as the process which evolves in the *absence of potential* starting from the initial data at time τ determined by the probability density $\rho(x,t)$ and the mean forward velocity field $v_+(x,t)$ associated with the process $Q(t)$ at time τ.

(ii) According to stochastical mechanical correspondence, the process $Q_f(t)$ can be studied by means of the associated solution ψ_f of the *free Schrödinger equation* which starts from the initial data $\psi_f(x,\tau) = \psi(x,\tau)$.

(iii) For almost every sample path $\omega(t)$ of the process $Q_f(t)$ the following limit exists:

$$\pi(\tau,\omega) = \lim_{T\to\infty} m\left[\frac{\omega(\tau+T) - \omega(\tau)}{T}\right]. \tag{9.47}$$

The existence of this limit is the observation that the mean forward velocity field $v_+^f(x,t)$ for $Q_f(t)$ behaves for large t as

$$v_+^f(x,t) = x/t + O[t^{-3/2}|\psi_f(x,t)|^{-1}]. \tag{9.48}$$

(iv) Granted the existence of the limit (9.47), the fact that

$$\lim_{T\to\infty}\left(\frac{T}{m}\right)^{1/2}\exp\left\{i\left(\frac{\pi}{4}-\frac{1}{2}\frac{p^2}{2m\hbar}T\right)\right\}\tilde{\psi}_f\left(\frac{T}{m}p;\tau+T\right)=\tilde{\psi}(p,\tau),$$

where $\tilde{\psi}(p,\tau)$ is the Fourier transform of $\psi(x,\tau)$, shows that the probability density of the *random variable*

$$\pi(\tau)\equiv\pi(\tau,\omega), \tag{9.49}$$

is $|\tilde{\psi}(p,\tau)|^2$. The random variable $\pi(\tau)$ is, therefore, to be identified with the *image of quantum mechanical momentum* in stochastic mechanical correspondence.

We now consider two simple examples which are interesting in canonical transformation for stochastic mechanics.

EXAMPLE 9.1. First of all we recall that what would seem the most natural definition of *momentum* in stochastic mechanics to be associated with the random motion of the particle, namely $m\,dQ(t)/dt$, is in fact *incorrect* because $dW=O[(dt)^{1/2}]$ in (9.41) prevents the existence of the relevant limit. Let us set

$$p_+(t)=mv_+(Q(t),t);\qquad p_-(t)=mv_-(Q(t),t).$$

Then it is easy to see that multiplying both sides of (9.45) by $m\rho$ and integrating over x, we obtain

$$\mathbb{E}\{p_+(t)\}=\mathbb{E}\{p_-(t)\}. \tag{9.50}$$

Thus, the two random variables $p_\pm(t)$ have the same expectation. Similarly, multiplying both sides of (9.45) by $m\rho x$, integrating over x, and using (9.50), we obtain

$$\mathbb{E}\left\{[Q(t)-\mathbb{E}(Q(t))]\left[\frac{p_-(t)-p_+(t)}{2}-\mathbb{E}\left(\frac{p_-(t)-p_+(t)}{2}\right)\right]\right\}=m\nu. \tag{9.51}$$

Schwarz's inequality then implies

$$\Delta Q\cdot\Delta\,\delta p\geqslant m\nu, \tag{9.52}$$

where ΔQ and $\Delta\,\delta p$ are the root mean square deviations of the random variables $Q(t)$ and $\delta p(t)=[p_+(t)-p_-(t)]/2$, respectively. Taking into account (9.44), we see that inequality (9.52) implies that the root mean square deviations of the random variables $Q(t)$ and $u(Q(t),t)$ satisfy

$$\Delta Q\cdot\Delta(mu)\gtrsim\hbar/2. \tag{9.53}$$

In order to connect this inequality with the *position-momentum uncertainty relations*, we observe that, because of (9.42), we have in fact

$$\Delta Q=\{\mathbb{E}[(Q(t)-\mathbb{E}(Q(t)))^2]\}^{1/2}=[\langle\psi(t),x^2\psi(t)\rangle-\langle\psi(t),x\psi(t)\rangle^2]^{1/2}.$$

Further, we see that the relevant expectation of the momentum operator

$p = (\hbar/i)\,\partial/\partial x$ in

$$\Delta p = [\langle \psi(t), p^2\psi(t)\rangle - \langle \psi(t), p\psi(t)\rangle^2]^{1/2},$$

can be easily read from the mean stochastic velocities of the process $Q(t)$. We have, indeed,

$$\mathbb{E}\{p_+(t)\} = \mathbb{E}\{p_-(t)\} = \int \mathrm{d}x \rho(x,t) \frac{\partial S(x,t)}{\partial x} = \langle \psi(t), p\psi(t)\rangle,$$

and, similarly

$$\mathbb{E}\{p_+^2(t)\} = \langle \psi, p^2\psi\rangle + 2\hbar \int \mathrm{d}x \rho(x,t) \frac{\partial S(x,t)}{\partial x}\frac{\partial R(x,t)}{\partial x}.$$

Namely,

$$\mathbb{E}\{p_+^2(t) + p_-^2(t)\}/2 = \langle \psi(t), p^2\psi(t)\rangle.$$

From these equalities, we have

$$\begin{aligned}\Delta p &= [\langle \psi(t), p^2\psi(t)\rangle - \langle \psi(t), p\psi(t)\rangle^2]^{1/2}\\ &= 2^{-1/2}\{\mathbb{E}[p_+^2(t)] - [\mathbb{E}(p_+(t))]^2 + \mathbb{E}[p_-^2(t)] - [\mathbb{E}(p_-(t))]^2\}^{1/2}. \qquad (9.54)\end{aligned}$$

Equation (9.54) is compared with

$$m\,\Delta u = \left\{\mathbb{E}\left[\left(\frac{p_+(t) - p_-(t)}{2}\right)^2\right]\right\}^{1/2}.$$

On the other hand, it is easy to verify that the following equality holds

$$\begin{aligned}(\Delta p)^2 - (\Delta mu)^2 &= \mathbb{E}\left\{\left[\frac{p_+(t) - p_-(t)}{2}\right]^2\right\} - \mathbb{E}\{p_+(t)\}\mathbb{E}\{p_-(t)\}\\ &= \int \mathrm{d}x \rho(x,t)[\partial S(x,t)/\partial x]^2 - \left\{\int \mathrm{d}x \rho(x,t)\,\partial S(x,t)/\partial x\right\}^2.\end{aligned}$$

From this, we can conclude that

$$(\Delta p)^2 = (\Delta mv)^2 + (\Delta mu)^2. \qquad (9.55)$$

In particular, $\Delta p \geqslant \Delta mu$. Thus, the *Heisenberg position-momentum uncertainty relations*

$$\begin{aligned}&[\langle \psi(t), x^2\psi(t)\rangle - \langle \psi(t), x\psi(t)\rangle^2]^{1/2} \times\\ &\times [\langle \psi(t), p^2\psi(t)\rangle - \langle \psi(t), p\psi(t)\rangle^2]^{1/2} \geqslant \tfrac{1}{2}\hbar \qquad (9.56)\end{aligned}$$

can be traced back to the position-stochastic velocity uncertainty relation

$$\{\mathbb{E}[Q^2(t)] - [\mathbb{E}(Q(t))]^2\}^{1/2} \times \left\{\mathbb{E}\left[\left(\frac{p_+(t) - p_-(t)}{2}\right)^2\right]\right\}^{1/2} \geqslant \tfrac{1}{2}\hbar. \qquad (9.57)$$

Finally, it should be noted that in the decomposition (9.55) of the root mean square deviation of the quantum mechanical momentum there is a part due to the current velocity and a part due to the stochastic velocity; it is just the *stochastic term* that forces the position-momentum uncertainty. This problem was first considered by de la Peña-Auerbach and Cetto (1972).

EXAMPLE 9.2. To estimate the rate of convergence in (9.47) we consider an explicit example due to Falso, Martino, and Siena (1983).

Let the wave function at time $\tau = 0$ be

$$\psi(x, 0) = (2\pi a)^{-1/2} \exp(-x^2/4a), \tag{9.58}$$

where $a = (\Delta x)^2$. The mean forward velocity field for the process $Q_f(t)$ evolving from the initial condition corresponding to equation (9.58) is

$$\begin{aligned} v_+^f(x, t) &= \frac{\hbar}{m}\left\{\mathrm{Re}\left[\psi_f^{-1}(x, t)\frac{\partial}{\partial x}\psi_f(x, \dot{t})\right] + \mathrm{Im}\left[\frac{\partial}{\partial x}\ln \psi_f(x, t)\right]\right\} \\ &= (\hbar^2 t - 2m\hbar a)x/(\hbar^2 t^2 + 4a^2 m^2). \end{aligned}$$

The process $Q_f(t)$ evolves therefore according to the stochastic differential equation

$$\mathrm{d}Q_f(t) = [(\hbar^2 t - 2m\hbar a)Q_f(t)/(\hbar^2 t^2 + 4a^2 m^2)]\,\mathrm{d}t + \mathrm{d}W, \tag{9.59}$$

where $\mathrm{d}t > 0$, and $\mathrm{d}W$ represents increments of Brownian motion with diffusion coefficient $\nu = \hbar/2m$.

From equation (9.59) it follows that the two-point function of the process $Q_f(t)$ satisfies for $t > 0$ the differential equation

$$\frac{\mathrm{d}}{\mathrm{d}t}\,\mathbb{E}\{Q_f(0)Q_f(t)\} = \frac{\hbar^2 t - 2m\hbar a}{\hbar^2 t^2 + 4a^2 m^2}\,\mathbb{E}\{Q_f(0)Q_f(t)\},$$

whose explicit solution, with the initial condition $\mathbb{E}\{Q_f^2(0)\} = a$ is

$$\mathbb{E}\{Q_f(0)Q_f(t)\} = a[1 + \hbar^2 t^2/4a^2 m^2]^{1/2} \exp\{-\operatorname{arc\,tg}(\hbar t/2ma)\}. \tag{9.60}$$

Defining

$$(\Delta\pi_t)^2 \equiv \mathbb{E}\{[m(Q_f(t) - Q_f(0))/t]^2\},$$

from equation (9.60), we obtain

$$\Delta x \cdot \Delta\pi_t = \tfrac{1}{2}\hbar\{1 + 2(\lambda/t)^2[1 - (1 + t^2/\lambda^2)^{1/2} \exp(-\operatorname{arc\,tg}(t/\lambda))]\}^{1/2}, \tag{9.61}$$

where $\lambda = 2m(\Delta x)^2/\hbar$. The main features of equation (9.61) are the following:

(a) As $t \to +\infty$, $\Delta x \cdot \Delta\pi_t$ goes to the quantum-mechanical uncertainty product $\Delta x \cdot \Delta p$ (which in the particular case at hand is $\hbar/2$);
(b) As $t \to 0^+$, $\Delta x \cdot \Delta\pi_t$ diverges as t^{-1};
(c) The limit (a) is reached from below;
(d) After a time of order $\lambda = 2m(\Delta x)^2/\hbar$, $\Delta x \cdot \Delta\pi_f$ is very close $\Delta x \cdot \Delta p$ (within a few percent).

As to the generality of the intriguing point (c), we observe that a necessary and sufficient condition for $\Delta\pi_t$ to reach its limit Δp from below is that for every t large enough

$$\mathbb{E}\left\{Q_f(0)\frac{1}{t}\int_0^t [v_+^f(Q_f(\tau),\tau) - v_+^f(Q_f(0),0)]\,\mathrm{d}\tau\right\} \geqslant \hbar/2m. \tag{9.62}$$

Condition (9.62) lends itself to the suggestive interpretation that, at least in some cases, such as the explicit example discussed above, part of the uncertainty in momentum builds up as a *cumulative effect* of small fluctuations over large times.

9.6. Appendix C. Concept of the 'Differential Form' and Directional Derivative

An important question in relativistic dynamics is associated with the evolution of the system on the transition from one hyperplane σ_1 to another σ_2. We suppose the hyperplane σ is characterized by a single time-like vector $\mathbf{v} = (v_0, \boldsymbol{v})$. The hyperplane σ is orthogonal to the world line $\mathscr{P}(s)$ of the particle, and the four-velocity $u^\mu = \mathrm{d}\mathscr{P}/\mathrm{d}s = \mathrm{d}x^\mu/\mathrm{d}s$ is the unit vector tangent to the world line of the particle.

Hitherto, we have considered the evolution of the system from the point of view of its proper time (see Chapter 8). For example, by means of this concept we have derived the probability density $\rho(x_\mu, s + \Delta s)$ for the 'time' $s + \Delta s$ from the probability density $\rho(x_\mu, s)$ for the earlier 'time' s, and we have also obtained a differential equation for $\rho(x_\mu, s)$.

Now we attempt to derive an equation for $\rho(x_\mu, s)$ from a different point of view, namely, on the basis of a geometrical object – a '*differential form*' or '1-form' (for more detail, see Misner *et al.*, 1973); then the evolution of the probability density ρ is determined by means of an *oriented family* of flat surfaces $\tilde{\mathbf{k}} = \mathbf{d}\rho$. Using only surfaces, it is impossible to characterize fully the 1-form $\tilde{\mathbf{k}}$; it is also necessary to specify an orientation. We suppose $\tilde{\mathbf{k}}$ is composed of the surfaces of the constant values, say, $\rho = 7, 8, 9, \ldots$; then the direction from surface to surface is assumed to be 'positive' if ρ increases in this direction.

We now define the concept of the derivative along the direction of the vector $\mathbf{v} = (v_0, \boldsymbol{v})$. We take a certain vector $\mathbf{v}$, construct the curve $\mathscr{P}(\lambda)$, which is determined by the equation $\mathscr{P}(\lambda) - \mathscr{P}_0 = \lambda \boldsymbol{v}$, and differentiate the function ρ along this curve:

$$\partial_{\mathbf{v}}\rho = (\mathrm{d}/\mathrm{d}\lambda)|_{\lambda=0}\,\rho(\mathscr{P}(\lambda)) = (\mathrm{d}\rho/\mathrm{d}\lambda)|_{\mathscr{P}_0}.$$

The 'differential operator' $\partial_{\mathbf{v}} = (\mathrm{d}/\mathrm{d}\lambda)$ for $\lambda = 0$ along the curve $\mathscr{P}(\lambda) - \mathscr{P}_0 = \lambda\mathbf{v}$, which carries out this differentiation, is called the '*operator of the derivative along the direction of the vector* $\mathbf{v}$'. The derivative along the direction $\partial_{\mathbf{v}}\rho$ and the gradient $\mathbf{d}\rho$ are intimately related. This is immediately seen by applying $\partial_{\mathbf{v}}$ to the equation $\rho(\mathscr{P}) = \rho(\mathscr{P}_0) + \langle \mathbf{d}\rho, \mathscr{P} - \mathscr{P}_0 \rangle$ + nonlinear terms, which determines the 'gradient', and calculating the result at the $\mathscr{P}_0$, which gives

$$\partial_{\mathbf{v}}\rho = \langle \mathbf{d}\rho, \mathrm{d}\mathscr{P}/\mathrm{d}\lambda \rangle = \langle \mathbf{d}\rho \cdot \mathbf{v} \rangle.$$

In terms of Misner, Thorne, and Wheeler (1973), this result reads: $\mathbf{d}\rho$ is a linear machine for calculating the rate of change of ρ along an arbitrary vector $\mathbf{v}$. Introducing $\mathbf{v}$ into $\mathbf{d}\rho$, we obtain at the output $\partial_{\mathbf{v}}\rho$, the 'number of intersected planes', a number which for sufficiently small $\mathbf{v}$ is simply equal to the increment of ρ between the base and tip of the vector $\mathbf{v}$.

In the co-ordinate representation, the operator $\partial_{\mathbf{v}}$ of the derivative along the direction of the vector $\mathbf{v}$ is determined by the expression $\partial_{\mathbf{v}} = v^\mu \, \partial/\partial x^\mu$. The notation and concepts are taken from Misner *et al.* (1973).

Thus, the method based on the use of the concept of the derivative along a direction warrants attention. For example, when this method is used equations (8.20) and (8.21) simplify significantly by virtue of the elimination of the dependence on $u^\mu_\pm$ in the expression for $P_\pm$, which now takes the form

$$P_\pm = (4\pi \nu_\pm \, \Delta s_\pm)^{-2} \exp[-(\Delta y_{\mathrm{E}})^2/4\nu_\pm \cdot \Delta s_\pm]; \quad (\nu_+ = \nu_- = \nu).$$

Here, s_+ and s_- are the parameters which characterize the derivative along the directions of the vectors u^μ_+ and u^μ_-, respectively. In this case, equations (8.22) take the form

$$\begin{aligned}
\partial u^\mu_+/\partial s_+ &= 2u^\lambda \partial_\lambda u^\mu_+ + \nu\Box u^\mu_+ + F^\mu_+/m;\\
\partial u^\mu_-/\partial s_- &= -2u^\lambda \, \partial_\lambda u^\mu_- - \nu\Box u^\mu_- + F^\mu_-/m;\\
\partial u^\mu_-/\partial s_+ &= 2u^\lambda \partial_\lambda u^\mu_- + \nu\Box u^\mu_- + F'^\mu_-/m;\\
\partial u^\mu_+/\partial s_- &= -2u^\lambda \, \partial_\lambda u^\mu_+ - \nu\Box u^\mu_+ + F'^\mu_+/m.
\end{aligned}$$

By definition, the derivatives along the directions of u^μ_+ and u^μ_- are

$$\partial u^\mu_\pm/\partial s_+ = u^\lambda_+ \, \partial_\lambda u^\mu_\pm; \quad \partial u^\mu_\pm/\partial s_- = u^\lambda_- \, \partial_\lambda u^\mu_\pm,$$

and we therefore obtain equations (8.23) and (8.24) with

$$D_c = v^\lambda \, \partial_\lambda; \qquad D_s = u^\lambda \, \partial_\lambda + \nu\Box.$$

Equations of form (8.23) with $\lambda = 1$ and $D_c = v^\lambda \, \partial_\lambda$ were obtained for the first time by Lehr and Park (1977) (see also de la Peña-Auerbach, 1971); they are, of course, equivalent to equations (8.23) with $D_c = \partial/\partial s + v^\lambda \, \partial_\lambda$ and $D_s = u^\lambda \, \partial_\lambda + \nu\Box$.

Thus, the derivative with respect to s which occurs in this appendix can be understood as the derivative along the direction of some arbitrary vector $\mathbf{v}$. In the special case when the particle's velocity u^μ is taken as the vector v^μ, we can interpret s as the *proper time* of the particle. In our approach, conservation of mass (of the probability density multiplied by the volume) signifies the absence of mass loss through any hyperplane characterized by the vector u^μ; it follows from this in particular that the probability density for finding the particle along the world trajectory is constant, i.e.,

$$\partial\rho/\partial s = (\mathbf{d}\rho\cdot\mathbf{u}) = (\partial\rho/\partial x_0)u^0 + (1-\beta^2)^{-1/2}\mathbf{v}\cdot(\partial\rho/\partial\mathbf{x}) = 0. \tag{9.63}$$

This last equation expresses the conservation of the mass (the probability density of the current) of the particle. Note that equation (9.63) recalls the equation that reflects the absence of heat transfer and requires constancy of the specific entropy (the entropy of unit mass) for a fluid particle (see Misner *et al.*, 1973):

$$\mathrm{d}\mu/\mathrm{d}t = 0 \quad \text{or} \quad \partial\mu/\partial t + (\mathbf{v}\nabla)\mu = 0,$$

and the last equation expresses the conservation of mass:

$$\partial\rho/\partial t + \nabla(\rho\mathbf{v}) = 0.$$

We now obtain the continuity equation for the probability density of the current in our case. We see from (8.18) that the condition (9.63) leads to the following equations:

$$\partial_\mu(\rho u^\mu_+) = \nu\Box\rho \quad \text{and} \quad \partial_\mu(\rho u^\mu_-) = -\nu\Box\rho.$$

Further, we determine the vectors of the four-dimensional current:

$$J^\mu_\pm = \rho u^\mu_\pm; \qquad J^\mu \equiv \tfrac{1}{2}(J^\mu_+ + J^\mu_-); \qquad v^\mu \equiv J^\mu/\rho,$$

and then J^μ satisfies the continuity equation

$$\partial_\mu J^\mu = 0. \tag{9.64}$$

Since $v_\mu v^\mu = c^2$, $|\rho|^2 = J_\mu J^\mu/c^2$ is a world scalar and, therefore, the equations

$$\partial_\mu J^\mu_+ = \nu\Box|\rho| \quad \text{and} \quad \partial_\mu J^\mu_- = -\nu\Box|\rho|,$$

are already expressed in covariant form.

Chapter 10

Further Developments in Stochastic Quantization

10.1. Introduction

The stochastic quantization programme of classical canonical systems, proposed by Fényes (1952), Nelson (1966, 1967) and their followers Cufaro Petroni and Vigier (1979); Vigier (1979, 1982); Yasue (1979); de la Peña-Auerbach and Cetto (1975); Davidson (1979–82); Guerra and Ruggiero (1973), etc. has been enriched by new methods in recent years. These methods (see Chapters 8 and 9; Baulieu and Zwanziger, 1981; Zwanziger, 1981; Parisi and Wu, 1981; Niemi and Wijewardhana, 1982; Mc'Clain *et al.*, 1982, 1983; etc.) rely on ideas of stochastic space-time (or vacuum fluctuation), and of the stochasticity of gauge fields, and the use of new techniques and concepts. A central role in these ideas is played by the fact that there exist stochastic solutions of a simple variant of the Yang–Mills theory, obtained by Matinyan *et al.* (1981a, b); Chirikov and Shepelyansky (1981).

We recall that the Nelson stochastic quantization procedure has already been generalized and used for the description of continuous systems and fields, in particular, in the case of the Euclidean–Markov scalar field (Guerra and Ruggiero, 1973; Davidson, 1980 and Moore, 1980), and of the electromagnetic (Guerra and Loffredo, 1980; Davidson, 1981), and gravitational (Davidson, 1982) fields. Moreover, the properties of physical vacuum in particular, the mechanism of the vacuum tunneling effect can be illustrated within the realm of the stochastic quantization (Yasue, 1979).

Recently, the relation between Euclidean quantum field theory and stochastic differential equations has attracted considerable attention with the development of new methods of stochastic quantization of any field theory, including gauge theories. In particular, they offer a new method of quantization, which sometimes presents conceptual advantages. For example, by considering a stochastic equation in real space, one can show that a supersymmetry and superspace naturally arise (Parisi and Sourlas, 1979, 1982; Mc'Clain *et al.*, 1983). In a more traditional approach (see Section 10.4 and references therein), one may introduce a fifth dimension in the form of a new 'time' τ and consider a stochastic equation in this variable (Parisi and Wu, 1981).

The stochastic quantization method proposed by Parisi and Wu is to use the solution of the Langevin equation to evaluate the *correlation functions* according to the following prescription:

(i) Let $\mathscr{L}[A(x)]$ $(x = x_4, \mathbf{x})$ be the Lagrangian density describing the dynamics of a physical system. $\mathscr{L}$ is a function of a set of fields in d-dimensional Euclidean space, which we denote collectively by $A(x)$ (internal indices are omitted here). In the Euclidean field theory (see Section 7.6) the Green functions are the vacuum expectation values of products of fields and are given by

$$\langle 0|A(x_1)\cdots A(x_n)|0\rangle = \frac{\int \delta A\, \mathrm{e}^{-S[A]}\, A(x_1)\cdots A(x_n)}{\int \delta A\, \mathrm{e}^{-S[A]}}, \tag{10.1}$$

where $S[A]$ is the action of the system.

(ii) Introduce an additional dimension and consider fields which depend on $d+1$ variables $A_\eta(x, \tau)$. The evolution in this extra 'time' τ is assumed to be governed by a *Langevin equation*:

$$\frac{\partial}{\partial\tau} A_\eta(x, \tau) = -\delta S[A_\eta]/\delta A_\eta(x, \tau) + \eta(x, \tau) \tag{10.2}$$

with the initial condition $A_\eta(x, 0) = 0$. It is shown (Floratos and Iliopoulos, 1983) that this condition can be replaced by a more general one $A_\eta(x, 0) = C(x)$, where $C(x)$ is an arbitrary function. Moreover, one can show that the large time $(\tau \to \infty)$ limits are independent of the choice of $C(x)$. In equation (10.2) $\eta(x, \tau)$ is a *random source* with Gaussian distribution, namely the random average over η has the following Wick decomposition property:

$$\langle \eta(x_1, \tau_1)\eta(x_2, \tau_2)\cdots\eta(x_{2n}, \tau_{2n})\rangle_\eta = \sum_{\substack{\text{possible pair}\\ \text{comb.}}}\prod \langle \eta(x_i, \tau_i)\eta(x_j, \tau_j)\rangle_\eta,$$

$$\langle \eta(x, \tau)\eta(x', \tau')\rangle_\eta = 2\,\delta^{(d)}(x - x')\,\delta(\tau - \tau'),$$

i.e., $\eta(x, \tau)$ is a *white, Gaussian noise*.

(iii) Correlation functions are now introduced as averages over η:

$$\begin{aligned}&\langle A_\eta(x_1, \tau_1)\cdots A_\eta(x_n, \tau_n)\rangle_\eta \\ &\quad = \int \delta\eta \exp\left[-\frac{1}{4}\int \mathrm{d}^d x\, \mathrm{d}\tau \cdot \eta^2(x, \tau)\right] \times A_\eta(x_1, \tau_1)\cdots A_\eta(x_n, \tau_n).\end{aligned} \tag{10.3}$$

A convenient way to rewrite (10.3) is to introduce the *probability density*

$P[A, \tau]$ which is defined by

$$P[A, \tau] = \int \delta\eta \exp\left\{ -\frac{1}{4} \int d^d x \, d\rho \cdot \eta^2(x, \rho) \right\} \prod_y \delta[A(y) - A_\eta(y, \tau)]. \tag{10.4}$$

In terms of P we can write the equal-time correlation function as

$$\langle A_\eta(x_1, \tau) \cdots A_\eta(x_n, \tau) \rangle_\eta = \int \delta A \cdot A(x_1) \cdots A(x_n) P[A, \tau]. \tag{10.5}$$

The probability density P satisfies the *Fokker–Planck equation*:

$$\frac{\partial}{\partial \tau} P[A, \tau] = \int d^d x \frac{\delta}{\delta A(x)} \left\{ \frac{\delta}{\delta A(x)} + \frac{\delta S}{\delta A(x)} \right\} P[A, \tau], \tag{10.6}$$

with the initial condition $P[A, 0] = \Pi_y \delta[A(y)]$. On the right-hand side of (10.6) we assume that a suitable regularization has been employed. In particular, we can use the nonlocal method proposed in Part I. Roughly speaking, in the given case our method of regularization leads to a *smeared* (*nonlocal*) *white noise*:

$$\eta_s(x, \tau) = \int d^d y K(x - y) \eta(y, \tau) \quad \text{or}$$

$$\eta_s(x, \tau) = \int d^d y \int d\rho \, K(x - y, \tau - \rho) \eta(y, \rho).$$

(iv) The correlation function (10.1) is then given by

$$\langle 0 | A(x_1) \cdots A(x_n) | 0 \rangle = \lim_{\tau \to \infty} \langle A_\eta(x_1, \tau) \cdots A_\eta(x_n, \tau) \rangle_\eta \tag{10.7}$$

or, equivalently,

$$\text{w.} \lim_{\tau \to \infty} P[A, \tau] = P^{\text{eq}}[A] = \frac{\exp\{-S[A]\}}{\int \delta A \, e^{-S[A]}}, \tag{10.8}$$

i.e., at large times the probability distribution reaches the *equilibrium* one given by (10.8). Here the limit is supposed to be taken 'weakly', as expressed in (10.7); in other words, $P[A, \tau]$ will always be applied to a string of fields (for details, see Floratos and Iliopoulos, 1983).

The property (10.7) or (10.8) has been shown to hold for several systems (Wick, 1981; Fox, 1978). It has been conjectured (Parisi and Wu, 1981) that it is true even for *gauge theories*, provided one looks for *gauge-invariant quantities*. In other words, it has been indicated that the limit (10.7) does not exist for arbitrary gauge non-invariant Green functions, but it exists and gives the correct result only for gauge-invariant quantities. This fact is very interesting because it allows one to formulate a new quantization method which does not require any gauge-fixing term and is therefore *free from gauge ambiguities* (Gribov, 1978; Singer, 1978). In Section 10.5 we shall prove this statement order by order in perturbation theory.

The purpose of this chapter is to report briefly on the progress in the stochastic interpretation of quantum field theory. These results are based on the studies (Guerra and Ruggiero, 1973; Davidson, 1979–81; Yasue, 1979; Baulieu and Zwanziger, 1981; and Floratos and Illiopoulos, 1983) of a generalization of the Nelson stochastic mechanics to systems with infinite number of degrees of freedom and of the stochastic quantization method introduced by Parisi and Wu (1981). In Sections 10.2 and 10.3 we present the stochastic quantization method for scalar and electromagnetic fields in the framework of Davidson's approach. Some ideas of stochastic quantization of the gauge theories are given in Sections 10.4 and 10.5. The last two sections are devoted to the study of the mechanism of vacuum tunneling phenomena and stochastic fluctuations of the classical Yang–Mills fields, respectively.

10.2. Davidson's Model for Free Scalar Field Theory

In Section 7.5 it is shown that Euclidean field theory results from the stochastic quantization of the free scalar field. We present here another approach due to Davidson (1980a). This model is useful in our scheme where a nonlocal Euclidean–Markov field is investigated (see Chapter 11). The basic idea of Davidson's stochastic theory is the following: Let $\varphi(\mathbf{x}, t)$ be the free scalar field with mass m. Then, consider the classical equation for the real free scalar field in the Minkowski space:

$$P_\mu P^\mu \varphi = m^2 \varphi, \qquad P_\mu = -i\hbar \partial/\partial x_\mu,$$

where c has been set to unite and $g^{00} = 1$. Let us require periodic boundary conditions on the field

$$\varphi(\mathbf{x}, t) = \varphi(\mathbf{x} + \mathbf{a}, t), \quad a_i = n_i L,$$

n_i is an integer, so that one may write

$$\varphi(\mathbf{x}, t) = (\hbar L^3)^{-1/2} \sum_{\mathbf{q}} \exp(i\mathbf{qx})\, \varphi_{\mathbf{q}}(t), \quad q_i = 2\pi m_i/L, \tag{10.9}$$

(m_i is an integer).

Further, as usual, it is assumed that each component of the Fourier decomposition (10.9) is a *random variable* satisfying the stochastic differential equation

$$\mathrm{d}\varphi_{\mathbf{q}}(t) = b_{\mathbf{q}}(\varphi_{\mathbf{q}}(t), t)\, \mathrm{d}t + \mathrm{d}W_{\mathbf{q}}(t), \tag{10.10}$$

where $b_{\mathbf{q}}$ is a smooth function of the type of a velocity field (see Chapter 7) and is determined by the ground state probability density $\rho_{\mathbf{q}}(\varphi_{\mathbf{q}}(t))$ for $\varphi_{\mathbf{q}}(t)$ (Davidson, 1980a). In equation (10.10) $W_{\mathbf{q}}(t)$ is a Wiener process satisfying

$$\mathbb{E}\{\mathrm{d}W_{\mathbf{q}}(t)\,\mathrm{d}W_{\mathbf{q}'}(t)\} = 4\nu\,\delta_{\mathbf{q},-\mathbf{q}'}\,\mathrm{d}t,$$

here $\mathbb{E}$ denotes the conditional expectation value with respect to the random variable $\varphi_{\mathbf{q}}(t)$, and ν is the diffusion coefficient. For illustration we write a solution to simple stochastic equation of the following type (see Nelson, 1967; Kac, 1959):

$$\mathrm{d}Q(t) = -\omega Q(t)\,\mathrm{d}t + \mathrm{d}W(t), \qquad Q(t) = \mathrm{e}^{-\omega t}\int_{-\infty}^{t} \mathrm{d}W(s)\,\mathrm{e}^{\omega s}. \tag{10.11}$$

For the Wiener process $\mathrm{d}W(s) \equiv y(s)$ we have then,

$$\mathbb{E}\{y(s_1)y(s_2)\mid y(\beta)=0\}$$

$$\underset{\mathrm{Df}}{=} \lim_{\varepsilon\to 0}\frac{\mathbb{E}\{y(s_1)y(s_2);\,-\varepsilon<y(\beta)<\varepsilon\}}{P\{-\varepsilon<y(\beta)<\varepsilon\}}$$

$$= \lim_{\varepsilon\to 0}\left\{\iint_{-\infty}^{\infty}\int_{-\varepsilon}^{\varepsilon} \mathrm{d}y_1\,\mathrm{d}y_2\,\mathrm{d}y_3\,y_1y_2(2\pi s_1)^{-1/2}\exp\left(-\frac{y_1^2}{2s_1}\right)[2\pi(s_2-s_1)]^{-1/2}\times\right.$$

$$\times\exp\left[\frac{-(y_2-y_1)^2}{2(s_2-s_1)}\right][2\pi(\beta-s_2)]^{-1/2}\frac{\exp[-(y_3-y_2)^2/2(\beta-s_2)]}{(2\pi\beta)^{-1/2}}\times$$

$$\left.\times\left[\int_{-\varepsilon}^{\varepsilon}\mathrm{d}y_3\exp\left(-\frac{y_3^2}{2\beta}\right)\right]^{-1}\right\}$$

$$= (2\pi\beta)^{1/2}\iint_{-\infty}^{\infty}\mathrm{d}y_1\,\mathrm{d}y_2\,y_1y_2(2\pi)^{-3/2}\times[s_1(s_2-s_1)(\beta-s_2)]^{-1/2}\times$$

$$\times\exp\left\{-\left[\frac{y_1^2}{2s_1}+\frac{(y_2-y_1)^2}{2(s_2-s_1)}+\frac{y_2^2}{2(\beta-s_2)}\right]\right\}$$

$$= s_1 - s_1s_2\beta^{-1}, \quad (0<s_1\leqslant s_2\leqslant\beta).$$

Here $s = 2\nu t$, $\nu = \frac{1}{2}$. Assuming $s_1 = s_2 = 2t\nu$ and $\beta = 2\nu(t+\mathrm{d}t)$ we obtain

$$\mathbb{E}\{\mathrm{d}W(t)\,\mathrm{d}W(t)\} = 2\nu t[1-(1-\mathrm{d}t/t)] = 2\nu\,\mathrm{d}t, \qquad \mathbb{E}\{\mathrm{d}W(t)\} = 0.$$

From the solution (10.11), we have

$$\mathbb{E}\{Q(t)\} = 0,$$

$$\mathbb{E}\{Q(t)Q(t')\} = \mathbb{E}\left\{\mathrm{e}^{-\omega t}\int_{-\infty}^{t}\exp(\omega s_1)\,\mathrm{d}W(s_1)\exp(-\omega t')\int_{-\infty}^{t}\exp(\omega s_2)\,\mathrm{d}W(s_2)\right\}$$

$$= (\nu/\omega)\exp(-\omega|t-t'|).$$

In the case of the Gaussian distribution, the ground state probability density $\rho_{\mathbf{q}}(\varphi_{\mathbf{q}}(t))$ and the function $b_{\mathbf{q}}(\varphi_{\mathbf{q}}(t))$ take the form

$$\rho_{\mathbf{q}}(\varphi_{\mathbf{q}}(t)) \sim \exp\{-(2\omega_{\mathbf{q}}/\hbar^2)\varphi_{\mathbf{q}}^2(t)\}, \quad \text{and} \quad b_{\mathbf{q}} = -4\nu(\omega_{\mathbf{q}}/\hbar^2)\varphi_{\mathbf{q}}(t),$$

respectively, where

$$\omega_{\mathbf{q}} = \tfrac{1}{2}\hbar(\mathbf{q}^2 + m^2/\hbar^2)^{1/2}.$$

Thus, in the Davidson (1980a) stochastic model, the process $\varphi_{\mathbf{q}}(t)$ is a Gaussian stochastic process characterized by the expectations:

$$\mathbb{E}\{\varphi_{\mathbf{q}}(t)\} = 0, \qquad \mathbb{E}\{\varphi_{\mathbf{q}}(t)\,\varphi_{\mathbf{q}'}(t')\} = 2\,\delta_{\mathbf{q},-\mathbf{q}'}\cdot(\hbar/4\omega_{\mathbf{q}})\exp[-\alpha|t-t'|], \tag{10.12}$$

where $\alpha = (4\nu/\hbar^2)\omega_{\mathbf{q}}$.

Here, the diffusion coefficient ν is an arbitrary quantity and undetermined with respect to $\nu = \hbar/2$ which is usually used in Nelson's stochastic mechanics. This situation allows the diffusion parameter to be continued to imaginary values; then the Markov expectations become the causal Green function of quantum field theory which are Lorentz invariants. From this one can conclude that if the diffusion parameter ν is not measurable (undetermined), then the *stochastic model for the field is consistent with special relativity*, although it is not manifestly Lorentz covariant.

Now we obtain this result.

THEOREM 10.1. *For non-coincident* x_i $(i = 1, 2, \ldots, n)$ $(x = \mathbf{x}_i, t_i)$:

$$\lim_{L\to\infty,\,\Lambda\to\infty} \mathbb{E}\{\varphi(\mathbf{x}_1,t_1)\cdots\varphi(\mathbf{x}_n,t_n)\} = S_n\left(\mathbf{x}_1,\frac{2\nu}{\hbar}t_1;\ldots;\mathbf{x}_n,\frac{2\nu}{\hbar}t_n\right), \tag{10.13}$$

where Λ *is a cutoff in* $\mathbf{q}^2$, $S_n(\cdot)$ *are the n-point* Schwinger functions *determined in Chapter* 7, *in particular*

$$S_2(\mathbf{x}_1,t_1;\mathbf{x}_2,t_2) = (4\pi)^{-4}\int d^4p\,\frac{\exp[i\mathbf{p}(\mathbf{x}_1-\mathbf{x}_2) + ip_4(t_1-t_2)]}{p_4^2 + \mathbf{p}^2 + m^2/\hbar^2}. \tag{10.14}$$

Proof. To show this result, it suffices to prove (10.13) for the two-point function only because the expectations of (10.13) will satisfy the following equation:

$$\begin{aligned} &S_{2n}(\mathbf{x}_1,t_1;\ldots;\mathbf{x}_{2n},t_{2n}) \\ &= \sum_{\pi} S_2(\mathbf{x}_1,t_1;\mathbf{x}_2,t_2)\cdots S_2(\mathbf{x}_{2n-1},t_{2n-1};\mathbf{x}_{2n},t_{2n}) \end{aligned} \tag{10.15}$$

(where the sum over π is a sum over distinct permutations of the arguments of the S's), since φ is a Gaussian process. Using equation (10.9), we have

$$\mathbb{E}\{\varphi(\mathbf{x}_1,t_1)\varphi(\mathbf{x}_2,t_2)\} = \frac{1}{\hbar L^3}\sum_{\mathbf{q},\mathbf{q}'}^{\Lambda}\exp(i\mathbf{q}\mathbf{x}_1 + i\mathbf{q}'\mathbf{x}_2)\mathbb{E}\{\varphi_{\mathbf{q}}(t_1)\varphi_{\mathbf{q}'}(t_2)\}, \tag{10.16}$$

where a cutoff in **q**-space has been included in equation (10.16). Substituting equation (10.12) into (10.16), we obtain

$$\mathbb{E}\{\varphi(\mathbf{x}_1,t_1)\varphi(\mathbf{x}_2,t_2)\} = L^{-3}\sum_{\mathbf{q}}^{\Lambda} \exp[i\mathbf{q}(\mathbf{x}_1 - \mathbf{x}_2)] \int \frac{dq_4}{2\pi} \frac{\exp[iq_4(t_1 - t_2)(2\nu/\hbar)]}{q_4^2 + \mathbf{q}^2 + m^2/\hbar^2}.$$

Taking the limits $L \to \infty$ and $\Lambda \to \infty$, this becomes

$$\lim_{L\to\infty,\,\Lambda\to\infty} \mathbb{E}\{\varphi(\mathbf{x}_1, t_1)\varphi(\mathbf{x}_2, t_2)\} = S_2\left(\mathbf{x}_1, \frac{2\nu}{\hbar}t_1; \mathbf{x}_2, \frac{2\nu}{\hbar}t_2\right)$$

and the result is shown. ■

It is interesting to note that if $\nu = \hbar/2$, which is the value used by Guerra and Ruggiero (1973), then the expectations of the Markov theory become equal to the Schwinger functions (see Section 7.5). Theorem 10.1 shows that the expectations of the Davidson model have the same analyticity properties in ν as the Euclidean field's Schwinger function does in t. It is believed that ν is not an observable and that measurable quantities are independent of ν. Therefore, one may consider continuing ν into the complex plane, as has been done by Davidson (1980a).

To illustrate the compatibility of the Davidson theory with special relativity, we continue expectations to the point

$$\nu = i\hbar/2. \tag{10.17}$$

We denote the analytically continued expectations by $\mathbb{E}_\nu$. One easily finds:

$$\mathbb{E}_{\nu = i\hbar/2}\{\varphi(\mathbf{x}, t)\varphi(0)\} = i(2\pi)^{-4} \int d^4q \frac{\exp(ix_\mu q^\mu)}{q_\mu q^\mu - (m^2/\hbar^2) + i\varepsilon}. \tag{10.18}$$

Now we compare (10.18) with the usual causal Green function of the ordinary quantum field theory (see Part I and Bogolubov and Shirkov, 1980):

$$\begin{aligned}\Delta_c(x) &= \langle 0|T\{\varphi(x)\varphi(0)\}|0\rangle \\ &= i^{-1}(2\pi)^{-4}\int d^4q \frac{\exp(iq_\mu x^\mu)}{m^2 - q^2 - i\varepsilon}, \quad (q^2 = q_0^2 - \mathbf{q}^2).\end{aligned} \tag{10.19}$$

From this we see that the two theories are related by

$$\mathbb{E}_{\nu = i\hbar/2}\{\varphi(x)\varphi(0)\} = \Delta_c(x) = \langle 0|T\{\varphi(x)\varphi(0)\}|0\rangle, \tag{10.20}$$

$$\mathbb{E}_{\nu = -i\hbar/2}\{\varphi(x)\varphi(0)\} = \Delta_c^*(x) = \langle 0|T^*\{\varphi(x)\varphi(0)\}|0\rangle, \tag{10.21}$$

where T and T^* denote, as usual, time ordering (later times to the left) and anti-time ordering, respectively.

Using the combinatorial rule of equation (10.15), which is satisfied by quantum expectations as well as the Markov expectations and Schwinger functions, we obtain immediately a generalization of (10.20) and (10.21) to higher-order Green's functions:

$$\mathbb{E}_{\nu = i\hbar/2}\{\varphi(x_1)\cdots\varphi(x_n)\} = \langle 0|T\{\varphi(x_1)\cdots\varphi(x_n)\}|0\rangle, \tag{10.22}$$

$$\mathbb{E}_{\nu = -i\hbar/2}\{\varphi(x_1)\cdots\varphi(x_n)\} = \langle 0|T^*\{\varphi(x_1)\cdots\varphi(x_n)\}|0\rangle, \tag{10.23}$$

where we have stopped writing the t arguments explicitly on the left-hand side. Since the right-hand side of equations (10.22) and (10.23) are Lorentz invariants, it follows that the Davidson model for the free scalar field theory is manifestly *Lorentz covariant* when continued in the diffusion parameter ν to $\pm i\hbar/2$.

It should be noted that the above results may be realized in the case of electromagnetic fields. We now go over to this problem.

10.3. The Electromagnetic Field as a Stochastic Process

We consider stochastic quantization of the electromagnetic field and apply the above obtained results for the scalar field to this case. Guerra and Ruggiero (1973, 1978) and Guerra and Loffredo (1980) first analyzed the relativistic scalar field and the free electromagnetic field using the quantization prescription of Nelson. Moore (1979) has also considered the Maxwell field as part of a general programme of the stochastic field theory. In these studies, only the fixed value of the diffusion coefficient ($\nu = \hbar/2$) of the original Fényes–Nelson theory was considered. As for the scalar field (Davidson, 1980a and in the preceding section), the moments of the electromagnetic field (EMF) in the ground state are equal to the Schwinger functions of the theory, but with the times scaled by a common factor $2\nu/\hbar$ which depends on the diffusion coefficient ν. In this section a stochastic quantization of EMF is carried out in terms of vector potentials rather than directly in terms of electric and magnetic fields as in Guerra and Loffredo's (1980) paper. Here we follow the approach developed by Davidson (1981).

So, in an appropriate system of units the *Maxwell equations* take the form

$$\nabla\cdot\mathbf{E} = \rho, \qquad \nabla\cdot\mathbf{B} = 0; \qquad \nabla\times\mathbf{E} = -\partial\mathbf{B}/\partial t; \qquad \nabla\times\mathbf{B} = \mathbf{J} + \partial\mathbf{E}/\partial t, \tag{10.24}$$

where it is assumed $c = 1$. In the *Coulomb gauge* one has

$$\nabla\cdot\mathbf{A} = 0, \qquad \mathbf{E} = -\nabla\phi - \partial\mathbf{A}/\partial t; \qquad \mathbf{B} = \nabla\times\mathbf{A}; \qquad \Delta\phi = -\rho.$$

The Maxwell equations become

$$(\partial^2/\partial t^2 - \Delta)\mathbf{A} = \mathbf{J} - \nabla\,\partial\phi/\partial\mathbf{t}. \tag{10.25}$$

The energy density of the field in these units is given by $u = \frac{1}{2}(\mathbf{E}^2 + \mathbf{B}^2)$.

Quantization as in the scalar case shall be carried out in a cubic volume of edge L with periodic boundary conditions in L. Consistency requires that $\mathbf{J}$ and ρ have this same periodicity, but little generality is lost by this, since ultimately the volume will tend to infinity. Because of the periodicity requirement, the following decompositions are allowed

$$\mathbf{A} = 2^{-1/2} L^{-3/2} \sum_{\lambda,\mathbf{k}} \varepsilon(\lambda, \mathbf{k})\, e^{i\mathbf{kx}}\, Q_{\lambda,\mathbf{k}}(t);$$

$$\phi = \sum_{\mathbf{k}} e^{i\mathbf{kx}}\, \rho(\mathbf{k}, t)/\mathbf{k}^2; \qquad \rho = \sum_{\mathbf{k}} e^{i\mathbf{kx}}\, \rho(\mathbf{k}, t); \qquad \mathbf{J} = \sum_{\mathbf{k}} e^{i\mathbf{kx}}\, \mathbf{J}(\mathbf{k}, t),$$

where the components of $\mathbf{k}$ are restricted to be integral multiples of $2\pi/L$ and

$$\varepsilon(\lambda, -\mathbf{k}) = \varepsilon(\lambda, \mathbf{k}), \qquad \varepsilon \cdot \mathbf{k} = 0, \qquad \lambda = 1, 2;$$
$$Q_{\lambda,-\mathbf{k}} = \mathbf{Q}^*_{\lambda,\mathbf{k}}; \qquad \rho(-\mathbf{k}, t) = \rho^*(\mathbf{k}, t); \qquad \mathbf{J}(-\mathbf{k}, t) = \mathbf{J}^*(\mathbf{k}, t);$$
$$\partial\rho/\partial t = -i\mathbf{k}\cdot\mathbf{J}. \tag{10.26}$$

It is customary to choose the transverse current by $\mathbf{J}_{\mathrm{T}} = \mathbf{J} - \boldsymbol{\nabla}\cdot\partial\phi/\partial t$, and it is easy to show that its divergence vanishes $\boldsymbol{\nabla}\cdot\mathbf{J}_{\mathrm{T}} = 0$ so that one may write

$$\mathbf{J}_{\mathrm{T}} = 2^{-1/2} L^{-3/2} \sum_{\lambda,\mathbf{k}} \varepsilon(\lambda, \mathbf{k})\, e^{i\mathbf{kx}}\, Z_{\lambda,\mathbf{k}}(t).$$

In this case we choose the real and imaginary parts of the Q's as the canonical co-ordinates. Equation (10.25), then, leads to the following equations for the Q's

$$(\partial^2/\partial t^2 + \mathbf{k}^2) Q_{\lambda,\mathbf{k}}(t) = Z_{\lambda,\mathbf{k}}(t).$$

A suitable Hamiltonian from which these equations may be derived is

$$H = \sum_{\lambda,\mathbf{k}} \{\tfrac{1}{4}(|\dot{Q}_{\lambda,\mathbf{k}}|^2 + \mathbf{k}^2 |Q_{\lambda,\mathbf{k}}|^2) - \tfrac{1}{2}\,\mathrm{Re}(Z^*_{\lambda,\mathbf{k}} Q_{\lambda,\mathbf{k}})\}, \tag{10.27}$$

which may also be written as

$$H = \int_V d^3x \left[\frac{1}{2}\dot{\mathbf{A}}^2 + \frac{1}{2}\sum_{i=1}^{3} (\boldsymbol{\nabla}\cdot A_i)^2 - \mathbf{J}_{\mathrm{T}}\cdot\mathbf{A}\right],$$

where V denotes the cubic volume of edge L.

Since the equations for the Q's (the canonical momenta are their time derivatives) are those for forced harmonic oscillators, in principle the quantization of the field can be achieved by solving a Schrödinger equation for uncoupled but forced harmonic oscillators. Thus, there is wave function in Q space which satisfies $H\psi = i\hbar(\partial\psi/\partial t)$, and where the Hamiltonian operator has the general form

$$H = \sum_i \left\{-\tfrac{1}{2}\hbar^2 \frac{\partial^2}{\partial Q_i^2} - Z_i Q_i + \tfrac{1}{2}k_i^2 Q_i^2\right\}. \tag{10.28}$$

Here the Q's and Z's are real and represent the real and imaginary parts of the

Q's and Z's in equation (10.27). The sum in equation (10.28) is only over independent Q's [in the sense of equation (10.26)], which explains the factor-of-two difference between equations (10.27) and (10.28).

So, we are ready to discuss the stochastic quantization procedure for the electromagnetic field (EMF). According to the previous section, the stochastic quantization of the EMF is straightforward. It is convenient to define

$$\boldsymbol{\delta}(\mathbf{x}, t) = 2^{-1/2} L^{-3/2} \sum_{\lambda, \mathbf{k}} \boldsymbol{\varepsilon}(\lambda, \mathbf{k})\, \mathrm{e}^{i\mathbf{k}\mathbf{x}} \left(\frac{\partial}{\partial \operatorname{Re} Q_{\lambda,\mathbf{k}}} + \frac{i\,\partial}{\partial \operatorname{Im} Q_{\lambda,\mathbf{k}}} \right).$$

The forward and backward time derivatives can be expressed in terms of $\boldsymbol{\delta}$ as

$$D_+ = \frac{\partial}{\partial t} - \int_V \mathrm{d}^3 x (\mathbf{E}_+ + \nabla\phi)\cdot\boldsymbol{\delta} + \nu \int_V \mathrm{d}^3 x\, \delta^2,$$

$$D_- = \frac{\partial}{\partial t} - \int_V \mathrm{d}^3 x (\mathbf{E}_- + \nabla\phi)\cdot\boldsymbol{\delta} - \nu \int_V \mathrm{d}^3 x\, \delta^2,$$

where $\mathbf{E}_+ = -D_+\mathbf{A} - \nabla\phi$, $\mathbf{E}_- = -D_-\mathbf{A} - \nabla\phi$. The Maxwell equations become

$$\nabla\cdot\mathbf{E}_+ = \nabla\cdot\mathbf{E}_- = \rho, \qquad \nabla\cdot\mathbf{B} = 0, \qquad \nabla\times\mathbf{E}_+ = -\mathbf{D}_+\mathbf{B},$$

$$\nabla\times\mathbf{E}_- = -D_-\mathbf{B},$$

$$\tfrac{1}{2}[D_+\mathbf{E}_- + D_-\mathbf{E}_+] + \tfrac{1}{8}\beta(D_+ - D_-)(\mathbf{E}_+ - \mathbf{E}_-) = \nabla\times\mathbf{B} - \mathbf{J}. \tag{10.29}$$

With the help of the following stochastic field,

$$\mathbf{W}_+(\mathbf{x}, t) = 2^{-1/2} L^{-3/2} \sum_{\lambda,\mathbf{k}} \boldsymbol{\varepsilon}(\lambda, \mathbf{k})\, \mathrm{e}^{i\mathbf{k}\mathbf{x}}\, \mathbf{W}^+_{\lambda,\mathbf{k}}(t), \tag{10.30}$$

the *stochastic differential equation* can be written in the compact and suggestive form

$$\mathrm{d}\mathbf{A} = -(\mathbf{E}_+ + \nabla\phi)\,\mathrm{d}t + \mathrm{d}\mathbf{W}_+, \quad \mathrm{d}\mathbf{A} = -(\mathbf{E}_- + \nabla\phi)\,\mathrm{d}t + \mathrm{d}\mathbf{W}_-, \tag{10.31}$$

where the second equation of (10.31) is the 'backward' version of the stochastic differential equations and $\mathbf{W}_-$ is another field of the form (10.30), but with different Wiener processes.

In the special case $\nu = \frac{1}{2}h$ and when there are no charges or currents present, equations (10.29) become precisely those derived by Guerra and Loffredo (1980). It is interesting to note, as is pointed out by Guerra and Loffredo, that the electric field does not exist as a well-defined stochastic process since the time derivatives of A do not exist. Despite this fact, the two functions $\mathbf{E}_+$ and $\mathbf{E}_-$ may be thought of as a stochastic replacement for the electric field.

For the free field, the scalar potential vanishes in the Coulomb gauge. When the Schrödinger equation is solved in the ground state and the Markov process for the field examined, it is found that the vector potential is a

Gaussian Markov field with the following *covariance*:

$$\mathbb{E}\{A_i(x)A_j(y)\} = \hbar \int \left(\frac{\mathrm{d}k}{2\pi}\right)^4 \frac{(\delta_{ij} - k_i k_j/k^2)}{k_0^2 + \mathbf{k}^2} \exp\left[i\mathbf{k}(\mathbf{x} - \mathbf{y}) + \left(\frac{ik_0 2v}{\hbar}\right)|x_0 - y_0|\right]$$

$$(x_0 = t_x, y_0 = t_y), \qquad (10.32)$$

where the infinite volume limit has been taken. The *covariance* for the magnetic field is

$$\mathbb{E}\{B_i(x)B_j(y)\} = \hbar \int \left(\frac{\mathrm{d}k}{2\pi}\right)^4 \frac{(\delta_{ij}k^2 - k_i k_j)}{k_0^2 + \mathbf{k}^2} \exp\left[i\mathbf{k}(\mathbf{x} - \mathbf{y}) + \right.$$

$$\left. + \left(\frac{ik_0 2v}{\hbar}\right)|x_0 - y_0|\right] \qquad (10.33)$$

As has been stated before, the electric field does not, strictly speaking, exist. However, because the following expressions are equal

$$\frac{\partial}{\partial x_0}\frac{\partial}{\partial y_0}\mathbb{E}\{A_i(x)A_j(y)\} = \mathbb{E}\{E^i_\pm(x)E^j_\pm(y)\}, \quad x_0 \neq y_0, \qquad (10.34)$$

it is natural to study the properties of the left-hand side of (10.34) and identify these with an electric field. From this we have

$$\frac{\partial}{\partial x_0}\frac{\partial}{\partial y_0}\mathbb{E}\{A_i(x)A_j(y)\}$$
$$= 2v\,\delta(x_0 - y_0)\,\delta^{ij}_{\mathrm{tr}}(\mathbf{x} - \mathbf{y}) + (2v/\hbar)^2\,\mathbb{E}\{B_i(x)B_j(y)\}, \qquad (10.35)$$

where $\delta(t)$ is the Dirac delta function and $\delta^{ij}_{\mathrm{tr}}(t)$ is the transverse delta function defined by

$$\delta^{ij}{}_{\mathrm{tr}}(\mathbf{x}) = (2\pi)^{-3}\int \mathrm{d}^3k(\delta_{ij} - k_i k_j/\mathbf{k}^2)\,\mathrm{e}^{i\mathbf{k}\mathbf{x}}. \qquad (10.36)$$

Finally, terms with both electric and magnetic fields may be evaluated. The result is

$$-\frac{\partial}{\partial x_0}\mathbb{E}\{A_i(x)B_j(y)\} = \mathbb{E}\{E^i_\pm(x)B^j(y)\}$$

$$= i\hbar(2v/\hbar)\varepsilon_{ijl}[1 - 2\theta(x_0 - y_0)](2\pi)^{-4} \times \qquad (10.37)$$

$$\times \int \mathrm{d}^4k\, k_l \exp[i\mathbf{k}(\mathbf{x} - \mathbf{y}) + (ik_0 2v/\hbar)|x_0 - y_0|],$$

where ε_{ijl} is the totally antisymmetric tensor. When both fields are at the same point in space, one obtains the results

$$\begin{aligned}
&\mathbb{E}\{B_i(x_0)B_j(y_0)\} = \hbar(\delta_{ij}/\pi^2)(\hbar/2\nu)^4|x_0 - y_0|^{-4},\\
&\mathbb{E}\{E^i_\pm(x_0)E^j_\pm(y_0)\} = \hbar(\delta_{ij}/\pi^2)(\hbar/2\nu)^2|x_0 - y_0|^{-4}, \qquad (10.38)\\
&\mathbb{E}\{E^i_\pm(x_0)B^j(y_0)\} = 0.
\end{aligned}$$

High-order correlations may be calculated from those above by means of the well-known combinatoric rule for Gaussian processes. For example,

$$\mathbb{E}\{A_{i_1}(x_1)\cdots A_{i_{2m}}(x_{2m})\} = \sum_{\pi} \mathbb{E}\{A_{i_1}(x_1)A_{i_2}(x_2)\}\cdots \times$$
$$\times\ \mathbb{E}\{A_{i_{2m-1}}(x_{2m-1})A_{i_{2m}}(x_{2m})\}, \qquad (10.39)$$

where the sum is over all distinct permutations and all odd correlations vanish.

As in the scalar field case (see Section 10.2) the *Green functions of the usual quantum field theory are obtained by analytically continuing the diffusion coefficient* ν *to imaginary values*, as has been done by Davidson (1980a). The analytically-continued expectations are denoted by a subscript. The following results are found when the usual quantization is carried out in the *Coulomb gauge*:

$$\mathbb{E}_{\nu = i\hbar/2}\{A_i(x)A_j(y)\} = \langle 0|T\{A_i(x)A_j(y)\}|0\rangle,$$
$$\mathbb{E}_{\nu = -i\hbar/2}\{A_i(x)A_j(y)\} = \langle 0|T^*\{A_i(x)A_j(y)\}|0\rangle.$$

The two-point function – Green function – for the quantum theory is given by

$$\langle 0|T\{A_i(x)A_j(y)\}|0\rangle = i\hbar\int\left(\frac{\mathrm{d}k}{2\pi}\right)^4 \frac{\delta_{ij} - k_i k_j/k^2}{k_\mu k^\mu + i\varepsilon}\,\mathrm{e}^{-ik(x-y)}, \qquad (10.40)$$

where $k_\mu k^\mu = k_0^2 - \mathbf{k}^2$, $x_0 > y_0$, the metric convention $g^{00} = 1$ is used and ε is positive and is taken to zero after the k_0 integration is performed (for details, see Chapters 2–4). The other time ordering is obtained by taking the complex conjugate of equation (10.40). According to the Gaussian combinatories of the quantum and stochastic theory, we may write in general

$$\mathbb{E}_{\nu = i\hbar/2}\{A_{i_1}(x_1)\cdots A_{i_n}(x_n)\} = \langle 0|T\{A_{i_1}(x_1)\cdots A_{i_n}(x_n)\}|0\rangle,$$
$$\mathbb{E}_{\nu = -i\hbar/2}\{A_{i_1}(x_1)\cdots A_{i_n}(x_n)\} = \langle 0|T^*\{A_{i_1}(x_1)\cdots A_{i_n}(x_n)\}|0\rangle.$$

The conclusion to be drawn is that the stochastic theory is physically equivalent to the Lorentz covariant quantum field theory, even though the stochastic theory is not itself manifestly Lorentz covariant because the expectations in equations (10.32)–(10.39) do not transform in a covariant way.

We notice that the generalized Fényes–Nelson model has been applied by Davidson (1982) to the weak field approximation of Einstein's general theory of relativity, the so-called linearized gravitational field (Einstein, 1979; Misner *et al.*, 1973; Weinberg, 1972). We have not considered this problem here. The stochastic quantization of supergravity has been discussed by Hori (1983).

10.4. Stochastic Quantization of the Gauge Theories

10.4.1. *Introduction*

Interest in the stochastic regularization and stochastic quantization of gauge theories has grown in recent years. The stochastic methods have already been applied to gauge theories. Such approaches are based on the fact that gauge theories possess stochastic properties. For example, it has been discovered that simple variants of the *Yang–Mills theory* have stochastic solutions (Matinyan *et al.*, 1981a, b). There have appeared a few stochastic models for gauge theories. For example, Parisi and Wu (1981) proposed a method of stochastic quantization of *gauge fields* whereby Euclidean expectation values are obtained by relaxation to equilibrium of a stochastic process depending on an artificial fifth time parameter (for further developments see Nakano, 1983; Alfaro and Sakita, 1983; Guha and Lee, 1982; Japan group, 1982). Within this approach Zwanziger (1981) has obtained the equilibrium distribution directly without reference to the artificial time, by a stationary condition which is an eigenfunction equation in the Euclidean Hilbert space. Zwanziger's model allows one to carry out a covariant quantization of gauge fields without *Gribov' ambiguity* (Gribov, 1977, 1978). On the basis of this model Baulieu and Zwanziger (1981) have proved the equivalence of stochastic quantization to the Faddeev–Popov (1967) ansatz for covariant and axial gauges. Equivalence of stochastic and canonical quantization in perturbation theory has been demonstrated by Floratos and Iliopoulos (1983) (see Section 10.5).

Niemi and his coworkers (1982) have analyzed the equivalence between a scalar quantum field theory (also gauge theories) in d-dimensions and its classical counterpart in $d + 2$ dimensions which is coupled to an external random source with Gaussian correlations. The main idea of this approach is the following: A classical six-dimensional Euclidean scalar field $\varphi(x)$ coupled to an external source $\eta(x)$ is considered. The Lagrangian of this system is given by

$$\mathscr{L}[\varphi] = \tfrac{1}{2}(\partial_\mu \varphi)^2 + \tfrac{1}{2}m^2\varphi^2 + \frac{g}{3!}\varphi^3 + \frac{\lambda}{4!}\varphi^4 + \eta(x)\varphi.$$

The classical equation of motion

$$\delta A[\varphi]/\delta\varphi(x) = -\eta(x) \tag{10.41}$$

is solved perturbatively. Here A is the classical action. This solution of equation (10.41) is denoted by $\varphi_\eta(x)$. It is assumed that the random sources $\eta(x)$ have Gaussian correlations

$$\mathbb{E}\{\eta(x)\} = 0,$$

$$\mathbb{E}\{\eta(x)\eta(y)\} = \int \delta\eta\ \eta(x)\eta(y)\exp\left[-\frac{1}{2}\frac{1}{\rho}\int \mathrm{d}^6 x\eta^2(x)\right] = \rho\cdot\delta^{(6)}(x-y),$$

$$\mathbb{E}\{\eta(x_1)\cdots\eta(x_{2n})\} = \sum_{\pi} \mathbb{E}\{\eta(x_1)\eta(x_2)\}\cdots\mathbb{E}\{\eta(x_{2n-1})\eta(x_{2n})\}. \tag{10.42}$$

In the momentum space (10.42) reads

$$\mathbb{E}\{\tilde{\eta}(k_1)\tilde{\eta}(k_2)\} = (2\pi)^6 \rho \cdot \delta^{(6)}(k_1 + k_2).$$

The parameter ρ defining the correlation strength has dimensions m^2 where m fixes the mass scale.

This stochastic quantization scheme was started by Aharony, Imry and Ma (1976). In this scheme the functional of the form

$$Z[J] = \int \delta\eta \exp\left\{\iint d^4x\, d^2y\left[\left[\varphi_\eta(x, y)J(x, y) - \frac{1}{2\rho}\eta^2(x, y)\right]\right]\right\}, \tag{10.43}$$

is investigated. Here x is four-dimensional and y is two-dimensional, and the source $J(x, y)$ of the form $J(x, y) = (1/m)\,\delta^{(2)}(y)j(x)$ is introduced (where $j(x)$ is arbitrary). It is proved that the four-dimensional n-point functions

$$\mathbb{E}\{\varphi(x_1)\cdots\varphi(x_n)\} = \int \delta\eta m^{-n} \varphi_\eta(x_1,0)\cdots\varphi_\eta(x_n,0) \times$$

$$\times \exp\left\{-\frac{1}{2\rho}\iint d^4x\, d^2y\, \eta^2(x,y)\right\} \tag{10.44}$$

generated by the expression on the right-hand side of equation (10.43), are equal to the full n-point Green functions in the corresponding four-dimensional quantum theory, order by order in perturbation theory (see Mc'Clain *et al.*, 1982; Niemi and Wijewardhana, 1982).

We briefly review the method of stochastic quantization due to Parisi and Wu (1981) and Zwanziger (1981) in this section, discuss the equivalence of stochastic and canonical quantization in perturbation theory, in the case of gauge theories, and the Euclidean path integral description of the vacuum tunneling phenomena from the point of view of the probability theory (Yasue, 1979) in Sections 10.5 and 10.6, respectively. We conclude with some remarks concerning stochastic solution of Yang–Mills-type theory in Section 10.7.

10.4.2. *Another Stochastic Quantization Scheme*

Recently, Parisi and Wu (1981) have suggested that the Euclidean expectation value $\mathbb{E}\{\phi\}$ of a functional $\phi(A)$ of the Euclidean gauge field $A^a_\mu(x)$ be defined as the limit to which the mean relaxes

$$\mathbb{E}\{\phi\} = \lim_{\tau\to\infty} \mathbb{E}\{\phi\}_\tau = \lim_{\tau\to\infty} \int \delta A\phi(A)P[A, \tau],$$

where the probability distribution $P[A, \tau]$ depending on an artificial fifth time parameter, is determined by a stochastic process, namely

$$\dot{P}[A,\tau] = \int \mathrm{d}^4 x \frac{\delta}{\delta A^a_\mu(x)} \left\{ \frac{\delta}{\delta A^a_\mu(x)} + \frac{\delta S[A]}{\delta A^a_\mu(x)} \right\} P[A,\tau],$$

$$S[A] = \frac{1}{4} \int \mathrm{d}^4 x \sum_{\mu,\nu,a} (F^a_{\mu\nu})^2,$$

$$F^a_{\mu\nu} = \partial_\mu A^a_\nu - \partial_\nu A^a_\mu + g f^{abc} A^b_\mu A^c_\nu \tag{10.45}$$

for some initial data

$$P[A,0] = P_0[A] \geqslant 0, \qquad \int \mathrm{d}A\, P_0[A] = 1.$$

Here $\dot{P} = \partial P/\partial\tau$, a is a colour index, $\mu = 1, \ldots, 4$ is a Lorentz index, and x is an Euclidean four-vector. Our consideration is based on the approach developed by Baulieu and Zwanziger (1981). As mentioned above it will frequently be convenient to replace $A^a_\mu(x)$ by *discrete variables* Q_i, $i = 1, \ldots, \infty$ according to

$$A^a_\mu(x) = \sum_i Q_i G^a_{i\mu}(x),$$

where $G^a_{i\mu}(x)$ are a complete orthonormal set. Thus, an arbitrary functional $\phi[A]$, for example

$$\phi[A] = A^a_\mu(x) A^b_\nu(y) \quad \text{or} \quad \phi[A] = \mathrm{tr}\, P \exp i \oint \lambda^a A^a(x)\, \mathrm{d}x$$

will be written $\phi[A] = \phi(Q)$. The stochastic method (Zwanziger, 1981) introduces an artificial fifth time τ and promotes $\phi(Q)$ to a stochastic variable $\phi(Q,\tau)$ governed by the diffusion equation with a drift force $\mathbf{f}(Q)$

$$\dot{\phi}(Q,\tau) = -L_Q \phi(Q,\tau), \tag{10.46}$$

$$L_Q = -[\Delta + \mathbf{f}\cdot\nabla] \tag{10.47}$$

supplemented by the initial condition

$$\phi(Q,0) = \phi(Q). \tag{10.48}$$

Here

$$\Delta \equiv \sum_i \frac{\partial^2}{\partial Q_i^2}, \qquad \mathbf{f}\cdot\nabla = \sum_i f_i(Q) \frac{\partial}{\partial Q_i}$$

and the drift force $f_i(Q)$ is obtained from the classical action $f_i(Q) = -\partial S(Q)/\partial Q_i$, which for a gauge theory is the Yang–Mills action (10.45). Stochastic quantization is completed by the ansatz for calculating Euclidean expectation values

$$\langle\phi\rangle = \lim_{\tau\to\infty} \phi(Q,\tau) \tag{10.49}$$

which is equivalent to the proposal of Parisi and Wu (1981). Clearly, for this formula to make sense, the limit must exist and be independent of Q. It is shown (Baulieu and Zwanziger, 1981) that in a non-gauge theory it is reduced to the standard functional integral formula

$$\langle\phi\rangle = \frac{\int \mathrm{d}Q\,\phi(Q)\exp[-S(Q)]}{\int \delta Q \exp[-S(Q)]}, \tag{10.50}$$

and that for a gauge theory it agrees with the Faddeev–Popov ansatz for *gauge-invariant* ϕ.

As a trivial example, let Q_i be the discrete Fourier components of A in a box quantization, let $S[A]$ be the free field action

$$S[A] = \frac{1}{2}\sum_i (k_i^2 + m^2)Q_i^2,$$

and take $\phi(Q)$ to be the two-point function $\phi(Q) = Q_i Q_j$. One may easily verify that the corresponding stochastic equation

$$\phi(Q,\tau) = \sum_i \left\{\frac{\partial^2}{\partial Q_i^2} - (k_i^2 + m^2)Q_i\frac{\partial}{\partial Q_i}\right\}\phi(Q,\tau),$$

with the initial condition $\phi(Q,0) = Q_i Q_j$ has the solution

$$\phi(Q,\tau) = \frac{\delta_{ij}}{k_i^2 + m^2}[1 - \exp(-2(k_j^2 + m^2)\tau)] + \\ + Q_i Q_j \exp[-(k_i^2 + k_j^2 + 2m^2)\tau].$$

One sees that $\lim_{\tau\to\infty} \phi(Q,\tau) = \delta_{ij}/(k_i^2 + m^2)$, so the stochastic prescription does give the familiar free field propagator $\langle Q_i Q_j\rangle$.

Now we consider the case of a guage theory. The infinitesimal gauge transformation $\delta A_\mu^a = D_\mu^{ab} w^b$, where $D_\mu^{ac} = \partial_\mu \delta^{ac} + f^{abc} A_\mu^b$, (here the coupling constant g has been absorbed in the structure constant f^{abc} of the gauge group), induces the change in ϕ given by

$$\delta\phi[A] = -\int \mathrm{d}^4x\, w^a D_\mu^{ab} \frac{\delta}{\delta A_\mu^b(x)}\phi[A],$$

so gauge-invariant functions are characterized by $G^a(x)\phi[A] = 0$, where $G^a(x) = -D_\mu^{ab}\,\delta/\delta A_\mu^b(x)$ satisfies the commutation relations of the local gauge group,

$$[G^a(x), G^b(y)] = \delta^{(4)}(x - y) f^{abc} G^c(x).$$

With L, equation (10.47), written in the functional notation

$$L = -\int d^4x \left\{ \frac{\delta}{\delta A_\mu^a(x)} - \frac{\delta S_{cl}[A]}{\delta A_\mu^a(x)} \right\} \frac{\delta}{\delta A_\mu^a(x)},$$

it is easily verified that L is strongly gauge invariant, namely $[G^a(x), L] = 0$. Consequently, for gauge invariant ϕ, the formal solution of equations (10.46) and (10.48), namely,

$$\phi[A, \tau] = \exp(-L\tau)\phi[A],$$

may also be written

$$\phi[A, \tau] = \exp(-L_v\tau)\phi[A], \tag{10.51}$$

where

$$L_v = L + \int d^4x v^a(x) G^a(x) = L + \int d^4x (D_\mu v)^a(x) \frac{\delta}{\delta A_\mu^a(x)}. \tag{10.52}$$

Here $v^a(x) = v^a(x, A)$ is an arbitrary functional of A which, in the full enjoyment of gauge freedom, we may specify at our pleasure. Reverting to the discrete notation, we see that L_v is again of the diffusion form $L_v = -(\Delta + \mathbf{f}_v \cdot \nabla)$ with a drift force $\mathbf{f}_v(Q)$ that depends on v determined from

$$\sum_i f_{v,i}(Q) G_{i\mu}^a(x) = -\delta S_{cl}[A]/\delta A_\mu^a(x) - (D_\mu v)^a(x, A). \tag{10.53}$$

Let $P(Q_2, Q_1, \tau)$ be the kernel of the time-translation operator (10.51)

$$\phi(Q_1, \tau) = \int dQ_2 \phi(Q_2) P(Q_2, Q_1, \tau). \tag{10.54}$$

According to equation (10.46), it satisfies

$$\dot{P}(Q_2, Q_1, \tau) = -L_{v,Q_1} P(Q_2, Q_1, \tau),$$

which is sometimes called the backward diffusion equation. From the composition law

$$\int dQ_2 P(Q_3, Q_2, \tau_3 - \tau_2) P(Q_2, Q_1, \tau_2 - \tau_1) = P(Q_3, Q_1, \tau_3 - \tau_1),$$

the forward equation follows:

$$\dot{P}(Q_2, Q_1, \tau) = -L_{v,Q_2}^* P(Q_2, Q_1, \tau),$$

where

$$L_{v,Q}^* = -[\Delta - \nabla \cdot \mathbf{f}_v], \qquad L_{v,Q}^* g = -[\Delta g - \nabla(\mathbf{f} g)]$$

and one has $P(Q_2, Q_1, 0) = \delta(Q_2 - Q_1)$. The form of the last three equations allows one to interpret $P(Q_2, Q_1, \tau)$ as the probability that a Brownian motion particle subject to a drift force $\mathbf{f}_v$ will arrive at Q_2 at time τ if it was originally at Q_1.

The relevant theorem on stochastic processes has been proved only for a finite number of degrees of freedom, and we suppose that our equations may be regularized by keeping only a finite number N of modes Q_i, $i = 1, \ldots, N$. In this case the following theorem applies.

THEOREM 10.2. *Given a drift force* $\mathbf{f}(Q)$, *if there exists a smooth positive function, call it* $P_{\text{eq}}(Q)$, *which satisfies*

$$L^* P_{\text{eq}}(Q) = 0, \qquad P_{\text{eq}}(Q) \geqslant 0, \qquad \int \mathrm{d}Q P_{\text{eq}}(Q) = 1, \tag{10.55}$$

where

$$L^* P_{\text{eq}}(Q) = -\Delta P_{\text{eq}}(Q) + \nabla \cdot (\mathbf{f}(Q) P_{\text{eq}}(Q)),$$

then $P_{\text{eq}}(Q)$ *is unique and, moreover* $P(Q, Q_1, \tau)$ *relaxes to it:*

$$\lim_{\tau \to \infty} P(Q, Q_1, \tau) = P_{\text{eq}}(Q). \tag{10.56}$$

We call $P_{\text{eq}}(Q)$ *the* equilibrium distribution. *More precisely, this theorem can be reformulated as follows: let* $L^* P_{\text{eq}}(Q) = 0$, *where* $P_{\text{eq}}(Q) \geqslant 0$ *is smooth and*

$$\int \mathrm{d}Q P_{\text{eq}}(Q) = 1.$$

Let K be any compact set. Then

$$\operatorname*{Sup}_{Q_1 \in K} \int \mathrm{d}Q \, |P(Q, Q_1, \tau) - P_{\text{eq}}(Q)| \to 0, \quad \textit{as } \tau \to \infty.$$

The existence of $P(Q, Q_1, \tau)$ *is assumed by Varadhan* (1980).

Thus, if there is a distribution satisfying (10.55) then, according to equation (10.54) the stochastic prescription (10.49) reduces to

$$\langle \phi \rangle = \int \mathrm{d}Q \, \phi(Q) P_{\text{eq}}(Q). \tag{10.57}$$

In a non-gauge theory ($v = 0$) one has $\mathbf{f}(Q) = -\nabla S(Q)$, and it is easy to verify that

$$P_{\text{eq}}(Q) = \frac{\exp[-S(Q)]}{\int \delta Q \exp[-S(Q)]},$$

(suitable regularization is needed) satisfies conditions (10.55) and so, in virtue of the theorem, the stochastic prescription (10.49) coincides with the standard formula (10.50).

The important point for a gauge theory is that the above theorem holds even if $\mathbf{f}(Q)$ is not derivable from a potential, and consequently formula (10.57) may be used, provided that for some $v^a(Q)$ there exists $P_{eq}(Q)$ satisfying conditions (10.55) [with $L = L_v$ and $\mathbf{f}(Q) = \mathbf{f}_v(Q)$ given by equations (10.52) or (10.53)]. Equivalence to the *Faddeev–Popov ansatz* follows because the equation

$$L_v^* P_{eq}(Q) = 0, \tag{10.58}$$

or

$$L^* P_{eq}[A] = -\int d^4 x D_\mu^{ac} \frac{\delta}{\delta A_\mu^c(x)} \{v^a(x, A) P_{eq}[A]\},$$

does, in fact, hold for

$$v^a(x, A) = -\int d^4 y \frac{\partial G^{ab}}{\partial y_\nu}(x, y; A) J_\nu^b(y, A),$$

with $P_{eq}[A]$ the *Faddeev–Popov distribution*

$$P_{eq}[A] = P_{FP}[A] = N \det(\partial \cdot D) \exp\left\{-S_{cl} - (2\alpha)^{-1} \int d^4 x (\partial \cdot A)^2\right\}.$$

Here $G^{ab}(x, y; A)$ is the *Green function or ghost field propagator* defined by

$$-\partial D_x^{ac} G^{cb}(x, y; A) = -(D \cdot \partial_y)^{bc} G^{ac}(x, y; A) = \delta^{(4)}(x - y) \delta^{ab},$$

and the effective induced current $J_\nu^b(y, A)$ is given by

$$J_\nu^b(y, A) = D_\mu^{bc} F_{\mu\nu}^c(y) + \alpha^{-1} \partial_\nu \partial \cdot A^b(y) + f^{bcd} \partial_{(z)\nu} G^{cd}(y, z; A)|_{z=y}.$$

It is possible to verify equation (10.58) directly using properties of determinants or the usual *ghost and antighost fields* (see Baulieu and Zwanziger, 1981).

10.5. Equivalence of Stochastic and Canonical Quantization in Perturbation Theory in the Case of Gauge Theories

In comparison with field theories without gauge invariance, in theories possessing a local invariance at the classical level, the proof of property (10.7) presents some difficulties since, already for the free field, there is no limiting distribution $P_0[A, \infty]$. It is hoped that the limit exists only for *gauge-invariant quantities.* Let $F[A]$ be a functional of the gauge potential $A_\mu^a(x)$. The average of F is determined by

$$\langle F \rangle_\eta = \int \delta A \cdot F[A] P[A, \tau], \tag{10.59}$$

where P is again the probability distribution. It satisfies the Fokker–Planck equation, (10.6), with $S[A]$ the gauge-invariant classical action. The claim is that, although $P[A, \tau]$ has no limit when $\tau \to \infty$, the average (10.59) does possess such a limit for gauge-invariant functional $F[A]$ which, furthermore, equals the result obtained by the usual method. We shall prove this statement, order by order in perturbation theory, following the method of Zwanziger (1981); Baulieu and Zwanziger (1981); Floratos and Iliopoulos (1983) (see preceding section). The very elegant proof has been done by Floratos and Iliopoulos (1983) and is based on a lemma which in the case of field theory without gauge invariance can be formulated as follows.

Let us consider a self-interacting scalar field $\varphi(x)$:

$$\mathscr{L} = \mathscr{L}_0 + g\mathscr{L}_{\text{in}}, \qquad \mathscr{L}_0 = -\tfrac{1}{2}(\partial_\mu \varphi)^2 - \tfrac{1}{2}m^2 \varphi^2. \tag{10.60}$$

In perturbation theory the Green functions are computed as power series in g. Similarly, we expand the probability density $P[\varphi, \tau]$:

$$P[\varphi, \tau] = \sum_{k=0} g^k P_k[\varphi, \tau].$$

The n-point equal time correlation function in the kth order of perturbation theory is given by

$$\langle \varphi_\eta(x_1, \tau) \cdots \varphi_\eta(x_n, \tau) \rangle_\eta = g^k \int \delta\varphi \cdot \varphi(x_1) \cdots \varphi(x_n) P_k[\varphi, \tau], \tag{10.61}$$

and is computed as the sum of the corresponding regularized Feynman diagrams. Regularization may be carried out by means of the nonlocal method introduced in Part I. This is the meaning we attach to $P_k[\varphi, \tau]$ and all our arguments should be interpreted in this way.

The Fokker–Planck equation (10.6) now takes the form

$$\frac{\partial}{\partial \tau} P_k[\varphi, \tau] = \int \mathrm{d}^4 x \frac{\delta}{\delta\varphi(x)} \left\{ \frac{\delta}{\delta\varphi(x)} + \frac{\delta S_0}{\delta \varphi(x)} \right\} P_k[\varphi, \tau] +$$

$$+ \int \mathrm{d}^4 x \frac{\delta}{\delta \varphi(x)} \frac{\delta S_{in}}{\delta \varphi(x)} P_{k-1}[\varphi, \tau], \tag{10.62}$$

with the initial conditions

$$P_0[\varphi, 0] = \prod_y \delta[\varphi(y)], \qquad P_k[\varphi, 0] = 0, \quad k = 1, 2, \ldots \tag{10.63}$$

We assume that, for $1 \leqslant l \leqslant k-1$, $P_l[\varphi, \tau]$ is uniformly bounded and satisfies

$$\text{w.}\lim_{\tau \to \infty} P_l[\varphi, \tau] = P_l^{eq}[\varphi]. \tag{10.64}$$

In (10.64) w. lim again denotes the limit in the weak sense, as explained in Introduction 10.1. We now go over to the following Lemma 10.1.

LEMMA 10.1.

$$\text{w.}\lim_{\tau\to\infty}\frac{\partial}{\partial\tau}P_k[\varphi,\tau]=0.$$

Proof. We first transform equation (10.62) in the integral form:

$$P_k[\varphi,\tau]=\int\delta\varphi'\int_0^\tau \mathrm{d}\tau' D_0[\varphi,\varphi';\tau-\tau']\times$$

$$\times\int \mathrm{d}^4x\frac{\delta}{\delta\varphi'(x)}\frac{\delta S_{\text{in}}}{\delta\varphi'(x)}P_{k-1}[\varphi',\tau'], \tag{10.65}$$

where $D_0[\varphi,\varphi',\tau-\tau']$ is the *Green functional* of the free Fokker–Planck equation

$$\frac{\partial}{\partial\tau}D_0[\varphi,\varphi';\tau-\tau']=\int \mathrm{d}^4x\frac{\delta}{\delta\varphi(x)}\left\{\frac{\delta}{\delta\varphi(x)}+\frac{\delta S_0}{\delta\varphi(x)}\right\}D_0[\varphi,\varphi';\tau-\tau']+$$

$$+\delta(\tau-\tau')\prod_x\delta[\varphi(x)-\varphi'(x)], \tag{10.66}$$

with the boundary condition

$$D_0[\varphi,\varphi';\tau-\tau']\Bigg|_{\tau=\tau'}=\prod_x\delta[\varphi(x)-\varphi'(x)]. \tag{10.67}$$

By solving equation (10.66) we can construct $D_0[\varphi,\varphi';\tau-\tau']$ explicitly. Similarly, we can construct $P_0[\varphi,\tau]$ by solving equation (10.62) for $g=0$. The results are (see Appendix *D*)

$$P_0[\varphi,\tau]=N_0^{-1}\exp\left\{-\frac{1}{2}\int \mathrm{d}^4k\varphi(k)\frac{k^2+m^2}{1-\exp[-2\tau(k^2+m^2)]}\varphi(-k)\right\}, \tag{10.68}$$

$$D_0[\varphi,\varphi';\tau]=\theta(\tau)\,\Delta_0^{-1}\exp\left\{-\frac{1}{2}\int \mathrm{d}^4k\varphi(k)D(\tau)\varphi(-k)\right\}, \tag{10.69}$$

where N_0 and Δ_0 are normalization factors such that

$$\int\delta\varphi\cdot P_0[\varphi,\tau]=1;\qquad \int\delta\varphi\cdot D_0[\varphi,\varphi';\tau]=\theta(\tau).$$

In (10.69) $\varphi(k)$ denotes $[\varphi(k),\varphi'(k)]$ and $D(\tau)$ is the following 2×2 matrix:

$$D_{11}(\tau)=(k^2+m^2)/\{1-\exp[-2\tau(k^2+m^2)]\},$$

$$D_{12}(\tau)=D_{21}(\tau)=-D_{11}(\tau)\exp[-\tau(k^2+m^2)],$$

$$D_{22}(\tau)=D_{11}(\tau)\exp[-2\tau(k^2+m^2)]. \tag{10.70}$$

We can easily verify that, in the limit $\tau \to \infty$, P_0 and D_0 satisfy the relations

$$P_0[\varphi, \tau] \underset{\tau \to \infty}{\to} P_0^{\text{eq}}[\varphi] = (N_0^{\text{eq}})^{-1} \exp\left\{-\frac{1}{2}\int \mathrm{d}^4 k \varphi(k)(k^2 + m^2)\varphi(-k)\right\}, \tag{10.71}$$

$$\frac{\partial}{\partial \tau} P_0[\varphi, \tau] \underset{\tau \to \infty}{\to} \mathrm{O}(\mathrm{e}^{-m^2\tau}), \qquad D_0[\varphi, \varphi'; \tau] \underset{\tau \to \infty}{\to} P_0^{\text{eq}}[\varphi],$$

$$\partial D_0/\partial\tau \underset{\tau \to \infty}{\to} \mathrm{O}(\exp(-m^2\tau)), \qquad \delta D_0/\delta\varphi'(x) \underset{\tau \to \infty}{\to} \mathrm{O}(\exp(-m^2\tau)).$$

We now look at $P_k[\varphi, \tau]$ given by equation (10.65). By partial functional integration we obtain

$$P_k[\varphi, \tau] = -\int \delta\varphi' \int \mathrm{d}^4 x \left[\frac{\delta S_{\text{in}}}{\delta\varphi'(x)}\right] \int_0^\tau \mathrm{d}\tau' \times$$
$$\times \left\{\frac{\delta D_0[\varphi, \varphi'; \tau - \tau']}{\delta\varphi'(x)}\right\} P_{k-1}[\varphi', \tau']. \tag{10.72}$$

We are also interested in the derivative $\partial P_k/\partial\tau$:

$$\frac{\partial P_k}{\partial \tau} = -\int \delta\varphi' \int \mathrm{d}^4 x \left[\frac{\delta S_{\text{in}}}{\delta\varphi'(x)}\right] \frac{\partial}{\partial\tau} \int_0^\tau \mathrm{d}\tau' \times$$
$$\times \left\{\frac{\delta D_0[\varphi, \varphi'; \tau - \tau']}{\delta\varphi'(x)}\right\} P_{k-1}[\varphi', \tau']. \tag{10.73}$$

The exchange of the orders of differentiation and integration over φ' and τ' is harmless because, as we have already emphasized, (10.72) and (10.73) represent the Green functions of the theory which are sufficiently regularized even in the presence of local vertex insertions. We must first consider equation (10.72) for $k = 1$, in order to verify the first step of the inductive hypothesis. The reason is that $P_0[\varphi, \tau]$ is not finite for $\tau \to 0$ because it satisfies the initial condition [the first equality of (10.63)]. However, this δ-function is multiplied by $\delta S_{\text{in}}/\delta\varphi'$, which is a monomial in φ'. Therefore, using the explicit expressions of P_0 and D_0 given by (10.68) and (10.69), it is straightforward to verify that $P_1[\varphi, \tau]$ exists, is bounded uniformly in φ for all values of τ, has a well-defined limit as $\tau \to \infty$, and vanishes at $\tau = 0$, as required by the second condition of (10.63). We must still prove that $P_1[\varphi, \infty] = P_1^{\text{eq}}[\varphi]$, but first we prefer to complete the proof of Lemma 10.1 for arbitrary k. This is simple, $P_{k-1}[\varphi', \tau']$ is bounded and $\delta D_0/\delta\varphi'$ is integrable for all φ. The functional integration over φ' in (10.72) is always possible because P_{k-1} contains the Gaussian measure of the free field theory. Therefore $P_k[\varphi, \tau]$, as given by (10.72), is bounded for all φ and has a limit $P_k[\varphi, \infty]$. A similar argument, applied to equation (10.73), shows that $\partial P_k/\partial\tau$ also exists for all τ and all φ

and has a well-defined limit when $\tau \to \infty$. It follows that $\partial P_k/\partial\tau \to 0$. This completes the proof of Lemma 10.1. ■

Thus, we now return to gauge theories. Under an infinitesimal gauge transformation with parameters $\theta^a(x)$ the fields $A_\mu^a(x)$ transform as

$$\delta A_\mu^a(x) = D_\mu^{ab}\theta^b(x) \equiv \delta^{ab}\,\partial_\mu\theta^b(x) - f^{abc}\theta^b(x)A_\mu^c(x) \tag{10.74}$$

and the functional $F[A]$ as

$$\delta F[A] = -\int d^4x\,\theta^a(x) D_\mu^{ab}[\delta/\delta A_\mu^b(x)]F[A]. \tag{10.75}$$

A gauge-invariant functional satisfies

$$D_\mu^{ab}[\delta/\delta A_\mu^b(x)]F[A] = 0. \tag{10.76}$$

The crucial observation (Zwanziger, 1981; Baulieu and Zwanziger, 1981) is that, if one wants to compute the average of such a gauge-invariant functional, one may replace the stochastic equation (10.2) with the following:

$$\partial A_\mu^a(x,\tau)/\partial\tau = -\delta S[A]/\delta A_\mu^a(x,\tau) - D_\mu^{ab}v^b + \eta(x,\tau), \tag{10.77}$$

where, in order to simplify the notation, we have omitted the subscript η in the stochastic fields. D^{ab} is given by (10.74) and $v^a = v^a[A,x]$ is an arbitrary gauge-noninvariant functional of A. Indeed, using (10.77), we can compute the *time evolution* of a *gauge-invariant functional* $F[A]$ which satisfies (10.76):

$$\frac{\partial F}{\partial\tau} = \int d^4x \left[\frac{\partial A_\mu^a(x,\tau)}{\partial\tau}\right]\left[\frac{\delta F}{\delta A_\mu^a(x,\tau)}\right], \tag{10.78}$$

and we can see that, by virtue of (10.76), the term proportional to v^a does not contribute. The average of F can be computed using a modified probability distribution $P_v[A,\tau]$ which satisfies the following *Fokker–Planck equation*

$$\frac{\partial P_v}{\partial\tau} = \int d^4x \frac{\delta}{\delta A_\mu^a(x)}\left\{\frac{\delta}{\delta A_\mu^a(x)} + \frac{\delta S}{\delta A_\mu^a(x)} + D_\mu^{ab}v^b\right\}P_v[A,\tau]. \tag{10.79}$$

We see that, because of (10.76), the average value of a gauge-invariant quantity F does not depend on the choice of v^a and is the same, irrespective of whether one uses the distribution $P[A,\tau]$ or the modified one $P_v[A,\tau]$. Of course, this is not true for gauge-noninvariant quantities.

We shall present here the special case $v^a(x) = (1/\alpha)\,\partial_\mu A_\mu^a(x)$ with α an arbitrary constant. Because S contains terms proportional to both g and g^2, the Fokker–Planck equation (10.62) will connect $P_{v,k}$ with $P_{v,k-1}$ and $P_{v,k-2}$ for $k \geqslant 2$. $P_{v,1}$ will only be connected to $P_{v,0}$. We again make the inductive hypothesis (10.64) which we shall verify for $P_{v,0}$. The analogs of equations (10.68) and (10.69) are

$$P_{v,0}[A,\tau] = N_0^{-1}\exp\left\{-\frac{1}{2}\int d^4k\, A_\mu^a(k)(\Delta_{\mu\nu}^{ab})^{-1}A_\nu^b(-k)\right\}, \tag{10.80}$$

with

$$\Delta^{ab}_{\mu\nu} = \delta^{ab} k^{-2}\{(\delta_{\mu\nu} - k_\mu k_\nu/k^2)[1-\exp(-2\tau k^2)] + \\ +\alpha(k_\mu k_\nu/k^2)[1 - \exp(-2\tau k^2/\alpha)]\}. \tag{10.81}$$

As expected, for $\alpha = 1$ we obtain the "Feynman propagator" for finite τ. For $\alpha \to \infty$ we recover the propagator given by Parisi and Wu (1981). We see that

$$\text{w.}\lim_{\tau\to\infty} P_{v,0}[A,\tau] = P^{\text{eq}}_{v,0}[A], \tag{10.82}$$

which verifies the inductive hypothesis. Similarly, we find for $D_0[A, A', \tau]$:

$$D_0 = \theta(\tau)\,\Delta_0^{-1}\exp\left\{-\frac{1}{2}\int \mathrm{d}^4k\; A^a_\mu(k) D^{ab}_{\mu\nu}(\tau) A^b_\nu(-k)\right\}, \tag{10.83}$$

$$D^{ab}_{\mu\nu} = \delta^{ab}\left\{\left(\delta_{\mu\nu} - \frac{k_\mu k_\nu}{k^2}\right)D(\tau, m=0) + \frac{1}{\alpha}\frac{k_\mu k_\nu}{k^2} D\left(\frac{\tau}{\alpha}, m=0\right)\right\}, \tag{10.84}$$

where again A^a_μ denotes $(A^a_\mu, A^{a'}_\mu)$ and the 2×2 matrix D is the scalar matrix given in equations (10.70) with $m = 0$.

From this point the proof goes as before with the estimations which correspond to a massless theory. From the explicit construction $P_{v,0}$ we first prove the theorem for $P_{v,1}$; then using both, we proceed to $P_{v,2}$, etc. The conclusion is that, in the limit $\tau \to \infty$, we obtain a regularized perturbation series which, however, is not the one we would derive from the usual theory with a gauge fixing term proportional to $(\partial_\mu A_\mu)^2$. This is obvious from the absence of explicit *Faddeev–Popov ghost terms* whose contribution is replaced by the more complicated structure of the vertices. There is a choice of v^a which precisely reproduces the Faddeev–Popov perturbation theory, but in the *stochastic language* it is a nonlocal functional of A (Zwanziger, 1981; Baulieu and Zwanziger, 1981).

10.6. The Mechanism of the Vacuum Tunneling Phenomena in the Framework of Stochastic Quantization

This problem has been considered by Yasue (1979). We give here his method of description of the *vacuum tunneling phenomena.* It is suggested (Polyakov, 1975a; 't Hooft, 1976, etc.) that a Euclidean path integral

$$\int \delta A \exp\left\{-\frac{1}{\hbar} S_{\mathrm{E}}[A] + \text{gauge-fixing term}\right\},$$

where $S_{\mathrm{E}}[A]$ is given by (10.45), provides a powerful tool to explore the structure of gauge theory vacuum. Classical Euclidean solutions, which minimize the Euclidean action $S_{\mathrm{E}}[A]$, were shown to manifest tunneling phenomena between *topologically inequivalent classical vacua* within the WKB approximation.

In the classical field theory, the SU(2) *Yang–Mills field* $A_\mu(\mathbf{x}, t)$ has an isovector expression

$$A_\mu(\mathbf{x}, t) = \sum_{a=1}^{3} A_\mu^a(\mathbf{x}, t)T^a,$$

where $\{T^a\}$, $a = 1, 2, 3$, is a basis of the Lie algebra of SU(2).

Dynamics of the Yang–Mills field is given by a Lagrangian

$$\mathscr{L} = -\frac{1}{4}\sum_{a=1}^{3}\sum_{\mu,\nu=0}^{3}{}' \int \mathrm{d}^3x\, F_{\mu\nu}^a F_{\mu\nu}^a, \tag{10.85}$$

where the terms in the primed sum for $\mu, \nu = 1, 2, 3$ are taken with reversed sign. In terms of 'electromagnetic fields'

$$E_i^a = F_{0i}^a, \qquad B_i^a = \frac{1}{2}\sum_{k,j=1}^{3} \varepsilon_{ijk} F_{jk}^a,$$

Lagrangian (10.85) can be written as

$$\mathscr{L} = \frac{1}{2}\sum_{i,a=1}^{3} \int \mathrm{d}^3x (E_i^a E_i^a - B_i^a B_i^a). \tag{10.86}$$

We shall work in $A_0 = 0$ gauge. Then E_i^a becomes identical with $\partial_0 A_i^a = \dot{A}_i^a$ and B_i^a does not contain $\dot{A}_i^a$. In this gauge A_i^a's are dynamical variables of the Yang–Mills field. Lagrangian (10.86) takes the form

$$\mathscr{L} = \frac{1}{2}\sum_{i,a=1}^{3} \int \mathrm{d}^3x (\dot{A}_i^a \dot{A}_i^a - B_i^a B_i^a),$$

which gives us the following equation motion for A_i^a:

$$\ddot{A}_i^a = -\frac{1}{2}\left(\frac{\delta}{\delta A_i^a}\right) \int \mathrm{d}^3y\, B_j^b B_j^b,$$

where the summation convention for all repeated Latin indices is assumed.

To *quantize the Yang–Mills field*, it is convenient to adopt the stochastic quantization procedure (Yasue, 1979). This is because not only the structure of the vacuum state but also the mechanism of the vacuum tunneling of the quantized Yang–Mills field can be illustrated within the stochastic quantization.

All the quantization procedures are the same as in Sections 7.4 and 7.5, provided that the field equation is replaced by the following one

$$\tfrac{1}{2}(D_+D_- + D_-D_+)A_i^a(t) = -\frac{1}{2}\left(\frac{\delta}{\delta A_i^a(t)}\right) \int \mathrm{d}^3y\, B_j^b B_j^b. \tag{10.87}$$

The quantized Yang–Mills field $A_i^a(t)$, $-\infty < t < \infty$, is of course, a generalized random process (for details, see Sections 10.3 and 10.4). The resulting *Schrödinger equation* is

$$i\hbar\frac{\partial}{\partial t}\Psi(A,t) = \frac{1}{2}\int d^3x\left\{-\hbar^2\left(\frac{\delta^2}{\delta A_i^a \delta A_i^a}\right) + B_i^a B_i^a\right\}\Psi(A,t), \tag{10.88}$$

where A is an abbreviation for $\{A_i^a\}$, $i, a = 1, 2, 3$.

Following Yasue we shall investigate a structure of the vacuum state by making use of stochastic quantization. Removing the infinite zero-point energy, we define a quantum theoretical vacuum state of the Yang–Mills theory by a ground state of wave functional $\Psi(\mathbf{A})$ which satisfies the Schrödinger equation

$$\int d^3x\{-\tfrac{1}{2}\hbar^2(\delta^2/\delta A_i^a\,\delta A_i^a) + \tfrac{1}{2}B_i^a B_i^a\}\Psi(\mathbf{A}) = 0.$$

In the classical field theory, vacuum states of the Yang–Mills theory are classical field configuration $A_i'^a(x)$ with zero potential energy

$$\frac{1}{2}\int d^3x\, B_i^a(\mathbf{A}')B_i^a(\mathbf{A}') = 0.$$

They are pure gauge fields $A_i' = (G')^{-1}\,\partial_i G' \in \mathrm{SU}(2)$ Lie module on $\mathbb{R}^3$, where G''s are unitary matrices such as $\lim_{|x|\to\infty} G'(x) = I$. As $\lim_{x\to\infty} \mathbf{A}'(x) = 0$, we can consider the pure gauge fields $\mathbf{A}'$ continuous mappings from $\dot{\mathbb{R}}^3$ to SU(2). Since $\dot{\mathbb{R}} \cong S^3$ (three-dimensional sphere) and also $\mathrm{SU}(2) \cong S^3$, A''s can be classified by the third *Homotopy group* of S^3: $\pi_3(S^3) \cong \mathbb{Z}$ (integers) with respect to a fixed point belonging to $\mathbb{R}^3$. Namely, classical vacuum states of the Yang–Mills theory consist of an infinite number of homotopy classes of S^3:

$$[A_i' = (G')^{-1}\,\partial_i G'],\quad [A_i'' = (G'')^{-1}\,\partial_i G''],\quad \ldots, \tag{10.89}$$

where $[\cdot]$ denotes a homotopy class to which the pure gauge field inside the brackets belongs. Two pure gauge fields which can be joined by a continuous gauge transformation in the manifold of SU(2) Lie module should be understood as the same classical vacuum state (Jackiw, 1977).

In order to classify the homotopy classes, it is convenient to introduce the *Pontryagin index*

$$q = -\frac{1}{24\pi^2}\sum_{i,j,k=1}^{3}\int d^3x\,\mathrm{tr}(A_i A_j A_k).$$

For pure gauge fields, q is an integer which belongs to the homotopy group $\pi_3(S^3) \cong \mathbb{Z}$. Then, the classical vacuum states (10.89) can be rearranged as

$$\{[A_i^q = G_q^{-1}\,\partial_i G_q]\}_{q\in\mathbb{Z}}, \tag{10.90}$$

where A_i^q is a pure gauge field with Pontryagin index q.

In the quantum field theory the classical vacuum states (10.90) are rendered unstable by the tunneling effect; they are metastable vacuum states. Let us study the vacuum tunneling phenomena between metastable vacuum states A_i^q's from the point of view of probability theory.

Within the conventional framework of quantum field theory, the wave function $\Psi(A)$ does not teach us the details of the vacuum tunneling. One can describe the tunneling behavior of the quantized Yang–Mills field only within semiclassical limits, i.e., within the WKB approximation.

It is assumed that the behavior of the quantized Yang–Mills field in the vacuum state $\Psi(\mathbf{A})$ is known to be a random process $A_i^a(t)$. Then, it can be found that the random process $A_i^a(t)$, $-\infty < t < \infty$, is a solution of the *stochastic differential equation*

$$\mathrm{d}A_i^a(t) = U_i^a(\mathbf{A}(t))\,\mathrm{d}t + \mathrm{d}W_i^a(t), \tag{10.91}$$

where the transformation $U_i^a(\cdot)$ is related to the vacuum state wave functional by

$$U_i^a(\mathbf{A}) = \hbar \frac{\delta}{\delta A_i^a(t)} \ln \Psi(\mathbf{A}), \tag{10.92}$$

and $W_i^a(t)$ denotes a (generalized) Wiener process with the diffusion constant $\frac{1}{2}\hbar$. Notice, that the stochastic differential equation (10.91) is an abbreviation for the relation

$$A_i^a(t) = A_i^a(s) + \int_s^t \mathrm{d}u\; U_i^a(A(u)) + W_i^a(t) - W_i^a(s), \quad t > s.$$

Therefore, a transition probability of the random process $A_i^a(t)$ manifests the tunneling process of the quantized vacuum field configuration between metastable vacuum states A_i^q. The transition probability is given by an elementary solution $\Omega[\mathbf{A}; s \mid \mathbf{A}'; u]$ of the Fokker–Planck equation of the type (7.21) or (7.22). Namely, we have

$$\frac{\partial}{\partial s}\Omega = -\int \mathrm{d}^3x \frac{\delta}{\delta A_i^a}[U_i^a(\mathbf{A})\Omega] + \frac{\hbar}{2}\int \mathrm{d}^3x \frac{\delta^2}{\delta A_i^a\,\delta A_i^a}\Omega, \tag{10.93}$$

and

$$\lim_{s\to u} \Omega[\mathbf{A}; s \mid \mathbf{A}'; u] = \delta(\mathbf{A} - \mathbf{A}'). \tag{10.94}$$

To illustrate the mechanism of the vacuum tunneling, one needs to solve the *Cauchy problem* of equations (10.93) and (10.94). This can be done by introducing a relative transition probability $F[\mathbf{A}; s \mid \mathbf{A}'; u]$ by

$$\Omega[\mathbf{A};s|\mathbf{A}';u] = \Psi(\mathbf{A})F[\mathbf{A};s|\mathbf{A}';u](\Psi(\mathbf{A}'))^{-1}. \tag{10.95}$$

Equation (10.93) is transformed into a self-adjoint form

$$-\hbar\left(\frac{\partial}{\partial s}\right)F = \int \mathrm{d}^3x\left\{-\tfrac{1}{2}\hbar^2\left(\frac{\delta^2}{\delta A_i^a\delta A_i^a}\right) + \tfrac{1}{2}B_i^a B_i^a\right\}F. \tag{10.96}$$

and the initial condition (10.94) into

$$\lim_{s \to u} F[\mathbf{A}; s \mid \mathbf{A}'; u] = \delta(\mathbf{A} - \mathbf{A}'), \tag{10.97}$$

by the substitution (10.95). Equation (10.96) is nothing but an Euclidean analog of the Schrödinger equation (10.88). The Feynman–Kac formula (6.32) and (6.33) asserts that a solution of the Cauchy problem (10.96) and (10.97) is given by the *Wiener integral*

$$F[\mathbf{A}; s \mid \mathbf{A}'; u] = \int \delta[\mathbf{A}(s) - \mathbf{A}]\, \mu_{\mathrm{W}}^{u,\mathbf{A}'}(\delta\mathbf{A}(\cdot)) \exp\left\{ -\frac{1}{2\hbar} \int_u^s \mathrm{d}t \times \right. \\ \left. \times \int \mathrm{d}^3x\, B_i^a[\mathbf{A}(t)] B_i^a[\mathbf{A}(t)] \right\}. \tag{10.98}$$

Here $\mu_{\mathrm{W}}^{u,\mathbf{A}'}$ denotes a (nonstandard) Wiener measure with the diffusion coefficient $\hbar/2$ defined on the totality of continuous paths $\mathbf{A}(\cdot)$ starting from $\mathbf{A}'$ at $t = u$ (for details, see Yasue, 1979). Correspondingly, the transition probability of the random process $A_i^a(t)$ is found to be

$$\Omega[\mathbf{A}; s \mid \mathbf{A}'; u] = (\Psi(\mathbf{A})/\Psi(\mathbf{A}')) \int \delta[\mathbf{A}(s) - \mathbf{A}]\, \mu_{\mathrm{W}}^{u,\mathbf{A}'}(\delta\mathbf{A}(\cdot)) \times \\ \times \exp\left\{ -\frac{1}{2\hbar} \int_u^s \mathrm{d}t \int \mathrm{d}^3x\, B_i^a[\mathbf{A}(t)] B_i^a[\mathbf{A}(t)] \right\}. \tag{10.99}$$

A tunneling probability of the quantized vacuun field configuration between metastable vacuum states A_i^q and A_i^p $(q \neq p)$ from a remote past to a remote future is $\Omega[\mathbf{A}^p; \infty \mid \mathbf{A}^q; -\infty]$. Noticing that $\Omega[\mathbf{A}^p; \infty \mid \mathbf{A}^q; -\infty]$ would coincide with the conventional expression for the tunneling probability, i.e., the ratio $|\Psi(\mathbf{A}^p)/\Psi(\mathbf{A}^q)|^2$, we find a vacuum tunneling amplitude of the quantized Yang–Mills field in $A_0 = 0$ gauge to be

$$\mathrm{Amp}\,[\mathbf{A}^p; \infty \mid \mathbf{A}^q; -\infty] = |\Psi(\mathbf{A}^p)/\Psi(\mathbf{A}^q)|^2 = \int \delta[\mathbf{A}(\infty) - \mathbf{A}^p]\, \mu_{\mathrm{W}}^{-\infty,\mathbf{A}^q}(\delta\mathbf{A}(\cdot)) \times \\ \times \exp\left\{ -\frac{1}{2\hbar} \int_{-\infty}^{\infty} \mathrm{d}t \int \mathrm{d}^3x\, B_i^a \times \right. \\ \left. \times [\mathbf{A}(t)] B_i^a[\mathbf{A}(t)] \right\}. \tag{10.100}$$

If we introduce a functional path integral expression of the *Wiener integral*

$$\int \delta[\mathbf{A}(\infty) - \mathbf{A}^p]\mu_{\mathrm{W}}^{-\infty,\mathbf{A}^q}(\delta\mathbf{A}(\cdot)\cdots = N\int_{\mathbf{A}^q}^{\mathbf{A}^p} \delta\mathbf{A}(\cdot) \times$$

$$\times \exp\left\{-\frac{1}{2\hbar}\int_{-\infty}^{\infty} dt \int d^3x \dot{A}_i^a(t)\dot{A}_i^a(t)\right\},$$

where $\delta\mathbf{A}(\cdot)$ means to take a functional path integral and N is a normalization constant, equation (10.100) becomes

$$\mathrm{Amp}\ [\mathbf{A}^p; \infty | \mathbf{A}^q; -\infty]$$

$$= N\int_{\mathbf{A}^q}^{\mathbf{A}^p} \delta\mathbf{A}(\cdot) \exp\left\{-\frac{1}{2\hbar}\int_{-\infty}^{\infty} dt \int d^3x(\dot{A}_i^a(t)\dot{A}_i^a(t) + B_i^a[\mathbf{A}(t)]B_i^a[\mathbf{A}(t)])\right\}.$$

In terms of the field strength tensor $F_{\mu\nu}^a$, this can be written as

$$\mathrm{Amp}\ [\mathbf{A}^p; \infty | \mathbf{A}^q; -\infty] = N\int_{\mathbf{A}^q}^{\mathbf{A}^p} \delta\mathbf{A}(\cdot) \exp\left\{-\frac{1}{2\hbar}\sum_{a=1}^{3}\sum_{\mu,\nu=0}^{3}\int d^4x F_{\mu\nu}^a F_{\mu\nu}^a\right\},$$

which provides an Euclidean path integral description of vacuum tunneling phenomena in $A_0 = 0$ gauge.

Thus, we have presented here Yasue's method of the Euclidean path integral description of the vacuum tunneling phenomena in SU(2) Yang–Mills theory, the validity of which has been proved from the point of view of probability theory.

10.7. Stochastic Fluctuations of the Classical Yang–Mills Fields

Recently, the stochastic solutions to the Yang–Mills equations and the possibility of derivation (based on these solutions) of effects connected with the basic states of quantum chromodynamics have been discussed intensively (Polyakov, 1975a; Gribov, 1977). In this connection, it is interesting to analyze the classical Yang–Mills equations without external sources. The result could be useful for the construction and study of a vacuum structure and a problem of asymptotic states of the theory.

In this section we briefly discuss the study of a *nonlinear oscillation in the classical Yang–Mills theory.* We consider the Yang–Mills equations without external sources, corresponding to SU(2) group:

$$\partial_\mu F_{\mu\nu}^a + gf^{abc}A_\mu^b F_{\mu\nu}^c = 0, \qquad (10.101)$$

in the co-ordinate system, where the *Poynting vector* $F_{0i}^a F_{ji}^a$ turns to zero. The field strength tensor $F^a_{\mu\nu}$ is given in Section 10.4. For this case it is enough (in the gauge $\partial_i A^a_i = 0$, $A^a_i = 0$) to satisfy one of the following relations:

(a) $A^a_{i,l} = 0, \quad (F_{i,l} \equiv \partial_l F_i;\ i, l = 1, 2, 3).$

(b) $\dot{A}^a_i = 0.$

(c) $A^a_{i,j} - A^a_{j,i} = 0.$

(the point denotes, as usual, a differentiation with respect to time). In the case of (a) when in the chosen co-ordinate system the potential depends only on time, the *Yang–Mills equations* take the form:

$$\ddot{A}^a_i + g^2(A^b_j A^b_j A^a_i - A^a_j A^b_j A^b_i) = 0. \tag{10.102}$$

The system described by equations (10.102) is naturally equivalent to the discrete nonlinear mechanical system with the Hamiltonian

$$H = \tfrac{1}{2}(\dot{A}^a_i)^2 + \tfrac{1}{4}g^2[(A^a_i A^a_i)^2 - (A^a_i A^a_j)^2], \tag{10.103}$$

which coincides with the energy of the field system (10.101).

It is natural to mention the (nonlinear) classical Yang–Mills mechanics determined by equations (10.102). Notice that the corresponding classical Maxwell mechanics is trivial and described by the equation $\ddot{A}_i = 0$, generating the constant electric field $A_i = -E_i t$, but due only to the nonlinearity of the Yang–Mills equations (as shown by Matinyan *et al.*, 1981), there appear complicated nonlinear oscillations of colour.

In the simple case that allows an analytic solution, a nonlinear plane wave arises. This plane wave is described by the elliptic cosine law with respect to time. Indeed, if one searches for a solution to system (10.102) in the nine-parameter form $A^a_i = Q^a_i R^{(a)}(t)/g$, (over a there is no summation), where Q^a_i is an orthogonal matrix independent of time, $Q^a_i Q^b_i = \delta^{ab}$, then from (10.102) we obtain the following system of equations

$$\ddot{R}^{(a)} + R^{(a)}(\mathbf{R}^2 - R^{(a)2}) = 0, \tag{10.104}$$

where

$$\mathbf{R}^2 = R^{(1)2} + R^{(2)2} + R^{(3)2}.$$

The simple case (allowing analytic solution), when all three colours change in the same form $R^{(1)} = R^{(2)} = R^{(3)} = R(t)$ with respect to time variable, was investigated by Baseyan *et al.* (1979). In this case system (10.104) is reduced to the nonlinear equation with one degree of freedom, the solution of which has the following form:

$$R(t) = (2g^2/3)^{1/4}\cdot\mu\cdot\mathrm{cn}((\tfrac{8}{3}g^2)^{1/4}\cdot\mu\cdot(t + t_0);\ 2^{-1/2}), \tag{10.105}$$

where $\mathrm{cn}(x; k)$ is the Jacob elliptic cosine of the argument x and module k; t_0 is an arbitrary origin of the time moment, and μ^4 is an energy density in the given co-ordinate system.

Corresponding to expression (10.105) *chromoelectric* $E_i^a = Q_i^a \dot{R}/g$ and *chromomagnetic*

$$B_i^a = \varepsilon_{ijk} f^{abc} Q_j^b Q_k^c R^2/g^2$$

intensities of fields in the 'rest' system of reference are parallel to each other, but three vectors $\mathbf{E}^1$, $\mathbf{E}^2$, $\mathbf{E}^3$ are orthogonal to each other. Such a picture differs from the one, realized in the plane wave of classical electrodynamics, where $\mathbf{E} \perp \mathbf{B}$. The intensities $\mathbf{E}^a$ and $\mathbf{B}^a$ change from time with the period

$$T = (3/8g^2)^{1/4}(4/\mu)K(1/\sqrt{2}),$$

where K is a total elliptic integral of the first kind. Under the Lorentz transformations the nonlinear wave [arising from (10.105)] with arguments $kx = k_\mu x_\mu$, where $k_0 = \mu\beta$, $k_i = \mu v_i \beta$ ($\beta^{-1} = (1 - v^2)^{1/2}$, $k^2 = \mu^2$) differ from the non-Abelian linear plane wave due to Coleman (1977) ($k^2 = 0$) by the fact that the value of the Poynting vector does not equal energy density ($k^2 = \mu^2$) (Baseyan *et al.*, 1979).

So, we see that investigating the mechanical system (10.104) by the Lorentz transformations, a solution to system (10.101) in an arbitrary co-ordinate system can be obtained. Such a solution always contains a plane wave. Therefore, our problem leads in fact, to the study of the classical Yang–Mills mechanics (10.102). In connection with the above-mentioned, it is interesting to study a further problem of the nonlinear oscillation of color in the classical Yang–Mills theory. The solution of these problems with the number of degrees of freedom $N > 1$ leads to the investigation of system (10.102) [or in the simple variant of system (10.104)].

Matinyan *et al.* (1981a, b); Chirikhov and Shepelyansky (1981) carried out a numerical integration of system equations (10.104) and obtained those *stochastic solutions* in the two and three dimensional cases. They investigated the nonlinear mechanical system with the Hamiltonian

$$n = 2 \quad H_2 = \dot{x}^2/2 + \dot{y}^2/2 + x^2y^2/2,$$

$$n = 3 \quad H_3 = \tfrac{1}{2}(\dot{x}^2 + \dot{y}^2 + \dot{z}^2) + \tfrac{1}{2}(x^2y^2 + y^2z^2 + x^2z^2),$$

and with the corresponding *equations of motion*

$$n = 2 \quad \ddot{x} + xy^2 = 0; \qquad \ddot{y} + yx^2 = 0;$$

$$n = 3 \quad \ddot{x} + x(y^2 + z^2) = 0; \quad \ddot{y} + y(x^2 + z^2) = 0; \quad \ddot{z} + z(x^2 + y^2) = 0.$$

The system when $n = 2$ is obtained from (10.104) for $R^{(3)} = 0$. Here, the following notation is introduced

$$x \equiv R^{(1)}, \qquad y \equiv R^{(2)}, \qquad z \equiv R^{(3)}.$$

In conclusion we notice that the obtained connection of the Yang–Mills mechanics with a stochastic process, is not surprising, since the Yang–Mills mechanics is a nonlinear system described by some nonlinear partial differential equations. Thus, among the set of solutions to the Yang–Mills equations generating a whole row of remarkable effects: *monopole* (Wu and Yang, 1969; Polyakov, 1975b; 't Hooft, 1974; Nielsen and Olesen, 1973), *instantons* (Polyakov, 1975a; Belavin *et al.*, 1975), and *merons* (de Alfaro *et al.*, 1976; Callan *et al.*, 1978; Baseyan and Matinyan, 1980) there should exist stochastic solutions responsible for the *stochastic origin* of all the *physical phenomena including the vacuum and space-time.*

10.8. Appendix D. Solutions to the Free Fokker–Planck Equation

In this appendix we calculate the quantities $P_0[\varphi, \tau]$ and $D_0[\varphi, \varphi'; \tau]$ in equations (10.62) and (10.66). By definition $P_0[\varphi, \tau]$ is

$$P_0[\varphi,\tau] = \int \delta\eta \exp\left\{ -\frac{1}{4} \int d^4 x \, d\rho \cdot \eta^2(x,\rho) \right\} \prod_y \delta[\varphi(y) - \varphi_\eta^0(y,\tau)], \tag{D.1}$$

where φ_η^0 satisfies the free *Langevin equation*

$$\partial\varphi_\eta^0/\partial\tau = (\partial^2 - m^2)\varphi_\eta^0 + \eta(x, \tau). \tag{D.2}$$

The solution of this equation with initial condition $\varphi_\eta^0(x, 0) = 0$ is the following

$$\varphi_\eta^0(x, \tau) = \int d^4 y \int_0^\tau d\rho \, G(x - y; \tau - \rho)\eta(y, \rho), \tag{D.3}$$

where the Green function $G(x - y; \tau - \rho)$ is given by (dimension of space $d = 4$)

$$G(x - y; \tau - \rho) = (2\pi)^{-4} \int d^4 k \exp[ik(x - y) - (\tau - \rho)(k^2 + m^2)]. \tag{D.4}$$

We are also interested in the correlation function D,

$$\begin{aligned} D(x,y;\tau,\tau') &\equiv \langle \varphi(x,\tau)\varphi(y,\tau')\rangle_\eta \\ &= 2\int d^4 z \int_0^{\min(\tau,\tau')} d\rho \, G(x - z; \tau - \rho)G(y - z; \tau' - \rho). \end{aligned} \tag{D.5}$$

In order to find $P_0[\varphi, \tau]$, we write equation (D.1) in the form

$$\begin{aligned} P_0[\varphi,\tau] = \int \delta\eta \, \delta\xi \exp\left\{ -\frac{1}{4} \int d^4 x \, d\rho \cdot \eta^2 (x, \rho) \right\} \times \\ \times \exp\left\{ i \int d^4 x \, \xi(x)[\varphi(x) - \varphi_\eta^0(x, \tau)] \right\}. \end{aligned} \tag{D.6}$$

Then, we introduce expression (D.3) for φ_η^0 and we do the Gaussian, η and ξ

integrations. The result is

$$P_0[\varphi, \tau] = N_0^{-1} \exp\left\{-\frac{1}{2}\iint d^4x\, d^4x'\, \varphi(x) D^{-1}(x, x'; \tau, \tau')\varphi(x')\right\}, \tag{D.7}$$

where D^{-1} is defined as

$$\int d^4z\, D(x, z; \tau, \tau) D^{-1}(z, x'; \tau, \tau) = \delta^{(4)}(x - x'). \tag{D.8}$$

From (D.5) and (D.4) we find

$$D(x, y; \tau, \tau) = (2\pi)^{-2} \int d^4k\, e^{ik(x-y)} \times$$
$$\times \{1 - \exp[-2\tau(k^2 + m^2)]\}(k^2 + m^2)^{-1}, \tag{D.9}$$

and so in momentum space (D.7) reduces to expression (10.68).

For the functional propagator $D_0[\varphi, \varphi'; \tau]$ we use its definition

$$D_0[\varphi, \varphi'; \tau - \tau'] \equiv P_0[\varphi, \varphi'; \tau, \tau']/P_0[\varphi', \tau'], \tag{D.10}$$

as the conditional probability density, where

$$P_0[\varphi, \varphi'; \tau, \tau'] = \int \delta\eta \exp\left[-\frac{1}{4}\int d^4x d\rho \cdot \eta^2(x, \rho)\right] \prod_y \delta[\varphi(y) - \varphi_\eta^0(y, \tau)] \times$$
$$\times \delta[\varphi'(y) - \varphi_\eta^0(y, \tau')]. \tag{D.11}$$

We follow the same method as for $P_0[\varphi, \tau]$ and forming the ratio (D.10), we find expression (10.69).

Chapter 11

Some Physical Consequences of the Hypothesis of Stochastic Space-Time and the Fundamental Length

11.1. Prologue

As was noted above (Chapter 7), there are three approaches to the construction of the theory of stochastic processes in physics. The traditional approach describing the diffusion phenomena (also Brownian motion) is based on the classical works due to Einstein (1956) and Smoluchowski (1923) (see the reviews by Skagerstam, 1975; Moore, 1979; Yasue, 1979; Nelson, 1979; Guerra, 1981). Further developments in this theory are connected with the names of Uhlenbeck and Ornstein (1930); Wang and Uhlenbeck (1945); Chandrasekhar (1943); Kac (1959); Rice (1944, 1945). Mathematical foundations of the stochastic theory for physical processes were laid down by A. Markov, N. Wiener, A. N. Kolmogorov, K. Itô, N. N. Bogolubov, etc. New stochastic methods in physics have been recently given in DeWitt-Morette and Elworthy (1981) (see Chapter 10). The other two approaches are connected with the properties of the electromagnetic vacuum (zero-point field) and with the postulate that the random behavior of a physical system is caused by the stochastic nature of physical space-time (see Chapters 7–9).

In this chapter, we would like to show that all the above-mentioned approaches to the stochastic theory are connected to each other with the hypothesis of stochastic space-time as the connection. On the basis of this connection, it is shown that the spectrum density of the zero-point electromagnetic field in stochastic electrodynamics changes at small distances. This fact gives a reinterpretation to the electromagnetic vacuum as a real physical 'medium' and allows us to conclude that it is an energy density, i.e., the density of matter due to electromagnetic interactions may be restricted. From our considerations presented below, one can therefore establish that the introduction of the fundamental length l into the stochastic theory leads automatically to a *restriction on the density of matter* by $u_l \sim \hbar\pi^{-2}l^{-4}/c$. Further, an attempt to unify the scales of physical quantities (masses, lengths, etc.) is presented in the framework of a hypothesis of the existence of a family of small 'black holes'. On the semi-empirical level this proposal provides the possibility of finding unique scale [characterized by the parameter $\frac{1}{2}\alpha$ (α being the fine structure constant)] which separates on different subscales including the scale of the *grand unification.* It transpires that all the typical physical scales (atomic, molecular, electromagnetic, weak, grand unification and gravitational) of lengths are covered by this unique scale.

In the framework of the new approach to gravitation (de Alfaro *et al.*, 1980) the *decay of the proton* is discussed (Section 11.5) on the basis of the *neutron oscillations* and of *superfields.* It is shown that a mechanism of the proton decay is very sensitive to a value of the fundamental length l. A possibility is discussed that if the fundamental length is not larger than the *Planck length* $l_{\mathrm{Pl}} \simeq l_{\mathrm{gr}} \sim 10^{-33}$ cm; then the proton decay is wholly due to the gravitational interaction and its *life-time* is of an order of 10^{43} years.

In Section 11.6 we presented some new results related to the hypothesis of *nonlocality and stochasticity of the space-time metric* and its consequences. As shown above (Chapter 1), the hypothesis of the stochastic nature of space-time at small distances leads to the space-time metric nonlocality (1.12). The last fact makes the theory nonlocal too, so the field equations and Green's functions of test fields (particles) are changed according to results obtained in Section 11.6. Moreover, in our proposal, *Einstein's relation between the energy and momentum of particles* is also changed. Speculation concerning the *creation* of elementary particles as some *perturbation of the space-time metric* resulting from the nonlocality, is presented. A generalized mass operator (momentum operator) of ground states of test bosons and fermions, and concrete forms of the space-time metric nonlocality are constructed, which allow one to describe the *equidistant behavior* of the mass spectra of observed particles in semi-empirical level only.

Further, we deal with some comments on the *physical mechanism responsible for the acceleration of cosmic rays.* It is suggested that this mechanism is caused by the stochastic (or fluctuational) structure of space-time at small distances. A method of introducing *fluctuations in a conformally flat Riemannian space-time metric due to ultra-high energy particles* is presented. In this framework, a *nonlinear dynamics of particles* and equations for the electromagnetic fields are obtained. The former admits the *acceleration mechanism* for cosmic-ray particles and the fact that their energy increases extremely during the evolution of the Universe. In our model the energy of cosmic-ray particles, say, protons and their radius (the effective Schwarzschild), the age of the Universe and the value of the fundamental length are interconnected, and are determined by a unified formula, Einstein's relation for the relativistic particle energy. It allows one to define *experimentally the bound on the value of the fundamental length* that is $l_{\mathrm{exp}} = 1.56 \times 10^{-33}$ cm for the maximum proton energy $E^{\mathrm{ob}} = 10^{20}$ eV observed in primary cosmic rays.

Within the new acceleration mechanism of cosmic rays proposed here the problem of the energy spectrum of the cosmic rays and the ratio of the intensities of the electron component to the proton component at the same energy level are discussed.

To conclude this chapter we speculate concerning an interrelationship between the properties of the space-time structure near moving particles and their dynamics. It is suggested that the space-time metric near particles becomes curved $\tilde{g}_{\mu\nu}(x, b_{\mathrm{E}})$ depending on a random vector $b_{\mathrm{E}} = (b_4, \mathbf{b})$ with a distribution $w(b_{\mathrm{E}}^2/l^2)$, the averaged space-time metric $\langle \tilde{g}_{\mu\nu}(x, b_{\mathrm{E}}) \rangle$ over this distribution gives a general effect on particle behavior. As a result the particle motion in our scheme is described by a nonlinear equation. It transpires that the nonrelativistic limit of this equation gives a simple connection between the space-time structure at small distances and the dynamic behavior of particles. Different types of the particle motion (nearly rectilinear, stochastic and soliton like) caused by some concrete forms of the averaged conformally flat space-time metric $\langle \tilde{g}_{\mu\nu}(x, b_{\mathrm{E}}) \rangle$ are considered.

11.2. Nonlocal-Stochastic Model for Free Scalar Field Theory

In this section we consider the free scalar field in the space-time $\mathbb{R}^4(\hat{x})$ in the framework of the Davidson (1980a) model (see Section 10.2). In the stochastic space-time $\mathbb{R}^4(\hat{x})$, the field (10.9) may be represented in the form:

$$\varphi(\hat{x}) = (\hbar L^3)^{-1/2} \sum_{\mathbf{q}} \mathrm{e}^{i\mathbf{q}(\mathbf{x}+\mathbf{b})} \varphi_{\mathbf{q}}(t + i\tau), \tag{11.1}$$

where τ is a real random variable. According to the above deduction (see Chapter 1) we must average the field (11.1) with the measure $w(b_{\mathrm{E}}^2/l^2)$. For this purpose, we introduce intermediate Euclidean variables $(\hbar/2v)b_4 = i\tau$, $x_4 = it$ (generally speaking, v is a complex number), and average the expression (11.1) with $w(b_{\mathrm{E}}^2/l^2)$, where $b_4 = i\tau(2v/\hbar)$; thus we obtain

$$\begin{aligned}\langle \varphi(\hat{x}) \rangle = \phi(x) &= (\hbar L^3)^{-1/2} \sum_{\mathbf{q}} \mathrm{e}^{i\mathbf{q}\mathbf{x}} \int \mathrm{d}^4 b_{\mathrm{E}} w\left(\frac{b_{\mathrm{E}}^2}{l^2}\right) \exp\left\{ i\mathbf{q}\mathbf{b} + \frac{1}{2}\frac{\hbar}{v} i b_4 \frac{\partial}{\partial x_4} \right\} \times \varphi_{\mathbf{q}}(t) \\ &= (\hbar L^3)^{-1/2} \sum_{\mathbf{q}} \mathrm{e}^{i\mathbf{q}\mathbf{x}} K(Q) \cdot \varphi_{\mathbf{q}}(t),\end{aligned} \tag{11.2}$$

where

$$K(Q) = \int \mathrm{d}^4 b_{\mathrm{E}}\, w\left(\frac{b_{\mathrm{E}}^2}{l^2}\right) \exp(i b_{\mathrm{E}}' \cdot Q) = 4\pi^2 l^3 a^{-1} \int_0^\infty \mathrm{d}y\, y^2 J_1(aly) w(y^2), \tag{11.3}$$

$$Q = \left[\left(\frac{\hbar}{2iv}\right)\frac{\partial}{\partial t}, \mathbf{q} \right], \qquad b_{\mathrm{E}}' = (2iv\tau/\hbar, \mathbf{b}); \qquad a = (Q^2)^{1/2} = \left[\mathbf{q}^2 - \left(\frac{\hbar^2}{4v^2}\right)\frac{\partial^2}{\partial t^2} \right]^{1/2}.$$

Here $J_1(z)$ is the Bessel function. As before, we are interested in such a class of measures for which $K(Q)$ are entire analytic functions of the variable Q^2. For example, in the cases of $w_0(y^2) = \lim_{m \to 0} w_m(y^2)$, $w_G(y^2)$ in Section 1.3, and

$w_1(y^2)$ in Section 11.6, we obtain

$$K_0(Q^2l^2) = 3(Q^2l^2)^{-1}[\sin(Q^2l^2)^{1/2}/(Q^2l^2)^{1/2} - \cos(Q^2l^2)^{1/2}],$$

$$K_G(Q^2l^2) = \exp[-Q^2l^2/8], \qquad K_1(Q^2l) = \sin^2\tfrac{1}{2}\sqrt{Q^2}/(\tfrac{1}{2}\sqrt{Q^2})^2.$$

We now come to the main result which can be formulated as the following theorem:

THEOREM 11.1. For noncoincident $\hat{x}_i[\mathbf{x}_i + \mathbf{b}_i, t_i + (b_4)_i(\hbar/2v)]$, $i = 1, 2, \ldots$

$$\lim_{L\to\infty} \mathbb{E}\{\langle\varphi(\hat{x}_1)\rangle\cdots\langle\varphi(\hat{x}_n)\rangle\} = S_n^M\left(\mathbf{x}_1, \frac{2v}{\hbar}t_1; \ldots; \mathbf{x}_n, \frac{2v}{\hbar}t_n\right), \tag{11.4}$$

where $S_n^M(\cdots)$ *are modified n-point Schwinger functions determined by formula (10.15) (see also Chapter 7), where for the scalar field the function* $S_2^M(\cdots)$ *takes the form*

$$S_2^M(\mathbf{x}_1, t_1; \mathbf{x}_2, t_2) = (2\pi)^{-4}\int d^4q \frac{\exp[i\mathbf{q}(\mathbf{x}_1 - \mathbf{x}_2) + iq_4(t_1 - t_2)]}{q_4^2 + \mathbf{q}^2 + m^2/\hbar^2} \times$$

$$\times K^2(l^2(\mathbf{q}^2 + q_4^2)). \tag{11.5}$$

The functions (11.5) with the form factor $V(z) = [K(z)]^2$, *physical meaning of which was discussed in Section 2.3, are called the* nonlocal Schwinger functions.

Proof. The proof is in the literal repetition of the results obtained above in Section 10.2. As a result, we obtain

$$\mathbb{E}\{\langle\varphi(\hat{x}_1)\rangle\langle\varphi(\hat{x}_2)\rangle\} = L^{-3}\sum_{\mathbf{q}} \exp[i\mathbf{q}(\mathbf{x}_1 - \mathbf{x}_2)]K\left(l^2\left[\mathbf{q}^2 - \left(\frac{\hbar^2}{4v^2}\right)\frac{\partial^2}{\partial t_1^2}\right]\right) \times$$

$$\times K\left(l^2\left[\mathbf{q}^2 - \left(\frac{\hbar^2}{4v^2}\right)\frac{\partial^2}{\partial t_2^2}\right]\right) \times$$

$$\times \int \frac{dq_4}{2\pi}\frac{\exp[iq_4(t_1 - t_2)2v/\hbar]}{q_4^2 + \mathbf{q}^2 + m^2/\hbar^2}. \tag{11.6}$$

Here we assume that $\langle\varphi(\hat{x})\rangle$ is a Gaussian process because, as discussed in Section 9.1, the method of averaging (11.2) cannot change the physical nature of the objects and may only change the spatial structure of the objects which are extended (nonlocal) in some domain characterized by length l.

Operating on the exponential in (11.6) by the operator $K(Q)$ and taking into account that $K(Q)$ is an entire analytical function of Q, we perform the limit $L \to \infty$ and obtain

$$\lim_{L\to\infty} \mathbb{E}\{\langle\varphi(\hat{x}_1)\rangle\langle\varphi(\hat{x}_2)\rangle\} = S_2^M\left(\mathbf{x}_1, \frac{2v}{\hbar}t_1; \mathbf{x}_2, \frac{2v}{\hbar}t_2\right). \qquad \square$$

Notice that what is most important in the proof is the order of differentiation with respect to the $\partial/\partial t$ in the operator $K(Q)$ and of averaging with the measure $w(b_E^2/l^2)$. It is necessary to perform first the averaging with w over the field (11.1) at every point of the space-time $\mathbb{R}^4(\hat{x})$ and then (at the last stage of the calculation) to act by the operator $K(Q)$. If this application of $K(Q)$ took place in some intermediate stage of the calculation, for example, if $K(Q)$ would act on the function $\exp\{-\alpha|t_1 - t_2|\}$, then the result obtained would correspond to the usual theory (i.e., local theory). Indeed

$$K\left(l^2\left[\mathbf{q}^2 - \left(\frac{\hbar^2}{4v^2}\right)\frac{\partial^2}{\partial t_1^2}\right]\right)K\left(l^2\left[\mathbf{q}^2 - \left(\frac{\hbar^2}{4v^2}\right)\frac{\partial^2}{\partial t_2^2}\right]\right)\exp\{-\alpha|t_1 - t_2|\}$$

$$\equiv K^2\left(\frac{-m^2l^2}{\hbar^2}\right)\exp(-\alpha|t_1 - t_2|),$$

but the function K is normalized so that $K(-m^2l^2/\hbar^2) = 1$.

One can easily see that for a particular value of the parameter $v = \hbar/2$, which is the value used by Guerra and Ruggiero (1973) we obtain the modified (or nonlocal) Schwinger function (11.5). In the limit $l \to 0$ this function becomes the local one therefore we reach the Guerra and Ruggiero result exactly.

In our case, equation (10.18) takes the form

$$\mathbb{E}_{v = i\hbar/2}\{\langle\varphi(\hat{x})\rangle\langle\varphi(\hat{y})\rangle\} = i(2\pi)^{-4}\int d^4qV(-q^2l^2)\frac{\exp[iq(x-y)]}{q^2 - m^2/\hbar^2 + i\varepsilon}, \tag{11.7}$$

where $V(-q^2l^2) = [K(q^2l^2)]^2$, $q^2 = q_0^2 - \mathbf{q}^2$. Comparing (11.7) with the usual Green's function of the nonlocal quantum field theory (Part I) we obtain

$$\mathbb{E}_{v = i\hbar/2}\{\langle\varphi(\hat{x})\rangle\langle\varphi(\hat{y})\rangle\} = D_c(x-y) = \langle 0|T\{\phi(x)\phi(y)\}|0\rangle.$$

Here

$$D_c(x) = i^{-1}(2\pi)^{-4}\int d^4q\,V(-q^2l^2)\frac{e^{iqx}}{m^2 - q^2 - i\varepsilon}$$

and $\phi(x)$ are the nonlocal causal Green function and nonlocal field constructed by the nonlocal distribution (generalized function) $K(x)$ given formula (1.11). Thus we come to the nonlocal theory investigated in Part I. In the nonlocal theory with the form factors $V(-q^2l^2)$ belonging to the class of entire analytical functions there exists an intermediate regularization procedure (Chapter 2), which permits the changing of the contour integration in (11.7) to a necessary domain in the complex plane q_0 (i.e., planes I and III).

In the nonlocal case the corrections similar to (10.22) and (10.23) are valid

$$\mathbb{E}_{v = i\hbar/2}\{\langle\varphi(\hat{x}_1)\rangle\cdots\langle\varphi(\hat{x}_n)\rangle\} = \langle 0|T\{\phi(x_1)\cdots\phi(x_n)\}|0\rangle,$$

$$\mathbb{E}_{v = -i\hbar/2}\{\langle\varphi(\hat{x}_1)\rangle\cdots\langle\varphi(\hat{x}_n)\rangle\} = \langle 0|T^*\{\phi(x_1)\cdots\phi(x_n)\}|0\rangle,$$

where T and T^* are the Wick time and antitime ordering of operators $\phi(x)$.

Thus, the free scalar field in our model is equivalent to the nonlocal Markov field whose covariances are obtained from the non-local Schwinger functions in which all times are scaled by a common factor $2\nu/\hbar$. The hypothesis concerning the indeterminate nature of ν in quantum field theory makes it possible to obtain the Efimov nonlocal theory. In the given scheme, the Euclidean–Markov field and quantum field theory may be constructed in which no cut-off procedure is involved.

11.3. Zero-Point Electromagnetic Field and the Connection Between the Value of the Fundamental Length and the Density of Matter

In the previous section, in the framework of the stochastic quantization method proposed by Fényes (1952), Nelson (1966, 1967) and Davidson (1980a), we have considered the free scalar field in the stochastic space-time $\mathbb{R}^4(\hat{x})$ and have obtained the two-point nonlocal Schwinger function for the Euclidean–Markov field. We shall consider here the electromagnetic field in $\mathbb{R}^4(\hat{x})$ (for the local case see Section 10.3 and Guerra and Loffredo, 1980). In the case of the electromagnetic field expression (11.4) takes the form (in the local case such generalization has been done by Moore, 1979):

$$\lim_{L\to\infty} \mathbb{E}\{\langle A_i(\hat{x})\rangle\langle A_j(\hat{y})\rangle\} = \delta_{ij}\,\Delta_l(x-y) = \delta_{ij}\int\frac{\mathrm{d}^4q}{(2\pi)^4}\,\frac{V(-q^2l^2)}{q_4^2+\mathbf{q}^2}\,\mathrm{e}^{iq(x-y)}$$

$$(i,j=1,2,3) \qquad (11.8)$$

Here it is assumed that $\nu = \hbar/2$. It appears that in the framework of our scheme the nonlocal Euclidean–Markov electromagnetic field (11.8) produces the zero-point electromagnetic (ZPEM) field.

THEOREM 11.2. *The* covariance *of the ith component of stochastic ZPEM potential* $\mathrm{A}_{l,i}^{(0)}$ (i = *1,2,3*) *for the whole space is*

$$\mathbb{E}\{A_{l,i}^{(0)}(\mathbf{x},t_1)A_{l,i}^{(0)}(\mathbf{y},t_2)\} = -2\,\Delta_l(\mathbf{x}-\mathbf{y},t_1-t_2), \qquad (11.9)$$

where

$$\mathbf{A}_l^{(0)}(\mathbf{x},t) = \langle \mathbf{A}^{(0)}(\hat{x})\rangle_{\mathbb{R}^4(\hat{x})}.$$

By using the dipole approximation in the case of a local theory ($l\to 0$) the spectrum density of the zero-point electromagnetic field has been obtained by Moore (1979) from equation (11.9). However, in this case frequency cutoffs are employed, since $\Delta_0(0)$ diverges. In our model $\Delta_l(0) < \infty$, the spectrum density may be calculated without the cutoff procedure, and the dipole approximation is not involved. Let us calculate the *spectrum density* of the ZPEM field for a specific form of the form factors $V(q_E^2 l^2)$, say $V_G = \exp(-q_E^2 l^2)$ see formula

(1.16). Then

$$\mathbb{E}\{A_{l,i}^{(0)}(\mathbf{x}, t_1)A_{l,i}^{(0)}(\mathbf{x}, t_2)\} = \tfrac{1}{2}\pi^{-2}\int_0^\infty d\omega\, Q_l(\omega)\cos\omega(t_1 - t_2),$$

where

$$Q_l(\omega) = \omega\,\mathrm{erfc}(l\omega) - \exp(-l^2\omega^2)/l\sqrt{\pi}. \tag{11.10}$$

Here erfc(z) is the integral of probability (see Section 2.3). The second term of (11.10) diverges in the limit $l \to 0$ which reflects the fact that a divergence appears in the local theory. In order to get rid of the second term of equation (11.10) in the nonlocal theory one also needs (finite) renormalization. Thus

$$\mathbb{E}\{E_{l,i}^{(0)}(\mathbf{x}, t)E_{l,i}^{(0)}(\mathbf{x}, t)\} = \int_0^\infty d\omega \cdot \rho_l(\omega)\cos\omega(t_1 - t_2),$$

where

$$\rho_l(\omega) = \pi^{-2}\omega^3\,\mathrm{erfc}(l\omega)/2, \mathrm{erfc}(z) = 1 - \mathrm{erf}(z). \tag{11.11}$$

As is to be expected, in the limit $l \to 0$ the nonlocal spectrum density (11.11) goes over to the spectrum density of the ZPEM field due to stochastic electrodynamics (SED) (see Chapter 7). We see that the introduction of the fundamental length into the spectrum density of the ZPEM field leads to a restriction of the density of matter (or more exactly, electromagnetic matter) by some value u_l. We now calculate a value of u_l. In our model the energy density is finite in accordance with (11.11):

$$u_l = \int_0^\infty d\omega \cdot \rho_l(\omega) = \tfrac{1}{2}\langle \mathbf{E}^2 + \mathbf{H}^2\rangle = \langle \mathbf{E}^2\rangle.$$

For $V = \exp(-q_E^2 l^2)$ we have $u_l = (3/32)\pi^{-2}l^{-4}\hbar/c$ in ordinary units. From high-energy experimental data (see Introduction 1.1 and references therein) it follows that $l \lesssim 10^{-16}$ cm and therefore $u_l \gtrsim \frac{1}{3}10^{25}$ g cm^{-3}. In view of this we notice that the estimate of l due to the astronomical approach (see also Section 11.7) proposed by Ginzburg (1975) and the search for superdense matter may be of great interest (see Shuryak, 1980).

Thus, in our scheme SED with the spectrum density (11.11) appears and we call it a *nonlocal SED*. Therefore the concept of nonlocality may be introduced by means of the properties of the vacuum fluctuations (or zero-point electromagnetic field).

It has been pointed out above that random or stochastic electrodynamics (SED) (i.e., the theory of the motion of charged particles in the presence of the electromagnetic vacuum) is constructed on the basis of the hypothesis about the existence of the *zero-point random electromagnetic radiation field* $\mathbf{E}^{(0)}$ with the spectrum density

$$\rho(\omega) = \hbar\omega^3/2\pi^2c^3.$$

If we neglect the force due to the magnetic radiation field; then the nonrelativistic basic equation of SED is

$$m\ddot{\mathbf{r}} = e\mathbf{E}^{(0)} + \mathbf{f} + m\tau\dddot{\mathbf{r}}, \qquad \tau = 2\alpha/3mc^3, \qquad \alpha = e^2/4\pi, \tag{11.12}$$

where m is the mass and e the charge of the particle, $\mathbf{f} = \mathbf{f}(\mathbf{r}, \dot{\mathbf{r}}, t)$ the external force.

In the nonlocal SED obtained above, the equation of motion for the charged particle has the same form as equation (11.12) with $\mathbf{E}^{(0)} \to \mathbf{E}_l^{(0)}$. Here $\mathbf{E}_l^{(0)}$ is the nonlocal electric component of the zero-point nonlocal electromagnetic field (modified by the presence of the fundamental length l) with the spectrum density given by (11.11). Study of equation (11.12) with $\mathbf{E}^{(0)} \to \mathbf{E}_l^{(0)}$ and its solution goes beyond the scope of the present book. We notice only that in the case of the nonlocal SED arising from the nonlocal Euclidean–Markov and Efimov theories, the fulfilment of the Lorentz covariance follows from the fact that the Lorentz and four-dimensional rotation groups are isomorphic in a complex domain (see Chapter 1).

In conclusion we would like to note that our hypothesis of existing superdense matter may provide some interesting consequences in the theory of gravitation. For example, as suggested by Markov (1982), this hypothesis allows the possibility of rewriting the gravitation equation in the following form

$$R_\mu^\nu - \tfrac{1}{2}R\,\delta_\mu^\nu = \frac{8\pi G}{c^4}\,T_\mu^\nu\left[1 - \left(\frac{u}{u_l}\right)^2\right]^n - \Lambda\left(\frac{u}{u_l}\right)^{2n}\delta_\mu^\nu, \quad (n \geqslant \tfrac{1}{2}). \tag{11.13}$$

Here it is assumed that a value of matter density u is always restricted by a limiting quantity

$$u_l \to u_{Pl} = \frac{c^5}{G^2\hbar} = \frac{\hbar}{c}\,l_{Pl}^{-4}, \qquad l_{Pl} = \left(\frac{\hbar G}{c^3}\right)^{1/2}.$$

Further, we distinguish two possibilities:

(1) In the case of $(u/u_{Pl})^2 \ll 1$, i.e., for $\hbar \to 0$ equation (11.13) turns to Einstein's classical equation (7.1).

(2) In the case of $(u/u_{Pl})^2 = 1$ equation (11.13) describes de Sitter's world

$$(u/u_{Pl})^2 = 1, \qquad R_\mu^\nu - \tfrac{1}{2}R\,\delta_\mu^\nu + \Lambda\cdot\delta_\mu^\nu = 0.$$

On the other hand, this limiting principle ($u \leqslant u_{Pl}$) may be expressed by geometrical language, for example, by the relation:

$$Q = R_{\mu\nu\delta\lambda}R^{\mu\nu\delta\lambda} \leqslant 1/l_{Pl}^4.$$

Now equation (11.13) can be rewritten in the form

$$R_\mu^\nu - \tfrac{1}{2}R\,\delta_\mu^\nu = \frac{8\pi G}{c^4}\,T_\mu^\nu(1 - Q l_{Pl}^4)^n - \Lambda\cdot(Q l_{Pl}^4)^n\delta_\mu^\nu, \quad n \geqslant \tfrac{1}{2}, \tag{11.14}$$

Denoting the left and right parts of (11.14) by G_μ^ν and B_μ^ν, respectively, we obtain an additional equation in the form of the covariant derivative of the right part of (11.14)

$$B^\nu_{\mu;\,\nu} = 0. \tag{11.15}$$

This equation results from the well-known property of Einstein's equation $G^\nu_{\mu;\,\nu} = 0$. Equation (11.15) leads to the fact that the covariant derivative of energy and momentum tensor does not turn to zero $T^\nu_{\mu;\,\nu} \neq 0$. The last equation describes the process of the disappearance of matter (tensor T_μ^ν) with generating de Sitter's world. In other words, equation (11.15) gives the transformation of matter into the vacuum state of the de Sitter world. As is known, this world is unstable. The state equation in it is $p + \varepsilon < 0$ that is a macroscopic analogy of overheated liquid or overcooled steam. Equations of the type of (11.13) and (11.14) are accommodated to the description of the oscillating Universe (see Markov, 1984 and Walstad, 1980).

11.4. Hierarchical Scales and 'Family' of Black Holes

In contemporary physics some hierarchical scales for the physical quantities (lengths, energies, times, etc.) are observed. (In particular, a problem of hierarchy in physics has been discussed by Recami, 1981 and in references therein). For example, atomic and molecular processes occur in the domain characterized by length $0(10^{-8})$ cm, but electromagnetic processes: $0(10^{-10}$–$10^{-11})$ cm. The length $0(10^{-13})$ cm is just a typical distance for nuclear physics. The majority of physicists believe that weak and quantum-gravitational effects will be manifested at distances of the order of $l_{\mathrm{w}} \sim 6.7 \times 10^{-17}$ cm and $l_{\mathrm{gr}} \sim 5.72 \times 10^{-33}$ cm, respectively. We recall that these lengths are connected with the Fermi G_{F} and Newton G_g constants by the formulas:

$$l_{\mathrm{w}} = (G_{\mathrm{F}}/\hbar c)^{1/2}, \qquad l_{\mathrm{gr}} = (4\pi G_g \hbar/c^3)^{1/2}.$$

Recently, some models (Pati and Salam, 1973; Georgi and Glashow, 1974; Buras *et al.*, 1978; Goldman and Ross, 1979; Marciano, 1979) have suggested that grand unification, i.e., the link of electroweak and strong interactions seems to take place at the energy scale of the order of $m_X \sim 6 \times 10^{14}$ GeV [at length $0(10^{-29})$ cm] with the gauge coupling constant $\alpha_{\mathrm{GUM}} = g^2/4\pi \sim 1/40$ at this scale.

The present section is devoted to obtaining the parameters of m_X and α_{GUM} for the *grand unification theories* within the framework of a hypothesis about the existence of a family of black holes. It turns out that because of this hypothesis all the above-mentioned typical lengths are confined to the unique scale measured by the parameter $\frac{1}{2}\alpha$, α being the fine structure constant. The analogous possibility has been pointed out by Salam (1980).

Following an idea of Tennakone (1974), we make the assumption that some elementary particles are black holes (or monopoles) in the strong gravitational field. Thus let us consider as a 'family' the chain of black holes;

$$\begin{matrix} & \begin{pmatrix} f_0 \\ f_1 \end{pmatrix} & \begin{pmatrix} f_1 \\ f_2 \end{pmatrix} & \begin{pmatrix} f_2 \\ f_3 \end{pmatrix} & \dots \\ i = & 1; & 2; & 3; & \dots \end{matrix} \tag{11.16}$$

(it is assumed that $m_{f_{i-1}} \ll m_{f_i}$, $i = 1, 2, \dots$) which are singularities in the Nordström–Reissner solutions of the Einstein equation for a particle of mass m_{f_i} and charge e. This solution is given by the metric

$$ds^2 = -\gamma^{-1}\,dr^2 - r^2\,d\theta^2 - r^2\sin^2\theta\,d\phi^2 + \gamma dt^2, \tag{11.17}$$

where

$$\gamma = 1 - 2G_g m_{f_i}/c^2 r + e^2 G_g/c^4 r^2. \tag{11.18}$$

If now we assume that the structure of space-time in the immediate neighborhood of the charged heavy particles f_i belonging to an ith pair of the family (11.16) is determined by strong gravitation, then the constant G_g in (11.18) should be replaced by the strong gravitational constant G_s^i:

$$G_s^i\, m_{f_i}^2/\hbar c \sim 1.$$

Then the singularities in the strong gravitational field of the charged particle f_i are given by the metric (11.17) with

$$\gamma = \gamma_s = 1 - 2G_s^i\, m_{f_i^0}/c^2 r + e^2 G_s^i/c^4 r^2 \tag{11.19}$$

where $m_{f_i^0}$ is the mass of an uncharged partner of the heavy particle f_i which is also assumed to exist.

Since $G_s^i m_{f_i}^2/e^2 \gg 1$, the equation $\gamma_s = 0$ has two real roots $r_{f_{i-1}}$ and r_{f_i}. For those one can easily obtain the following ratio (for details, see Tennakone, 1974):

$$m_{f_{i-1}}/m_{f_i} = r_{f_{i-1}}/r_{f_i} \tag{11.20}$$

or

$$2G_s^i m_{f_{i-1}}/c^2 r_{f_{i-1}} = 1, \qquad 2G_s^i\, m_{f_i}/c^2 r_{f_i} = 1, \tag{11.21}$$

where

$$r_{f_{i-1}} = e^2/2c^2 m_{f_i^0}. \tag{11.22}$$

Here r_{f_i} may be interpreted as the *Schwarzschild radius* of the singularity corresponding to particle f_i; m_{f_i} and $m_{f_{i-1}}$ are masses of heavy and light particles (black holes) belonging to an ith pair of (11.16), respectively.

It should be noted that the strong gravitational constant G_s^i emerges as a by-product of the theory. Indeed, from (11.21) and (11.22) we have

$$G_s^i = e^2/4m_{f_{i-1}} m_{f_i^0}. \tag{11.23}$$

Now we determine a parameter which characterizes the hierarchical scales for the physical quantities in 'black holes' physics. According to Tennakone (1974), identifying f_0 with the electron and f_1 with the proton (in given case $f_i^0 \equiv$ neutron) we get from (11.20) (since $m_p \sim m_n$)

$$r_1 = \frac{e^2}{2\hbar c}\frac{\hbar}{m_e c} = \tfrac{1}{2}\alpha\frac{\hbar}{m_e c} = 1.4 \times 10^{-13}\,\text{cm},$$

or in terms of the energy scale

$$m_{w_1} = \hbar/cr_1 = 140\,\text{MeV}.$$

Further, assuming $f_1 = p$ in the next pair ($i = 2$) of the family (11.16) we have

$$m_p = e^2 m_{f_2}/2c^2 r_2 m_{f_2^0}.$$

By assumption $m_{f_2} \approx m_{f_2^0}$. Then

$$r_2 = \alpha\hbar/2m_p c = 7.7 \times 10^{-17}\,\text{cm} \quad \text{or} \quad m_{w_2} = \hbar/r_2 c = 274 m_p = 257\,\text{GeV},$$

etc.

Thus, we see that number $\frac{1}{2}\alpha$ plays the role of scale in the transition from one pair to another in the family of black holes, and it may be interpreted as a parameter of hierarchy in physics. Further it is postulated that the mass ratio of each hole belonging to the family (11.16) is given by parameter $\frac{1}{2}\alpha$:

$$m_{f_i} = \frac{2}{\alpha} m_{f_{i-1}} \quad \text{or} \quad m_{f_i} = (2/\alpha)^{i-1} m_p, \quad (i = 2, 3, \ldots). \tag{11.24}$$

Typical scales for $r_i\,[m_{w_i}]$ corresponding to this parameter $\alpha/2$ are shown in Table 11.1. Now with the same scale $\alpha/2$ and formally continuing the hierarchical scales shown in Table 11.1, to the left we obtain the typical lengths 3.86×10^{-11} cm and 1.06×10^{-8} cm for the electromagnetic and atomic-molecular processes. Further continuation of the hierarchical scales to the right is impossible and the chain of black holes breaks at $i = 9$. Since in the case $i > 9$ (we obtain an object with the mass larger than the Planck mass $m_{Pl} \sim 1.2 \times 10^{19}$ GeV) classical solutions of the Einstein equations are not acceptable and quantum gravitational effects should be taken into account.

Let us calculate the gauge coupling constant at the scale of order $m_X \sim 4 \times 10^{14}$ GeV. For this assuming $m_{f_7^0} \sim m_{f_7}$ and taking into account (11.24), from (11.23) we obtain

$$\alpha_{\text{GUM}} = \frac{g^2}{4\pi\hbar c} = \frac{G_s^7 m_{f_7}^2}{4\pi\hbar c} = \frac{\alpha}{16}\left(\frac{m_{f_7}}{m_{f_6}}\right) \sim \frac{1}{25}.$$

From Table 11.1 we see that $m_X/m_W \sim 1.5 \times 10^{12}$.

Table 11.1

Black holes	$\binom{e}{p}$	$\binom{p}{f_2}$	$\binom{f_2}{f_3}$	$\binom{f_3}{f_4}$	$\binom{f_4}{f_5}$
No. (i)	$i = 1$	$i = 2$	$i = 3$	4	5
r_i (cm)	1.4×10^{-13}	7.7×10^{-17}	2.8×10^{-19}	1.0×10^{-21}	3.7×10^{-24}
m_{W_i} (GeV)	0.14	257	7.0×10^4	1.9×10^7	5.3×10^9
Comparison	m_π	$\sim m_W$	m_{mon}		

So, we notice that the scale corresponding to the seventh pair of black holes may claim to be a scale of the grand unification (or, at least, a lower bound to the grand unification scale). In distinction to the paper by Tennakone (1974) where one pair (e, p) of black holes was considered, in this section we have assumed that there exists a family of black holes in the strong gravitation field. Within the framework of this hypothesis the number of pairs of black holes is restricted and equals $9 = 1 + 8$. Thus, in our case the 'natural' emergence of mass hierarchies is $[m_{\mathrm{P}l}, (\alpha/2)m_{\mathrm{P}l}, (\alpha/2)^2 m_{\mathrm{P}l}, \ldots, m_W]$ or $[m_p, (2/\alpha)m_p, (2/\alpha)^2 m_p, \ldots, m_{\mathrm{P}l}]$.

Speculative extension of the unique scale discussed in the case of the microworld to the large scale of the Universe gives the following hierarchy in lengths $L_i = (1/\alpha)^i L_0$ $(i = 0, 1, 2, \ldots)$, where it is naturally assumed that $L_0 = r_{\mathrm{ear}}$ is the radius of the Earth. Quantities of this macroscale are presented in Table 11.2.

Table 11.2

Number of scales	Scales in lengths (cm): Theoretical calculating lengths	Scales in lengths (cm): Expected lengths from Astrophysical Data
0	6.35×10^8	The radius of Earth r_{ear}.
1	8.7×10^{10}	$r_\odot = 7 \times 10^{10}$ cm
2	1.2×10^{13}	Average distance from Earth to Sun, i.e., $r_{\mathrm{AU}} = 1.5 \times 10^{13}$ cm
3	1.6×10^{15}	Radius of the solar system $r_{\mathrm{ss}} = 0.6 \times 10^{15}$ cm
4	2.2×10^{17}	Light year 9.5×10^{17} cm
5	3.1×10^{19}	Nucleus of Galaxy $r_{\mathrm{NG}} = 10$ parsec.
6	4.2×10^{21}	Distance from Sun to Galaxy centre $r_{\mathrm{GC}} = 8$ kilo parsec.
7	5.7×10^{23}	Distance between centres of Galaxies $R \sim 18 \times 10^{23}$ cm
8	7.9×10^{25}	Average diameter of Meta-Galaxy $R \sim 10^{25}$ cm
9	1.1×10^{28}	Radius of the Universe, i.e., Hubble radius $R_{\mathrm{H}} \sim 10^{28}$ cm

Table 11.1—(Continued).

Black holes	$\binom{f_5}{f_6}$	$\binom{f_6}{f_7}$	$\binom{f_7}{f_8}$	$\binom{f_8}{f_9}$
No. (i)	6	7	8	9
r_i (cm)	1.3×10^{-26}	4.9×10^{-29}	1.8×10^{-31}	6.6×10^{-33}
m_{W_i}(GeV)	1.4×10^{12}	3.9×10^{14}	1.0×10^{17}	2.9×10^{19}
Comparison		m_X		m_{Pl}

11.5. The Decay of the Proton and the Fundamental Length

Recently the problem of proton decay was discussed intensively within the framework of *grand unified theories* (see, for example, Ellis *et al.*, 1980; Weinberg, 1981; Langacker, 1981). The possibility that the proton decays is not only important for grand unified theories but it is of great interest from the point of view of cosmology and of the general philosophical problem of matter structure and its stability. The hypothesis of neutron oscillation and the effective many-fermion operators for the proton decay have been considered by Kazarnovsky (1980); Mohapatra and Marshak (1980); Pati *et al.* (1981); L. Chang and N. Chang (1980); Sawada and Fukugita (1980); Weinberg (1979); Wilczek and Zee (1979); Smirnov (1980), etc. High sensitivity experiments in search for proton decay are undertaken and projected in many laboratories.

In this section we study proton decay within the framework of the hypothesis concerning the existence of the fundamental length in Nature, i.e., we try to connect a mechanism of proton decay to the properties of space-time. It is shown (from the grand unified theories) that if the proton decays, then it occurs at least in an energetic domain determined by a scale $M \sim 10^{14}$–10^{15} GeV (or $l = \hbar/Mc \sim 10^{-28} - 10^{-29}$ cm). Therefore we make the assumption that the structure of space-time (i.e., an effect of gravitation) can play an essential role in the mechanism of proton decay.

Moreover, starting from the nonlocal stochastic electrodynamics constructed by the hypothesis of stochastic space-time, it has been shown above that the introduction of the fundamental length into the quantity of spectrum density for zero-point electromagnetic field reduces the restriction on the density of matter to be $u_l \sim \pi^{-2} l^{-4} \hbar/c$. From this, the conclusion can be made that the existence of a superdense object may expose the space-time structure in the general theory of relativity (see also Ginzburg, 1975). In this connection, it seems to me, that study of the properties of the proton deserves great attention. This is related too to the fact that there exists some indication (Miettinen and Thomas, 1980) of the possibility of the existence of a high density matter core (nucleus) in the proton with respect to its charge density.

Furthermore, according to our present structure concept of matter, the *proton represents a superstable compound object from which our world is built.*

Our basic assumptions are the following:

(i) Fundamental length exists in Nature.
(ii) The mechanism of proton decay is highly sensitive to the value of fundamental length.
(iii) The structure of space-time appears in the phenomena of the microworld through an oscillation and acceleration mechanism of particles in space-time (the acceleration mechanism will be discussed in Section 11.7).

Now we consider the *oscillation mechanism of particles*, say a *neutron*. Our approach is based on the new method of gravitation proposed by de Alfaro, Fubini, and Furlan (1980). Let us consider one component the scalar field $\varphi(x)$. The invariant action of $\varphi(x)$ interacting with gravitation is

$$A = \int d^4x \sqrt{-g}\, \mathscr{L}(x), \qquad \mathscr{L}(x) = \mathscr{L}_E(x) + \mathscr{L}_M(x), \tag{11.25}$$

where $\mathscr{L}_E = -R/4$ and $\mathscr{L}_M = \frac{1}{2} g^{\mu\nu}\, \partial_\nu \varphi\, \partial_\mu \varphi$ are the Lagrangians of gravitation and matter field $\varphi(x)$, respectively. Here $g^{\mu\nu}$ is the metric tensor. The total Green function for the scalar field $\varphi(x)$ coupled with gravitation is

$$D_c(x - y) = \langle 0 | T\{\varphi(x)\varphi(y) S_M\} | 0 \rangle / \langle 0 | S_M | 0 \rangle, \tag{11.26}$$

$$S_M = T \exp\left\{ i \int d^4x \sqrt{-g}\; \mathscr{L}_M(x) \right\}.$$

Following the idea of de Alfaro *et al.* (1980), we study the gravitation effects in the flat space limit

$$g^{\mu\nu} \Rightarrow [g^{\mu\nu}]_{\text{flat}} = l^2\, \delta^{\mu\nu}, \tag{11.27}$$

where $\delta^{\mu\nu}$ is the Galilean metric and l is some universal constant of the dimension of a length. We call it the fundamental length. For the present we do not identify l with l_{gr} determined by de Alfaro *et al.* (1980):

$$l_{gr} = (4\pi\hbar G_g/c^3)^{1/2} = 5.72 \times 10^{-33}\ \text{cm}. \tag{11.28}$$

We notice again that there exists an enormous range between experimental restrictions $l_{exp} \lesssim 10^{-16}$ cm (see Chapter 1) and l_{gr} which must be investigated in physics. It is quite possible that starting from some small distances l but $l \gg l_{gr}$ the space-time structure may play an important role in particle physics processes in the microworld. Such a possibility can be interpreted formally as the appearance of some strong gravity with coupling constant $G_s = l^2 c^3/4\pi\hbar$.

Notice that it is necessary to make a change $\varphi \to m\varphi$ in expression (11.26) at the same time with the transition to limit (11.27). Then the S-matrix for $\varphi(x)$

coupled with gravity takes the form

$$S_M = T\exp\left\{i\tfrac{1}{2}m^2l^2\int d^4x\,\delta^{\mu\nu}\,\partial_\nu\varphi\,\partial_\mu\varphi\right\}. \tag{11.29}$$

Making use of the modified Wick theorem (see Chapter 2 and Bogolubov and Shirkov, 1980) we obtain from (11.26) and (11.29)

$$D_c(x-y) = \Delta_0(x-y) + m^2l^2\,\Delta_1(x-y) + m^4l^4\,\Delta_2(x-y) + \cdots, \tag{11.30}$$

where

$$\Delta_0(x) = i^{-1}(2\pi)^{-4}\int d^4p\,\frac{e^{-ipx}}{m^2-p^2-i\varepsilon}$$

is the usual Green function for a free scalar field $\varphi(x)$ and

$$\Delta_1(x) = i^{-1}(2\pi)^{-4}\int d^4p\,p^2\,e^{-ipx}(m^2-p^2-i\varepsilon)^{-2}, \tag{11.31}$$

etc. Here $(m^2l^2)^n\,\Delta_n(x)$ $(n = 1, 2, \ldots)$ are terms of higher order with respect to $\Delta_0(x)$ which are due to the gravitational interaction of the scalar field $\varphi(x)$. It is interesting to calculate a contribution due to gravitation to the Yukawa potential of two scalar particles coupled with gravitation. In the static limit this potential is given by (11.31):

$$\varphi_g(r) = \frac{m^2l^2}{(2\pi)^3}\int d^3p\,e^{-i\mathbf{pr}}\,\mathbf{p}^2(m^2+\mathbf{p}^2)^{-2} = -\frac{m^2l^2}{r}\frac{1}{8\pi}(2-mr)\,e^{-mr}, \tag{11.32}$$

Setting $l = l_{\mathrm{gr}}$, we obtain a modified *Newton potential*

$$\varphi_g(\mathrm{r})|_{l=l_{\mathrm{gr}}} = -\frac{G_g m^2}{r}\left(1-\frac{mr}{2}\right)\exp(-mr). \tag{11.33}$$

We recall that the *Yukawa potential* of two scalar particles is given by the first term of (11.30) (in the static limit):

$$\varphi_Y(r) = (2\pi)^{-3}\int d^3p\,\frac{e^{-i\mathbf{pr}}}{m^2+\mathbf{p}^2} = \frac{1}{4\pi r}e^{-mr}.$$

We see that sum of the potentials $\varphi_g(r)$ and $\varphi_Y(r)$ is finite at the point $r = 0$ in the case of the *Compton wavelength* $l_c = \hbar/mc$.

Strictly speaking, the expressions (11.32) and (11.33) for potentials (and the concept of potential at all) are valid at distances $r \gg l$ and $r \gg l_{\mathrm{gr}}$, respectively. We see that the real corrections to the physical processes of the microworld due to space-time structure (i.e., gravitation effects) are very small, of the order m^2l^2 or even m^4l^4.

Now we consider a mechanism of proton decay within our hypothesis. For this purpose we define first a generalized concept of particle oscillation. A motion of a particle from one point to another in the space may be formally considered as repeated (multi) oscillations of a particle. The matrix elements for transitions-into-itself of the particle coupled with gravitation during its time evolution, are given by

$$M(p \to p') = \langle 0|a^-(p)S_M a^+(p')|0\rangle \sim \delta^{(4)}(p - p')(1 + im^2l^2).$$

Then, by definition, the *oscillation time* is

$$\tau \sim 1/[m(1 + m^4l^4)] = (1 - m^4l^4)/m.$$

The oscillation period of a free particle is understood as the time during which this particle travels (or succeeds to go through) a distance equal to its Compton wave-length, i.e., $\tau = \lambda/c$, $\lambda = \hbar/mc$.

Such a formal definition of the oscillation period is generalized easily to a more general case, when the particle during its time evolution in space accomplishes transitions to completely different particles. For example, $n \to \tilde{n}$, $\nu_e \to \nu_\mu \to \nu_\tau$ and even $n \to \pi^0$, $n \to K^0$, etc. For description of such strange transitions it is necessary to introduce the concept of a *superfield* (concerning superfields, supersymmetry and supergravity see the review by Mezincescy and Ogievetsky, 1975; the book edited by Van Nieuwenhuizen and Freedman, 1979) responsible for the above-mentioned transitions. The particles (and their antiparticles) with different spins and masses, etc. enter a superfield multiplet. For example,

$$\Psi_s = \sum_i (\alpha_i \varphi_i + \beta_i M^{-1/2} \psi_i),$$

where φ_i and ψ_i are boson and fermion field operators, respectively, and α_i and β_i are some real numbers (generally speaking, they are Grassmann variables). The appearance of the factor $M_i^{-1/2}$ in the second term of the superfield Ψ_s follows from a dimensional argument. A simple Lagrangian for the superfield coupled with gravity is

$$\mathcal{L}_s(x) = \tfrac{1}{2} g^{\mu\nu}\, \partial_\nu \overline{\Psi}_s(x) \partial_\mu \Psi_s(x). \tag{11.34}$$

Then the total Green function of the type of (11.26) for the transition $n \to \tilde{n}$ takes the form

$$D_{n\tilde{n}}(x - y) = \langle 0|T[\psi_{\tilde{n}}(x)\bar{\psi}_n(y)S_s]|0\rangle \, \langle 0|S_s|0\rangle^{-1}$$

$$= iM_p^2 l^2 \int d^4z\, \delta^{\mu\nu}[\partial\, \Delta_{\tilde{n}s}(x - z)/\partial z_\mu][\partial \Delta_{sn}(z - y)/\partial z_\nu] + \cdots$$

$$= \frac{M_p l^2}{(2\pi)^4 i} \int d^4 p\, p^2 e^{-ip(x - y)} (m_n - \hat{p} - i\varepsilon)^{-2} + O(M_p^4 l^4), \tag{11.35}$$

where $M_p = m_n$ and

$$\Delta_{sn}(x - z) \underset{\mathrm{Df}}{=} \langle 0| T[\Psi_s(x)\bar{\psi}_n(z)|0\rangle = i^{-1} M_p^{-1/2} \int \frac{\mathrm{d}^4 p}{(2\pi)^4} \frac{\mathrm{e}^{-ip(x-z)}}{m_n - \hat{p} - i\varepsilon}.$$

Here we have assumed $\beta_n = \beta_{\tilde{n}} = 1$. Note that since, in a free state, direct transitions between n and $\tilde{n}$ are impossible the term $\Delta_{n\tilde{n}}(x - y)$ of the type $\Delta_0(x - y)$ in (11.30) does not appear in expression (11.35) for the $n\tilde{n}$-transitions.

In the case of neutron oscillations it is easy to see that the oscillation period for the $n\tilde{n}$-transitions is

$$\tau_{n\tilde{n}} \sim 1/M_p^5 l^4. \tag{11.36}$$

We see that if the $n \to \tilde{n}$ oscillation exists, then it is wholly responsible for the space-time structure (i.e., gravity). The neutron oscillation period $\tau_{n\tilde{n}}$ must be $\gtrsim 10^{30}$ years, since it would otherwise contradict the stability of matter, as was determined experimentally (Reines *et al.*, 1979; Reines and Crouch, 1974; see also the review by Goldhaber and Sulak, 1981). This is because, if the oscillation mechanism in the presence of gravity changes the neutron into an antineutron, the $\tilde{n}$ can be destroyed with other nucleons in matter into multipions resulting in an effective-baryon-number nonconserving decay. For example, $(A, Z) \to (A - 2, Z) + \pi^+\pi^-$.

Let us obtain restrictions on the value of the fundamental length l. From inequality $\tau_{n\tilde{n}} \gtrsim 10^{30}$ years and (11.36) it follows that $l \lesssim 10^{-29}$ cm. On the other hand, according to our semi-empirical approach (see previous section) the next possible candidates for quantities of l are $l_8 \sim 1.8 \times 10^{-31}$ cm and $l_9 \sim l_{\mathrm{gr}} = 5.72 \times 10^{-33}$ cm. Corresponding to these quantities of l the *proton life-time* is $\tau_p^8 \sim 10^{38}$ years and $\tau_p^9 \sim \tau_{\mathrm{gr}} \sim 10^{43}$ years, respectively. The last possibility means that the *proton decay is wholly due to the gravitation interaction with the Newton constant* G_g. In particular, the true value of the fundamental length $l \sim 10^{-33}$ cm results from our consideration carried out in Section 11.7. So that the value $\tau_p \sim 10^{43}$ years is *more probable.*

11.6. A Hypothesis of Nonlocality of Space-Time Metric and its Consequences

In this section we consider some consequences of introducing nonlocality into the metric of space-time given by formulas (1.4) and (1.12) in Chapter 1. We again recall that introduction of the proposal about the stochasticity of space-time inevitably leads to the change of the space-time metric, say (1.4):

$$s_0 = x_0^2 - \mathbf{x}^2 \Rightarrow s_l = \int \mathrm{d}^4 b_{\mathrm{E}}\, w\!\left(\frac{b_{\mathrm{E}}^2}{l^2}\right)[(x_0 + i\tau)^2 - (\mathbf{x} + \mathbf{b})^2] = s_0 - l^2, \tag{11.37}$$

where $\tau = b_4$. A change of this form of metric was discussed by Markov (1959) and Takano (1961) too. But there is another possibility

$$s_0 = x_0^2 - \mathbf{x}^2 \Rightarrow s_l = \langle [(x_0 + \tau)^2 - (\mathbf{x} + i\mathbf{b})^2] \rangle_{\mathbb{R}^4(\hat{x})}$$
$$= \int d^4 b_E\, w\left(\frac{b_E^2}{l^2}\right)[(x_0 + \tau)^2 - (\mathbf{x} + i\mathbf{b})^2] = s_0 + l^2. \tag{11.38}$$

Now we consider another case when the space-time metric is determined by the formula $ds^2 = g_{\mu\nu}(x)\, dx^\mu\, dx^\nu$. In such a curved space-time we have the following expression, according to (1.12):

$$ds^2 = g_{\mu\nu}(x) dx^\mu dx^\nu \Rightarrow (ds_l^i)^2 = G_{\mu\nu}^i(x) dx^\mu dx^\nu, \quad i = A, C;$$

$$G_{\mu\nu}^A(x) = \int d^4 b_E w\left(\frac{b_E^2}{l^2}\right) g_{\mu\nu}(x_0 + i\tau, \mathbf{x} + \mathbf{b}) = \int d^4 y K^A(x - y) g_{\mu\nu}(y), \tag{11.39}$$

$$G_{\mu\nu}^C(x) = \int d^4 b_E w\left(\frac{b_E^2}{l^2}\right) g_{\mu\nu}(x_0 + \tau, \mathbf{x} + i\mathbf{b}) = \int d^4 y K^C(x - y) g_{\mu\nu}(y),$$

where $K^A(x)$ and $K^C(x)$ are some distributions of type (1.11), Fourier transforms of which are given by

$$\tilde{K}^A(-p^2 l^2) = (2\pi)^2 \int_0^\infty dr\, r^2 w\left(\frac{r^2}{l^2}\right) J_1(\sqrt{-r^2 p^2})(-p^2)^{-1/2}, \tag{11.40}$$

$$\tilde{K}^C(p^2 l^2) = (2\pi)^2 \int_0^\infty dr\, r^2 w\left(\frac{r^2}{l^2}\right) J_1(\sqrt{r^2 p^2})(p^2)^{-1/2}, \quad (p^2 = p_0^2 - \mathbf{p}^2) \tag{11.41}$$

Furthermore, as usual, we shall investigate only the class of distributions – measures $w(b_E^2/l^2)$ for which the form factors (11.40) and (11.41) will be entire analytic functions of the variables $z = p^2 l^2$ with a finite order of growth $\infty > \rho \geqslant \frac{1}{2}$. Then the metric tensor $G_{\mu\nu}^i$ in (11.39) can be rewritten in the form

$$G_{\mu\nu}^i = K^i(l^2 \Box) g_{\mu\nu}. \tag{11.42}$$

For $K^i(l^2\Box)$ the representation of type (1.11) holds.

From expressions (11.40) and (11.41) it follows that the functions $\tilde{K}^A$ and $\tilde{K}^C$ have a completely different behavior in the limit $p^2 \to \pm\infty$, i.e., in the pseudo-Euclidean and Euclidean directions, respectively. In order to present clearly a general characteristic of this behavior, we now shall calculate these functions for a specific form of the measures $w(b_E^2/l^2)$ under consideration (see Chapter 1). Let

1. $w_G(y^2) = 4\pi^{-2} l^{-4} \exp(-2y^2), \quad 0 \leqslant y < \infty,$

2. $w_1(y^2) = \begin{cases} \frac{1}{2}\pi^{-2} l^{-4} y^{-2}(1 - y^2)^{-1/2}, & 0 < y < 1 \\ 0 & y \geqslant 1 \end{cases}$

3. $$w_2(y^2) = \begin{cases} \pi^{-2} l^{-4}(1+\mu)(2+\mu)(1-y^2)^\mu & \mu > -1, \quad 0 < y < 1 \\ 0 & , \quad y \geqslant 1 \end{cases}$$

Then the form factors $\tilde{K}^A$ and $\tilde{K}^C$ corresponding to these measures are equal to

$$\tilde{K}_G(\mp p^2 l^2) = \exp(\pm p^2 l^2/8), \tag{11.43}$$

$$\tilde{K}_1(\mp p^2 l^2) = 4\sin^2[(\mp p^2 l^2/4)^{1/2}][(\mp p^2 l^2)^{1/2}]^{-2}, \tag{11.44}$$

$$\tilde{K}_2(\mp p^2 l^2) = 4.2^\mu \cdot \Gamma(3+\mu) J_{2+\mu}((\mp p^2 l^2)^{1/2})[(\mp p^2 l^2)^{1/2}]^{-2-\mu}. \tag{11.45}$$

Here the upper (lower) sign corresponds to $\tilde{K}^A(\tilde{K}^C)$. Whence, it is obvious that the functions $\tilde{K}^{A,C}$ have a different behavior, say

$$\tilde{K}_1^A(-p^2 l^2) = \begin{cases} O(\exp(l\sqrt{|p^2|})) & \text{at } p^2 \to +\infty, \\ O(1/|p^2|) & \text{at } p^2 \to -\infty, \end{cases}$$

$$\tilde{K}_1^C(p^2 l^2) = \begin{cases} O(\exp(l\sqrt{|p^2|})) & \text{at } p^2 \to -\infty, \\ O(1/|p^2|) & \text{at } p^2 \to +\infty. \end{cases}$$

Such a different behavior of the form factors $\tilde{K}^{A,C}$ plays, as seen below, an important role in constructing a finite theory of interacting quantized (test) fields from the *hypothesis of nonlocality of the space-time metric point of view*.

Now the following questions arise: How to realize the abovementioned idea of nonlocality of the space-time metric in physical applications and what is a physical quantity which represents a ‘carrier’ of this nonlocality? We shall discuss these questions in detail. Indeed, it is quite possible that in Nature there exist exclusive exotic objects (we call them *test particles*) with respect to the space-time structure at small distances. These test particles, if they exist, would give rise to a transition mechanism from the nonlocality of the space-time metric to fluctuations in it, and, therefore, would act as a resonant factor in the process of the creation of elementary particles by some perturbation (or fluctuation) of the space-time metric (see below). However, the question of the possible existence of such objects requires, it seems, a more fundamental physical principle connected to the quantum theory of gravitation, which is not clear in the present formulation. We assume that as such an object (test particle), black holes (small) and other hypothetical particles like quarks, bags (for example, MIT bag), monopoles (Dirac’s, ’t Hooft-Polyakov’s and GUT’s super-heavy), maximons (particles with the Planck mass $m_{\text{Pl}} \sim 1.2 \times 10^{19}$ GeV), etc. may be taken. The role of these test particles in the process of the creation of elementary particles and their specific differences from usual local particles will be discussed below.

Further, we postulate that the square-energy-momentum operator of particles is constructed by means of the metric tensor $G^i_{\mu\nu}$ of the type of (11.42) and acquires the form:

$$\hat{p}^2 \Rightarrow \hat{\mathscr{P}}^2 = -G^i_{\mu\nu}\frac{\partial^2}{\partial x^\mu\,\partial x^\nu}, \tag{11.46}$$

and in the flat space case one can assume that

$$\hat{\mathscr{P}}^2 = -K^i(l^2\Box)g_{\mu\nu}\frac{\partial^2}{\partial x^\mu\,\partial x^\nu}, \tag{11.47}$$

where $g_{\mu\nu}$ is the Minkowski metric. We assume that such a form of the operator $\hat{\mathscr{P}}^2$ results from the hypothesis of nonlocality of the space-time metric at small distances or in a singular region of confinement of particles which are represented by solitary waves (Efinger, 1981). In the last case one could then hypothesize that particles under consideration are geometrical objects on a *Riemannian space-time metric* $g_{\mu\nu}(x)$ which is nearly singular (or nonlocal) in a region corresponding to the width of the solitary wave-amplitude (or in a region of confinement of particles).

Moreover, the form we have proposed of operator (11.46) or (11.47) naturally generalizes the idea of quantization of three-dimensional space according to Fujiwara (1980). His scheme is derived as a particular case for our consideration. Indeed, the new momentum operator $\hat{\mathscr{P}}$[according to (11.46)] constructed by the generalized function K^C_1, the Fourier transform of which is determined by the form factor (11.44) in the static limit $p_0 \to 0$, coincides explicitly with the momentum operator proposed by Fujiwara (1980). The difference in both the schemes is that in our model the relativistic invariance is strictly conserved.

Thus, our postulation concerning the nonlocality of the space-time metric leads to a profound connection between the particle-wave dualism and the structure of space-time. In other words, such a structure of the space-time-stochasticity or nonlocality (after averaging over a large scale) as a self-memory would affect the ω–k relation for the waves and E–p relations for particles and makes the theory nonlocal. Notice that in the three-dimensional case this problem has been discussed by Fujiwara in detail.

Now we consider some consequences of the nonlocal theory of quantized fields based on the assumption of the nonlocality of the space-time metric. Since the important physical quantities such as the Hamiltonian (or Lagrangian) and field equations of *test particles* are constructed by means of the operator (11.46) or (11.47), these quantities are changed in accordance with the change of the metric forms (11.42). For example, the field equation and the Green function for a scalar field acquire the form:

$$(\hat{\mathscr{P}}^2 - m^2)\phi(x) = [K^i(l^2\Box)\Box - m^2]\phi(x) = 0, \tag{11.48}$$

$$G(x) = i^{-1}(2\pi)^{-4}\int d^4p\, e^{ipx}[m^2 - p^2\tilde{K}^{A,C}(p^2l^2)]^{-1}. \tag{11.49}$$

Here the Green function $G(x)$ satisfies the following equation

$$(\hat{\mathscr{P}}^2 - m^2)G(x) = -\delta^{(4)}(x).$$

The contour integration in (11.49) is chosen as in the case of local theory, so that in the limit $l \to 0$, the value of $G(x)$ gives the usual Green function for the scalar field with mass m. Similarly, in our model the propagators of photons and spinor fields read:

$$D^{c}_{\mu\nu}(x) = -i(2\pi)^{-4} g_{\mu\nu} \int d^4p \frac{e^{ipx}}{p^2 \tilde{K}^{A,C}(p^2 l^2)}, \tag{11.50}$$

and

$$S_c(x) = i^{-1}(2\pi)^{-4} \int d^4p \, e^{ipx} \frac{m + \hat{p}[\tilde{K}^{A,C}(p^2 l^2)]^{1/2}}{m^2 - p^2 \tilde{K}^{A,C}(p^2 l^2)}. \tag{11.51}$$

Now we discuss the choice of the form factors $\tilde{K}^A$ or $\tilde{K}^C$ in (11.40) and (11.41). It is well-known that (see Section 2.3) the introduction of nonlocality into theories in accordance with (11.46), (11.49)–(11.51) changes the form of potentials between interacting fields (for example, the Coulomb and Yukawa laws) at small distances. For example, the *Coulomb potential* of two interacting electrons is given by expression (11.50) in the static limit:

$$\varphi(r) = e(2\pi)^{-3} \int d^3p \frac{e^{-i\mathbf{pr}}}{\mathbf{p}^2 \tilde{K}^{A,C}(-\mathbf{p}^2 l^2)}. \tag{11.52}$$

From the condition $\varphi(0) < \infty$ it follows that only form-factors of the type $\tilde{K}^C(p^2 l^2)$ given by expression (11.41) are acceptable. In particular, for form-factor K^C_G (11.43) the equality (11.52) reads

$$\varphi(r) = e(2\pi)^{-3} \int d^3p \, \mathbf{p}^{-2} \exp\left(-i\mathbf{pr} - \frac{\mathbf{p}^2 l^2}{8}\right) = \left(\frac{e}{4\pi r}\right) \operatorname{erf}\left(\frac{\sqrt{2}r}{l}\right).$$

One verifies that for the form-factors $\tilde{K}^C(p^2 l^2)$ the propagators (11.49)–(11.51) are finite in the Euclidean metric. Thus, in our model all matrix elements of the S-matrix are studied, as in the case of the nonlocal theory of quantized fields (see Chapters 1 and 2), in the Euclidean domains $p_i^2 < 0$ of all the momentum variables p_i. The passage to the case $p_i^2 > 0$ is done by an analytic continuation.

A concrete scheme for the quantization of fields and the construction of a nonlocal theory of electromagnetic and weak, etc. interactions with the use of the propagators (11.49)–(11.51) will be presented in a separate work. Actually, in our scheme the ultraviolet divergences are absent.

We now consider the more interesting consequences of the idea of the space-time metric nonlocality. First, it transpires that in our scheme the Einstein relation between the energy and momentum of a particle is changed. For this, consider a simple case when the metric is defined by formulas (11.37) and

(11.38). According to formula (11.42), these equalities can be written in the unified way

$$s_0 \Rightarrow s_l = (1 - al^2 \Box) s_0, \tag{11.53}$$

where a is some number depending on the form of measures $w(b_E^2/l^2)$ and taking two values, positive and negative, in accordance with the choice of metric formulas (11.37) and (11.38) [or the form of form factors (11.40) and (11.41)]. Then the equality $\mathscr{P}^2 = m^2$ from equation (11.48) in the given (free) case yields the following equation

$$p^2 - m^2 - al^2 p^4 = 0,$$

that has two solutions:

$$p_1^2 = m^2(1 + am^2 l^2), \qquad p_2^2 = 1/al^2 - m^2(1 + am^2 l^2). \tag{11.54}$$

Here it is assumed that $ml \ll 1$. If $a > 0$, then both the solutions are positive, which corresponds to the appearance (or creation) of a *supplementary particle* with mass of an order of $1/al^2 - m^2$, at which the initial mass m acquires a small additional term of an order of $m^2 l^2$ resulting from the nonlocality of the space-time metric. If $a < 0$, then a particle of the tachyon type corresponds to the second solution, i.e., a particle with complex mass.

From the first equation (11.54) it follows that

$$E_l = [\mathbf{p}^2 + m^2(1 + am^2 l^2)]^{1/2} \approx E_0 + (1/2E_0)am^4 l^2,$$
$$E_0 = (\mathbf{p}^2 + m^2)^{1/2}.$$

In particular, for the rest energy we have

$$E_l^0 = mc^2(1 + \tfrac{1}{2}am^2 l^2)$$

and in the classical limit

$$E_l^{cl} = \frac{\mathbf{p}^2}{2m}(1 - \tfrac{1}{2}am^2 l^2).$$

Further, for the concrete form of form factors, say for $\tilde{K}_1^C$ given by (11.44) we have

$$p_0^2 = \mathbf{p}^2 + 4l^{-2}[\arcsin(\tfrac{1}{2}ml)]^2 \quad \text{or} \quad E_l = [\mathbf{p}^2 + 4l^{-2}(\arcsin(\tfrac{1}{2}ml))^2]^{1/2}.$$

In view of this we notice that the test of Einstein's formula $E = mc^2$ may be of a great interest.

Another exceptionally important fact is that the space-time metric nonlocality, as can be seen from expression (11.54), admits the existence of a corresponding partner with mass $1/al^2 - m^2$ for any particle with mass m, even in the case of the flat-space-time-metric nonlocality (11.53).

It turns out that the *equidistant behavior of observable mass* spectra of the elementary particles may be understood within our hypothesis of the space-time metric nonlocality, if the parameter l used here is assumed to be an order of the size of the domain in which the quark confinement is reached, i.e., $l \to l_{\text{eff}} \sim 10^{-13}$ cm. On the other hand, this quantity may be identified with the effective *Schwarzschild radius of the proton* (see Li, 1982; Salam and Strathdee, 1976). As mentioned above (Chapter 1), in our scheme such a possibility is based on the assumption that the parameter l is not a universal fundamental constant but characterizes only the domain of the nonlocal interaction of test particles. In connection with this we notice that there is an another interesting approach (Brooke *et al.*, 1982) in which elementary particles are regarded as quantum space-time *excitons* within *Born's reciprocity principle* (Born, 1938).

Now we give some speculation concerning the possibility of obtaining the equidistant behavior of the observable mass spectra of elementary particles. It should be noted that in Nature there exist *two basic different objects* with respect to the space-time description. Those of the first sort behave as point-like particles and are described by the local quantum field theory, i.e., their equations of motion satisfy finite order differential equations (for example, Klein–Gordon and Dirac equations). In this case, according to the above-deduction, the connection between the properties of these particles and of the space-time metric nonlocality is expressed by the change of the energy-momentum relation at small distances, in particular, for the rest energy of the particles we have obtained

$$E_l^0 = mc^2(1 + \tfrac{1}{2}am^2l^2).$$

Here we assume $m^2l^2 \ll 1$ for any free point-like observable particles and $|a| \sim 1$ for the measures (distributions) $w_i(y^2)$ listed above.

The situation is radically changed for the *extended* (nonlocal) objects. A role of such objects, as mentioned above, may be played by bags, strings, monopoles, different compound states of quarks and gluons etc. For definiteness, we have called them above *test particles*. The features of these test particles are the following: first they are described by integral (strictly speaking, integro-differential) equations, i.e., differential equations of infinite order; secondly, they have some internal structure and those physical quantities (charge, mass, etc.) are characterized by some distributions. Namely, the presence of the internal structure and distributions for the test particles plays an essential role in their space-time description. In this nonlocal case, we should distinguish between the internal and external mediums (in particular, space-time) around the nonlocal objects and take into account their structures. The interrelation of these mediums or their reaction on nonlocality with respect to the change of space-time properties may be expressed by means of the concept of the *perturbative effect* of test particles on the space-time metric nonlocality. Our basic assumption is that this perturbative effect is described by some differential equation of infinite order and gives rise to equidistant behavior of observable particle mass spectra, if we would regard the observable particles as some definite states (excitons) of excited test particles. Below we consider a few excited test particles which generate spectra of masses of observable particles.

In the case under consideration, the parameter l and the measure $w(b_E^2/l^2)$ characterize the physical properties of the perturbative effect of test particles on the space-time metric nonlocality. Roughly speaking, now the parameter l *loses its universal character and may be understood as an effective size* l_{eff} of the test particles (henceforth by l we denote just this value l_{eff} in this section). The measure $w(b_E^2/l^2)$ corresponds to an energetic (or mass) distribution of excited test particles. Further, we assume that the effective size of the considered test particles are comparable with the size of the domain in which the *quark confinement* is reached, i.e., $l \to l_{\text{eff}} \sim 10^{-13}$ cm. Therefore, in the test particle case, $M^2 l_{\text{eff}}^2 \sim 1$ and $E_l^0 = Mc^2$, where M is the characteristic mass of test particles.

Thus, we assume that basis (ground) states ϕ_B and ψ_F of test particles for bosons and fermions satisfy the following differential equation of infinite order

$$(K_B(l_B^2\Box)\Box - m_0^2)\phi_B = 0, \qquad \left(iK_F^{1/2}(l_F^2\Box)\gamma_\mu \frac{\partial}{\partial x_\mu} - M_0\right)\psi_F = 0, \tag{11.55}$$

and their eigenvalues are determined by the formulas

$$p^2\tilde{K}_B(p^2 l_B^2) - m_0^2 = 0, \qquad p^2\tilde{K}_F(p^2 l_F^2) - M_0^2 = 0, \tag{11.56}$$

where $l \to l_B$ and l_F denote the proper radii of test particles for bosons and fermions, respectively. The higher order equations (11.55) are reduced to the Klein–Gordon and Dirac equations as $l_{B,F} \to 0$. Quantities m_0 and M_0 correspond to some lowest (or unperturbed) values of the masses in elementary particle mass spectra for bosons and fermions. On the other hand, equations (11.56) can be understood as conditions for the *quantization of masses*.

As is seen above, only form-factors of type (11.41), say, like K^C's in (11.43)–(11.45), can describe the equidistant behavior of the elementary particle mass spectra. It turns out that the parameters m_0, M_0, l_B, l_F and the form of form-factors $\tilde{K}_i^C(p^2 l_i^2)$ $(i = \text{B}, \text{F})$ are sufficient to obtain some theoretical mass formulas which roughly correspond to the equidistant behavior of the spectra masses of observed particles. For this, consider the simple cases (11.44) and (11.45). The first equation of (11.56) in the case of (11.44) takes the form

$$m_0^2 - (2/l_B^2)[1 - \cos(p^2 l_B^2)] = 0.$$

Assuming $m_0 = \sqrt{2}/l_B$, we get the quantized condition $\cos(p^2 l_B^2) = 0$. From this we have a family of particles:

$$m_n = (p_n^2)^{1/2} = (\pi/l_B)(\tfrac{1}{2} + n), \quad n = 0, 1, 2, \ldots$$

In the second case (11.45) the second equation of (11.56) yields

$$M_0^2 - 4.2^\mu \cdot \Gamma(3 + \mu)p^2(\sqrt{p^2 l_F^2})^{-2-\mu} J_{2+\mu}(\sqrt{p^2 l_F^2}) = 0.$$

Setting $M_0 = 0$ here we obtain the following condition for the fermion mass quantization $J_{2+\mu}(p^2 l_F^2) = 0$, that corresponds to some family of particles with masses:

$$M_n = (1/l_F)j_{2+\mu,n}, \quad n = 1, 2, \ldots$$

where $j_{2+\mu,n}$ are zeros of the Bessel function $J_{2+\mu}(x)$.

We assume that the basic (initial) metric perturbation gives rise to some family of particles, but newly created particles of this family may in turn generate the space-time metric, and in consequence, that there are possible second (daughter) metric perturbations which are able to generate other families of particles, etc. Such a process of metric perturbation is analogous to the chain reaction of atomic nuclei (or analogous to the cosmic tree diagrams according to 't Hooft, 1979). Of course, the mechanism of perturbations (or fluctuations) of space-time metric proposed here differs from the *quantum gravitational concept of fluctuations in metric*, this should be on length scales which are far smaller than any at present probed. As a result of such consecutive perturbations, the observed equidistant behavior of the particle mass spectra may be explained. It appears that for a satisfactory description of the equidistant behavior of the particle mass spectra, it is sufficient to know two basic metric perturbations connected with the generation of the π-meson and proton and their corresponding partners. Next, a few second (daughter) metric perturbations are needed which arise from the K-meson and Λ-, Σ-, Ξ-hyperons generated already by the initial metric perturbations.

So, assume that two metric perturbations are defined by the following formulas:

$$m_n^{(1)} = \frac{\pi}{l_B}(\tfrac{1}{2} + n), \qquad \frac{1}{l_B} = \frac{c}{\hbar} m_0^{(1)} = 90 \text{ MeV}, \quad n = 0, 1, 2, \ldots$$

$$M_n^{(2)} = \frac{1}{l_F} j_{20\frac{1}{2},n}, \qquad \frac{1}{l_F} = 36.138 \text{ MeV}, \quad n = 1, 2, \ldots$$

where $j_{20\frac{1}{2},n}$ denotes the nth positive zero of the Bessel function J of the order $20\frac{1}{2}$ (Olver (ed.), 1960). The corresponding theoretical spectra of particle masses are presented in Table 11.3 (the experimental data were taken from Particle Data Group, 1982).

Daughter metric perturbations caused by the K-meson and Λ-, Σ-, Ξ-hyperons are given by

$$M_n^{(i)} = (1/l_i)J_{20\frac{1}{2},n}, \quad i = 1, 2, 3, 4,$$

where $1/l_1 = 19.07$ MeV, $1/l_2 = 43$ MeV; $1/l_3 = 46$ MeV; $1/l_4 = 50.66$ MeV; for K-, Λ-, Σ- Ξ-family of particles, respectively, and the corresponding theoretical mass spectra are shown in Table 11.4.

Since masses of the K-meson, proton and Λ-, Σ-, Ξ-hyperons are equal in the order of magnitude, we assume that their form factors must be the same and the corresponding values of the parameter $1/l$ must grow with increasing particle mass. It is natural because the generation of a particle with a larger

Table 11.3

n	$m_n^{(1)}$	Experimental values	$m_n^{(2)}$	Experimental values
0	141	π		
1	424	K	938	p
2	707	$\sim\rho$	1103	Λ
3	990	$S^*(975)$	1249	Σ
4	1272	$f(1270), \eta(1275)$	1386	Ξ
		$D(1285)$		$N(1440)$
5	1555	$\rho'(1600)$	1518	$N(1520)$
6	1838	$\phi(1850)$	1647	Ω^-, $N(1650)$, $N(1675)$, $N(1680)$
7	2120	$\pi(2100)$	1773	$N(1700)$, $N(1710)$, $N(1720)$
8	2403	$\delta(2450)$	1897	$N(1990)$
9	2686	?	2020	$N(2080)$
10	2969	$\eta_c(2980)$	2141	$N(2190)$
11	3251	$\psi(3100)$	2662	Λ_c^+, $N(2200)$, $N(2220)$, $N(2250)$
12	3534	$\psi(3415)$	2381	
		$\psi(3510), \psi(3555)$		$\Delta(2380)$–$\Delta(2450)$
13	3817	$\psi(3685), \psi(3770)$	2501	
14	4100	$\psi(4030), \psi(4160)$	2619	$N(2600)$
15	4382	$\psi(4415)$	2737	$\Delta(2750)$, $N(2700)$?
16	4665	?	2855	$\Delta(2850)$, $N(2800)$?
17	4948	?	2972	$\Delta(2950)$?
18	5230	$B(5200)$	3090	$N(3030)$?
19	5513	?	3206	$\Delta(3230)$, $N(3245)$?
20	5796	?	3323	$\Delta(3350)$?
...	...	...	...	
33	9472	$\Upsilon(9460)$		
34	9755	?		
35	10037	$\Upsilon(10020)$		
36	10320	$\Upsilon(10350)$		
37	10603	$\Upsilon(10570)$		
...	...	...		
285	80723	m_W		
290	82137			
295	83551			
320	90619			
325	92033	m_{Z^0}		

mass requires a larger space-time 'curvature' $R = 1/l$ or smaller l. Further we notice that the choice of the form-factors (11.44) and (11.45) describing the equidistant behavior of the particle-mass-spectra in the given case need not be unique, and other forms of the metric perturbation (correspondingly, of the measure $w(b_E^2/l^2)$ or the chosen form-factor) are possible. Thus, in our approach the fundamental open problem is the question of a possible unique choice of form-factors, which requires, it seems, another fundamental physical

Table 11.4

n	$m_n^{(K)}$	Experimental values	n	$m_n^{(i)}$	Experimental values
1	495	K			
2	582	η		$m_n^{(\Lambda)}$	
3	659	ρ	1	1115	Λ
4	732		2	1312	$\Lambda(1405)$
5	801	ω	3	1486	$\Lambda(1520)$
6	869	$K^*(892)$	4	1650	$\Lambda(1600, 1670, 1690)$
7	938	$\eta'(958)$	5	1807	$\Lambda(1800, 1820, 1830)$
8	1001	$\delta(980)$	6	1960	$\Lambda(1890)$
9	1066	$\phi(1020)$	7	2110	$\Lambda(2100, 2110)$
10	1130	$H(1190)$	8	2257	?
11	1193	$B(1235)$	9	2403	$\Lambda(2350)$
12	1257	$\rho'(1250), A_1(1270),$	10	2548	$\Lambda(2583)$
		$Q_1(1280)$		...	
13	1320	$\pi(1300), A_2(1320)$			
		$\varepsilon(1300)$		$m_n^{(\Sigma)}$	
14	1382	$k(1350)$	1	1194	Σ
15	1444	$Q_2(1400), K'(1400),$	2	1404	$\Sigma(1385)$
		$K^*(1430), E(1420)$	3	1590	$\Sigma(1560, 1580, 1620)$
16	1507	$f'(1515)$			$\Sigma(1660, 1670)$
17	1569	$L(1580)$	4	1765	$\Sigma(1750, 1775)$
18	1630	$\theta(1640), K^*(1650)$	5	1933	$\Sigma(1915, 1940)$
19	1692	$\omega(1670), A_3(1680)$	6	2096	$\Sigma(2030)$
		$\phi'(1680), g(1690)$	7	2257	$\Sigma(2250)$
20	1753	$L(1770), K^*(1780)$	8	2415	$\Sigma(2455)$
21	1815	$X(1850)$	9	2571	$\Sigma(2620)$
22	1876	$D(1870)$	10	2725	?
23	1937	$S(1915)$		...	
24	1999	$D^*(2007, 2010),$		$m_n^{(\Xi)}$	
		$\delta(2030), F(2020), F(2021)$	1	1315	Ξ
25	2060	$h(2040), \rho(2050)$	2	1546	$\Xi(1530)$
		$K^*(2060)$	3	1751	$\Xi(1820)$
26	2121	$F^*(2140)$	4	1944	$\Xi(2030)$
27	2181	$\rho(2150), \varepsilon(2150)$			
28	2242	$K^*(2200), \rho(2250)$			
29	2303	$\varepsilon(2300)$			
30	2364	$\rho(2350)$			

principle. As pointed out above, formulation of this principle is not known and needs deeper studies in field theories. Moreover, the mechanism of metric perturbation (allowing one to create the particles) resulting from the space-time metric nonlocality is not clear.

In conclusion, notice that the mass splitting connected with the internal quantum numbers such as charge, isospin, charm, etc. would not be surprising. In our model, we do not consider these quantum numbers, and consequently, the metric perturbation is not sensitive to the changing values of these numbers. Therefore, we associate some mass values of theoretical calculation with experimental values that are closer to them.

From Table 11.3 it follows that if the equidistant behavior of the mass spectra caused by the nonlocality of space-time metric is valid, then our scheme predicts the existence of a family of bosons with masses: 2700, 4700, 4900, 5500, 5800, ..., 9750 MeV, etc.

Finally, it should be noted that an attempt to explain the equidistant behavior of observed particle mass spectra in the framework of our proposal listed above, is indeed not of principle interest, but it plays an important role when it deals with hypothetical particles (for example, maximons by term of M. A. Markov) of mass of the order of the Planck mass M_{Pl}. In this case, it seems to us that mass spectra of these maximons, if they exist, should be caused by fluctuations in metric and quantized according to the above deduction, and obeyed also the equidistance behavior as in the case of usual particles observed recently.

11.7. On the Origin of Cosmic Rays and the Value of the Fundamental Length

The problem of the origin of cosmic rays plays an important part in high-energy astrophysics. To solve this problem one should answer a set of questions (see Ginzburg and Ptuskin, 1976) among which the *acceleration mechanism* is of great interest. If the main sources (their spatial distribution) of cosmic rays are pointed out and the intensity and spectra of cosmic rays emitted by the sources are given, the problem of the cosmic-ray origin can be divided into an external (acceleration mechanism) and an internal (source theory) problem. The former, i.e., the acceleration mechanism that carries cosmic rays (especially, protons) to the energies over 10^{20}–10^{22} eV extent in the primary cosmic radiation still remains unsolved (Ginzburg and Ptuskin, 1976; Hillas, 1975).

Here we discuss this important problem in the framework of the hypothesis of *space-time stochasticity and fluctuations in metric* (see Chapter 1). It is well known that a high isotropy and a rather larger content of secondary nuclei in cosmic rays indicate an effective 'mixing' and a long wandering of high-energy particles in the Galaxy. We assume that such a mixing and isotropization are caused by the stochastic (or fluctuational) structure of space-time in which the cosmic-ray propagation takes place. More exactly, we suggest that the physical mechanism responsible for the acceleration of cosmic rays during their propagation in the Galaxy has the *stochastic origin* connected with the space-time properties at small distances.

As shown above, the idea of using the stochasticity of space-time and random fluctuations in metric as an origin of the stochastic and nonlocal interpretation of quantum mechanics plays an important role in constructing the nonlocal theory of quantized fields. Now we obtain the dynamics of cosmic-ray particles within the hypothesis of fluctuations in metric. The basic idea is the following: there exists a profound connection between the structure of space-time and the propagation mechanism of ultra-high energy particles. To realize this connection we assume that *freely moving particles perturb space-time around themselves,* and at the same time the space-time metric is generated and becomes a Riemannian one $\tilde{g}_{\mu\nu}(b_E, x)$ depending on a random vector $b_E = (b_4 = \tau, \mathbf{b})$ with a measure $w(b_E^2/l^2)$ obeying conditions (1.2) in Chapter 1.

As a result of such fluctuations in metric, as shown below, a nonlinear dynamics of particles arises and the maximum energy of particles, say protons in the primary cosmic-rays depends on the *value of the fundamental length* at which fluctuations in metric take place. This allows one to define the value of the fundamental length from high-energy astrophysical experiments.

A simple form of $\tilde{g}_{\mu\nu}$ is chosen, that is $\tilde{g}_{\mu\nu} = \phi^2(b_E^2, x^2)\eta_{\mu\nu}$, and the *averaged space-time metric* acting on the behavior of particles is obtained by the formula

$$g_{\mu\nu}(l^2/x^2) = \langle \tilde{g}_{\mu\nu}(b_E, x) \rangle_{\mathbb{R}^4(\hat{x})} = \int d^4b_E w(b_E^2/l^2)\phi^2(b_E^2, x^2)\eta_{\mu\nu}$$

$$= \phi^2(l^2/x^2)\eta_{\mu\nu}, \tag{11.57}$$

where $\eta_{\mu\nu}$ is the Minkowski metric.

Now we pass to obtaining cosmic-ray particle dynamics in this *conformally flat space-time.* The action for a free particle takes the form

$$S = -mc \int_a^b ds, \tag{11.58}$$

$$ds = [g_{\mu\nu}(l^2/x^2)\, dx^\mu\, dx^\nu]^{1/2} = \phi\, ds_0, \quad ds_0 = (dx_\nu\, dx^\nu)^{1/2}.$$

According to the action principle

$$\delta S = -mc \int_a^b \delta ds = -mc \int_a^b \frac{\delta ds^2}{2ds} = -mc \int_a^b \frac{\delta(g_{\mu\nu}\, dx^\mu\, dx^\nu)}{2ds}$$

$$= -mc \int_a^b \left\{ \frac{1}{2} \frac{dx^\nu}{ds} \frac{dx^\mu}{ds} \frac{\partial g_{\nu\mu}}{\partial x^\lambda} \delta dx^\lambda + g_{\nu\mu} \frac{dx^\nu}{ds} \frac{d\delta x^\mu}{ds} \right\} ds = 0.$$

Integration by parts in the second term gives

$$\delta S = -mcg_{\nu\mu} \frac{dx^\nu}{ds} \delta x^\mu \Big|_a^b - mc \int_a^b ds \left\{ \frac{1}{2} \frac{dx^\nu}{ds} \frac{dx^\mu}{ds} \frac{\partial g_{\nu\mu}}{\partial x^\lambda} \delta x^\lambda - \frac{d}{ds}\left(g_{\nu\mu} \frac{dx^\nu}{ds} \right) \delta x^\mu \right\}.$$

From this we have the following nonlinear equation

$$\frac{Du_\nu}{Ds} \equiv g_{\lambda\nu} \frac{du^\lambda}{ds} + \Gamma_{\nu,\mu\lambda} u^\mu u^\lambda = 0, \quad (u^\mu = dx^\mu/ds)$$

or taking into account that $g_{\mu\nu} = \phi^2\eta_{\mu\nu}$, $ds = \phi\, ds_0$, we get

$$du^0_\nu/ds_0 - \partial \ln \phi/\partial x^\nu + u^0_\nu(u^\lambda_0 \partial \ln \phi/\partial x^\lambda) = 0. \tag{11.59}$$

Here $u^\nu_0 = dx^\nu/ds_0$,

$$\Gamma_{\nu,\mu\lambda} = \tfrac{1}{2}(\partial g_{\mu\nu}/\partial x^\lambda + \partial g_{\lambda\nu}/\partial x^\mu - \partial g_{\mu\lambda}/\partial x^\nu)$$

is the *Christoffel symbol.*

In the nonrelativistic case equation (11.59) will be discussed in the next section. We now solve it in the two-dimensional Minkowski space-time (x, t) and in the simple case when $\phi(x)$ depends only on the spatial variable x (now the variable x is regarded as a parameter). Since $u^0_x = -v_x(1-\beta^2)^{-1/2}$, $ds_0 = c\,dt(1-\beta^2)^{1/2}$, $\beta^2 = v_x^2/c^2$, from (11.59) we get $(v_x = v)$:

$$\frac{dq}{dt} = \varepsilon(1-q^2), \quad \left(q = \frac{v}{c}, \ \varepsilon = \frac{-c\,\partial \ln \phi}{\partial x}\right). \tag{11.60}$$

This is the *Riccati-type equation* (for example, see Bender and Orszag, 1978). The substitution $q = \dot{w}/\varepsilon w$, $\dot{w} = dw/dt$ converts it into a second-order linear equation for $w(t)$: $\ddot{w} - \varepsilon^2 w = 0$. The solution to the last equation is $w = c_1 \operatorname{ch}(\varepsilon t + c_2)$, and therefore $q = \operatorname{th}(\varepsilon t + c_2)$. Here the integration constant c_2 is given by an initial condition, say $q(0) = v_0/c$.

So, the solution of the *Cauchy problem* for the Riccati-type equation (11.60) is

$$q(t) = \operatorname{th}(\varepsilon t + \operatorname{Arth}(v_0/c)), \tag{11.61}$$

or

$$\dot{q}(t) = \varepsilon \operatorname{sech}^2(\varepsilon t + \operatorname{Arth}(v_0/c)).$$

The next question concerns the choice of the metric form $g_{\mu\nu}(l^2/x^2)$, i.e., the function $\phi(l^2/x^2)$ in conformally flat space-time. A simple form may be obtained if one assumes that:

(i) Due to the small parameter l it is sufficient to determine the metric form $g_{\mu\nu}(l^2/x^2)$ in (11.57) by the few first moments of the random variable b_E^2 of the function $\phi^2(b_E^2/l^2)$.

(ii) According to an old concept by Einstein (1967) and recently by Efinger (1981) the averaged Riemannian metric (11.57) may be remarkably peaked (even nearly singular) in the region of the confinement of a particle and outside this region one expects

$$g_{\mu\nu} \approx \eta_{\mu\nu} = \operatorname{diag}(1, -1, -1, -1)$$

(or the boundary condition $g_{\mu\nu} \to \eta_{\mu\nu}$ at infinity).

(iii) $\qquad g_{\mu\nu} \to \eta_{\mu\nu}, \quad$ for $l \to 0$.

(iv) The size of the domain of particle confinement corresponds to the particle radius r which may be interpreted as the effective *Schwarzschild radius* (Li, 1982; Salam and Strathdee, 1976), for example, the experimental value of the proton radius is $r = r_p = 0.8 \times 10^{-13}$ cm. Thus, the simple form

$$\phi(l^2/x^2) \Rightarrow \phi(l^2/\mathbf{x}^2) = 1 + l^2(r^2 + \mathbf{x}^2)^{-1} \tag{11.62}$$

satisfies the above-mentioned conditions automatically.

The perturbative effect of the particles on the metric leads to fluctuations such that the deviation from $\eta_{\mu\nu}$ has a nonzero mean. In view of (11.61) and the assumption (11.62) the particles are accelerated to large velocities $v \sim c$ during the evolution of the Universe. At this, the energy of the particle acquires the form

$$E \Rightarrow E_l = mc^2(1 - \beta_l^2)^{-1/2}, \tag{11.63}$$

where $\beta_l^2 = q^2(t)$ is given by equation (11.61).

Thus, measuring the maximum energy of the particle (the largest proton energy value observed in cosmic-rays is about $E_{l,p}^{\text{ob}} = 10^{20}$ eV, see Hillas, 1974) in primary cosmic-rays, we can determine experimentally the value of the fundamental length from formulas (11.61) and (11.63). We assume that the age of primary cosmic-rays T_{cr} in the Galaxy is of the order of the age of the Universe $T_u = 1/2H_0 = 8.85 \times 10^9$ yr. Recently model-dependent calculations (see Ginzburg and Ptuskin, 1976) have shown that a 'large' $T_{\text{cr}} \gtrsim 10^8$ yr is more possible. Here the value of the *Hubble constant* is $H_0 = 55$ km/sec $\times$ $\times$ (megaparsec).$^{-1}$

The simple extremum problem for equation (11.61) in the case of (11.62) gives the maximum value of $q(t)$ at $x = r/\sqrt{3}$ $(l \ll r)$. At this the argument $c\varepsilon t$ of $q(t)$ takes the form:

$$c\varepsilon t \Rightarrow c\varepsilon_{\max} T_{\text{cr}} \Rightarrow c \cdot \varepsilon_{\max} T_u = \frac{9}{8\sqrt{3}} cT_u \frac{l^2}{r^3} = 10.68\Delta^2,$$

for $r = r_p$, where $\Delta = l/l_0$, $l_0 = 10^{-33}$ cm. So, assuming $t = T_u$, $v_0 = 0$, $r = r_p = 0.8 \times 10^{-13}$ cm, $x = r_p/\sqrt{3}$ and $E_l = E_{l,p}^{\text{ob}}$, we obtain from equations (11.61) and (11.63)

$$E_{l,p}^{\text{ob}} = 10^{20}\ \text{eV} = m_p c^2 \tfrac{1}{2} \exp(10.68\,\Delta^2).$$

From this it is easy to find that $\Delta = 1.56$, therefore

$$l = 1.56 \times 10^{-33}\ \text{cm}, \tag{11.64}$$

which is very close to the *Planck length* $l_{Pl} = (G_g \hbar/c^3)^{1/2} = 1.62 \times 10^{-33}$ cm discussed in the quantum gravitational theory, where G_g is the Newtonian constant.

It transpires that the very slighest change in the value of the fundamental length from the quantity (11.64) (or l_{Pl}) gives rise to the enormous range of energy of the cosmic-ray proton (see Table 11.5).

Table 11.5

l/l_0	1.56	1.62	1.69	1.82	1.93	2.04	2.14	2.24	2.34	2.43	2.51	2.60
E_l(eV)	10^{20}	7.1×10^{20}	10^{22}	10^{24}	10^{26}	10^{28}	10^{30}	10^{32}	10^{34}	10^{36}	10^{38}	10^{40}

Thus, we see that if the *true fundamental length* that exists in Nature is of $l_{Pl} = 1.62 \times 10^{-33}$ cm, then the maximum energy of the proton is *restricted* by the value $E_{l,p}^{\text{lim}} = 7.1 \times 10^{20}$ eV. We note that another value $T'_u = 1/H_0$ for T_u is discussed in literature, that gives $l = 1.11 \times 10^{-33}$ cm for $E_{l,p}^{\text{ob}} = 10^{20}$ eV. In this case $E_{l,p}^{\text{lim}} \sim 10^{33}$ eV.

Now we consider the motion of a charged particle in an electromagnetic field characterized by a potential A_μ within our scheme. After some calculations analogous to the usual (gravitational) case (for example, see Landau and Lifshitz, 1971) we have the following equation

$$mc\frac{Du_\nu}{Ds} = \frac{e}{c}F_{\nu\rho}u^\rho, \tag{11.65}$$

where

$$F_{\nu\rho} = A_{\rho;\nu} - A_{\nu;\rho} \equiv \partial A_\rho/\partial x^\nu - \partial A_\nu/\partial x^\rho.$$

It turns out that in our model the equations for the electromagnetic field acquire the forms:

$$\partial F_{\nu\mu}/\partial x^\lambda + \partial F_{\lambda\nu}/\partial x^\mu + \partial F_{\mu\lambda}/\partial x^\nu = 0$$

and

$$F^{\nu\mu}_{\mu} = (1/\sqrt{-g})\partial/\partial x^\mu(\sqrt{-g}\,F^{\nu\mu}) = -(4\pi/c)j^\nu, \tag{11.66}$$

where j^ν is the four-vector current satisfying the equation of continuity

$$j^\nu_\nu = (1/\sqrt{-g})\,\partial_y(\sqrt{-g}j^\nu) = 0.$$

We recall again that according to conformally flat space-time considered above, the determinant of the metric tensor $g_{\mu\nu}$ is calculated from equation (11.57).

Further, we write the equation of motion for the charge in an external force f_μ in accordance with equation (11.65):

$$mc\frac{Du_\nu}{Ds} = \frac{e}{c}F_{\nu\mu}u^\mu + f_\nu, \tag{11.67}$$

where f_v has components:

$$f_v = \{(\mathbf{f}\cdot\mathbf{v}/c\sqrt{1-\beta^2}),\ -(\mathbf{f}/c^2\sqrt{1-\beta^2})\},$$

in particular, for bremsstrahlung $\mathbf{f} = \tau\ddot{\mathbf{v}}$, $\tau = \frac{2}{3}e^2c^{-3}$. Then, the equation of motion in the case of $v \ll c$, $\phi = \phi(\mathbf{x})$ has the form (in the absence of the electromagnetic field):

$$\dot{\mathbf{v}} = (\tau/m)\ddot{\mathbf{v}} - (1-\beta^2)\left\{c^2(1-\beta^2)\frac{\partial \ln\phi}{\partial \mathbf{x}} + \mathbf{v}\left(\mathbf{v}\cdot\frac{\partial \ln\phi}{\partial \mathbf{x}}\right)\right\}, \tag{11.68}$$

or in the two-dimensional case

$$\dot{q} = (\tau/m)\ddot{q} + \varepsilon(1-q^2), \quad (q = v/c). \tag{11.69}$$

This is an *autonomous equation* if it is assumed that t is an independent variable and x is a parameter. Then a standard trick is to express $u = \dot{q}(t)$ as a function of q and to find an equation for u and its derivatives. We thus let (in the two-dimensional case)

$$\dot{q} = u(q), \qquad \ddot{q} = u'(q)u(q), \qquad u'(q) = \mathrm{d}u/\mathrm{d}q,$$

and for $u(q)$ we have the following equation:

$$u(q) - (\tau/m)u'(q)u(q) = \varepsilon(1-q^2).$$

This equation is still too difficult to solve in a closed form. This is easy to show if one assumes that $\varepsilon \sim l^2$ ($x \ll 1$) is a small parameter, then a solution is sought in the form

$$u(q) = u_0(q) + \varepsilon u_1(q) + \varepsilon^2 u_2(q) + \cdots$$

First two terms of $u(q)$ satisfy the following equations:

$$u_0(q) - (\tau/m)u_0'(q)u_0(q) = 0,$$
$$u_1(q) - (\tau/m)u_1'(q)u_0(q) - (\tau/m)u_0'(q)u_1(q) = 1 - q^2. \tag{11.70}$$

Solutions of the former are

$$u_0^{(1)} = 0, \qquad u_0^{(2)} = (m/\tau)q + c_1.$$

Corresponding to these $u_0^{(1)}$ and $u_0^{(2)}$ solutions to the last equation (11.70) are

$$u_1^{(1)}(q) = 1 - q^2$$

and

$$u_1^{(2)}(q) = \tfrac{1}{2}q^2 - c_1\tau\cdot q/m + (c_1^2\tau^2/m^2 - 1)\ln(c_1 + mq/\tau) + c_2.$$

So, for equation (11.69) we have the following approximate solutions:

$$q^{(1)} = \mathrm{th}(\varepsilon(t + c_4)), \qquad \int^{q(2)} \mathrm{d}x[mx/\tau + c_1 + \varepsilon u_1^{(2)}(x)]^{-1} = t + c_3,$$

where $c_i (i = 1, \ldots, 4)$ are integration constants which may be defined by the initial and other physical conditions for the given problem.

We notice that the conditions of propagation of the electron-positron component are different from those for protons and nuclei since relativistic electrons (and positrons) undergo considerably larger synchrotron and Compton energy losses. Therefore, the fundamental equation (11.60) describing the cosmic-ray proton-nuclear component propagation is changed for the electron-positron component but for its propagation, it seems, equations of the type of (11.67) and (11.69) are more suitable (see below). Thus, we assume that the electron acceleration mechanism differs from the acceleration mechanism for protons and nuclei. As is mentioned by Ginzburg and Ptuskin (1976), it is most probable that the electron-positron component is completely *secondary*, i.e., it is generated by cosmic-rays (protons and nuclei) in the interstellar medium and perhaps in the sources.

There are a few approaches devoted to explanation of the particle acceleration mechanism in stochastic electrodynamics (Rueda, 1977; 1978; Rueda and Lecompte, 1979) and to the motion of a particle with the shadow of the vacuum fluctuation (Cheon, 1982). Unlike these approaches, in our model the particle-acceleration mechanism is caused by fluctuations in metric and depends on the value of the fundamental length l that may be obtained from the experimental data on measuring ultra-high energy protons. In other words, ultra-high energy particles are 'carriers' of space-time properties at small distances, i.e., they *perturb* space-time around themselves and make the metric stochastic (fluctuational). This perturbative effect of the particles on the metric gives rise to fluctuations such that the deviation from the Minkowski metric $\eta_{\mu\nu}$ near the particles has a nonzero mean. As a result of the last fact, the particles move with acceleration due to self-interacting gravitational field and acquire very high energies during the evolution of the Universe.

The most important distinctive outline of our acceleration mechanism presented here is its smooth character of acceleration by which a high final energy of particles is achieved little by little – continuously or in small portions. This circumstance is especially essential for the explanation of existing ultrarelativistic heavy nuclei components in cosmic rays. It is clear that compound nucleus may conserve its structure as a whole in the acceleration process if and only if it gains energy in sufficiently small portions. From this point of view it is naturally assumed that acceleration has a macroscopic character and is not the consequence of some elementary act or scattering process as it might be, for example, if it occurred in ultrarelativistic percussion waves.

Now we discuss other unsolved questions of interest including the problem of the energy spectrum of the cosmic rays and the ratio of the intensities of the electron component to the proton component at the same energy level, within the framework of our scheme proposed above. It should be noted that these two problems play an important role in the construction and choice of the concrete model for the propagation mechanism of the cosmic rays, which must reproduce all the experimental data. Thus, in order to characterize our mechanism with respect to the others and the experimental data, we will consider, first, the problem of the energy spectrum of the cosmic rays. It appears that our acceleration mechanism leads to the generation of particles with the *power-law spectrum*

$$N(E)\,dE = KE^{-\gamma}\,dE, \tag{11.71}$$

where N is a particle concentration and K is some constant. Indeed, by the same proposition discussed by Ginzburg and Syrovatskii (1964) it is assumed that the particles start to move with acceleration with equal probability at any time moment, and the probability density of finding the particles with 'age' in the interval t, $t + dt$, is

$$dW = (1/T)\exp(-t/T)\,dt, \tag{11.72}$$

where T is the average lifetime of the particles during which the particles undergo acceleration mechanism (for discussion of T see below there T denotes T_{ef}). Then, according to the formula (11.63), we have

$$\frac{dE}{dt} = E\frac{1}{1-q^2}q\dot{q}, \quad (q = v/c). \tag{11.73}$$

Here $\dot{q}$ and q are given by equations (11.60) and (11.61). Thus,

$$\frac{dE}{dt} = E\varepsilon\,\mathrm{th}(\varepsilon t), \tag{11.74}$$

where we assume $q(0) = v_0/c = 0$. Solution of this equation is simple, that is

$$E = E_0\,\mathrm{ch}(\varepsilon t), \tag{11.75}$$

here E_0 is the initial energy. But in the ultrarelativistic case we can take $v \sim c$ or $\varepsilon t \to \infty$. i.e., $q \sim 1$ or $\mathrm{th}(\varepsilon t) \to 1$ in equations (11.73) and (11.74). In this approximate case, instead of (11.74) and (11.75), we have

$$\frac{dE}{dt} = \varepsilon E, \qquad E = E_0\exp(\varepsilon t), \qquad E_{max} = E_0\exp(\varepsilon t_{max}), \tag{11.76}$$

where t_{max} is the maximum acceleration time.

Taking into account equation (11.76) it is easy to verify that

$$dt = dE/\varepsilon E, \qquad t = (1/\varepsilon)\ln(E/E_0)$$

and

$$dW \sim N(E)\,dE \sim (1/\varepsilon T)E_0^{1/\varepsilon T}E^{-(1+1/\varepsilon T)}\,dE, \quad (\gamma = 1 + 1/\varepsilon T). \tag{11.77}$$

A more exact equation (11.74) gives

$$dW \sim (1/\varepsilon T)2^{-1/\varepsilon T}E_0^{-1+1/\varepsilon T}E^{-1/\varepsilon T}(E/E_0 - 1)\,dE. \tag{11.78}$$

According to the above deduction and reference of Ginzburg and Syrovatskii (1964), we obtain for the proton component

$$\varepsilon_p = (9/8\sqrt{3})cl^2 r_p^{-3} \quad \text{and} \quad T_p \sim (1 \div 3) \times 10^8 \text{ year.}$$

Thus, $\varepsilon_p T_p \sim 0.645$ for $T_p = 2.2 \times 10^8$ year and therefore

$$\gamma_p = 2.55 \tag{11.79}$$

which coincides with the experimental value of γ_{exp}.

Within our scheme, it is naturally assumed that the acceleration mechanism and the value of $\varepsilon_N T_N$ for the nuclei component are the same as the proton component, so that in this case the energy spectrum and the value of γ are determined by the analogous formulas of (11.77), (11.78), and (11.79).

With regard to the acceleration mechanism of the electron-positron component, energy losses should be taken into account, and it should be assumed that, as mentioned above, the electron-positron component is secondary, i.e., it is generated by means of the processes $\pi \to \mu \to e$ [for detail see Ginzburg and Syrovatskii, 1964)]. Here we study the equation of the type of (11.67) and (11.69) in the relativistic case, when the electron undergoes energy losses connected with bremsstrahlung in the external magnetic field

$$H_{\text{G}} \sim (3 \div 10) \times 10^{-6} \text{ Oe (Oersted)}$$

or (11.80)

$$H_{\text{MG}} \sim (3 \div 10) \times 10^{-8} \text{ Oe}$$

of the Galaxy or of the MetaGalaxy (Ginzburg and Syrovatskii, 1964). In this case the force (Landau and Lifshitz, 1971)

$$\mathbf{f} = -\frac{2}{3}\frac{e^4}{m^2 c^5}(F_{\mu\nu}u_\nu)^2 \mathbf{v} \tag{11.81}$$

is directed to the opposite direction of the particle velocity $\mathbf{v}$; choosing the latter as x-axis and exposing the expression (11.81), we have

$$f_x = -\frac{2}{3}\frac{e^4}{m^2c^4}\frac{qH_\perp^2}{(1-q^2)}, \quad (q = v/c).$$

Here it is assumed that the electric component of $F_{\mu\nu}$ is zero. Then, from the equation (11.67) we obtain the following analogous equation of (11.69):

$$\dot{q} = \varepsilon_e(1-q^2) - \frac{2}{3}\left(\frac{e^2}{mc^2}\right)^2 H_\perp^2 \left(\frac{q}{mc}\right)(1-q)^{1/2}.$$

Thus, equation (11.73) in this case acquires the form

$$\frac{\mathrm{d}E}{\mathrm{d}t} = \varepsilon_e Eq - \delta E^2 q^2, \tag{11.82}$$

where

$$\delta = \tfrac{2}{3}c\left(\frac{e^2}{mc^2}\right)^2 H_\perp^2(mc^2)^{-2} = 0.98 \times 10^{-3} H_\perp^2 (mc^2)^{-2}\ \text{eV/sec}.$$

Here the energy E and H are expressed in electron volts and oersteds, respectively. In the ultrarelativistic case $q \sim 1$ we have the approximate equation

$$\frac{dE}{dt} = \varepsilon_e E - \delta E^2. \tag{11.83}$$

This is the *Bernoulli equation.* The substitution

$$y(t) = [E(t)]^{-1} \tag{11.84}$$

leads to the new differential equation for $y(t)$,

$$\dot{y} = -\varepsilon_e y + \delta, \tag{11.85}$$

which is soluble because it is linear in $y(t)$. The solution of this inhomogeneous linear differential equation is

$$y(t) = I^{-1}(t)\left[c_1 + \delta \int^t d\rho \exp(\varepsilon_e \rho)\right] = I^{-1}(t)\left[c_1 + \frac{\delta}{\varepsilon_e}\exp(\varepsilon_e t)\right],$$

where

$$I(t) = \exp\left\{\varepsilon_e \int^t d\rho\right\} = \exp(\varepsilon_e t)$$

is the integrating factor for equation (11.85). Further, according to formula (11.84), we have

$$E(t) = \exp(\varepsilon_e t)[c_1 + (\delta/\varepsilon_e)\exp(\varepsilon_e t]^{-1}, \tag{11.86}$$

here the integration constant c_1 is given by an initial condition, that is $E(0) = E_0$. Thus,

$$E(t) = \exp(\varepsilon_e t)[(1/E_0) + (1/E_{\max})(\exp(\varepsilon_e t) - 1)]^{-1}. \tag{11.87}$$

From this it is easy to see that

$$E(t)|_{t\to\infty} \sim E_{\max} = \varepsilon_e/\delta.$$

We observe here the interesting fact that within our hypothesis concerning the acceleration mechanism of cosmic-ray particles, the *electron energy is restricted by the maximum value of* $E_{\max}$. This means that for some time T_e during which our acceleration mechanism gives rise to increasing the electron energy up to $E_{\max}$, after which T_e the electron energy is not increased and remains constant at the order of $E_{\max}$. In other words, it may be said that our acceleration mechanism is switched off, when the electron energy reaches $E_{\max}$, i.e., for

$t \gtrsim T_e$. This effect is caused by the ratio of energy gains and losses due to the space-time structure near particles, and bremsstrahlung in the external magnetic field. When, the electron energy $E(t)$ reaches E_{max}, this ratio becomes unit, and energy gained compensates for energy losses. We now calculate the value of E_{max} which depends on ε_e and $H_\perp$. As the proton case, the former is given by $\varepsilon_e = (9/8\sqrt{3})\, cl^2 r_e^{-3}$, where r_e may be understood as an effective Schwarzschild radius connected with the internal structure of the leptons. Some theoretical prediction (Li, 1982) gives $r_e \sim 4 \times 10^{-17}$ cm. Therefore, $\varepsilon_e \sim 8 \times 10^{-7}$ sec^{-1} and by expression (11.80) for the magnetic field $H_\perp$, we have

$$E_{max} \sim \begin{cases} (24 \div 2.1) \times 10^{18}\,\text{eV} & \text{for } H_G^\perp \sim (3 \div 10) \times 10^{-6}\,\text{Oe}, \\ (24 \div 2.1) \times 10^{22}\,\text{eV} & \text{for } H_{MG}^\perp \sim (3 \div 10) \times 10^{-8}\,\text{Oe}. \end{cases}$$

Now we turn to the calculation of the energy spectrum for the electron component. Taking into account equation (11.83) and formula (11.87), we have

$$\mathrm{d}t = \frac{\mathrm{d}E}{\varepsilon_e \cdot E(1 - E/E_{max})}, \qquad t = \frac{1}{\varepsilon_e} \ln\left\{\frac{E_{max}/E_0 - 1}{E_{max}/E - 1}\right\}.$$

Substituting these expressions into (11.72), we get

$$\mathrm{d}W_e \sim N_e(E)\,\mathrm{d}E \sim \frac{\mathrm{d}E}{T_e\varepsilon_e} E_0^{1/T_e\varepsilon_e}\left(1 - \frac{E_0}{E_{max}}\right)^{-1/T_e\varepsilon_e} \times$$

$$\times\, E^{-1-1/T_e\varepsilon_e}(1 - E/E_{max})^{-1+1/T_e\varepsilon_e}. \tag{11.88}$$

Here we again observe the power-law spectrum for the electron-position component. Assuming as above $T_e\varepsilon_e \sim 1$, it is easy to satisfy the experimental value of $\gamma \sim 2 \div 3$.

Now it is relevant to discuss the ratio of the intensities $I_i (i = e, p)$ of the electron component to the proton component at the same energy level within our scheme. In the case of the isotropic radiation the particle concentration N_i and the intensity I_i are related by the formula (see Ginzburg and Syrovatskii, 1964).

$$I_i(E) = \frac{1}{4}\frac{v_i}{\pi} N_i(E), \qquad v_i = c\left(1 - \left(\frac{m_i c^2}{E}\right)^2\right)^{1/2}$$

Here the intensity of the particle with energy larger than the given value E is determined by

$$I_i(>E) = \int_E^\infty I_i(x)\mathrm{d}x = \frac{c}{4\pi}\int_E^\infty \mathrm{d}x\, N(x)\left[1 - \left(\frac{m_i c^2}{x}\right)^2\right]^{1/2}$$

Thus, the intensities of the electron and proton components are given by the following formulas

$$I_p \sim \frac{c}{4\pi}\left(\frac{1}{T_p\varepsilon_p}\right)(E_0^p)^{1/T_p\varepsilon_p}\int_E^\infty \mathrm{d}x\, x^{-1-1/T_p\varepsilon_p}\left[1 - \left(\frac{Mc^2}{x}\right)^2\right]^{1/2}$$

and

$$I_e \sim \frac{c}{4\pi}(E_0^e)^{1/T_e\varepsilon_e}\left(\frac{1}{T_e\varepsilon_e}\right)\left(1-\frac{E_0^e}{E_{\max}}\right)^{-1/T_e\varepsilon_e}\int_E^{E_{\max}} \mathrm{d}x\, x^{-1-1/T_e\varepsilon_e} \times$$

$$\times\,(1-x/E_{\max})^{-1+1/T_e\varepsilon_e}[1-(mc^2/x)^2]^{1/2},$$

respectively
It is easy to verify that the following asymptotic estimations may be obtained for both cases:

$$I_p \lesssim \frac{c}{4\pi}(E_0^p/E)^{1/T_p\varepsilon_p}$$

and

$$I_e \lesssim \frac{c}{4\pi}\left(\frac{E_0^e}{E}\right)^{1/T_e\varepsilon_e}\left(1-\frac{E_0^e}{E_{\max}}\right)^{-1/T_e\varepsilon_e}.$$

Taking into account that

$$\gamma - 1 = 1/T_p\varepsilon_p \sim 1/T_e\varepsilon_e \sim 1.4 \div 1.5 \quad \text{and} \quad E_0^e \ll E_{\max},$$

we get

$$I_e/I_p \lesssim (E_0^e/E_0^p)^{\gamma-1}.$$

It is naturally suggested that $E_0^p = M_p c^2$ and $E_0^e = 34$ MeV. The latter case is obtained by the assumption that the electron component is secondary and is generated by the processes $\pi \to \mu \to e$. Therefore, the average energy of electrons generated by $\mu \to e + \nu + \tilde{\nu}$ decay is just $E_0^e = 34$ MeV. Thus,

$$I_e/I_p \lesssim (E_0^e/E_0^p)^{\gamma-1} = \begin{cases} 0.96\% & \text{for } \gamma = 2.4 \\ 0.69\% & \text{for } \gamma = 2.5 \end{cases}$$

which coincides with the average measured ratio being about 1% obtained by experimentalists (see Meyer *et al.*, 1974).

Finally, we discuss some questions concerning the value of the effective time $T_{\text{ef}} \sim t_{\max}$ of acceleration during which the cosmic-ray particle's (mainly protons) energy reaches 10^{19}–10^{20} eV. It is usually suggested that one could expect $T_{ef} \lesssim T_u$, where T_u is the Hubble time. In particular, this assumption is based on the fact that in early cosmological epochs, particles would also apparently *lose energy adiabatically*, due to the *expansion of the Universe*, so that energy gained at very early times would not be retained. The effective time T_{ef} of acceleration will therefore be appreciably less than the Hubble time. We notice that this problem may be solved within the framework of our acceleration mechanism proposed above. Indeed, according to formula (11.63) the maximum energy of the particle is $E = (mc^2/2)\exp(\varepsilon t_{\max}\cdot c)$, where $t_{\max} \sim T_{\text{ef}}$ and $c\cdot\varepsilon T_{\text{ef}} = (9/8\sqrt{3})cl^2 r_c^{-3} T_{\text{ef}}$. Above we have assumed $r_c = r_0 \sim 0.8 \times 10^{-13}$ cm which gives $E \sim 10^{20}$ eV for $T_{\text{ef}} = T_{\text{cr}} \sim T_u$. However, generally speaking the value of the radius r_c is a semi-empirical one in our approach and may be changed. From this it is easy to see that the *very slightest change* in the value of r_c gives rise to the *enormous range of the energy* of the cosmic-ray particles. Thus, by some choice of r_c, say $r_c \sim 10^{-14} \div 10^{-15}$ cm we can obtain $E \sim 10^{19} - 10^{20}\, eV$ for $T_{ef} \ll T_u$. Here it is assumed that $l = l_{Pl} = 1.62 \times 10^{-33}$ cm.

11.8. Space-Time Structure near Particles and its Influence on Particle Behavior

11.8.1. *Introduction*

In this section we consider equation (11.59) and discuss the possibility of an interrelationship between the properties of space-time structure and the dynamics of particles. This problem is not new to scientific literature. For example, Einstein's (1924) idea to use the random fluctuations of the metric field $g_{\mu\nu}(x)$ as the origin of the real quantum forces which justify the stochastic interpretation of quantum mechanics were considered by Frederick (1976); Vigier (1982) and others (see, the review due to Vigier, 1982). Moreover, an analogous idea on the question has been discussed by Efinger (1981), who wrote: 'freely moving particles are represented by solitary waves which, by definition, preserve their shape, then one could hypothesize that associated with these waves is a Riemannian metric $g_{\mu\nu}(x)$ which is nearly singular in a region corresponding to the width of the solitary wave-amplitude (an idea reminiscent of an old concept by Einstein, 1967)'. According to this point it is assumed that the particles in question are geometrical objects on a Riemannian space-time.

It is well known that in Einstein's theory of gravitation, space-time structure near matter is changed and differs from flat space-time (for illustration see Figure 16). The validity of this assumption in a macroscale was demonstrated by the experiment measuring the deviation of light near the Sun, i.e., a ray of light going past the Sun is accordingly undergoing deflection at the rate of 1.7 seconds of arc. However, the question of what effect would be expected in the microworld is still open to discussion within this assumption, and in particular the influence of space-time structure near particles on their dynamic behavior.

In accordance with the previous section, the next assumption is that a form of $g_{\mu\nu}(l^2/x^2)$ in (11.57) determines the space-time structure near particles and gives a general effect on the particle behavior. In other words, the character and type of the particle motion is essentially dependent on the space-time structure near moving particles. For example, we propose that the flat space-time structure gives rectilinear motion, and soliton-like, stochastic and other types of particle motion may be caused by different forms of space-time structures around the particle.

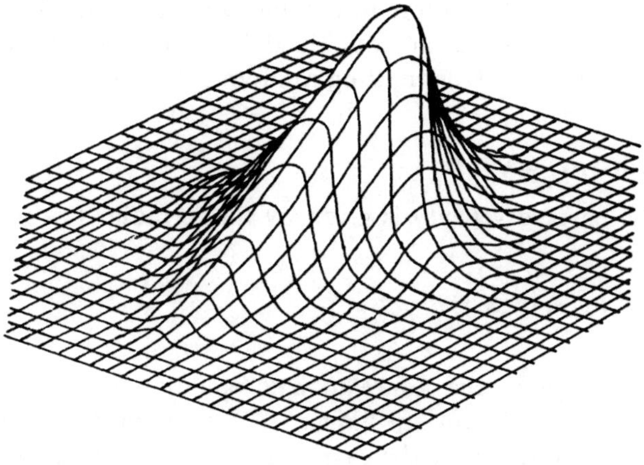

Fig. 16. The illustration of changing space-time structure near the particle.

Our aim is now to discuss equation (11.59) in the nonrelativistic limit by the appropriate choice of the form of the function $\phi(x, l)$. It is important to notice that the interrelationship between properties of space-time structure at small distance and the dynamic behavior of the particles appears at a deeper level and requires careful investigations. In this respect our approach is modest and belongs to the semi-empirical level.

11.8.2. *Stochastic Behavior of Particles and its Connection with Stochastic Mechanical Dynamics*

It is convenient to study the particle dynamic behavior arising from equation (11.59) and depending on the concrete form of the function $\phi(x, l)$ in the *nonrelativistic limit*. Before approaching this limit for equation (11.59) we make some remarks. Generally speaking, according to the hypothesis of the curved space-time structure around particles, the concept of their trajectory and velocity should be changed and generalized appropriately. In other words, it is quite possible that owing to the metric fluctuations in the presence of the particles, the description of the particles has a universal character in Nature and requires probabilistic methods. In particular, the particle velocity depends not only on the time variable t but also on the spatial variables $\mathbf{x}$: $\mathbf{v}(t) \to \mathbf{v}(\mathbf{x}, t)$. We suggest that for a complete description of the particle motion within our approach, and at the same time with particle velocity $\mathbf{v}(\mathbf{x},t)$, two more quantities $\rho(\mathbf{x},t)$ and $\mathbf{u}(\mathbf{x},t)$ should be introduced, where $\rho(\mathbf{x},t)$ is the probability density of finding the particle at point $\mathbf{x}$ and at time t, $\mathbf{u}(\mathbf{x}, t)$ is the stochastic velocity given by the formula $\mathbf{u}(\mathbf{x}, t) = D\nabla \ln \rho(\mathbf{x}, t)$ (see Chapter 8).

Thus, our basic idea is the following: fluctuations in metric take place *everywhere* (increasing in the presence of the particles), and may play the role of the origin of the real 'quantum' forces and lead to the random behavior of the particles; their dynamics are described by nonlinear partial differential equations of a type of (11.59) admitting random solutions and by the equation of continuity for $\rho(\mathbf{x}, t)$:

$$\partial\rho/\partial t + \mathrm{div}(\rho\mathbf{v}) = 0. \tag{11.89}$$

In this case the particle velocity is given by the formula

$$\mathbf{v}(t) = \int \mathrm{d}^3x\, \rho(\mathbf{x}, \mathbf{t})\mathbf{v}(\mathbf{x}, t).$$

In order to obtain the nonrelativistic equation of motion in our approach we go over to the formal limit $c \to \infty$ in equation (11.59). The time component of this equation in the case of $c \to \infty$ acquires the form

$$\mathbf{v}\cdot\mathbf{\nabla} \ln \phi = 0. \tag{11.90}$$

The passage to the limit $c \to \infty$ in the spatial component of (11.59) gives the following relation:

$$\frac{1}{2}\frac{\partial}{\partial t}\mathbf{v}^2 + \mathbf{v}^2\frac{\partial \ln \phi}{\partial t} = 0. \tag{11.91}$$

Here taking into account relation (11.90), from this equation we see that if the function ϕ does not depend on the time variable, the particle velocity is constant and the particle moves along the rectilinear trajectory. This situation also takes place at the limit, when the value of the fundamental length l is neglected, thus by our assumption $\phi(x, l) = 1$ at $l \to 0$ $(x = \mathbf{x}, t)$.

After simple integration of (11.91) we have

$$v^2\phi^2 = \text{const}. \tag{11.92}$$

Here the constant may be obtained by the initial condition, say $v^2\phi^2|_{t=0} = v_0^2$. Thus, the particle velocity $\mathbf{v}(\mathbf{x}, t)$ and the form of the function $\phi(x, l)$ are related by the formula

$$v^2 = v_0^2/\phi^2. \tag{11.93}$$

Physically this relationship means that by knowing the space-time structure near the particle we can calculate its velocity (generalized) and, on the contrary, by the value of the particle velocity one can try to build the space-time structure near the moving particle. Thus, it seems, there exists a profound connection between these two concepts and they enter as a *single inseparable entity into our scheme*. Therefore, generally speaking, the function $\phi(x, l)$ should be dependent on the properties of the particle: on its mass, an effective size λ (in the quantum mechanical case $\lambda = \hbar/mc$) and the initial velocity v_0 and so on. In this book we consider three cases for the function

$\phi(x, l) \Rightarrow \phi(x, t; l, \lambda, v_0)$ (here the space-time dimension is two; x, t):

$$\text{I.}\quad \phi_1(x, t; l, \lambda, v_0) \to \infty \quad \text{for } t \to \infty, \quad \text{and} \qquad (11.94)$$
$$\phi_1(x, t; l, \lambda, v_0) \to 0 \quad \text{at } |x| \to \infty,$$
$$\text{II.}\quad \phi_2(x, t; l, \lambda, v_0) \to 1, \quad \text{for } t, |x| \to \infty, \qquad (11.95)$$
$$\text{III.}\quad \phi_3(x, t; l, \lambda, v_0) \to \infty \quad \text{for } t, |x| \to \infty. \qquad (11.96)$$

We assume here the equality $\phi_i(x, t; l, \lambda, v_0) = 1$ at $l \to 0$, is fulfilled for all three cases. Now we consider the first case and simple form

$$\phi_1(x, t; l, \lambda, v_0) = (1 + bt^2)[1 + (b/v_0)xt]^{-1} \qquad (11.97)$$

gives

$$v = (v_0 + btx)(1 + bt^2)^{-1}, \qquad (11.98)$$

here $b = 4D(l/\lambda^3)$, D is a universal constant dimension of [cm^2/sec], obviously that is $D \sim \hbar/m$. Solution of the equation of continuity (11.89) with the velocity (11.98) and the initial condition

$$\rho(x, t = 0) = a^{-1}\pi^{-1/2} \exp(-x^2/a^2)$$

has the form

$$\rho(x, t) = a^{-1}\pi^{-1/2}(1 + bt^2)^{-1/2} \exp\{-(x - v_0 t)^2/a^2(1 + bt^2)\}, \qquad (11.99)$$

where $a = (l/\lambda^3)^{-1/2}$. According to the above deduction in our case the stochastic velocity of the particle is given by

$$u(x, t) = D\nabla \ln \rho(x, t) = -b^{1/2}(x - v_0 t)(1 + bt^2)^{-1}. \qquad (11.100)$$

Mean value and dispersion of the quantities $v(x, t)$ and $u(x, t)$ are determined by the formulas:

$$\langle v(x, t)\rangle = \int dx\, v(x, t)\, \rho(x, t) = v_0, \qquad (11.101)$$

$$\langle u(x, t)\rangle = \int dx\, u(x, t)\, \rho(x, t) = 0,$$

$$\text{Dis. } v^2(x, t) = \tfrac{1}{2}a^2 b^2 t^2 (1 + bt^2)^{-1}.$$

Moreover, by definition

$$\langle X(t)\rangle = \int dx\, x\, \rho(x, t) = v_0 t,$$

$$\text{Dis. } X^2(t) = \tfrac{1}{2}a^2(1 + bt^2), \qquad (11.102)$$

$$\langle v^2(x, t)\rangle = v_0^2 + \tfrac{1}{2}a^2 b^2 t^2 (1 + bt^2)^{-1}.$$

From formulas (11.98)–(11.102) we conclude immediately that the simple space-time structure near moving particles which is determined by expression (11.97) gives completely stochastic mechanical results obtained in Section 9.3 if the constant D is taken equal to the diffusion coefficient $\nu = \hbar/2m$. In this case the space-time metric near stochastic particles takes the form

$$ds^2 = \phi_1^2(x, t; l, \lambda, v_0)(c^2\,dt^2 - dx^2), \tag{11.103}$$

where $\phi_1(x, t; l, \lambda v_0)$ is given by formula (11.97).

It should be noted that this metrical form gives rise to the random behavior of particles (like Brownian motion), i.e., the particle is forced to move stochastically during its propagation in space-time. But, on the other hand, particles always take care of the space-time structure around themselves in order to move further. It is simple the essence of the dialectical unity of motion and space-time.

11.8.3. *Soliton-Like Behavior of Particles*

Now we consider the cases (11.95) and (11.96), and choose the following simple forms:

$$\phi_2(x, t; l, \lambda, v_0) = 1 + l^2[\lambda^2 + (x - v_0 t)^2]^{-1}, \tag{11.104}$$

$$\phi_3(x, t; l, \lambda, v_0) = \left(1 + \frac{l}{\lambda}\right)\frac{\operatorname{sh} b}{\operatorname{ch}^3 b}\operatorname{ch}^3\left(l\frac{x - v_0 t}{\lambda^2} + b\right)\operatorname{sh}^{-1} \times$$
$$\times \left(l\frac{x - v_0 t}{\lambda^2} + b\right) \tag{11.105}$$

where b is a positive constant. These expressions satisfy conditions (11.95), (11.96) and $\phi_i = 1$ $(i = 2, 3)$ at $l = 0$, for any x and t. The following generalized velocities of the particle:

$$v_2 = v_0/\phi_2 = v_0[\lambda^2 + (x - v_0 t)^2][\lambda^2 + l^2 + (x - v_0 t)^2]^{-1} \quad \text{and} \quad v_3 = v_0/\phi_3,$$

correspond to formulas (11.104) and (11.105), respectively. From these velocities v_2 and v_3 we have generalized 'trajectories':

$$[v_0 t - X_2(x, t)] = \frac{l^2}{(\lambda^2 + l^2)^{1/2}}$$
$$\operatorname{arc\,tg}\left(\frac{v_0 t}{(\lambda^2 + l^2)^{1/2}}\frac{1}{1 + [x(x - v_0 t)]/(\lambda^2 + l^2)}\right) \tag{11.106}$$

$$X_3(x, t) = \frac{\lambda^2}{2l}\left(1 + \frac{l}{\lambda}\right)^{-1}\operatorname{ch}^3 b\operatorname{sh}^{-1} b \times$$
$$\times\left\{\operatorname{sech}^2\left(\frac{l}{\lambda^2}(x - v_0 t) + b\right) - \operatorname{sech}^2\left(\frac{lx}{\lambda^2} + b\right)\right\}. \tag{11.107}$$

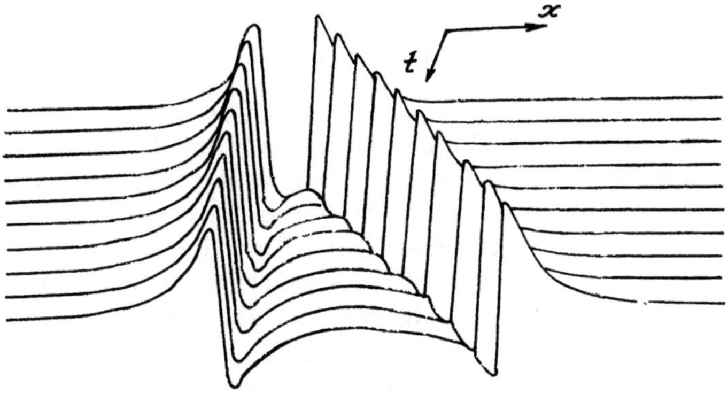

Fig. 17. Solitary type wave corresponding to formula (11.106). We see here two precise maximum-waves, the right part of which moves at a constant velocity without change of shape, while the left part remains near the point $x = 0$. From this plot and formula (11.106) one can suggest that in nature there may exist a type of particle motion which consists of two parts: rectilinear and soliton-like; however, because of the small value of the fundamental length l the latter type motion is not observed in practice.

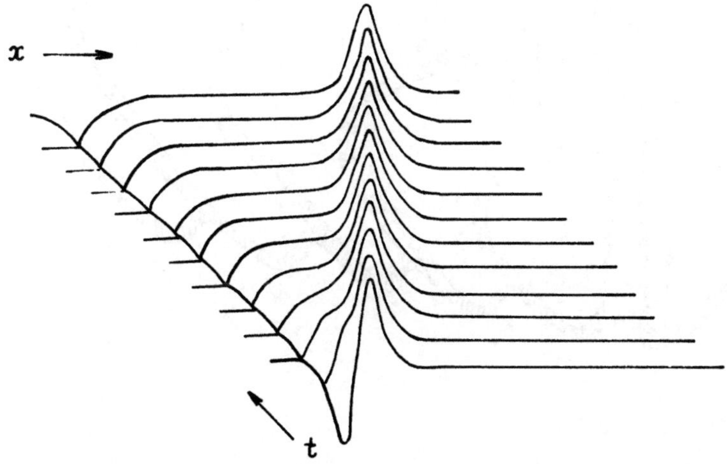

Fig. 18. Behavior of the 'trajectory' (11.107) at the value of $\lambda = l$. This plot is a hump-shaped wave exactly as Scott Russell observed (see Dodd *et al.*, 1983), which moves along the classical trajectory $x = v_0 t$ at a constant velocity without change of shape. For larger values of λ the wave is broader and higher, but becomes thinner and shorter the smaller λ becomes. The left part of this figure, it appears, has no physical meaning which is caused by the initial condition of the problem and corresponds to some minimum located near the point $x = 0$. This motionless minimum gives a 'canal' over the time variable, a plot of which is not seen in the figure.

It is easy to verify that

$$X_2(x, t) \to x_2(t) = v_0 t, \qquad X_3(x, t) \to x_3(t) = v_0 t$$

in the limit $l \to 0$. Behaviors of the functions $[v_0 t - X_2(x, t)]/l$ and $X_3(x, t)/l$ are shown in Figures 17 and 18, respectively. From these figures we observe that the 'trajectories' obtained by the functions ϕ_2 and ϕ_3 correspond to the soliton-like motion of the particle. However, in the case of (11.106) because of the small value of the fundamental length l we do not, in fact, observe the deviation from the rectilinear particle trajectory for any quantity λ, even up to $\lambda - l$. But the situation is different in the case of (11.107). When $\lambda \lesssim 10^{-15} \div 10^{-16}$ cm, the amplitude of the process (11.107) is an order of unit, therefore we can, in principle, observe experimentally soliton-like particle behavior which is caused by the space-time structure near the moving particle.

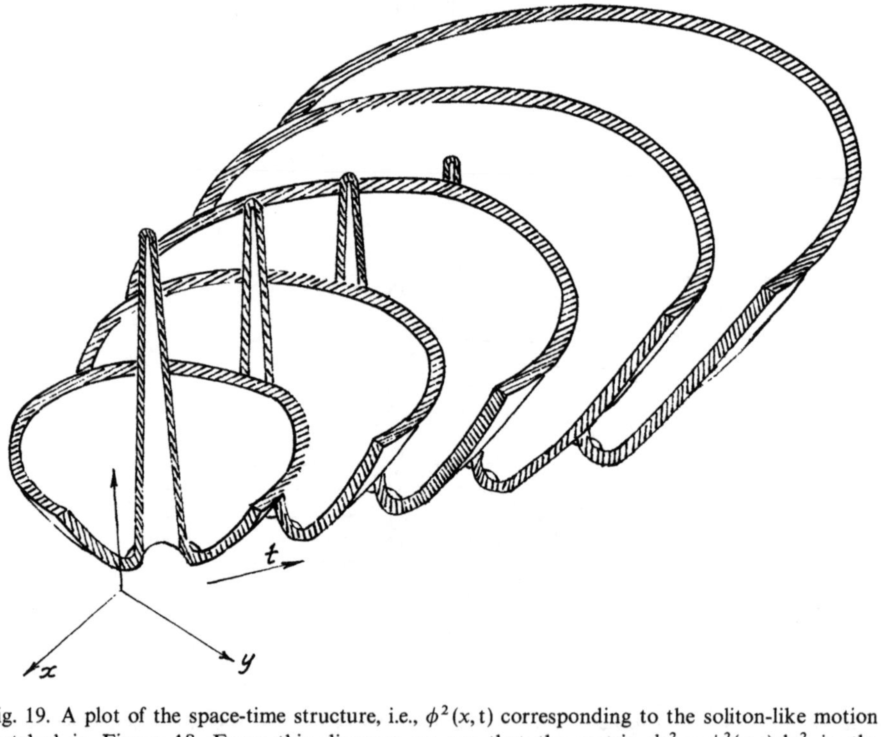

Fig. 19. A plot of the space-time structure, i.e., $\phi^2(x, t)$ corresponding to the soliton-like motion sketched in Figure 18. From this diagram we see that the metric $ds^2 = \phi^2(x, t)\,ds_0^2$ in the conformally flat space-time is singular in the region corresponding to the classical trajectory $x = v_0 t$. According to the diagram one can conclude that in order to hold a particle in a localized region during its propagation in space-time, or, for energy to be propagated in localized stable 'packets' without being dispersed, it needs enormous efforts for space-time, i.e., space-time sets up an infinitely higher barrier-wall around a particle.

It should be noted that case (11.94) essentially differs from the other two cases (11.95) and (11.96). Since in the first case the generalized trajectory $X_1(x, t)$ corresponding to the velocity $v_1(x, t)$ $[v_1(x, t) = \dot{X}_1(x, t)]$ becomes to infinity at $x \to \pm\infty$, i.e., the concept of trajectory in this case loses its significance and has no physical meaning, while the functions $[v_0 t - X_2(x, t)]$ and $X_3(x, t)$ are located along the rectilinear classical trajectory $x(t) = v_0 t$.

Finally, for illustration, the space-time structure $ds^2 = \phi^2(x, t)\eta_{\mu\nu}\, dx^\nu\, dx^\mu$, i.e., the function $\phi^2(x, t)$ corresponding to the particle behavior determined by formula (11.107) is shown in Figure 19. In our semi-empirical approach the question of a possible unique choice of the functions $\phi(x, t; l, \lambda, v_0)$ is not solved and seems to require another fundamental physical principle. At present, the formulation of this principle is not known and needs deeper study. However, we assume that this important problem may be solved alternatively, namely by the character of particle motion. It means that one can try to construct the space-time structure near the moving particle by means of the value of its velocity. For example, if we observe the rectilinear trajectory of the particle, we can conclude that the space-time structure near such a particle is flat. The role of that particle may be played by the light quantum-photon since according to the relation (11.92) if the constant in the right-hand side of (11.92) is equal to the velocity of light c then the current velocity $v(x, t)$ should be equal to c, therefore $\phi = 1$ everywhere at any t. Of course, we propose here that the light velocity in space-time is truly constant. If the last assumption is valid at small distances (or at very-high energies) then light does not disturb space-time around itself, which always remains flat. In contradistinction to the rectilinear motion other types of particle motion correspond to a curved space-time structure near particle under consideration.

In conclusion we notice that fluctuations in metric which give rise to stochastic and other types of particle motion, if they exist, would be detected by ultra-high energy particles if their wavelength λ were comparable to, or shorter than, the value of the fundamental length $l \sim 10^{-33}$ cm, in other words the properties of the particle with very small mass $m \ll M_{Pl} = \hbar/cl_{Pl} = 10^{-5}$ gm ($\lambda = \hbar/mc \to \infty$) would be essentially insensitive to the fluctuational structure of space-time.

Bibliography

Aharony, A., Imry, Y., and Ma, S.-K. (1976) 'Lowering of Dimensionality in Phase Transitions with Random Fields', *Phys. Rev. Lett.* **37**, 1364–1367.

Akama, K. (1981) 'Pregeometry Including Fundamental Gauge Bosons', *Phys. Rev.* **D24**, 3073–3081.

Akhiezer, A. I. and Berestetskii, V. B. (1965) *Quantum Electrodynamics*, Interscience, New York.

Akhiezer, A. I. and Berestetskii, V. B. (1981) *Quantum Electrodynamics* (in Russian), Nauka, Moscow.

Albeverio, S. and Høegh-Krohn, R. (1974) 'A Remark on the Connection Between Stochastic Mechanics and the Heat Equation', *J. Math. Phys.* **15**, 1745–1747.

Albeverio, S. and Høegh-Krohn, R. (1976) *Mathematical Theory of Feynman Path Integrals*, Springer, Berlin.

Alebastrov, V. A. and Efimov, G. V. (1973). 'A Proof of the Unitarity of S-Matrix in a Nonlocal Quantum Field Theory', *Comm. Math. Phys.* **311**, 1–24.

Alebastrov, V. A. and Efimov, G. V. (1974) 'Causality in Quantum Field Theory with Nonlocal Interactions', *Comm. Math. Phys.* **38**, 11–28.

Alebastrov, V. A., Efimov, G. V. and Seltzer, Sh. Z. (1973) 'Nonlocal Theory of the Electromagnetic and Weak Interactions with *W*-Boson', *Ann. Phys.* (*N.Y.*) **76**, 251–280.

Alexandrov, P. and Hopf, H. (1935) *Topologie*, Vol. 1, Springer, Berlin.

Alfaro, J. and Sakita, B. (1983) 'Derivation of Quenched Momentum Prescription by Means of Stochastic Quantization', *Phys. Lett.* **121B**, 339–344.

de Alfaro, V., Fubini, S. and Furlan, G. (1980) 'A New Approach to the Theory of Gravitation', *Nuovo Cimento* **B57**, 227–252.

de Alfaro, V., Fubini, S. and Furlan, G. (1976) 'A New Classical Solution of the Yang–Mills Field Equations', *Phys. Lett.* **65B**, 163–166.

Andryushin, V. I., Bilenky, S. M. and Gershtein, S. S. (1971) 'On a Possible Method of a Measurement of Neutrino Electromagnetic Formfactor' *Pisma JETP* **13**, 573–576.

Arbuzov, B. V. (1975) 'High-Energy Weak Interactions', in *Procedings of the 1975 JINR-CERN School of Physics*, Alushta, USSR, Preprint of JINR, E2-9086, Dubna.

Aron, J. C. (1981) 'Stochastic Foundation of Microphysics. A Critical Analysis', *Found. Phys.* **11**, 699–720.

Aron, J. C. (1966) 'Stochastic Processes in Microphysics in Connection with Relativity', *Progr. Theoret. Phys.* **35**, 147–171.

Aspect, A., Dalibard, J. and Roger, G. (1982a) 'Experimental Test of Bell's Inequalities Using Time-Varying Analysis', *Phys. Rev. Lett.*, **49**, 1804–1807.

Aspect, A., Grangier, P. and Roger, G. (1982b) 'Experimental Realization of Einstein–Podolsy–Rosen–Bohm Gedankenexperiment: A New Violation of Bell's Inequalities', *Phys. Rev. Lett.* **49**, 91–94.

Bailey, J. *et al.* (CERN-Mainz-Daresbury collaboration) (1979) 'Final Report on the CERN Muon

Storage Ring Including the Anomalous Magnetic Moment and the Electric Dipole Moment of the Muon, and a Direct Test of Relativistic Time Dilation', *Nuclear Phys.* **B150**, 1–75.

Ballentine, L. E. (1970) 'The Statistical Interpretation of Quantum Mechanics', *Rev. Modern Phys.*, **42**, 358–380.

Baltay, C. (1979) 'Recent Results from Neutrino Experiments in Heavy Preon Bubble Chambers', in *Proceedings of the 1979 JINR-CERN School of Physics, Dobogokö, Hungary.*

Banner, M. *et al.* (UA2 Collaboration) (1983) 'Observation of Single Isolated Electrons of High Transverse Momentum in Events with Missing Transverse Energy at the CERN pp Collider', *Phys. Lett.* **122B**, 476–485.

Barber, D. P. *et al.* (1979a) 'Test of Quantum Electrodynamics at $s^{1/2} = 13$ and 17 GeV', *Phys. Rev. Lett.* **42**, 1110–1113.

Barber D. P. *et al.* (1979b) 'Test of Universality of Charged Leptons', *Phys. Rev. Lett.* **43**, 1915–1918.

Barber, D. P. *et al.* (1980) 'Experimental Study of Heavy Charged Leptons and Search for Scalar Partners of Muons at Petra (12 GeV $\leqslant E_{\text{c.m}} \leqslant$ 36 GeV)', *Phys. Rev. Lett.* **45**, 1904–1907.

Barbieri, R., Nanopoulos, D. V., Morchio, G. and Strocchi, F. (1980a) 'Neutrino Masses in Grand Unified Theories', *Phys. Lett.* **90B**, 91–97.

Barbieri, R., Ellis, J. and Gaillard, M. K. (1980b) 'Neutrino Masses and Oscillations in SU(5)', *Phys. Lett.*, **90B**, 249–252.

Bardin, D. Yu. and Mogilevsky, O. A. (1974). 'On the Neutrino Charge Radius', *Lett. Nuovo Cimento* **9**, 549–553.

Barger, V., Whisnant, K., Cline, D. and Phillips, R. J. N. (1980a) 'Possible Indications of Neutrino Oscillations', *Phys. Lett.* **93B**, 194–198.

Barger, V., Whisnant, K. and Phillips, R. J. N. (1980b) 'Three-Neutrino Oscillations and Present Experimental Data', *Phys. Rev.* **D22**, 1636–1646.

Bargmann, V. and Wigner, E. P. (1948). 'Group Theoretical Discussion of Relativistic Equations', *Proc. Nat. Acad. Sci. U.S.A.* **34**, 211–.

Bartel, W. *et al.* (JADE collaboration) (1980) 'Test of Quantum Electrodynamics at Petra', *Phys. Lett.* **92B**, 206–210.

Baseyan, G. Z., Matinyan, S. G. and Savvidi, G. K. (1979) 'Nonlinear Plane Waves in Yang-Mills Massless Theory', *Pisma JETP* **29**, 641–644.

Baseyan, G. Z. and Matinyan, S. G. (1980) 'Solutions of Classical Yang–Mills Equations Containing Instantons and Merons, *Pisma JETP* **31**, 76–77.

Baulieu, L. and Zwanziger, D. (1981) 'Equivalance of Stochastic Quantization and the Faddeev–Popov Ansatz', *Nuclear Phys.* **B193**, 163–172.

Belavin, A. A., Polyakov, A. M., Schwartz, A. S. and Tyupkin, Yu. S. (1975) 'Pseudo-Particle Solutions of the Yang-Mills Equations', *Phys. Lett.* **59B**, 85–87.

Bender, C. M. and Orszag, S. A. (1978) *Advanced Mathematical Methods for Scientists and Engineers* McGraw-Hill, New York.

Berezin, F. A. (1966) *Method of Second Quantization*, Academic Press, New York and London.

Berger, Ch. *et al.* (PLUTO Collaboration) (1980) 'Test of QED in the Reactions $e^+e^- \to e^+e^-$ and $e^+e^- \to \mu^+\mu^-$ at CMS Energy from 9.4 to 31.6 GeV', *Z. Physik* **C4**, 269–276.

Beron, B. L. *et al.* (1978) 'Experimental Tests of Quantum Electrodynamics Through the Measurement of the Reactions $e^+e^- \to e^+e^-$, $e^+e^- \to \gamma\gamma$ and $e^+e^- \to \mu^+\mu^-$ at a Center-of-mass Energy of 5.2 GeV', *Phys. Rev.* **D17**, 2187–2198.

Berrondo, M. (1973) 'The Path Integral Method as Derived from a Stochastic Variational Principle', *Nuovo Cimento* **18B**, 95–109.

Bervillier, C., Drouffe, J. M. and Zinn-Justin, J. (1978) 'Comparison Between Large-Order Estimates and Perturbation Series in a Scalar Field Theory with Gaussian Propagator', *Phys. Rev.* **D17**, 2144–2159.

Bilenky, S. M., Hosek, J. and Petcov, S. T. (1980) 'On the Oscillations of Neutrinos with Dirac and Majorana Masses', *Phys. Lett.* **94B**, 495–498.

Bilenky, S. M. and Pontecorvo, B. (1980a) 'Majorana and Dirac Masses, Neutrino Oscillations and the Number of Charged Leptons', *Phys. Lett.* **95B**, 233–236.

Bilenky, S. M. and Pontecorvo, B. (1980b) 'On "Fast" and "Slow" Neutrinos', *Lett. Nuovo Cimento* **28**, 601–602.

Bilenky, S. M. and Pontecorvo, B. (1978) 'Lepton Mixing and Neutrino Oscillations', *Phys. Reports* **41C**, 225–276.

Bloch, C. (1952) 'On Field Theories with Non-localized Interaction', *Danske Vid. Selsk. Mat-Fys. Medd.* **27(8)**, 3–55.

Blokhintsev, D. I. (1973a) 'Stochastic Space and Nonlocal Fields', *Teoret. Math. Fiz.* **17**, 153–159.

Blokhintsev, D. I. (1973b) *Space and Time in the Microworld*, D. Reidel, Dordrecht.

Blokhintsev, D. I. (1975) 'The Stochastic Spaces', *Particles and Nuclei* **5**, 242– (ed. by APS, New York).

Blokhintsev, D. I. (1947) 'The Wave Field with Mass Spectrum', *JETP* 17, 115–120 (English Trans. *J. Phys.* **11**, 72–).

Bogolubov, N. N. and Shirkov, D. V. (1980) *Introduction to the Theory of Quantized Fields*, 3rd ed., Wiley-Interscience, I., New York.

Bogolubov, N. N., Medvedev, B. V. and Polivanov, M. K. (1958) *Problems in the Theory of Disperision Relations*, Institute for Advanced Studies, Princeton.

Bogolubov, N. N., Logunov, A. A. and Todorov, I. T. (1975) *Introduction to Axiomatic Quantum Field Theory*, Benjamin, New York.

Bogoliubov, N. N. and Parasiuk, O. A. (1957) 'Uber die Multiplikation der Kausalfunktionen in der Quantentheorie der Felder', *Acta Math.* **97**, 227–.

Bohm, D. (1952) 'Suggested Interpretation of Quantum Theory in terms of "Hidden Variables"', (Parts I and II) *Phys. Rev.*, **85**, 166–193.

Bohm, D. and Vigier, J.-P. (1954) 'Model of the Causal Interpretation of Quantum Theory in Terms of a Fluid with Irregular Fluctuations', *Phys. Rev.* **96**, 208–216.

Born, M. (1949) 'Reciprocity Theory of Elementary Particles' *Rev. Modern Phys.* **21**, 463–473.

Born, M. (1938) 'A Suggestion for Unifying Quantum Theory and Relativity', *Proc. Roy. Soc. London* **A165**, 291–303.

Bourrett, R. C. (1964) 'Quantized Fields as Random Classical Fields', *Phys. Lett.* **12**, 323–325.

Boyer, T. H. (1975) 'Random Electrodynamics: The Theory of Classical Electrodynamics with Classical Electromagnetic Zero-Point Radiation', *Phys. Rev.* **D11**, 790–808.

Boyer, T. H. (1975) 'General Connection Between Random Electrodynamics and Quantum Electrodynamics for Free Electromagnetic Fields and for Dipole Oscillator Systems', *Phys. Rev.* **D11**, 809–830.

Bracci, L., Fiorentini, G., Mezzorani, G. and Quarati, P. (1983) 'Bounds on a Hypothetical Fundamental Length', *Phys. Lett.* **133B**, 231–233.

Braffort, P. and Tzara, C. (1954) 'Energie de L'oscillateur Harmonique Dans le Vide', *C. R. Acad. Sci. Paris* **239**, 1779–1780.

Braffort, P. *et al.* (1965). L'energie Moyenne d'un Oscillateur Harmonique Non-Relativiste en Electromagnetique Aleatoire. *C. R. Acad. Sci. Paris* **261**, 4339–4341.

Bricman, C. *et al.* (Particle Data Group) (1980) 'Review of Particle Properties', *Rev. Modern Phys.*, **52**, No. 2 s1–s286.

Brodsky, S. J. and Drell, S. D. (1970) 'The Present Status of Quantum Electrodynamics', *Annual Review of Nuclear Science*, Vol. 20, E. Segre (ed.), Stanford Univ. Press, Palo Alto, California, pp. 147–194.

de Broglie, L. (1964) 'La Thermodynamique "Cachée" des Particules', *Ann. Inst. H. Poincaré* **1**, 1–19.

de Broglie, L. (1967) 'Notes des Membres et Correspondants et Notes Présentées ou Transmises par Leurs Soins', *C. R. Acad. Sci. Paris* **B264**, 1041–1044.

Brooke, J. A., Guz, W. and Prugovečki, E. (1982) 'The Reciprocity Principle in Stochastic Quantum Mechanics', *Hadronic J.* **5**, 1717–1733.

Brout, R., Englert, F., Frere, J.-M., Gunzig, E., Nardone, P., Truffin, C. and Spindei, Ph. (1980)

'Cosmogenesis and the Origin of the Fundamental Length Scale', *Nuclear Phys.* **B170**, 228–264.

Bunge, M. (1956). 'Survey of the Interpretations of Quantum Mechanics', *Amer. J. Phys.* **24**, 272–286.

Buras, A. J., Ellis, J., Gaillard, M. K. and Nanopoulos, D. V. (1978) 'Aspects of the Grand Unification of Strong, Weak and Electromagnetic Interactions', *Nuclear Phys.* **B135**, 66–92.

Busser, F. W. (1980 'High-Energy Neutrino Experiments', in *Field Theory and Strong Interactions*, Paul Urban Graz (ed)., Springer-Verlag, Vienna and New York.

Callan, C. G., Dashen, R. and Gross, D. J. (1978) 'Towards a Theory of the Strong Interactions', *Phys. Rev.* **D17**, 2717–2763.

Calmet, J., Narison, S., Perrottet, M. and Rafael, E. (1977) 'The Anomalous Magnetic Moment of the Muon: A Review of the Theoretical Contributions' *Rev. Modern Phys.*, **49**, 21–29.

Cameron, R. H. and Martin, W. T. (1949) 'Transformations of Wiener Integrals by Non-Linear Transformations', *Trans. Amer. Math. Soc.* **66**, 252–283.

Caubet, J.-P. (1976) *Le Mouvement Brownien Rėlativiste*, Lecture Notes in Mathematics, N. 559, Springer–Verlag, Berlin.

Cavalleri, G. (1981) 'The Propagator of Stochastic Electrodynamics', *Phys. Rev.* **D23**, 363–372.

Cerofolini, G. F. (1980) 'Quantum and Subquantum Mechanics', *Nuovo Cimento* **B58**, 286–300.

Chaichian, M. and Nelipa, N. F. (1984) *Introduction to Gauge Field Theories*, Springer-Verlag, Berlin, Heidelberg and New York.

Chandrasekhar, S. (1943) 'Stochastic Problems in Physics and Astronomy', *Rev. Modern Phys.*, **15**, N. 1, 1–89.

Chang, L.-N. and Chang, N.-P. (1980) 'B-L Nonconservation and Neutron Oscillation', *Phys. Lett.* **92B**, 103–106.

Cheng, T. P. and Ling-Fong Li. (1980) 'Neutrino Masses, Mixings, and Oscillations in SU(2) × U(1) Models of Electroweak Interactions', *Phys. Rev.* **D22**, 2860–2868.

Cheng, T. P. and Ling-Fong Li. (1977) 'Nonconservation of Separate μ- and e-Lepton Numbers in Gauge Theories with V + A Currents', *Phys. Rev. Lett.* **38**, 381–384.

Cheon, IL-T. (1978) 'Hypothesis of the Fundamental Length and Quantum Electrodynamics', *Internat. J. Theoret. Phys.* **17**, 611–629.

Cheon IL-T. (1979) 'Special Relativity on the Discrete Space-Time and a Fundamental Length', *Lett. Nuovo Cimento* **26**, 604–608.

Cheon, IL-T. (1982) 'On the Motion of a Particle with the Shadow of Vacuum Fluctuation', *Lett. Nuovo Cimento* **34**, 513–519.

Chirikov, B. V. and Shepelyansky, D. L. (1981) 'Stochastic Oscillation of Yang–Mills Classical Fields', *Pisma JETP* **34**, 171–175.

Chretien, M. and Peierls, R. E. (1953) 'Properties of Form Factors in Non-local Theories', *Nuovo Cimento*, **10**, 668–676.

Claverie, P. and Diner, S. (1976) '"Statistical and Stochastic Aspects of the Delocalization Problem in Quantum Mechanics" and "The Classical Limit in the Framework of Stochastic Mechanics"', in *Localization and Delocalization in Quantum Chemistry. Ionized and Excited States*, Vol. 2, O. Chalvet, and R. Daudel, *et al.* (eds.), D. Reidel, Dordrecht, pp. 395–448 and 449–460.

Claverie, P. and Diner, S. (1978) 'Stochastic Electrodynamics and Quantum Theory', *Internat. J Quantum Chem.*, **12**, Suppl. 1, 41–82.

Claverie, P. and Diner. S. (1973, 1975) 'Nates des Membres et Correspondants et Notes Présentées ou Transmises par Leurs Soins', *C.R. Acad. Sci. Paris* **B277**, 579–582, **B280**, 1–4.

Cole, E.A.B. (1972) 'The Classification of Displacements and Rotations in a Cellular Space-Time', *Internat. J. Theoret. Phys.* **5**, 437–446.

Coleman, S. (1977) 'Non-Abelian Plane Waves', *Phys. Lett.* **70B**, 59–60.

Cufaro Petroni, N. and Vigier, J.-P. (1979) 'Causal Superluminal Interpretation of the Einstein–Podolsky–Rosen Paradox', *Lett. Nuovo Cimento* **26**, 149–154.

Cufaro Petroni, N. and Vigier, J.-P. (1983a) 'Causal Action-At-Distance Interpretation of the Aspect–Rapisarda Experiments', *Phys. Lett.* **93A**, 383–387.

Cufaro Petroni, N. and Vigier, J.-P. (1983b) 'Dirac's Aether in Relativistic Quantum Mechanics', *Found. Phys.*, **13**, 253–286.

Dankel, T.-J. (1970) 'Mechanics on Manifold and the Incorporation of Spins into Nelson's Stochastic Mechanics', *Arch. Rational Mech. Anal.* **37**, 192–221.

Dankel, T.-J. (1977) 'Higher Spin States in the Stochastic Mechanics of the Bopp–Haag Spin Model', *J. Math. Phys.* **18**, 253–255.

Daum, M., Eaton, G. H., Frosch, R., Hirschmann, H., McCulloch, J., Minehart, R. C., and Steiner, E. (1978) 'Precision Measurement of the Muon Momentum in Pion Decay at Rest', *Phys. Lett.*, **74B**, 126–129.

Davidson, M. (1978). 'On the Equivalence of Quantum Mechanics and a Certain Class of Markov Processes', *J. Math. Phys.*, **19**, 1975–1978.

Davidson, M. (1979) 'A Generalization of the Fényes–Nelson Stochastic Model of Quantum Mechanics', *Lett. Math. Phys.* **3**, 271–277, 367–376.

Davidson, M. (1980a) 'The Generalized Fényes–Nelson Model for Free Scalar Field Theory', *Lett. Math. Phys.* **4**, 101–106.

Davidson, M. (1980b) 'Generalized Stochastic Mechanics on Riemannian Maniforld', *Lett. Math. Phys.* **4**, 475–483.

Davidson, M. (1981) 'Stochastic Quantization of the Electromagnetic Field' *J. Math. Phys.* **22**, 2588–2593.

Davidson, M. (1982) 'Stochastic Quantization of the Linearized Gravitation Field', *J. Math. Phys.* **23**, 132–137.

DeWitt-Morette, C. (1972) 'Feynman's Path Integrals. Definition Without Limiting Procedure', *Comm. Math. Phys.* **28**, 47–67.

DeWitt-Morette, C. and Elworthy, K. D. (eds.), (1981) 'New Stochastic Methods in Physics', *Phys. Reports* **C77**, 122–375.

DeWitt-Morette, C., Maheshwai, A. and Nelson, B. (1979) 'Path Integration in Non-Relativistic Quantum Mechanics', *Phys. Reports* **50**, 255–372.

Dineykhan, M. and Namsrai, Kh. (1977) 'Formulation of Gauge Invariant Quantum Electrodynamics in the Stochastic Space', *Teoret. Mat. Fiz.* **33**, 32–41 [*Theoret. Math. Phys. (U.S.A.)* **33**, 862–].

Dineykhan, M., Namsrai, Kh. and Omboo, Z. (1977) 'Construction of Gradient-Invariant Quantum Electrodynamics for Particles with Spin 0 and 1 in Stochastic Space', Communication of the JINR, P2-10963, Dubna.

Dodd, R. K., Eilbeck, J. C., Gibbon, J. D. and Morris, H. C. (1983) *Solitons and Nonlinear Wave Equations*, Academic Press, New York and London.

Dohrn, D. and Guerra, F. (1978) 'Nelson's Stochastic Mechanics on Riemannian Manifold', *Lett. Nuovo Cimento* **22**, 121–127.

Doob, J. (1953) *Stochastic Processes*, Wiley, New York.

Duffin, R. (1938) 'On the Characteristic Matrices of Covariant Systems', *Phys. Rev.* **54**, 1114–.

Dunford, N. and Schwartz, J. (1958) *Linear Operators*, Vol. 1, Interscience, New York.

Dynkin, E. B. (1965) *Markov Processes*, Springer-Verlag, Berlin and New York.

Ebel, G. *et al.* (1970) 'Compilation of Coupling Constants and Low-Energy Parameters', *Springer Tracts in Modern Physics* **55**, 239–290.

Eden, R. J. and Landshoff, P. V. (1965) 'The Problem of Causality in S-matrix Theory'. *Ann. Phys. (N.Y.)* **31**, 370–390.

Efimov, G. V. (1963) 'On the Development of a Nonlocal Quantum Field Theory Without Ultraviolet Divergences', *JETP* **44**, 2107–2117.

Efimov, G. V. (1968) 'On a Class of Relativistic Invariant Distributions', *Comm. Math. Phys.* **7**, 138–151.

Efimov, G. V. (1970) 'On Representation of Entire Functions of an Arbitrary Order of Growth', *Dokl. Akad. Nauk SSSR* **195**, 536–539.

Efimov, G. V. (1977a) *Nonlocal Interactions of Quantized Fields*, Nauka, Moscow.

Efimov, G. V. (1977b) 'Strong Coupling in the Quantum Field Theory with Nonlocal Nonpolynomial Interaction', *Comm. Math. Phys.* **57**, 235–258.

Efimov, G. V. (1972) 'On the Construction of Nonlocal Quantum Electrodynamics', *Ann. Phys.* (*N.Y.*) **71**, 466–485.

Efimov, G. V. (1979) 'Vacuum Energy in $g\varphi_d^4$-Theory for $g \to \infty$', *Comm. Math. Phys.* **65**, 15–44.

Efimov, G. V. (1974) 'Quantization of Nonlocal Field Theory', *Internat. J. Theoret. Phys.*, **10**, 19–37.

Efimov, G. V. and Ivanov, M. A. (1981). 'Vacuum Energy Y_d in the Yukawa Model in the Strong Coupling Limit', Preprint of the JINR, P2-81-707, Dubna.

Efimov, G. V., Malyshkin, V. G., Mogilevsky, O. A., Namsrai, Kh. and Yumatov, A. Yu. (1973) 'Anomalous Magnetic Moments of Leptons and the Lamb Shift in Nonlocal Theory of Current–Current Weak Interactions', *Nuclear Phys.*, **B59**, 1–22.

Efimov, G. V. and Namsrai, Kh. (1982). The Schrödinger Equation in the Quantum Field Theory with Nonlocal Interactions, *Teoret. Mat. Fiz.* **50**, 221–229.

Efimov, G. V. and Namsrai, Kh. (1975). 'To the Construction of Nonlocal Theory of Quantum Electrodynamics of Particles with spin 0 and 1', *Teoret. Mat. Fiz.* **22**, 186–202 (*Theoret. Math. Phys. U.S.A.* **22**, 129–141).

Efimov G. V. and Seltzer, Sh. Z. (1971) 'Gauge Invariant Nonlocal Theory of the Weak Interactions', *Ann. Phys.* (*N.Y.*) **67**, 124–144.

Efinger, H. J. (1981). 'Solitary Waves on a Curved Space-Time', *Found. Phys.*, **11**, 791–795.

Ehrlich, R. (1978) 'Are There an Elementary Length $l_0 = 0.66$ fm and an Elementary Time $\tau_0 = 0.66$ fm/c Associated with the Strong Interactions?' *Phys. Rev.* **D18**, 320–325.

Ehrlich, R. (1976) 'Possible Evidence for the Quantization of Particle Lifetimes', *Phys. Rev.* **D13**, 50–55.

Einstein, A. (1924) *Schweiz. Natur. Gesell. Verhand.* **85**, 85–.

Einstein, A. (1950, 1979) *The Meaning of Relativity*, 3rd and 4th edn, Princeton Univ. Press, Princeton, New York.

Einstein, A. (1956) *Investigations on the Theory of Brownian Movement*, Dover, New York.

Einstein, A. (1967) *The Meaning of Relativity*, 6th edn. rev., Appendix ii (1956), Science Paperbacks and Methuen and Co.

Einstein, A., Podolsky, B. and Rosen, N. (1935); 'Can Quantum-Mechanical Description of Physical Reality Be Considered Complete?' *Phys. Rev.*, **47**, 777–780.

Ellis, J., Gaillard, M. K., Nanopoulos, D. V. and Rudaz, S. (1980) 'Uncertainties in the Proton Lifetime', *Nuclear Phys.* **B176**, 61–99.

Exner, P. (1984) *Open Quantum Systems and Feynman Integrals*, D. Reidel, Dordrecht.

Faber, S.M. and Gallagher, J. S. (1979) 'Masses and Mass-to-Light Ratios of Galaxies', *Ann. Rev. Astronom. Astrophys.* **17**, 135–187.

Faddeev, L. D. and Popov, V. N. (1967) 'Feynman Diagrams for the Yang–Mills Field', *Phys. Lett.* **25B**, 29–30.

Faddeev, L. D. and Slavanov, A. A. (1980) *Introduction to the Theory of Gauge Theories*, Benjamin, Reading, Mass.

Fainberg, V. Ya. and Soloviev, M. A. (1978) 'How Can Local Properties Be Described in Field Theories Without Strict Locality?', *Ann. Phys.* (*N.Y.*) **113**, 421–447.

Fainberg, V. Ya, and Soloviev, M. A. (1977) 'Causality, Localizability and Holomorphically Convex Hills', *Comm. Math. Phys.* **57**, 149–159.

de Falso, D., de Martino, S. and de Siena, S. (1982) 'Position-Momentum Uncertainty Relation in Stochastic Mechanics', *Phys. Rev. Lett.* **49**, 181–183.

de Falso, D., de Martino, S. and de Siena, S. (1983) 'Momentum from Sample Paths in Stochastic Mechanics', *Lett. Nuovo Cimento* **36**, 457–460.

Faustov, R. N. (1972) 'Energy Levels and Electromagnetic Properties of the Hydrogen-Like Atoms', *Fizika Element. Chast. Atomnogo Yadra* **3**, 238–268.

Feynman, R. P. (1948) 'Space-Time Approach to Non-Relativistic Quantum Mechanics', *Rev. Modern Phys.* **20**, 367–387.

Feynman, R. P. (1950) 'Mathematical Formulation of the Quantum Theory of Electromagnetic Interactions', *Phys. Rev.* **80**, 440–457.

Feynman, R. P. (1951) 'An Operator Calculus Having Applications in Quantum Electrodynamics', *Phys. Rev.* **84**, 108–128.

Feynman, R. P. and Hibbs, A. R. (1965) *Quantum Mechanics and Path Integrals*, McGraw-Hill, New York.

Fényes, I. (1952) 'Eine Wahrscheinlichkeitstheoretische Begründund und Interpretation der Quantenmechanik', *Z. Physik* **132**, 81–106.

Finkelstein, D. (1969) 'Space-Time Code', *Phys. Rev.* **184**, 1261–1271.

Finkelstein, D. (1972) 'Space-Time Code, II and III', *Phys. Rev.* **D5**, 320–328, 2922–2931.

Finkelstein, D. (1974) 'Space-Time Code, IV', *Phys. Rev.* **D9**, 2219–2230.

Finkelstein, D., Frye, G. and Susskind, L. (1974) 'Space-Time Code, V', *Phys. Rev.* **D9**, 2231–2236.

Fiorini, E. (1979) 'Low Energy Weak Interactions', in *Proceedings Neutrino 1979 Conference, Bergen, Norway*, Vol. 1, A. Haatuft and C. Jarlskog (eds.), p. 236.

Floratos, E. and Iliopoulos, J. (1983) 'Equivalence of Stochastic and Canonical Quantization in Perturbation Theory', *Nuclear Phys.* **B214**, 392–404.

Flügge, G. (1980) 'Recent e^+e^- Physics', in *Proceedings of the 8th Internatinal Winter Meeting in Fundamental Physis, Ronda, Spain*, p. 169.

Fox, R. F. (1978) 'Gaussian Stochastic Processes in Physics', *Phys. Reports* **48**, 179–283.

Fradkin, E. S. (1954) 'The Green Function for Interacting Nucleons with Mesons', *Doklad. Akad. Nauk SSSR* **98**, 47–50.

Fradkin, E. S. (1954) 'On the Renormalization of Quantum Electrodynamics', *JETP* **26**, 751–754.

Frederick, C. (1976) 'Stochastic Space-Time and Quantum Theory', *Phys. Rev.* **D13**, 3183–3191.

Fridman, A. A. (1965) *The World as Space-Time*, Nauka, Moscow.

Friedberg, R. and Lee, T. D. (1983) 'Discrete Quantum Mechanics', *Nuclear Phys.* **B225** [FS9], 1–52.

Fröhlich, J. (1974) 'Verification of Axioms for Euclidean and Relativistic Fields and Haag's Theorem in a Class of $P(\phi)_2$ Models', *Ann. Inst. H. Poincaré* **A21**, 271–317.

Fubini, S. (1974) 'Summary Talk: Present Trends in Particle Physics', in *Proceedings of the XVII International Conference on High Energy Physics, London*, Chilton, London.

Fujiwara, K. (1980) 'Is the Light Velocity in Vacuum Really a Constant: Possible Break-Down of the Linear ω-k Relation at Extremely High Frequencies', *Found. Phys.* **10**, 309–332.

Fujikawa, K. and Shrock, R. E. (1980) 'Magnetic Moment of a Massive Neutrino and Neutrino-Spin Rotation', *Phys. Rev. Lett.* **45**, 963–966.

Fürth, R. (1933) 'Uber Einige Bezeihungen Zwischen Klassischer Statistik and Quantenmechanik', *Z. Physik* **81**, 143–162.

Gel'fand, I. M. and Shilov, G. E. (1964) *Generalized Functions*, Vol. 1, Academic Press, New York.

Gel'fand, I. M. and Shilov, G. E. (1968) *Generalized Functions*, Vol. 2, Academic Press, New York.

Gel'fand, I. M. and Yaglom, A. M. (1956) 'Integration in Functional Spaces and Its Application in Quantum Physics', *S. Uspehi Mat. Nauk* **67**, 77–114 (English Translation (1960), *J. Math. Phys.* **1**, 48–69).

Gell-Mann, M., Ramond, P. and Slansky, R. (1979) 'Complex Spinors and Unified Theories', in *Supergravity*, D. Z. Freedmann and P. van Nieuwenhuizen (eds), North-Holland, Amsterdam, pp. 315–321.

Georgi, H. and Glashow, S. L. (1974) 'Unity of all Elementary Particle Forces', *Phys. Rev. Lett.* **32**, 438–441.

Georgi, H. and Nanopoulos, D. V. (1979) 'Ordinary Predictions from Grand Principles: t-quark Mass in 0(10)', *Nuclear Phys.* **B155**, 52–74.

Ghirardi, G. C., Omero, C., Rimini, A. and Weber, T. (1978) 'Stochastic Interpretation of Quantum Mechanics: A Critical Review', *Riv. Nuovo Cimento* **1**, 1–34.

Gihman, I. I. and Skorokhod, A. V. (1972) *Stochastic Differential Equations*, Springer-Verlag, Berlin.

Gihman, I. I. and Skorokhod, A. V. (1975) *The Theory of Stochastic Processes*, Springer-Verlag, New York.

Ginzburg, V. L. (1975), 'Evaporation of Black Holes and the Fundamental Length', *Pisma JETP* **22**, 514–515.

Ginzburg, V. L. and Ptuskin, V. S. (1976) 'On the Origin of Cosmic Rays: Some Problems in High-Energy.Astrophysics', *Rev. Modern Phys.*, **48**, 161–189.

Ginzburg, V. L. and Syrovatskii (1964) *The Origin of Cosmic Rays*, Pergamon Press, London.

Glaser, V. (1974) 'On the Equivalence of the Euclidean and Wightman Formulation of Field Theory', *Comm. Math. Phys.* **37**, 257–272.

Glashow, S. L. (1961) 'Partial Symmetrics of Weak Interactions', *Nuclear Phys.* **22**, 579–588.

Glashow, S. L. (1980) 'Towards a Unified Theory: Threads in a Tapestry', *Rev. Modern Phys.* **52**, 539–543.

Glashow, S. L., Illiopoulos, J. J. and Maiani, L. (1970). 'Weak Interactions with Lepton-Hadron Symmetry', *Phys. Rev.*, **D2**, 1285–1292.

Glimm, J. and Jaffe, A. (1981) *Quantum Physics* (*A Functional Integral Point of View*), Springer-Verlag, New York, Heidelberg and Berlin.

Goldhaber, M. and Sulak, L. R. (1981) 'An Overview of Current Experiments in Search of Proton Decay', *Comment in Nuclear and Particle Physics* **10**, 215–225.

Goldman, T. J. and Ross, D. A. (1979) 'A New Estimate of the Proton Lifetime', *Phys. Lett.* **84B**, 208–210.

Grabert, H., Hänggi, P. and Talkner, P. (1979) 'Is Quantum Mechanics Equivalent to a Classical Stochastic Process?', *Phys. Rev.* **A19**, 2440–2445.

Gribov, V. N. (1977) 'Unstability of the Non-Abelian Gauge Theories and Impossibility of Choice of the Coulomb Gauge', in *Materials for the XII Winter School of the Leningrad Nuclear Research Institute*, pp. 147–162.

Gribov, V. N. (1978) 'Quantization of Non-Abelian Gauge Theories', *Nuclear Phys.* **B139**, 1–19.

Gueret, Ph. and Vigier, J.-P. (1982) 'De Broglie's Wave Particle Duality in the Stochastic Interpretation of Quantum Mechanics: A Testable Physical Assumption', *Found. Phys.* **12**, 1057–1083.

Guerra, F. (1981) 'Structural Aspects of Stochastic Mechanics and Stochastic Field Theory', *Phys. Reports* **77**, 263–312.

Guerra, F. and Loffredo, M. I. (1980) 'Stochastic Equations for the Maxwell Field', *Lett. Nuovo Cimento* **27**, 41–45.

Guerra, F. and Ruggiero, P. (1978) 'A Note on Relativistic Markov Processes', *Lett. Nuovo Cimento* **23**, 529–534.

Guerra, F. and Ruggiero, P. (1973) 'New Interpretation of the Euclidean–Markov Field in the Framework of Physical Minkowski Space-Time', *Phys. Rev. Lett.* **31**, 1022–1025.

Guha, A. and Lee, S.-C. (1982) 'Stochastic Quantization of Matrix and Lattice Gauge Models', Preprint of State Univ. of New York, ITP-SB-82-71, Stony Brook, New York, 11794.

Guth, A. H. (1981) 'Inflationary Universe: A Possible Solution to the Horizon and Flatness Problems', *Phys. Rev.* **D23**, 347–356.

Guth, A. H. (1982) '10^{-35} Seconds After the Big Bang', in *The Birth of Universe*, J. Audouze and J. Tran Thank Van (eds.). (Proceedings of the Seventeeth Rencontre de Moriond Astrophysics Meeting Les Arcs-Savoie-France, March, 14–26, 1982), Gif Sur Yvette Editions Frontieres, pp. 25–43.

Harish-Chandra (1946) 'The Correspondence Between the Particle and the Wave Aspects of the Meson and the Photon', *Proc. Roy. Soc.* **A186**, 502–.

Heisenberg, W. (1936), 'Zur Theorie der "Schauer" in der Höhenstrahlung', *Z. Phys.* **101**, 533–540.

Hillas, A. M. (1974) 'Survey of Data on Primary Cosmic-Ray Nuclei above 10^{14} eV', *Philos. Trans. Roy. Soc. Lond.* **A277**, 413–428.

Hillas, A. M. (1975) 'Some Recent Developments in Cosmic Rays', *Phys. Reports* **20C**, 59–136.

't Hooft, G. (1976) 'Symmetry Breaking Through Bell-Jackiw Anomalies', *Phys. Rev. Lett.* **37**, 8–11.

't Hooft, G. (1979) 'Quantum Gravity: A Fundamental Problem and Some Radical Ideas', in *Recent Development of Gravitation*, M. Levy and S. Deser (eds.), (NATO Advanced Study Institutes Series B: *Physics*, Vol. 44), Plenum Press, New York and London, pp. 323–345.

't Hooft, G. (1974) 'Magnetic Monopoles in Unified Gauge Theories', *Nuclear Phys.* **B79**, 276–284.

Hori, T. (1983) 'BRST Invariance and Stochastic Quantization of Supergravity' *Phys. Lett.* **123B**, 391–395.

Hsu, J. P. and Mac, E. (1979) 'Fundamental Length, Bubble Electrons and Nonlocal Quantum Electrodynamics', *Nuovo Cimento*, **B49**, 55–67.

Ibragimov, I. I. (1962) *External Properties of Entire Functions of the Finite Order of Growth*, Academic Press of Az. SSR, Baky.

Ingraham, R. I. (1967) *Renormalization Theory of Quantum Field with a Cutoff*, Gordon and Breach, New York.

Ioffe, B. L. (1973) 'Weak Interactions at Small Distances', *Uspehi Fiz. Nauk* **110**, 357–404.

Itô, K. (1961) *Lectures on Stochastic Processes*, Tata, Bombay.

Jackiw, R. (1977) 'Quantum Meaning of Classical Field Theory', *Rev. Modern Phy.* **49**, 681–706.

Jaffe, A. M. (1966) 'Form Factors at Large Momentum Transfer', *Phys. Rev. Lett.* **17**, 661–663.

Jaffe, A. M. (1967) 'High-Energy Behavior in Quantum Field Theory. I Strictly Localizable Fields', *Phys. Rev.* **158**, 1454–1461.

Jammer, M. (1974) *The Philosophy of Quantum Mechanics* J. Wiley, New York.

Jannussis, A. D. and Papatheou, V. (1982) 'Remarks on the Complex Time of Physical Phenomena', *Lett. Nuovo Cimento* **35**, 485–487.

Japan group (Fukai, T., Nakazato, N., Namiki, M., Ohba, I., Okano, K., Yamanaka, Y.) (1982) 'Stochastic Quantization Method...', Preprints of the Univ. of Waseda, WU-HEP-82-4, 82-5, 82-6, 82-7, Tokyo, 160, Japan; *Progr. Theoret. Phys.* (1983), **69**, 1580–1616.

Jost, R. (1965) *The General Theory of Quantized Fields*, American Mathematical Society, Providence, Rhode Island.

Kac, M. (1959) *Probability and Related Topics in Physical Sciences*, Interscience, New York.

Kadyshevsky, V. G. (1980) 'A New Approach to the Theory of Electromagnetic Interactions', *Fiz. Element. Chast. Atomnogo Yadra* **11**, 5–39.

Kang, K., Kim, J., Kim, J. E., Soh, K. S. and Song, H. S. (1980) 'Natural Mass Scale and Neutrino Oscillation', *Phys. Rev.* **D22**, 2869–2875.

Kantorovich, L. V. and Akilov, G. P. (1964) *Functional Analysis in Normed Spaces*, MacMillan, New York.

Kazarnovsky, M. V., Kuzmin, V. A., Chetyrkin, K. G. and Shaposhnikov, M. E. (1980) 'On the Neutron-Antineutron Oscillation', *Pisma JETP* **32**, 88–91

Kemmer, N. (1939) 'The Particle Aspect of Meson Theory', *Proc. Roy. Soc.* **A173**, 91–116.

Kershaw, D. (1964) 'Theory of Hidden Variables', *Phys. Rev.* **B136**, 1850–1856.

Kim, R., Mathur, V. and Okubo, S. (1974) 'Electromagnetic Properties of the Neutrino from Neutral-Current Experiments', *Phys. Rev.* **D9**, 3050–3053.

Kinoshita, T. (1979) 'Anomalous Magnetic Moment of an Electron and High Precision Test of Quantum Electrodynamics', in *Luminy CNRS Collogium.*

Kinoshita, T. (1950) 'On the Interaction of Mesons with the Electromagnetic Field. I', *Progr. Theoret. Phys.* **5**, 473–488.

Kinoshita, T. and Nambu, Y. (1950) 'On the Interaction of Mesons with Electromagnetic Field. II', *Progr. Theoret. Phys.* **5**, 749–768.

Kirzhnits, D. A. (1966) 'A Non-local Quantum Field Theory', *Uspehi Fiz. Nauk* **90**, 129–142.

Kirzhnits, D. A. and Linde, A. D. (1972) 'Macroscopic Consequences of the Weinberg Model', *Phys. Lett.*, **42B**, 471–474.

Kogut, J. and Susskind, L. (1975) 'Hamiltonian Formulation of Wilson's Lattice Gauge Theories', *Phys. Rev.* **D11**, 395–408.

Kolmogorov, A. N. (1933) *Grundbegriffe der Wahrscheinlichkeitsrechnung*, Berlin; (1956) *Foundations of the Theory of Probability*, Chelsea Publishing Co., New York.

Komar, A. A. and Markov, M. A. (1959) 'An Example of a Field Theory with Indefinite Metric in Hilbert Space II', *Nuclear Phys.* **12**, 190–203.

Kozik, V. S., Lubimov, V. A., Novikov, E. G., Nozik, V. Z. and Tretiakov, I. F. (1980) 'An Estimate of the $\tilde{\nu}_e$-Mass from the β-Spectrum of Tritium in the Valine Molecule', *Jadernaya Fiz.* **32**, 301–303; *Phys. Lett.* **94B**, 266–268.

Kracklauer, A. F. (1974) 'Comment on Derivation of the Schrödinger Equation from Newtonian Mechanics', *Phys. Rev.*, **D10**, 1358–1360.

Kraus, J. (1975) 'Delbrück Scattering and Tests of the Electron Propagator', *Nuclear Phys.* **B89**, 133–154.

Kroll, N. M. (1966) 'Ad Hoc Modifications of Quantum Electrodynamics', *Nuovo Cimento* **A45**, 65–92.

Lacroix, R. (1979) 'Sur L'existence d'une Longueur Elementaire et d'un Intervalle de Temps 'Elementaire. II', *Canad. J. Phys.* **57**, 1681–1685.

Lai, C. H. (ed.) (1983) *Gauge Theory of Weak and Electromagnetic Interactions* (Selected Papers), World Scientific Publishing Co. Pte Ltd, Singapore.

Landau, L. D. and Lifschitz, E. M. (1971) *The Classical Theory of Fields* (3rd edn) Pergamon Press, Oxford.

Langacker, P. (1981) 'Grand Unified Theories and Proton Decay', *Phys. Rep.* **C72**, 185–385.

Langouche, F., Roekaerts, D. and Tirapegui, E. (1979) 'Functional Integral Methods for Stochastic Fields', *Physica* **95A**, 252–274.

Lautrup, B. E., Peterman, A. and de Rafael, E. (1972) 'Recent Developments in the Comparison Between Theory and Experiments in Quantum Electrodynamics', *Phys. Rep.* **C3**, 193–259.

Lavenda, B. H. (1980) 'On the Equivalence Between Classical Markov Processes and Quantum Mechanics', *Lett. Nuovo Cimento* **27**, 433–436.

Lee, B. W., Pakvasa, S., Shrock, R. E. and Sugawara, H. (1977) 'Muon and Electron Number Nonconservation in a V–A Gauge Model', *Phys. Rev. Lett.* **38**, 937–939.

Lee. T. D. (1981) 'Is the Vacuum a Physical Medium?', in *Statistical Mechanics of Quarks and Hadrons*, H. Satz, (ed.), (Proceedings of the International Symposium Held at the Univ. of Bielfeld, FRG), Nord-Holland, Amsterdam, New York and Oxford, pp. 3–15.

Lee, V. J. (1980) 'Physical Foundation of Quantum Theory: Stochastic Formulation and Proposed Experimental Test', *Found. Phys.* **10**, 77–107.

Lehr, W. J. and Park, J. L. (1977) 'A Stochastic Derivation of Klein–Gordon Equation', *J. Math. Phys.* **18**, 1235–1240.

Leschke, H. and Schmutz, M. (1977) 'Operator Orderings and Functional Formulations of Quantum and Stochastic Dynamics', *Z. Physik* **B27**, 85–94.

Levine, M. J., Park, H. Y. and Roskies, R. Z. (1082) 'High-Precision Evalution of Contributions to g-2 of the Electron in Sixth-Order', *Phys. Rev.* **D25**, 2205–2207.

Li, F. (1982) 'The Effective Schwarzschild Radius and the Internal Interaction of Leptons', *Lett. Nuovo Cimento* **34**, 507–508.

Linde, A. D. (1982) 'A New Inflationary Universe Scenario: A Possible Solution of Horizon, Flatness, Homogeneity, Isotropy and Primordial Monopole Problems', *Phys. Lett.* **108B**, 389–393.

Magg, M., Ringhofer, K. and Salecker, H. (1972) 'μ-Mesic Hydrogen and g-2 of the Muon as Possible Tests of the Electron Propagator', *Nuclear Phys.* **B40**, 367–374.

March, A. Z. (1934) 'Die Geometrie Kleinster Räume. I, II', *Z. Physik* **104**, 93–99, 161–168.

March, A. Z. (1937) 'Zur Grundlegung und Anwendung Einer Statistischen Metrik', *Z. Physik* **105**, 620.

Marciano, W. J. (1979) 'Weak Mixing Angle and Grand Unified Gauge Theories', *Phys. Rev.* **D20**, 274–288.

Markov, M. A. (1958) *Gyperons and K-Mesons*, Fizmatgiz, Moscow.

Markov, M. A. (1959) 'An Example of a Field Theory with Indefinite Metric in Hilbert Space. I', *Nuclear Phys.*, **10**, 140–150.

Markov, M. A. (1982) 'Maximum Density of Matter as an Universal Law of Nature', *Pisma JETP* **36**, 214–216.

Markov, M. A. (1984) 'Some Problems of the Recent Theory of Gravitation', *Priroda*, No. 4 (824), 3–10.

Marshall, T. W. (1963) 'Random Electrodynamics', *Proc. Roy. Soc.* **A276**, 475–491.

Marshall, T. W. (1965) 'Stochastic Electrodynamics', *Proc. Cambridge Philos. Soc.* **61**, 537–546.

Marzke, J. and Wheeler, J. A. (1964) *Gravitation and Relativity*, H.-Y. Chiu and W. F. Hoffman (eds.), W. A. Benjamin, New York.

Matinyan, S. G., Savvidi, G. K. and Ter-Arutinyan-Savvidi, N. G. (1981a) 'Classical Yang–Mills Mechanics. Nonlinear Colour Oscillations', *JETP* **80**, 830–838.

Matinyan, S. G., Savvidi, G. K. and Ter-Arutinyan-Savvidi, N. G. (1981b) 'Stochasticity of Classical Yang–Mills Mechanics and its Elimination by Higgs Mechanism', *Pisma JETP* **34**, 613–616.

Mc'Clain, B., Niemi, A. and Taylor, C. (1982) 'Stochastic Quantization of Gauge Theories', *Ann. Phys. (N.Y.)* **140**, 232–246.

Mc'Clain, B., Niemi, A., Taylor, C. and Wijewardhana, L. C. R. (1983) 'Superspace, Dimensional Reduction, and Stochastic Quantization', *Nuclear Phys.* **B217**, 430–460.

Mc'Kean (Jr.), H. P. (1969) *Stochastic Integrals*, Academic Press, New York and London.

Meiman, N. N. (1964) 'The Causality Principle and Asymptotic Behavior of the Scattering Amplitude', *JETP* **47**, 1966–1983.

Menger, K. (1942) 'Statistical Metrics', *Proc. Nat. Acad. Sci. U.S.A.* **28**, 535.

Menger, K. (1949) 'Probability Geometry', *Proc. Nat. Acad. Sci. U.S.A.* **37**, 226.

Meyer, P., Ramaty, R. and Webber, W. R. (1974) *Phys. Today*, **27**, 23.

Mezincescy, L. and Ogievetsky, V. I. (1975) 'A Symmetry Between Bosons and Fermions and Superfields', *Uspehi Fiz. Nauk*, **117**, 637–683.

Michelson, D. M. and Sivashinsky, G. I. (1977) 'Nonlinear Analysis of Hydrodynamic Instability in Laminar Flames. II. Numerical Experiments', *Acta Astronautica* **4**, 1207–1221.

Mielnik, B. and Tengstrand, G. (1980) 'Nelson-Brown Motion: Some Question Marks', *Internat. J. Theoret. Phys.* **19**, 239–250.

Miettinen, H. I. and Thomas, G. H. (1980) 'Evidence that Hadronic Interiors have a Denser Matter than Charge Distribution', *Nuclear Phys.* **B166**, 365–377.

Migdal, A. B. (1981) 'Instability of Quantum Yang–Mills Fields and Vacuum Fluctuations', in *Statistical Mechanics of Quarks and Hadrons*, H. Satz (ed.), (Proceedings of the International Symposium Held at the Univ. of Bielfeld FRG), North-Holland, Amsterdam, New York and Oxford, pp. 349–353.

Misner, Ch. W., Thorne, K. S. and Wheeler, J. A. (1973) *Gravitation*, W. H. Freeman and Co., San Francisco.

Miura, T. (1979) 'Relativistic Path Integrals', *Progr. Theoret. Phys.* **61**, 1521–1535.

Mohapatra, R. N. and Marshak, R. E. (1980) 'Phenomenology of Neutron Oscillations', *Phys. Lett.* **94B**, 183–186.

Moore, S. M. (1979) 'Can Stochastic Physics a Complete Theory of Nature?', *Found. Phys.* **9**, 237–259.

Moore, S. M. (1980) 'Stochastic Fields from Stochastic Mechanics', *J. Math. Phys.* **21**, 2102–2110.

Morrison, D. R. O. (1980) 'Review of Neutrino Oscillations and Hunt for the Tay Neutrino', Preprint of CERN, EP-80 -190.

Morse, Ph. M. and Feschbach, H. (1953) *Methods of Theoretical Physics*, McGraw-Hill, New York.

Nady, K. L. (1966) *State Vector Spaces with Indefinite Metric in Quantum Field Theory*, Akademia Kiado, Budapest.

Nagels, M. M. *et al.* (1976) 'Compilation of Coupling Constants and Low-Energy Parameters', *Nuclear Phys.* **B109**, 1–90.

Nakano, T. (1959) 'Quantum Field Theory in Terms of Euclidean Parameters', *Prog. Theoret. Phys.* **21**, 241–259.

Nakano, Y. (1983) 'One-Time Characteristic Functional in the Stochastic Quantization', *Progr. Theoret. Phys.* **69**, 316–365.

Namsrai, Kh. (1980a) 'Relativistic Dynamics of Stochastic Particles. *Found. Phys.* **10**, 353–361.

Namsrai, Kh. (1980b) 'A Stochastic Derivation of the Sivashinsky Equation for the Self-Turbulent Motion of a Free Particle', *Found. Phys.* **10**, 732–742.

Namsrai, Kh. (1980c) 'Relativistic Feynman-Type Integrals', *Internat. J. Theoret. Phys.* **19**, 397–404.

Namsrai, Kh. (1981a) 'The Hypothesis of Stochastic Space and General Connection of Stochastic Theory', *Phys. Lett.* **82A**, 103–106.

Namsrai, Kh. (1981b) 'A Stochastic Model for the Motion of Two Relativistic Particles', *J. Phys. A: Mathematical General*, **14**, 1307–1311.

Namsrai, Kh. (1981c) 'A Nonlocal Stochastic Model for the Free Scalar Field Theory', *Internat. J. Theoret. Phys.*, **20**, 365–375.

Namsrai, Kh. (1981d) 'Stochastic Mechanics', *Fiz. Element. Chast. Atomnogo Yadra*, **12**, 1116–1156 (*Theoret. Math. Phys.* (*U.S.A.*) **12**, 449–464).

Namsrai, Kh. (1981e) 'Hierarchical Scales and Family of Black Holes', *Internat. J. Theoret. Phys.*, **20**, 749–754.

Namsrai, Kh. (1982) 'Proton Decay in the Theory of Gravitation and the Hypothesis of Fundamental Length', *Phys. Lett.* **88A**, 269–271.

Namsrai, Kh. and Dineykhan, M. (1983) 'Electromagnetic and Weak Interactions in Stochastic Space-Time: A Review', *Internat. J. Theoret. Phys.* **22**, 131–192.

Nash, C. (1978) *Relativistic Quantum Fields*, Academic Press, New York and San Francisco.

Nelson, E. (1964) 'Feynman Integrals and the Schrödinger Equation', *J. Math. Phys.* **5**, 332–343.

Nelson, E. (1966) 'Derivation of the Schrödinger Equation from Newtonian Mechanics', *Phys. Rev.*, **150**, 1079–1085.

Nelson, E. (1967) *Dynamical Theories of Brownian Motion*, Princeton Univ. Press, Princeton, New Jersey.

Nelson, E. (1971) 'Quantum Fields and Markoff Fields', in *American Mathematical Society, Summer Institute on Partial Differential Equations held at Berkeley.*

Nelson, E. (1973) 'Construction of Quantum Fields from Markoff Fields', *J. Funct. Anal.* **12**, 97–112.

Nelson, E. (1973) 'The Free Markoff Field', *J. Funct. Anal.* **12**, 211–227.

Nelson, E. (1979) 'Connection Between Brownian Motion and Quantum Mechanics', in *Einstein Symposium*, H. Nelkowski (ed.), Berlin.

Neutrino 82, Proceedings of the International Conference, 14–19 June, 1982), A. Frenkel and L. Jenik (eds.), Vols I and II, Supplement, Balatonfüred, Budapest.

Nielsen, H. B. and Olesen, P. (1973) 'Vortex-Line Models for Dual Strings', *Nuclear Phys.* **B61**, 45–61.

Niemi, A. and Wijewardhana, L.C.R. (1982) 'Stochastic Regularization of Quantum Field Theory', *Ann. Phys.* (*N.Y.*) **140**, 247–265.

Olver, F. W. J. (ed.) (1960) *Royal Society's Mathematical Tables*, Vol. 7, Bessel Functions; Part III, Zeros and Associated Values, Univ. Press, Cambridge.

Onofri, E. (1979) 'The Stochastic Interpretation of Quantum Mechanics: A Reply to Ghirardi *et al.*', *Lett. Nuovo Cimento* **24**, 253–254.

Osterwalder, K. and Schrader, R. (1973) 'Axioms for Euclidean Green's Functions', *Comm. Math. Phys.* **31**, 83–112.

Osterwalder, K. and Schrader, R. (1975) 'Axioms for Euclidean Green's Functions. II', *Commun. Math. Phys.* **42**, 281–305.

Pais, A. and Uhlenbeck, G. E. (1950) 'On Field Theories with Non-Localized Action', *Phys. Rev.*, **79**, 145–165.

Paley, R. E. A. C. and Wiener, N. (1934) *Fourier Transforms in the Complex Domain*, American Mathematical Society, New York.

Paley, R. E. A. C., Wiener, N. and Zygmund, A. (1933) 'Note on Random Functions', *Math. Z.* **37**, 647–668.

Parisi, G. (1977) 'The Perturbative Expansion and the Infinite Coupling Limit', *Phys. Lett.*, **69B**, 329–331.

Parisi, G. and Sourlas, N. (1979) 'Random Magnetic Fields, Supersymmetry, and Negative Dimensions', *Phys. Rev. Lett.* **43**, 744–745.

Parisi, G. and Sourlas, N. (1982) 'Supersymmetric Field Theory and Stochastic Differential Equations', *Nuclear Phys.* **B206**, 321–332.

Parisi, G. and Wu, Y.-S. (1981) *Sci. Sinica* **24**, 483–

Particle Data Group (1982) 'Review of Particle Properties', *Phys. Lett.* **111B**, i-xxii, 1–294.

Pati, J. C. and Salam, A. (1973a) 'Is Baryon Number Conserved?', *Phys. Rev. Lett.* **31**, 661–664.

Pati, J. C. and Salam, A. (1973b) 'Unified Lepton-Hadron Symmetry and a Gauge Theory of the Basic Interactions', *Phys. Rev.* **D8**, 1240–1251.

Pati, J. C., Salam, A. and Strathdee, J. (1981) 'Probings Through Proton Decay and n–$\bar{n}$ Oscillations', *Nuclear Phys.* **B185**, 445–472.

de la Peña-Auerbach, L. (1969) 'New Formulation of Stochastic Theory and Quantum Mechanics', *J. Math. Phys.* **10**, 1620–1630.

de la Peña-Auerbach, L. (1971) 'Stochastic Theory of Quantum Mechanics for Particles with Spin', *J. Math. Phys.*, **12**, 453–461.

de la Peña-Auerbach, L. and Cetto, A. M. (1972) 'Stronger Form for the Position-Momentum Uncertainty Relation', *Phys. Lett.* **39A**, 65–66.

de la Peña-Auerbach, L. and Cetto, A. M. (1974) 'Stochastic Treatment of the Quantum Mechanical Harmonic Oscillator: Mass Renormalization and Lamb Shift', *Phys. Lett.* **A47**, 183–184.

de la Peña-Auerbach, L. and Cetto, A. M. (1977) 'Derivation of Quantum Mechanics from Stochastic Electrodynamics', *J. Math. Phys.* **18**, 1612–1622.

de la Peña-Auerbach, L. and Cetto, A. M. (1978) 'Why Schrödinger's Equation?', *Internat. J. Quant. Chem.* **12**, Suppl. 1, 23–37, Discussion, 39–40.

de la Peña-Auerbach, L. and Cetto, A. M. (1982) 'Does Quantum Mechanics Accept a Stochastic Support?', *Found. Phys.* **12** 1017–1037.

de la Peña-Auerbach, L. and Cetto, A. M. (1975) 'Stochastic Theory for Classical and Quantum Mechanical Systems', *Found. Phys.* **5**, 355–370.

Petrina, D. Ya., Ivanov, S. S. and Rebenko, A. L. (1979) *Equations for Coefficient Functions of the Scattering Matrix*, Nauka, Moscow.

Poincaré, H. (1898) *La Mesure du Temps Rev. de Metaphysique et de Morale V, VI.*

Polyakov, A. M. (1975a) 'Compact Gauge Fields and the Infrared Catastrophe', *Phys. Lett.* **59B**, 82–84.

Polyakov, A. M. (1975b) 'Isomeric States of Quantum Fields', *JETP* **68**, 1975–1990.

Pontecorvo, B. (1967) 'Neutrino Experiments and the Question of Lepton Charge Conservation', *JETP* **53**, 1717–1725.

Prugovečki, E. (1978) 'Consistent Formulation of Relativistic Dynamics for Massive Spin-Zero Particles in External Fields', *Phys. Rev.* **D18**, 3655–3675.

Prugovečki, E. (1978) 'Relativistic Quantum Kinematics on Stochastic Phase Spaces for Massive Particles', *J. Math. Phys.* **19**, 2260–2270.

Prugovečki, E. (1981) 'Quantum Space-Time Operationally Based on Propagators for Extended Test Particles', *Hadronic J.* **4**, 1018–1104.

Prugovečki, E. (1984) *Stochastic Quantum Mechanics and Quantum Spacetime*, D. Reidel, Dordrecht.
Quigg, C. and Jackson, J. D. (1968) 'Decays of Neutral Pseudoscalar Mesons into Lepton Pairs', Preprint of California Univ., Berkeley, Lawrence Radiation Laboratory, UCRL, Report No. 18487.
Recami, E. (1981) 'Elementary Particles as Micro-Universes', Preprint of Institute Nazional di Fisica Nucleare INFN/AE-81/3. (Gennaio, Sezione de Catania, Italia).
Reed, M. and Simon, B. (1972) *Methods of Modern Mathematical Physics*, Vol. 1, *Functional Analysis*, Academic Press, New York.
Reed, M. and Simon, B. (1975) *Methods of Modern Mathematical Physics*, Vol. 2, *Fourier Analysis. Self-Adjointness*, Academic Press, New York.
Reines, F. and Crouch, M. F. (1974) 'Baryon-Conservation Limit', *Phys. Rev. Lett.* **32**, 493–494.
Reines, F., Learned, J. and Soni, A. (1979) 'Limits on Nonconservation of Baryon Number', *Phys. Rev. Lett.* **43**, 907–910.
Reines, F., Sobel, H. W. and Pasierb, E. (1980) 'Evidence for Neutrino Instability', *Phys. Rev. Lett.* **45**, 1307–1311.
Rice, S. O. (1944) 'Mathematical Analysis of Random Noise, I, II', *Bell System Tech. J.*, **23**, 282–332.
Rice, S. O. (1945) 'Mathematical Analysis of Random Noise, III: Statistical Properties of Random Noise Currents', *Bell System Tech. J.* **24**, 46–108.
Ringhofer, K. and Salecker, H. (1980) 'What Can Be Tested in Quantum Electrodynamics?' *Found. Phys.* **10**, 185–196.
Robiscoe, R. T. (1968) 'Reconciliation of Experimental Lamb Shifts', *Phys. Rev.* **168**, 4–11.
Robiscoe, R. T. and Shyn, T. W. (1970) 'Kinematics Corrections to Atomic Beam Experiments', *Phys. Rev. Lett.* **24**, 559–562.
Rosen, N. (1974) 'A Classical Picture of Quantum Mechanics', *Nuovo Cimento* **19B**, 90–98.
Ross-Bonney, A. A. (1975) 'Does God Play Dice? A Discussion of Some Interpretation of Quantum Mechanics', *Nuovo Cimento* **30B**, 55–79.
Roy Choudhury, R. K. and Roy, S. (1980) 'Stochastic Space-Time and the Concept of Potential in Classical Mechanics', *Phys. Rev.* **D22**, 2384–2386.
Roy, S. (1979) 'Relativistic Brownian Motion and the Foundations of Quantum Mechanics', *Nuovo Cimento* **B51**, 29–44.
Rueda, A. (1977) 'Particle Acceleration Mechanism in Stochastic Electrodynamics', (Preprint of the Univ. de los Andes, Departamento de Fisica, AR-1, Apartado Aero 4976), Bogota, D. E. Colombia, S. A.
Rueda, A. (1978) 'Model of Einstein and Hopf for Protons in Zero-Point Field and Cosmic-Ray Spectrum', *Nuovo Cimento* **48A**, 155–182.
Rueda, A. (1981) 'Behavior of Classical Particles Immersed in the Classical Electromagnetic Zero-Point Field', *Phys. Rev.* **A23**, 2020–2040.
Rueda, A. and Lecompte, A. (1979) 'On Approximations in the Model of Einstein and Hopf', *Nuovo Cimento* **52A**, 264–275.
De Rujula, A. and Glashow, S. L. (1980) 'Neutrino Weight Watching', *Nature*, **286**, No. 5775, 755–756.
De Rujula, A., Lusignoli, M., Maiani, L., Petcov, S. T. and Petronzio, R. (1980) 'A Fresh Look at Neutrino Oscillations', *Nuclear Phys.* **B168**, 54–68.
Salam, A. (1968) in *Elementary Particle Physics* (Nobel Symposium, No. 8), N. Svartholm (ed.), Almquist and Wiksell, Stockholm, p. 367.
Salam, A. (1980) 'Gauge Unification of Fundamental Forces', *Rev. Modern Phys.* **52**, 525–538.
Salam, A. and Strathdee, J. (1976) 'Black Holes as Solitons', *Phys. Lett.*, **61B**, 375–376.
Santamato, E. and Lavenda, B. H. (1981) 'The Underlying Brownian Motion of Nonrelativistic Quantum Mechanics', *Found. Phys.* **11**, 653–678.
Santos, E. (1972) 'Brownian Motion and Stochastic Theory', in *Irreversibility in the Many-Body Problem*, J. Biel, and J. Rae (eds.) Plenum Press, New York, pp. 459–470.

Santos, E. (1974) 'The Harmonic Oscillator in Stochastic Electrodynamics', *Nuovo Cimento* **B19**, 57–89.

Santos, E. (1974) 'Foundations of Stochastic Electrodynamics, 1: General Formalism and the Free Radiation Field', *Nuovo Cimento* **22B**, 201–214.

Sawada, O. and Fukugita, M. (1980) 'Neutron Oscillation as a Course of Cosmic Ray Antinucleons', (Preprint of National Laboratory of High Energy Physics), KEK. TH-19, Tsukuba.

Scadron, M. D. (1980) *Advanced Quantum Theory and Its Applications. Through Feynman Diagrams*, Springer-Verlag, New York, Heidelberg and Berlin.

Schramm, D. N. (1979) 'Current Outlook for Neutrino Astronomy', in *Proceedings of Neutrino 1979 Conference*, Vol. 1, A. Haatuft and C. Jarlskog, (eds.), Bergen, Norway, p. 503.

Schramm, R. and Steigman, G. (1980) 'Gravitiy Foundation Prize Essay' (unpublished).

Schrödinger, E. (1931) *Uber die Umkehrung der Naturgesetze*, Berliner Sitzungsberichte, pp. 144–153.

Schwartz, L. (1957, 1959) *Theorie des Distributions*, Vols. I–II. Hermann, Paris.

Schweizer, B. (1967) 'Probability Metric Space', in *The First 25 Years of the University of Massachusetts.*

Schwinger, J. (1959) 'Euclidean Quantum Electrodynamics', *Phys. Rev.* 115, 721–731.

Sehgal, L. M. (1969) 'Electromagnetic Contribution to the Decays $K_s \to l\bar{l}$ and $K_L \to l\bar{l}$', *Phys. Rev.* **183**, 1511–1513.

Sherstnev, A. N. (1963) 'On a Notion of a Random Normalized Space', *Dokl. Akad. Nauk SSSR*, **149**, 280–283.

Shucker, D. S. (1978) 'Stochastic Mechanics', Princeton Thesis.

Shucker, D. S. (1980a) 'Davidson's Generalization of the Fényes–Nelson Stochastic Model of Quantum Mechanics', *Lett. Math. Phys.* **4**, 61–65.

Shucker, D. S. (1980b) 'Stochastic Mechanics of Systems with Zero Potential', *J. Funct. Anal.* **38**, 146–155.

Shuryak, E. V. (1980) 'Quantum Chromodynamics and the Theory of Superdense Matter', *Phys. Reports* **C61**, 71–158.

Simon, B. (1974) *The $P(\varphi)_2$ Euclidean (Quantum) Field Theory*, Princeton Univ. Press, Princeton, New Jersey.

Simon, B. (1979) *Functional Integration and Quantum Physics*, Academic Press, New York, San Francisco and London.

Singer, I. M. (1978) 'Some Remarks on the Gribov Ambiguity', *Comm. Math. Phys.* **60**, 7–12.

Sivashinsky, G. I. (1977) 'Nonlinear Analysis of Hydrodynamic Instability in Laminar Flames. I. Derivation of Basic Equations', *Acta Astronautica* **4**, 1177–1206.

Sivashinsky, G. I. (1978) Self-Turbulence in the Motion of a Free Particle', *Found. Phys.* **8**, 735–744.

Sivashinsky, G. I. (1980) 'Turbulence in the Motion of a Free Particle and de Broglie Waves', *Lett. Nuovo Cimento* **27**, 504–508.

Skagerstam, Bo-S. K. (1975) 'Stochastic Processes on a Classical Configuration Space', Preprint of Institute for Theoretical Physics, Report 75-21, Gothenburg, Sweden.

Slavnov, D. A. (1974) 'The Generalized Pauli–Villars Regularization', (Ph.D. Thesis), Moscow State Univ.

Smirnov, A. Ya. (1980) 'Superheavy Fermions and Proton Lifetime', *Pisma JETP* **31**, 781–783.

Smoluchowsky, M. V. (1923) *Abhandlung Über dir Brownsche Bewegung und Verwandte Erscheinugen*, Akademii-Verlag, Leipzig.

Snyder, H. S. (1947) 'Quantized Space-Time', *Phys. Rev.* **71**, 38– .

Snyder, H. S. (1947) 'The Electromagnetic Field in Quantized Space-Time', *Phys. Rev.* **72**, 68–71.

Soloviev, M. A. (1971) 'On the Choice of the Class of Generalized Functions Compatible with Microcausality Condition and Spectrality', *Teoret. Mat. Fiz.* **7**, 183–194.

Soloviev, M. A. (1980) 'Relativistic Invariant Formulation of Causality in Nonlocal Field Theory of Exponential Growth', *Teoret. Mat. Fiz.* **43**, 202–209.

Stapp, H. P. (1965) 'Space and Time in *S*-Matrix Theory', *Phys. Rev.* **B139**, 257–270.
Streater, R. and Wightman, A. S. (1964) *CPT, Spin and Statistics and All That*, Benjamin, New York.
Surdin, M. (1971) 'Le Champ Electromagnetique Fluctuant de L'univers (Essai d'une Electrodynamique Stochastique)'. *Anns. Inst. H. Poincaré* **15**, 203–241.
Surdin, M. (1978) 'The Steady State Universe Revisited, with Stochastic Electrodynamics as a Guide', *Found. Phys.* **8**, 341–357.
Sutton, Ch. (1980) 'Neutrinos: Do They Rule the Universe', *New Scientist*, **86**, No. 1206, 308– .
Symanzik, K. (1966) 'Euclidean Quantum Field Theory. I. Equations for a Scalar Model', *J. Math. Phys.* **7**, 510–525.
Symanzik, K. (1969) 'Euclidean Quantum Field Theory', in *Proceedings of the International School of Physics "Enrico Fermi"*', Varenna Course XLV, R. Jost (ed.), Academic Press, New York.
Takano, Y. (1961) 'The Singularity of Propagators in Field Theory and the Structure of Space-time', *Prog. Theoret. Phys.* **26**, 304–314; **27**, 212–213.
Takano, Y. (1967) 'Fluctuation of Space-Time and Elementary Particles, I, II', *Progr. Theoret. Phys.* **38**, 1185–1187.
Tamm, I. E. (1965) 'On the Curved Momentum Space', in *Proceedings of the International Conference on Elementary Particles, Kyoto*, Kyoto, Japan, pp. 314–326.
Taylor, J. G. (1966) 'On the Existence of Field Theory, I. The Analytic Approach', *J. Math. Phys.* **7**, 1720–1729.
Tennakone, K. (1974) 'Electron, Muon, Proton, and Strong Gravity', *Phys. Rev.* **D10**, 1722–1725.
Terazawa, H. (1981) 'Pregeometry', Preprint of the Institute for Nuclear Study, INS-Report No. 429, Univ. of Tokyo.
Thomsen, D. E. (1980) 'Ups and Downs of Neutrino Oscillation', *Sci. News* **117**, No. 24, 377.
Ting, S. C. C. (1982) 'Test of Quantum Electrodynamics and the Study of Heavy Leptons', *Riv. Nuovo Cimento* **5**, 1–36.
Titchmarsh, E. (1939) *The Theory of Functions*, Oxford Univ. Press.
Turner, M. S. and Wilczek, F. (1982) 'Is Our Vacuum Metastable?' *Nature* **298**, 633–634.
Uhlenbeck, G. E. and Ornstein, L. S. (1930) 'On the Theory of the Brownian Motion', *Phys. Rev.* **36**, 823–841.
Van Dyck, R. S. *et al.* (1979) 'Progress of the Electron Spin Anomaly Experiment', *Bull. Amer. Phys. Soc.* **24**, 758.
Van Dyck, R. S., Schwinberg, P. B. and Dehmelt, H. G. (1977) 'Precise Measurements of Axial Magnetron, Cyclotron, and Spin-Cyclotron-beat Frequencies on a Isolated 1-MeV Electron', *Phys. Rev. Lett.* **38**, 310–314.
Van Nieuwenhuizen, P. and Freedman, D. Z. (eds.) (1979) *Supergravity* North-Holland, Amsterdam.
Varadham, S. R. S. (1980) *Diffusion Problems and Partial Differential Equations*, Published for Tata Institute of Fundamental Research by Springer-Verlag, p. 251.
Vialtsev, A. I. (1965) *Discrete Space-Time*, Nauka, Moscow.
Vigier, J.-P. (1979) 'Model of Quantum Statistics in Terms of a Fluid with Irregular Stochastic Fluctuations Propagating at the Velocity of Light: A Derivation of Nelson's Equations', *Lett. Nuovo Cimento* **24**, 265–272.
Vigier, J.-P. (1982a) 'Nonlocality, Causality and Aether in Quantum Mechanics', Astronom. Nach. **303**, I, 55–80.
Vigier, J.-P. (1982b) Louis de Broglie – Physicist and Thinker', *Found. Phys.* **12**, 923–930.
Vladimirov, V. S. (1979) *Generalized Functions in Mathematical Physics*, Nauka, Moscow.
Vuilleumier, J. L. (1979) 'Search for Neutrino Oscillations – A Progress Report', in *Proceedings of Neutrino 1979 Conference*, Vol. 1, A. Haatuft and C. Jarlskog (eds.), Bergen, Norway, p. 185.
Wachsmuch, H. (1979) 'Recent Data on Prompt Single Lepton Production in Hadron-Nucleus Collisions', in *Proceedings of the 1979 International Symposium on Lepton and Photon Interactions at High Energy*, Fermi National Acceleration Laboratory, Batavia, Illinois, USA, pp. 541–552.

Walstad, A. (1980) 'Time's Arrow in an Oscillating Universe', *Found. Phys.* **10**, 743–750.

Wanders, G. (1959) 'On the Problem of Causality', *Nuovo Cimento*, **14**, 168–184.

Wang, M. C. and Uhlenbeck, G. E. (1945) 'On the Theory of the Brownian Motion. II', *Rev. Modern Phys.* **17**, 323–342.

Wataghin, G. (1934) 'Bemerkung über die Selbstenergie der Electronen', *Z. Physik* **88**, 92–98.

Weinberg, S. (1967) 'A Model of Leptons', *Phys. Rev. Lett.* **19**, 1264–1266.

Weinberg, S. (1972) *Gravitation and Cosmology*, Wiley, New York.

Weinberg, S. (1979) 'Baryon- and Lepton-Nonconserving Processes', *Phys. Rev. Lett.* **43**, 1566–1570.

Weinberg, S. (1980) 'Conceptual Foundations of the Unified Theory of Weak and Electromagnetic Interactions', *Rev. Modern Phys.* **52**, 515–532.

Weinberg, S. (1981) The Decay of Proton. *Sci. Amer.* **244**, 64–75.

Weizel, W. (1953) 'Ableitung der Quantemtheorie aus einem Klassischen, Kausal Determinierten Modell, I, II', *Z. Physik* **134**, 264–285; **135**, 270–273.

Welton, T. A. (1948) 'Some Observable Effects of the Quantum-Mechanical Fluctuations of the Electromagnetic Field', *Phys. Rev.* **74**, 1157–1167.

Weisskopf, V. F. (1949) 'Recent Developments in the Theory of the Electron', *Rev. Modern Phys.* **21**, 305–321.

Weyl, H. (1929) 'Electron and Gravitation I', *Z. Physik* **56**, 330– .

Weyl, H. (1950) 'A Remark on the Coupling of Gravitation and Electron', *Phys. Rev.* **77**, 699–701.

Wheeler, J. A. (1964) 'Geometrodynamics and the Issue of the Final State', in *Relativity, Groups and Topology*, C. DeWitt and B. S. DeWitt (eds.), Gordon and Breach, New York.

Wick, W. D. (1981) 'Convergence to Equilibrium of the Stochastic Heisenberg Model', *Comm. Math. Phys.* **81**, 361–377.

Wiener, N. (1923) 'Differential Space', *J. Mathematical and Physical Sci.* **2**, 132–174.

Wightman, A. S. (1964) *Introduction to Some Aspects of the Relativistic Dynamics of Quantized Fields*, Lectures at the French Summer School of Theoretical Physics, Cargêse, Corsica.

Wightman, A. S. (1972) *Constructive Field Theory. Introduction to the Problems* Coral Gables Lectures.

Wilczek, F. and Zee, A. (1979) 'Operator Analysis of Nucleon Decay', *Phys. Rev. Lett.* **43**, 1571–1573.

Wilson, K. G. (1974) 'Confinement of Quarks', *Phys. Rev.* **D10**, 2445–2459.

Winter, K. (1979) 'Neutrino Physics', in *Proceedings of the 1979 JINR-CERN School of Physics, Dobogokö, Hungary.*

Witten, E. (1980) 'Neutrino Masses in the Minimal 0(10) Theory', *Phys. Lett.* **91B**, 81–84.

Wolf, G. (1980 'Selected Topics on e^+e^- Physics', Preprint DESY 80/13, Hamburg.

Wu, T. T. and Yang, C. N. (1969) 'Some Solutions of the Classical Isotopic Gauge Field Equations', in *Properties of Matter Under Unusual Conditions*, H. Mark and S. Fernbach (eds.), Interscience, New York, pp. 349–354.

Yanagida, T. and Yoshimura, M. (1980) 'Neutrino Mixing in a Class of Grand Unified Theories' *Phys. Lett.*, **97B**, 99–102.

Yang, C. N. and Mills, R. L. (1954) 'Conservation of the Isotopic Spin and the Isotopic Gauge Invariance', *Phys. Rev.* **96**, 191–194.

Yasue, K. (1979) 'Stochastic Quantization: A Review', *Internat. J. Theoret. Phys.* **18**, 861–913.

Yukawa, H. (1950) 'Quantum Theory of Non-local Fields, I, II', *Phys. Rev.* **77**, 219–226, 1047–1052.

Yukawa, H. (1966), Preprint of Research Institute for Fundamental Physics, Kyoto Univ., RIFP-55.

Zwanziger, D. (1981) 'Covariant Quantization of Gauge Fields Without Gribov Ambiguity', *Nuclear Phys.* **B192**, 259–269.

Index

GPSR Compliance

The European Union's (EU) General Product Safety Regulation (GPSR) is a set of rules that requires consumer products to be safe and our obligations to ensure this.

If you have any concerns about our products, you can contact us on ProductSafety@springernature.com

In case Publisher is established outside the EU, the EU authorized representative is:

Springer Nature Customer Service Center GmbH
Europaplatz 3
69115 Heidelberg, Germany

Batch number: 10815974

Printed by Printforce, the Netherlands